Nitric Oxide
and the Kidney

Nitric Oxide and the Kidney

Physiology and Pathophysiology

Edited by

Michael S. Goligorsky
Department of Medicine
State University of New York
Stony Brook, NY

and Steven S. Gross
Department of Pharmacology
Cornell University Medical College
New York, NY

CHAPMAN & HALL

ITP International Thomson Publishing

New York • Albany • Bonn • Boston • Cincinnati • Detroit • London • Madrid • Melbourne
Mexico City • Pacific Grove • Paris • San Francisco • Singapore • Tokyo • Toronto • Washington

Library of Congress Cataloging-in-Publication Data

Nitric oxide and the kidney: physiology and pathophysiology/edited by M.S. Goligorsky and S.S.
Gross.
     p.   cm.
     Includes bibliographical references and index.
     ISBN 0-412-08061-3 (HB : alk. paper)
     1. Nitric oxide-Physiological effect. 2. Nitric oxide-Pathophysiology.
3. Kidneys-Pathophophysiology. I. Goligorsky, Michael S. II. Gross, S.S. (Steven S.)
(DNLM: 1. Nitric Oxide-physiology. 2. Kidney-physiology. 3. Kidney Diseases-physiopathology.
QD 181.N1 N73093 1997
QP535.N1N546   1997
612.4'63-dc21
DNLM/DLC   96-48721
for Library of Congress   CIP

British Library Cataloguing in Publication Data available

Azote is one of the most abundant elements; combined with caloric it forms azotic gas, or mephitis, which composes nearly two thirds of the atmosphere . . . When combined with oxygen, azote forms the nitrous and nitric oxides and acids . . .

Antoine Laurent Lavoisier, "Elements of Chemistry"

One treadle sets a thousand threads a-going,
And to and fro the shuttle flies;
Quite unperceived the threads are flowing,
One stroke effects a thousand ties.

Johann Wolfgang von Goethe, "Faust," part I, 1924–8

Contents

Part III.   NO and the Regulation of Renal Hemodynamics

Part IV.   Role of NO system in Renal Pathophysiology

Preface

This book is unusual for two reasons. First, the very mingling of one of the most archaic signaling molecule, nitric oxide, with one of the most recent fields of Internal Medicine, Nephrology, is unorthodox. Second, both comingled fields of knowledge are advancing so swiftly that describing them is akin to counting sheep from a window of a moving train. Even a snapshot will provide certain aberrations in the fast-changing landscape. Anyone glancing outside for the first time will obtain a different prospective from that seen just a few moments ago. This was exactly the reason why we felt an urging necessity to artificially *carpe diem*, "freeze the moment," and scan the field with high-power binoculars.

There is a school of thought, favored by historians, that an objective description of events requires a certain distance in time, when the dust has settled. Indeed, Sir Walter Raleigh was once amazed by the difference in individual accounts on a commotion in a courtyard that was actually observed by a host of prisoners. Nonetheless, he doubted the accuracy of historic description of events from the nonwitnessed past.

Hence, this book captures the first decade of nitric oxide history as it has been evolving until the day when a group of investigators, all leaders in their respective fields, settled to reflect on the subject "How advent in nitric oxide biology and physiology fertilizes and fuels the development of Renal Medicine." The task was accomplished briskly, within a few months; however, the duration was sufficient for accumulation of some novel results. Therefore, some chapters are accompanied by "notes added in proof" to reflect on such unincorporated new data. This intensity notwithstanding, we hope that the produced snapshot would offer the reader an appropriate starting point into diverse subjects, whereas the high professionalism of the contributors should secure adequate objectivity and depth of presentation. With this, we wish to express our profound gratitude to all contributors who, frequently at the expense of their other calls, have dedicated their invaluable time to this work produced in a timely fashion. We are also most obliged to Professor Louis J. Ignarro for the Introduction to the book and to Mrs. Lisa LaMagna for excellent editorial assistance.

Contributors

Yoram Agmon, MD
Division of Nephrology
Hadassah University Hospital
Jerusalem 91240
Israel

Rexford S. Ahima, MD
Albert Einstein Medical College
Montefiore Hospital
111 East 210 Street
Bronx, NY 10467

Sebastian Bachmann, PhD
Professor
A.G. Anatomie & Zellbiologie
Klinikum Charlottenburg,
Spandauer Damm 130
Haus 31
14050 Berlin
Germany

Norman Bank, MD
Professor of Medicine
Albert Einstein Medical College
Montefiore Hospital
111 East 210 Street
Bronx, NY 10467

Christine Baylis, PhD
Professor
Department of Physiology, POB 9229
West Virginia University
Morgantown, WV 26506-9229

Timothy R. Billiar, MD
Professor
Department of Surgery
University of Pittsburgh
497 Scaife Hall
Pittsburgh, PA 15261

Mayer Brezis, MD
Head, Division of Nephrology
Hadassah University Hospital, Mount-
   Scopus
Jerusalem 91240, Israel

Josie P. Briggs, MD
Professor
Department of Physiology and Medicine
University of Michigan
7712 Medical Science Bldg, II
Ann Arbor, MI 48109

Rudi Busse, MD
Professor and Head, Department of
   Physiology
Johann Wolfgang Goethe University
   Clinic
Theodor-Stern-Kai 7
60590 Frankfurt am Main
Germany

Victoria Cattell, MD
Professor
St. Mary's Medical Center
Norfolk Place
London, W2 1PG
UK

John Conger, MD
Professor and Head, Division of
   Nephrology
VA Medical Center
1055 Clermont Street
Denver, CO 80220

Terry Cook, BA, MB, BS, MRCP,
   FRCPath
Reader in Renal Pathology
Royal Postgraduate Medical School
Hammersmith Hospital
London, W12 ONN
UK

Dennis A. Diederich, MD
Professor
Department of Medicine
Division of Nephrology
Kansas Medical Center
3901 Rainbow Blvd
Kansas City, KS 66160

Ingrid Fleming, PhD
Department of Physiology
Johann-Wolfgang-Goethe University
Frankfurt am Main D-60590
Germany

Michael S. Goligorsky, MD, PhD
Professor of Medicine and Physiology
Department of Medicine
State University of New York
Stony Brook, NY 11794-8152

Owen W. Griffith, MD
Head, Department of Biochemistry
Medical College of Wisconsin
8701 Watertown Plank Road
Milwaukee, WI 53226

Steven S. Gross, PhD
Associate Professor
Department of Pharmacology
Cornell University Medical Center
New York, NY 10021

Neil Hogg, PhD
Assistant Professor
Department of Biophysics
Medical College of Wisconsin
8701 Watertown Plank Road
Milwaukee, WI 53226

Professor Louis J. Ignarro
Department of Pharmacology
UCLA
Los Angeles, CA 90095

Young-Myeong Kim, PhD
Department of Surgery
University of Pittsburgh
497 Scaife Hall
Pittsburgh, PA 15261

Saulo Klahr, MD
Simon Professor of Medicine
Barnes-Jewish Hospital, North Campus
216 S. Kings Highway
St. Louis, MO 63110-1092

Armin Kurtz, PhD
Professor and Chairman
Department of Physiology
University of Regensburg, Postfach
   101042
D-93040 Regensburg
Germany

Thomas F. Lüscher, MD
Professor and Head
Department of Cardiology
University Hospital of Zurich
Ramistrasse 100
CH-8091 Zurich
Switzerland

Colin G. M. Millar, MD
The William Harvey Research Institute
St. Bartholomew's Hospital Medical
  College
Charterhouse Square
London EC1M 6BQ
UK

Heiko Mühl, PhD
Biozentrum
University of Basel
Klingelbergstr. 70
CH-4056 Basel
Switzerland

Eduardo Nava, MD
Kardiologie
Inselspital Hopital de L'Ile
CH-3010 Bern
Switzerland

Eisei Noiri, MD
Department of Medicine
State University of New York
Stony Brook, NY 11794-8152

Georg Noll, MD
Kardiologie
Inselspital Hopital de L'Ile
CH-3010 Bern
Switzerland

Marina Noris, MD
Institute di Ricerche Farm.
'Mario Negri'
Via Gavazzeni 11
24125 Bergamo
Italy

Suzette Y. Osei, MD
Montefiore Hospital
111 East 210 Street
Bronx, NY 10467

Andreas Papapetropoulos, PhD
Boyer Center for Molecular Medicine
Yale University Medical School
295 Congress Avenue
New Haven, CT 06536

Joseph Pfeilschifter, MD
Professor and Director
Department of Pharmacology
Johann Wolfgang Goethe University
60590 Frankfurt am Main
Germany

Giuseppe Remuzzi, MD
Professor and Chairman
Inst di Ricerche Farm "Mario Negri"
Via Gavazzeni, 11
24125 Bergamo
Italy

Jurgen Schnermann, PhD
Professor
Department of Physiology and Medicine
University of Michigan
7712 Medical Science Bldg, II
Ann Arbor, MI 48109

Karin Schricker
Department of Physiology
University of Regensburg
D-93040 Regensburg
Germany

David A. Schulsinger
Department of Urology
The New York Hospital–Cornell Medical
  Center
525 East 68th Street
New York, NY 10021

William Sessa, PhD
Associate Professor, Pharmacology
Boyer Center for Molecular Medicine
Yale University Medical School
295 Congress Avenue
New Haven, CT 06536

Shannon D. Smith
Section of Urology
Yale University School of Medicine
333 Cedar Street
New Haven, CT 06510

Christoph Thiemermann, PhD
Senior Scientist
The William Harvey Research Institute
St. Bartholomew's Hospital Medical
   College
Charterhouse Square
London, EC1M 6BQ
UK

Edith Tzeng, PhD
Department of Surgery
University of Pittsburgh
Pittsburgh, PA 15261

E. Darracott Vaughan, Jr., MD
Professor and Chairman
Department of Urology
The New York Hospital–Cornell Medical
   Center
525 East 68th Street
New York, NY 10021

Robert M. Weiss, MD
Professor
Section of Urology
Yale University School of Medicine
333 Cedar Str., PO Box 3333
New Haven, CT 06510

William Welch, PhD
Division of Nephrology
Georgetown University Medical Center
3800 Reservoir Rd NW
PHC F6003
Washington, DC 20007

Marcia A. Wheeler, PhD
Section of Urology
Yale University School of Medicine
New Haven, CT 06510

Christopher Wilcox, MD
Head, Division of Nephrology
Georgetown University Medical Center
3800 Reservoir Road NW
PHC F6003
Washington, DC 20007

Annotation

During the 1980s, an endothelium-derived relaxing factor (EDRF) described by Furchgott and Zawadski has been extensively studied. Its identification with nitric oxide heralded a new era in physiology and pathophysiology. Indeed, there hardly exists a cell type in the body which does not either produce this smallest and simplest autacoid or respond to its action. Not surprisingly, investigations on the different aspects of EDRF-nitric oxide system are mounting. It has been recognized that the kidney is not only the ultimate synthetic source for the substrate for nitric oxide synthase L-arginine, and not only renal dysfunction leads to the accumulation of an endogenous inhibitor of this enzyme, but that the kidney and its various structural units represent also sensitive targets for nitric oxide's multiple actions. All three known isoforms of nitric oxide synthase have been disclosed within different segments of the nephron. Their complex topography sheds light on the intricate and delicate regulatory network, which, upon disarray, results in renal dysfunction. It is most appropriate, in this context, to recall Shakespeare's lines:

> Take but degree away, untune that string,
> And, hark, what discord follows.

Despite a short history of research on the EDRF-nitric oxide system, the amount of information is overwhelming and controversies are abundant. It has become necessary, therefore, to summarize the existing knowledge on chemistry of nitric oxide, molecular biology of the enzymes generating it, physiological targets of nitric oxide in the kidney, and the role of this system in the pathogenesis and pathophysiology of various renal diseases, including glomerulonephritis, hypertension, renal failure, diabetic nephropathy and preeclampsia, to name a few. According to these goals, the book consists of five major parts: "Introduction" by L.J. Ignarro, which summarizes the first decade of NO research, "Biochemistry of nitric oxide", "Renal expression of nitric oxide synthases and production of nitric oxide", "Functions of nitric oxide in regulating renal hemodynamics", and "Role of nitric oxide in renal pathophysiology".

This venture was accomplished by an international team of experts in specific areas of EDRF-nitric oxide research. The book will be of interest to physiologists, biochemists and pharmacologists, as well as many nephrologists working on the pathophysiology of renal diseases. The book should provide a solid background for those nephrologists and physiologists who are just being attracted to the field of nitric oxide, should systematize and prioritize the issues for those who are already working on various aspects of nitric oxide, and should sketch the future directions of this field of research for those who are still considering whether or not to join investigative efforts on nitric oxide, thereby ensuring a lasting impact on the development of this branch of science.

Introduction

Nitric oxide (NO), one of the ten smallest molecules in nature according to Linus Pauling, has evolved from being an important molecule in the atmosphere, microbes and plants to being one of the most important and ubiquitous biological molecules in mammalian cells. It all started with Joseph Priestly more than 200 years ago, who discovered not only oxygen but also NO. Nitrogen oxides in general have played an important part in the development of our understanding of atmospheric pollution, the nitrogen cycle in microbes and plants, and in food preservation. For example, nitrites and nitrates have been added to certain meats not only to preserve them against microbial-induced spoilage but also to provide a deep red color and thereby make the meat more appealing to the consumer. This latter well-known use is attributed to the microbial-mediated reduction of nitrite or nitrate to NO, which reacts with the hemoproteins abundant in meat to form nitrosyl-hemoproteins having a distinctly red color. Indeed, NO has been used since the 1860s as a probe to study ligand properties in metalloproteins such as hemoglobin and myoglobin.

Interest in NO as a molecule that could affect mammalian cell function came in the mid 1970s when NO was found to activate guanylate cyclase and stimulate the formation of cyclic GMP in a variety of tissues. Following these observations came the discovery that NO is an exceedingly potent vascular smooth muscle relaxant and inhibitor of platelet aggregation, which elicits these pharmacological effects via the intracellular second messenger actions of cyclic GMP. The question arose as to why mammalian cells respond in this way to an air pollutant with such great potency and efficacy. In 1981, the mechanism of vasodilator action of nitroglycerin was attributed to the liberation of NO gas from nitroglycerin in vascular smooth muscle. Although this was an important discovery that explained the prior 100 years of documented clinical effects of nitroglycerin, the question still remained as to why mammalian cells have receptors for the exogenous chemical, NO. As a scientist very much interested in learning more about the pharmacology of NO, I truly believed that mammals must have an endogenous NO or NO carrier such as nitroglycerin for the purpose of facilitating local tissue blood flow and protecting against thrombosis. Designing experiments to test this hypothesis proved to be quite a difficult task, but then it all came together in my mind in 1985, just after relocating from New Orleans to Los Angeles.

While studying factors affecting the production and action of endothelium-derived relaxing factor (EDRF), which was discovered five years earlier in 1980, we noted that EDRF generated in arterial rings in response to acetylcholine and bradykinin caused the activation of purified preparations of guanylate cyclase. The EDRF had a very short half-life that was equivalent to that of authentic NO, about 5 seconds. Moreover, the cyclic GMP stimulating effects of EDRF and NO were antagonized by the guanylate cyclase inhibitor, methylene blue. These observations were cautiously interpreted as signifying that EDRF might be NO or a labile nitroso precursor of NO. As prior experiments from this laboratory had revealed that NO activates guanylate cyclase by heme-dependent mechanisms, new experiments were conducted to ascertain whether EDRF-elicited activation of guanylate cyclase might also be heme-dependent. Upon learning that such was indeed the case, we realized the impact of what we had found, namely, that EDRF must be NO, and we presented the data in support of this hypothesis at several meetings in 1986.

Clearly, these observations explained the long-appreciated potency of nitroglycerin as a vasodilator drug and quenched my search for the endogenous nitroglycerin-like molecule. The discovery made independently by two groups that EDRF as NO triggered an explosion in the field of NO research that is unparalleled in modern science. This discovery took the form of a tidal wave, sweeping the interest and imagination of so many basic and clinical researchers. The number of annual publications on NO has surged from several dozen to several thousand in the last decade. Who would have ever thought just 15 years ago, even for a brief moment, that the free radical gas, NO, would turn out to become one of the most universal and ubiquitous biological signaling molecules involved in both physiology and pathophysiology.

The kidney represents one of many important target organs for NO. Renal physiology and pathophysiology involve a multitude of actions of NO ranging from the regulation of blood flow, renin secretion and glomerular filtration to glomerulonephritis and renal failure. As is the case with other organs, there appears to be both a good side and a bad side regarding the actions of NO in the kidney. This book represents a state-of-the-art exposition of our current understanding of NO in the kidney, and should serve as a springboard for future research in renal physiology and pathophysiology.

Louis J. Ignarro

Nitric Oxide
and the Kidney

PART I

General Biochemistry of NOSs and Cellular Actions of NO

1

The Biological Chemistry of NO

Neil Hogg and Owen W. Griffith

1. Introduction

The most well-characterized biological function of nitric oxide (•NO) is the elevation of cyclic guanosine monophosphate (cGMP) levels by the activation of guanylyl cyclase [1–3]. Simplistically, this pathway involves diffusion of •NO from its site of synthesis to a target enzyme in an adjacent cell or tissue. In this respect, •NO is acting as a conventional paracrine messenger. However, the known chemistry of •NO suggests that this journey is not at all simple and that there is the potential for a plethora of side reactions to occur. The chemical reactions between •NO and other biological molecules impact not only on the efficiency of guanylyl cyclase activation but also contribute to the cytotoxic potential of activated immune cells. Moreover, if uncontrolled, •NO synthesis may be responsible for a variety of pathological processes.

It has become clear that in order to fully understand the biological chemistry of •NO, the spontaneous, nonenzymatic reactions of •NO have to be considered. Such reactions lead to the generation of secondary species which have been variously (and perhaps confusingly) termed "reactive nitrogen intermediates" and "reactive nitrogen species." Included in this group are peroxynitrite (ONOO⁻), nitroxyl anion (NO⁻), and nitrosonium cation (NO⁺). Appreciation of the chemistry, biochemistry, pharmacology, and pathology of such species is critical to a complete understanding of the often contradictory biological effects of •NO.

2. Molecular Properties of •NO

2.1. Nitric Oxide Is a Hydrophobic Gas

At room temperature •NO is in the gaseous state [boiling point (bp) = $-151°C$] [4]. In this form •NO is primarily monomeric, although small quantities of the dimeric form (N_2O_2) have been detected [5]. In aqueous solution, •NO is also mainly monomeric and has limited solubility (about 2 mM at room temperature [6]). •NO is more soluble in organic solvents and has an octanol partition coefficient of about 6 [7]. The hydrophobic nature of •NO implies that biological membranes will not represent a barrier to •NO diffusion [7–9] and that •NO may partition preferentially into the hydrophobic interior of phospholipid bilayers and lipoproteins. The diffusion coefficient of •NO is 3300 μm^2 s^{-1} in water [10].

2.2. Nitric Oxide Is a Free Radical

Free radicals are molecules that contain one or more unpaired electrons (conventionally indicated by a • symbol) in s or p orbitals. Free radical reaction mechanisms are generally dominated by the propensity of unpaired electrons to find a partner. Many free radicals are transient, highly reactive species that are able to abstract electrons from other nonradical molecules, generating secondary radicals and possibly initiating chain reactions. An example relevant to biology is the initiation of lipid peroxidation by the highly reactive hydroxyl radical:

$$\text{OH•} + \text{LH} \rightarrow \text{H}_2\text{O} + \text{L•} \tag{1}$$

The hydroxyl radical (OH•) abstracts a hydrogen atom from an unsaturated lipid (LH) to form a water molecule and a lipid radical. The lipid radical can then proceed to initiate the lipid peroxidation chain reaction. Other types of free radicals are generally not reactive enough to abstract a hydrogen atom or an electron from a nonradical source, but their electrons can be paired by radical–radical reactions. This can lead to dimerization as shown in Eq. (2).

$$\text{A•} + \text{A•} \rightarrow \text{A} - \text{A} \tag{2}$$

NO belongs to this second class of radicals and its biological chemistry is dominated by reactions with other free radicals. For example, the reaction of •NO with oxygen (a biradical), superoxide ($O_2^{•-}$), and nitrogen dioxide ($•NO_2$) are all radical–radical reactions. Dimerization of •NO may occur, but the resultant N_2O_2 is unstable and rapidly dissociates back to •NO.

2.3. Nitric Oxide Is a Reductant/Oxidant

The redox potentials for the reduction of •NO is within the biological range ($E^{0'}$ (•NO/NO$^-$) = 0.39 V [11]) indicating that, at least thermodynamically, it is possible for •NO to be reduced in biological systems. In fact, such reactions occur; for example, •NO undergoes redox-type reactions with glutathione [12,13].

One electron reduction of •NO leads to the formation of the nitroxyl anion (NO$^-$). Although the reactions of NO$^-$ in biological systems are incompletely understood, it is clear that in anaerobic systems dimerization of NO$^-$ leads to the formation of nitrous oxide (N$_2$O, "laughing gas") and detection of this species is often taken as evidence for NO$^-$ formation. Under aerobic conditions, the reaction of NO$^-$ with oxygen can form peroxynitrite (ONOO$^-$) [14], a species more commonly formed from the reaction between •NO and superoxide (see Section 3.2).

NO$^-$ has been shown to posses Endothelium-Derived Relaxing Factor (EDRF)-like bioactivity [15]. It is difficult to be certain, however, that NO$^-$ was not oxidized to •NO, which has EDRF activity. To this end, Pino and Feelish have used L-cysteine to distinguish between the effects of •NO and NO$^-$ [16]. L-cysteine enhanced •NO-dependent relaxation of aortic rings, whereas it inhibited the effects of NO$^-$.

One-electron oxidation of •NO (E$^{0'}$ (NO$^+$/•NO) = 1.21 V [11]) yields the nitrosonium cation NO$^+$ [17,18]. In aqueous solution, this molecule will ultimately give nitrite (NO$_2^-$) by the addition of water. However, •NO$^+$ is a highly reactive species and there exists the potential for other reactions in biological systems [19]. In fact, evidence for the formation of free NO$^+$ is lacking, and the biological chemistry of NO$^+$ may be restricted to the transfer of "bound" NO$^+$, as occurs, for example, during transnitrosation reactions (Section 3.4).

2.4. Nitric Oxide as a Metal Ligand

One of the most widely appreciated properties of •NO is its ability to bind to heme iron. Many years before the discovery of •NO as a mammalian biological product, •NO was used to probe the ligand binding site of heme proteins [20,21]. Such studies proved highly relevant, as one of the major biological functions of •NO involves binding to the heme prosthetic group of guanylyl cyclase, activating the enzyme [22,23]. Upon binding of •NO to ferrous heme, the histidine on the opposite side of the heme ring is displaced, and by this mechanism, the •NO binding signal can be transformed into a protein conformational change [24]. •NO can bind to both the ferric and ferrous forms of heme proteins although the affinity for the ferrous form is usually much greater [24].

•NO can also bind to nonheme iron, and it has been suggested that such an interaction with low-molecular-weight iron complexes may inhibit the oxidative stress associated with the presence of such complexes; that is, •NO will bind to the iron complex and prevent iron redox cycling and the consequent oxygen radical formation [25]. The reduction, by •NO, of nonheme iron in the active site of lipoxygenase has been suggested as a mechanism of inhibition [26].

Electron paramagnetic resonance (EPR) spectroscopy has identified dinitrosyl iron complexes of •NO in biological tissues [27]. Such complexes spontaneously form from low-molecular-weight thiols, such as cysteine, ferric or ferrous iron, and nitric oxide, and have potent vasodilator activity [28]. Protein-bound dinitrosyl complexes have also been identified and it has been proposed that such high-molecular-weight •NO complexes may act as a store of •NO that can be released by low-molecular-weight thiols [29]. The biological consequences of dinitrosyl iron complex formation in vivo has yet to be established.

3. Biological Reactions of Nitric Oxide

3.1. Oxygen

•NO reacts with oxygen in aqueous solution as shown in Eq. (3).

$$4•NO + O_2 + 2H_2O \longrightarrow 4H^+ + 4NO_2^- \tag{3}$$

One molecule of oxygen consumes four molecules of •NO and generates nitrite as the only nitrogen-containing product. The rate-limiting step for this reaction is the third-order reaction between •NO and oxygen shown in Eq. (4) [30,31].

$$2•NO + O_2 \xrightarrow{k = 6\times10^6 \ M^{-1} \ s^{-1}} 2•NO_2 \tag{4}$$

$$•NO_2 + •NO \xrightarrow{k = 1.1\times10^9 \ M^{-1} \ s^{-1}} N_2O_3 \tag{5}$$

$$N_2O_3 + H_2O \xrightarrow{k=1\times10^3 \ s^{-1}} 2H^+ + 2NO_2^- \tag{6}$$

The nitrogen dioxide formed from this reaction reacts rapidly with nitric oxide to give dinitrogen trioxide [N_2O_3, Eq. (5)]. Hydrolysis of N_2O_3 finally yields nitrite [Eq. (6)]. Dimerization of nitrogen dioxide gives dinitrogen tetraoxide (N_2O_4) (not shown), which can hydrolyze to give equal quantities of nitrite and nitrate. However, kinetic simulation of the above reaction scheme has indicated that very low, nonbiological rates of •NO production are required before nitrate is formed in significant amounts [32,33].

The kinetics of Eq. (4) imply that the rate of decomposition of •NO is proportional to the square of the •NO concentration. This dependency on $[•NO]^2$ has the important consequence that at high concentrations of •NO, the reaction with oxygen is rapid, whereas at low concentrations of •NO, the reaction is very slow. For example, if 1 mM solutions of these gases are mixed, the initial rate of oxidation will be approximately 6 mM s^{-1}. However, if physiological concentrations are considered (100 nM •NO and 20 µM oxygen), the initial rate of oxidation will be 1.2 pM s^{-1}. It is arguable whether the reaction between •NO and oxygen plays any role in vivo. It clearly cannot account for the measured biological half-life of endothelial-derived nitric oxide of about 4 s [34].

In cell culture systems, the concentration of oxygen is usually significantly higher than that found in vivo. Additionally, •NO synthesized by cells in culture or added to such a system, will largely remain dissolved in the cell culture medium; only a small proportion is likely to be lost by partition into the gas phase. Under such conditions, the reaction between •NO and oxygen may be a significant determinant of the biochemistry and toxicology of •NO. Lewis et al. [35] have attempted to analyze the fate of •NO, generated by activated macrophages, in cell culture by end-product determination. They concluded that even in cell culture conditions, only about 50% of the measured NO^-_2 can be accounted for by the reaction between •NO and oxygen.

There is evidence, however, that suggests the biological half-life of •NO is oxygen dependent even though the kinetics of the reaction with oxygen are too slow to explain •NO consumption. In an attempt to reconcile this paradox, it has been suggested that the initial step of the reaction of •NO with oxygen is the association of •NO and oxygen, to form the intermediate free radical ONOO•, and that further reactions of this radical may represent a route of •NO decomposition [36]. This idea is as still highly speculative.

The reaction of •NO with oxygen as shown in Eqs. (4) and (5) forms N_2O_3 as an intermediate. N_2O_3 is a powerful nitrosating agent and can react with amines and thiols to form *N*-nitrosamines and *S*-nitrosothiols, respectively [37]. The former are usually highly toxic mutagens, and this reaction has been implicated in the mutagenic and toxic potential of •NO [38]. *S*-Nitrosothiols have been suggested to be physiological storage and transport molecules for nitric oxide and will be discussed later.

3.2. Superoxide

The reaction of •NO with superoxide was shown by Blough and Zifarou [39] to form peroxynitrite:

$$•NO + O_2^- \xrightarrow{k=6\times10^9\ M^{-1}\ s^{-1}} ONOO^- \qquad (7)$$

The chemistry of peroxynitrite had been sporadically studied since it was first described in 1901 [40]. However, after Beckman et al. proposed in 1990 that this molecule may account for much of the cytotoxic potential of •NO [41], the reactions between peroxynitrite and biological molecules has been extensively investigated, and peroxynitrite has been implicated in the mechanism of a number of disease processes. The biological chemistry of peroxynitrite has been reviewed by Pryor and Squadrito [42].

Peroxynitrite has a pK_a of 6.8 [42] and is, therefore, partially protonated at physiological pH. The state of protonation is of major consequence, as peroxynitrite and peroxynitrous acid undergo distinct chemical reactions with biological molecules. The peroxynitrite anion will oxidize thiols such as glutathione with a rate constant of about 3–6000 M^{-1} cm^{-1} [43] and will nitrate tyrosine and tryptophan residues in the presence of a suitable catalyst (i.e., metal ions or superoxide dismutase) [44]. These nitrated amino acids, especially nitrotyrosine, have been used as an immunological marker of peroxynitrite-mediated damage

[45]. Peroxynitrous acid is a potent oxidant and undergoes both one- and two-electron oxidation reactions [42].

There is evidence that the peroxynitrite-dependent oxidation of methionine [46] and α-tocopherol [47] occur mainly via two electron oxidation reactions. For example, peroxynitrite will oxidize α-tocopherol predominantly to α-tocopherol quinine and only a small quantity of α-tocopheryl radical is formed [47]. Free radical intermediates have also been observed in the oxidation of ascorbate [48] and glutathione [49,50] and it is not yet established if they are a major or minor product of the oxidation reaction.

The one-electron oxidation reactions of peroxynitrite are mediated either by molecule-assisted homolysis reactions, where the interaction between peroxynitrous acid and the reactant molecule promotes homolytic dissociation, or by a noninduced hydroxyl radical-like reactivity [42]. Although it was initially thought that peroxynitrite could undergo homolysis to give hydroxyl radical and nitrogen dioxide, such cleavage is not thermodynamically favorable. It has been more recently proposed that the hydroxyl radicallike activity of peroxynitrite is due to an excited-state intermediate that occurs during internal rearrangement to nitric acid [51]. This modality of peroxynitrite chemistry includes many of the reactions previously thought to be diagnostic for the hydroxyl radical, including the oxidation of deoxyribose to malondialdehyde and the hydroxylation of benzoic acid [41]. Moreover, typical hydroxyl-radical scavengers such as mannitol and ethanol can inhibit product formation in these systems [41].

Many of the reactions of peroxynitrite such as deoxyribose degradation, α-tocopherol oxidation, and the initiation of lipid peroxidation also occur in systems where •NO and superoxide are generated simultaneously [52–54]. Such co-formation can be achieved conveniently by the used of SIN-1 [52], a sydnonimine derivative that decomposes in oxygenated buffers to give stoichiometric amounts of •NO and superoxide [55], or by using a separate source of both superoxide and nitric oxide [54,56].

The biological consequences of the reaction between •NO and superoxide remain controversial and are likely to depend on the system studied. For example, in a number of studies, the reaction between •NO and superoxide has been implicated as a protective mechanism of superoxide removal [57,58]. From another perspective, it may be argued that the removal of •NO by this reaction is of greater functional consequences than the formation of peroxynitrite. It has also been suggested that the reaction has a profound impact on endothelial cell function by regulating NFκB-dependent gene expression [59]. In some circumstances, the reaction of peroxynitrite with molecules in the experimental system, such as glutathione and glucose, can lead to the generation of nitric oxide donor compounds that will slowly release •NO [60,61]. Exposure of cells or biological solutions to peroxynitrite can thus result in a long-lasting effect closely resembling that seen with •NO.

3.3. Heme Proteins

Heme biochemistry and •NO are intimately related. As noted earlier, the best established physiological target of •NO is the heme group of guanylyl cyclase. It has also been suggested that nitric oxide synthase is product-inhibited by •NO binding to that enzyme's heme prosthetic group [61–62].

Although it has been known for many years that •NO is a ligand for the heme a_3-Cu b (oxygen binding site) of cytochrome c oxidase, the functional consequences of such a reaction have only recently been explored. •NO may not only act as an inhibitor of cytochrome c oxidase (and hence mitochondrial respiration) [63] but may also be reduced to NO$^-$ in the process [64]. It is intriguing to speculate that mitochondria may represent a site of •NO catabolism in vivo.

Another set of reactions that has been known for many years [65] is the complex chemistry of •NO with hemoglobin and myoglobin. •NO reacts rapidly with the oxygen-ligated ferrous forms of these proteins to form nitrate and the ferric "met" form of the heme protein:

$$Hb(Fe^{2+}) - O_2 + \bullet NO \longrightarrow Hb(Fe^{3+}) + NO_3^- \qquad (8)$$

This reaction is rapid and is accompanied by a dramatic change in the visible spectrum of the heme protein; the spectral change has be used to quantitate •NO production [66]. This method is a very useful way of determining the concentration of an aqueous solution of •NO [67]. •NO will also bind to both ferric and ferrous hemoglobin [68]. The ferrous–•NO complex has a characteristic electron spin resonance (ESR) signature that has been observed in the blood of endotoxin-treated animals [69]. The corresponding myoglobin signal has been recently observed in cardiac tissue [70]. The ferrous–•NO complex is relatively stable and its decay in the presence of oxygen is limited by the rate at which •NO dissociates from the complex. The binding of •NO to ferric hemoglobin leads to an ESR silent complex that undergoes slow autoreduction by an unknown mechanism to eventually give the ferrous–•NO complex [71].

The rapid kinetics of these reactions and the high concentration of hemoglobin in blood make it likely that once •NO diffuses into a vessel it is rapidly destroyed. This may represent the major route of •NO decomposition in vivo. Mathematical simulations have indicated that the presence of blood-containing vessel can have a major impact on the sustainable steady-state concentration of •NO in a vessel wall [34].

3.4. Thiols

As nitric oxide diffuses within cells, one of the molecules it is most likely to collide with is glutathione, which is present inside cells at very high concentrations

(1–8 mM) [72]. There has been much speculation about the functional consequences of reactions between •NO and thiols. It has been suggested that the formation of S-nitrosothiols from this reaction represents a mechanism of either storage or transport of •NO [73].

The biological chemistry of •NO and thiols is complex. •NO will oxidize glutathione (GSH) to glutathione disulfide (GSSG) generating NO$^-$ [Eq. (9)] [74].

$$2GSH + 2•NO \longrightarrow GSSG + 2NO^- + 2H^+ \qquad (9)$$
$$2NO^- + 2H^+ \longrightarrow N_2O + H_2O \qquad (10)$$
$$NO^- + O_2 \longrightarrow ONOO^- \qquad (11)$$

The chemistry of NO$^-$ in biological systems is incompletely understood. Dimerization of NO$^-$ leads to the formation of nitrous oxide [N$_2$O, Eq. (10)] and under anaerobic conditions, 70% of •NO can be detected as N$_2$O after reaction with glutathione [74]. NO$^-$ may also be reduced by glutathione to yield hydroxylamine and GSSG [17]. In the presence of oxygen, NO$^-$ can form peroxynitrite [Eq. (11)] [14]. The reaction between protein thiols and nitric oxide also leads to the formation of nitrous oxide, and if disulfide formation is not possible, the protein cysteinyl residue is oxidized to the sulfenic acid (RSOH) [75]. The reaction of peroxynitrite with glutathione generates mainly GSSG [43] and also small yields of glutathione radical [50] and S-nitrosoglutathione (GSNO) [61]. It has been reported that S-nitroglutathione (GSNO$_2$) is also generated from the reaction between peroxynitrite and glutathione [76].

As mentioned previously, the reaction of •NO with oxygen leads to the formation of N$_2$O$_3$. This potent nitrosating agent can react with thiols to generate S-nitrosothiols as shown in Eq. (12).

$$GSH + N_2O_3 \longrightarrow GSNO + NO_2^- + H^+ \qquad (12)$$

Hence the reaction of •NO with GSH in well-oxygenated solutions leads predominantly to the formation of GSNO. However, in vivo, the reaction of •NO with oxygen is thought to be of minor consequence, and the importance of this mechanism of S-nitrosothiol synthesis remains to be established.

Although there is as yet limited direct evidence of endogenous low-molecular-weight S-nitrosothiols in vivo, these compounds have been implicated in a number of biochemical processes. For example, it has been suggested that EDRF more closely resembles S-nitrosocysteine than •NO [77], though this has been contested [78]. There is good evidence for the presence of S-nitroso-proteins in plasma (e.g., S-nitrosoalbumin) [79] and in erythrocytes (e.g., S-nitrosohemoglobin) [80]. These and other examples establish that posttranslational modification of protein cysteinyl residues can result in the formation of an S-nitrosocysteinyl residue [81]. Such modification is easy to achieve in vitro, as incubation of any thiol-containing protein with excess low-molecular-weight S-nitrosothiol generally results in the formation of an *S*-nitroso-protein. The mechanism for this reaction involves transnitrosation: the transfer of NO^+ from an *S*-nitrosothiol to a thiol:

$$RSNO + R'SH \longrightarrow RSH + R'SNO \tag{13}$$

These reactions are fairly rapid ($k \sim 10–100 \ M^{-1} \ s^{-1}$), reversible, and represent a distribution of the nitrosyl functional group among all available thiols [82–84]. The high concentration of glutathione in the intracellular space may make transnitrosation reactions effectively unidirectional and favors formation of GSNO above all other *S*-nitrosothiols. One exception is perhaps in the erythrocyte, where the concentration of hemoglobin matches that of glutathione. Significant quantities of *S*-nitrosohemoglobin have been measured in venous and arterial blood [79]. In the extracellular space, the thiol group of serum albumin may represent the thermodynamic sink for the nitrosyl moiety [78]. Transnitrosation reactions have been suggested to be the predominant mechanism of action of exogenously added GSNO in vivo [85]. De Groot et al. [86] demonstrated that the cytostatic action of GSNO on *Salmonella typhimurium* appears to be associated with the transfer of the *S*-nitroso functional group to the cytoplasmic space rather than external •NO release.

Most biologically relevant *S*-nitrosothiols are remarkably stable compounds in the absence of light and contaminating transition metal ions, both of which stimulate *S*-nitrosothiol decomposition [84,87,88]. Light stimulates homolytic decomposition of *S*-nitrosothiols, resulting in the production of •NO and the corresponding thiyl free radical [89]. Interestingly, transition metal ion-dependent decomposition of *S*-nitrosothiols does not appear to involve a thiyl radical intermediate [83]. As both light and free metal ions are unlikely to be present in biological systems under normal conditions, it is not known how, or if, *S*-nitrosothiols decompose to give •NO in vivo. Transnitrosation reactions will only redistribute the nitroxyl moiety according to the thermodynamic parameters of the various reactions but will not lead to •NO release. In the absence of a faster reaction, GSNO will slowly react with GSH to give GSSG and NO^- as shown in Eq. (14) [74]:

$$GSNO + GSH \longrightarrow GSSG + NO^- + H^+ \tag{14}$$

As mentioned previously, NO^- can result in the formation of N_2O and $ONOO^-$. The mechanisms of *S*-nitrosothiol decomposition in vivo remain to be elucidated.

3.5. Peroxyl Radicals

Lipid peroxidation may be a consequence or a cause of many pathological states. Peroxidation of lipid bilayers not only disrupts membrane integrity at the level of the lipid but may transmit the damage to integral membrane proteins. Moreover, the products of lipid peroxidation can have adverse consequences on the normal eicosanoid signaling pathways. As noted above, reactive free radicals can initiate lipid peroxidation by a hydrogen abstraction reaction and $ONOO^-$ can also initiate lipid peroxidation by a radical-like reaction. The propagation of lipid peroxidation is represented in Eqs. (15) and (16):

$$L\bullet + O_2 \longrightarrow LOO\bullet \tag{15}$$
$$LOO\bullet + LH \longrightarrow LOOH + L\bullet \tag{16}$$

The lipid radical (L•) from Eq. (1) reacts rapidly with oxygen to give a lipid peroxyl radical (LOO•). The lipid peroxyl radical can then abstract a hydrogen atom from another molecule of lipid (LH) to give a lipid hydroperoxide (LOOH) and a further lipid radical. Equation (16) is the rate-limiting step of this chain reaction, and conventional chain-breaking antioxidants, such as α-tocopherol, function by scavenging the peroxyl radical intermediate, generating a lipid hydroperoxide and a stable antioxidant radical.

Recently, it has been discovered that •NO can inhibit the propagation of lipid peroxidation by scavenging peroxyl radicals in both low-density lipoprotein [90], free fatty acid [56], and liposome systems [91]. The reaction between •NO and peroxyl radicals [Eq. (17)] is diffusion controlled [92] and forms nitrogen-containing lipid adducts that have been detected by mass spectrometry [91,93].

$$LOO\bullet + \bullet NO \longrightarrow LOONO \tag{17}$$

•NO, delivered using a slow-release •NO donor, is an extremely potent inhibitor of lipid oxidation and will suppress the propagation chain reaction at physiological concentrations [94]. A corollary of this observation is that any process that inhibits the production of •NO may be regarded as pro-oxidant due to the removal of the radical scavenging activity of •NO.

3.6. Iron–Sulfur Complexes

The combination of iron, thiol, and •NO leads to the formation of dinitrosyl iron complexes (DNIC). These complexes are paramagnetic and have a characteristic ESR spectrum. There is evidence that such complexes are formed in biological systems and represent either a degradation product of ferritin or iron–sulfur centers, or a stabilized form of •NO, having either a storage or transport role. Low-molecular-weight DNIC can interact with serum albumin to generate protein-bound DNIC and other nitrosyl–iron complexes, and also generate protein S-nitrosothiols [95].

The reaction between •NO and iron–sulfur complexes of proteins leads to similar protein-bound complexes with characteristic ESR spectra. Such complexes can be observed in whole macrophages stimulated to produce •NO [96].

The functional consequences of the reaction between •NO and iron–sulfur complexes are under investigation and have unearthed a surprising relationship between •NO and iron regulation. The cytoplasmic form of aconitase, an enzyme of the tricarboxylic acid cycle, contains a cubane iron–sulfur cluster. However, a form of cytoplasmic aconitase that is deficient in the iron–sulfur cluster is a powerful translational regulator of proteins involved in iron metabolism such as ferritin and transferrin receptor [97]. It has been reported that elevated •NO can act as a functional switch between the aconitase activity and the iron-regulatory protein activity of this enzyme [98]. By this mechanism, elevated •NO mimics iron deficiency and results in increased translation of transferrin receptor and decreased translation of ferritin.

3.7. Other Reactions

Proteins

Nitric oxide-dependent inhibition of ribonucleotide reductase is thought to occur via the direct reaction of •NO with a protein tyrosyl radical that is an essential intermediate in the catalytic cycle of this enzyme [99]. This reaction, which would block DNA synthesis, represents a possible mechanism for the well documented cytostatic effects of •NO.

Several studies have indicated that •NO is able to modify nonthiol amino acid residues of proteins. Moriguchi et al. [100] implicated deaminination of the *N*-terminus of hemoglobin and carbonic anhydrase by nitric oxide, presumably through N_2O_3 formation and nitrosamine formation.

Antioxidants

Although there are a number of reports indicating that •NO, added in gaseous form, directly oxidizes α-tocopherol to the α-tocopheryl radical and ultimately the quinone form [101–103], other studies indicate that if •NO is released slowly from donor compounds, no reaction with α-tocopherol is observed [53,94]. It has recently been demonstrated that •NO gas is also incapable of oxidizing α-tocopherol if appropriate steps are taken to remove contaminating oxides of nitrogen [104].

4. Conclusion

There are many potential reactions of •NO as it journeys from its site of synthesis to its site of action. Some of these reactions are clearly deleterious, some are clearly advantageous, but most lie in the ambiguous middle ground. This can be illustrated by the differential toxicity of SIN-1, the molecule that simultaneously generates •NO and superoxide on various cell types. SIN-1 is toxic to cortical neurons, in culture, and Superoxide Dismutase (SOD) inhibits this toxicity by preventing ONOO⁻ production [18]. However, the opposite effect is observed in a human epithelial ovarian cancer cell line, where SOD enhanced the toxicity of SIN-1 by causing the generation of hydrogen peroxide [105]. The combination of •NO with hydrogen peroxide appears to be the toxic event in this system. It is therefore dangerous to extrapolate the effects of •NO from system to system and impossible to generalize about the biological consequences of changes in •NO production.

References

1. Furchgott, R.F. The role of the endothelium in the responses of vascular smooth muscle to drugs. *Ann. Rev. Toxicol.* **24**, 175–197 (1984).

2. Ignarro, L.J., Lippton, H., Edwards, J.C., Baricos, W.H., Hyman, A.L., Kadowitz, P.J., and Gruetter, C.A. Mechanism of vascular smooth muscle relaxation by organic nitrates, nitrites, nitroprusside and nitric oxide: evidence for the involvement of *S*-nitrosothiols as active intermediates. *J. Pharm. Exp. Ther.* **218**, 739–749 (1981).

3. Palmer, R.M.J., Ferrige, A.G., and Moncada, S. Nitric oxide release accounts for the biological activity of endothelium-derived relaxing factor. *Nature* **327**, 524–526 (1987).

4. Merck Index, Eleventh Edition (Ed.) S. Budavari, Merck Co. Inc., 1989.

5. Dinnerman, C.E. and Ewing, G.E. Infrared spectrum, structure, and heat of formation of gaseous $(NO)_2$. *J. Phys. Chem.* **53**, 626–631 (1970).

6. Feelisch, M. The biochemical pathways of nitric oxide formation from nitrovasodilators: Appropriate choice of exogenous NO donors and aspects of preparation and handling of aqueous NO solutions. *J. Cardiovasc. Pharm.* **17** (Suppl. 3), S25–S33 (1991).

7. Subczynski, W.K., Lomnicka, M., and Hyde, J.S. Permeability of nitric oxide through lipid bilayer membranes. *Free Rad. Res.* **24**, 343–349 (1996).

8. Denicola, A., Souza, J.M., Radi, R., and Lissi, E. Nitric oxide diffusion in membranes determined by fluorescence quenching. *Arch. Biochem. Biophys.* **328**, 208–212 (1996).

9. Singh, R.J., Hogg, N., Mchaourab, H.S., and Kalyanaraman, B. Physical and chemical interaction between nitric oxides and nitroxides. *Biochim. Biophys. Acta* **1201**, 437–441 (1994).

10. Malinski, T., Taha, Z., Grunfeld, S., Patton, S., Kapturczak, M., and Tomboulian, P. Diffusion of nitric oxide in the aorta wall monitored in situ by porphyrinic microsensors. *Biochem. Biophys. Res. Commun.* **193**, 236–244 (1993).

11. Koppenol, W.H., Pryor, W.A., Moreno, J.J., Ischiropoulos, H., and Beckman, J.S. Peroxynitrite, a cloaked oxidant formed by nitric oxide and superoxide. *Chem. Res. Toxicol.* **5**, 834–842 (1992).

12. Hogg, N., Singh, R.J., and Kalyanaraman, B. The role of glutathione in the transport and catabolism of nitric oxide. *FEBS Lett.* **382**, 223–228 (1996).

13. Pryor, W.A., Church, D.F., Govindan, C.K., and Crank, G. Oxidation of thiols by nitric oxide and nitrogen dioxide: Synthetic utility and toxicological implication. *J. Organ. Chem.* **47**, 159–161 (1982).

14. Donald, C.E., Hughes, M.N., Thompson, J.M., and Bonner, F.T. Photolysis of the NZ–SN bond in trioxodinitrate: Reaction between triplet NO⁻ and O_2 to form peroxynitrite. *Inorg. Chem.* **25**, 2676–2677 (1988).

15. Fukuto, J.M., Chiang, K., Hszieh, R., Wong, P., and Chaudhuri, G. The pharmacological activity of nitroxyl: a potent vasodilator with activity similar to nitric oxide and/or endothelium-derived relaxing factor. *J. Pharm. Exp. Ther.* **263**, 546–551 (1992).

16. Pino, R.Z. and Feelisch, M. Bioassay discrimination between nitric oxide (•NO) and nitroxyl (NO⁻) using L-cysteine. *Biochem. Biophys. Res. Commun.* **201**, 54–62 (1994).

17. Arnelle, D. and Stamler, J.S. NO⁺, •NO, and NO⁻ donation by *S*-nitrosothiols: Implications for regulation of physiological functions by *S*-nitrosylation and acceleration of disulfide formation. *Arch. Biochem. Biophys.* **318**, 279–285 (1995).

18. Lipton, A.S., Chol, Y.-B., Pan, Z-H, Lei, S.Z., Chen, H-S.V., Sucher, N.J., Loscalzo, J., Singel, D.J., and Stamler J.S. A redox based mechanism for the neuroprotective and neurodestructive effects of nitric oxide and related nitroso-compounds. *Nature* **364**, 626–631 (1993).

19. Bonner, F. and Stedman, G. The chemistry of nitric oxide and redox related species. In: *Methods in Nitric Oxide Research.* Eds. Feelisch, M. and Stamler, J.S. John Wiley and Sons Ltd., London, 1996, pp. 3–18.

20. Antonini, E. Science, 158, 1417–1425 (1967).

21. Stevens, T.B., Brudvig, G.W., Bocian, D.F., and Chan, S.I. Structure of cytochrome a2-Cua3 couple in cytochrome c oxidase as revealed by nitric oxide binding studies. *Proc. Natl. Acad. Sci. USA* **76**, 3320–3324 (1979).

22. Ignarro, L.J., Wood, K.S., and Wollin, M.S. Regulation of purified soluble guanylate cylase by porphyrins and metalloporphyrins: a unifying concept. *Adv. Cyclic Nucleotide Protein Phosphoryl. Res.* **17**, 267–274 (1984).

23. Stone, J.R. and Marletta, M.A. Soluble guanylate cyclase from bovine lung: Activation with nitric oxide and carbon monoxide and spectral characterizations of the ferrous and ferric forms. *Biochemistry* **33**, 5636–5640 (1994).

24. Traylor, T.G. and Sharma, V.S. Why NO? *Biochemistry* **31**, 2847–2849 (1992).

25. Kanner, J., Harel, S., and Granit, R. Nitric oxide as an antioxidant. *Arch. Biochem. Biophys.* **289**, 130–136 (1991).

26. Kanner, J., Harel, S., and Granit, R. Nitric oxide, an inhibitor of lipid oxidation by lipoxygenase, cyclooxygenase and hemoglobin. *Lipids* **27**, 46–49 (1992).

27. Vanin A.F., Mordvintcev, P.I., Hauschildt, S., and Mulsch, A. The relationship between L-arginine-dependent nitric oxide synthesis, nitrite release and dinitrosyl–iron complex formation by activated macrophages. *Biochim. Biophys. Acta* **1177**, 37–42 (1993).

28. Verdernikov, Y.P., Mordvintcev, P.I., Malenkova, I.V., and Vanin A.F. Similarity between the vasorelaxing activity of dinitrosyl iron cysteine complexes and endothelium-derived relaxing factor. *Eur. J. Pharmacol.* **211**, 313–317 (1992).

29. Mulsch, A., Mordvintcev, P.I., Vanin A.F., and Busse R. The potent vasodilating and guanylyl cyclase activating dinitrosyl iron(II) complex is stored in a protein bound orm in vascular tissue and is released by thiols. *FEBS Lett.* **294**, 252–256 (1991).

30. Wink, D.A., Darbyshire, J.F., Nims R.W., Saavedra J.E., and Ford P.C. Reaction of the bioregulatory agent nitric oxide in oxygenated aqueous media: Determination of the kinetics for the oxidation and nitrosation by intermediates generated in the •NO/O₂ reaction. *Chem. Res. Toxicol.* **6**, 23–27 (1993).

31. Kharitonov, V.G., Sundquist A.R., and Sharma, V.J. Kinetics of nitric oxide autoxidation in aqueous solution. *J. Biol. Chem.* **269**, 5881–5883 (1994).

32. Hogg, H., Singh, R.J., Joseph, J., Neese, F., and Kalyanaraman B. Reaction of nitric oxide with nitronyl nitroxides and oxygen: Prediction of nitrite and nitrate formation by kinetic simulation. *Free Rad. Res.* **22**, 47–56 (1995).

33. Bonner, F.T. and Stedman, G. The chemistry of nitric oxide and redox-related species. In: *Methods in Nitric Oxide Research.* Eds. Feelisch, M. and Stamler, J.S. John Wiley and Sons, Ltd., London, 1996.

34. Lancaster, J.R. Simulation of the diffusion and reaction of endogenously produced nitric oxide. *Proc. Natl. Acad. Sci. USA* **91**, 8137–8141 (1994).

35. Lewis, R.S., Tamir, S., Tannenbaum, S.R., and Deen W.M. Kinetic analysis of the fate of nitric oxide synthesized by macrophages in vitro. *J. Biol. Chem.* **270**, 29350–29355 (1995).

36. Czapski, G. and Goldstein, S. The role of the reactions of •NO with superoxide and oxygen in biological systems: A kinetic approach. *Free Rad. Biol. Med.* **19**, 785–794 (1995).

37. Wink, D.A., Nims, R.W., Darbyshire, J.F., Christodoulou, D., Hanbauer, I., Cox, G.W., Laval, F., Laval, J., Cook, J.A., Krishna, M.C., LeGraff, W.G., and Mitchell, J.B. Reaction kinetics for nitrosation of cysteine and glutathione in aerobic nitric oxide solutions at neutral pH. Insights into the fate and physiological effects of intermediates generated in the NO/O₂ reactions. *Chem. Res. Toxicol.* **7**, 519–525 (1994).

39. Miwa, M., Stuehr, D.J., Marletta, M.A., Wishnok, J.S., and Tannenbaum, S.R. Nitrosation of amines by stimulated macrophages. *Carcinogenesis* **8**, 955–958 (1987).

39. Blough, N.V. and Zafiriou, O.C. Reaction of superoxide with nitric oxide to form peroxynitrite in alkaline solution. *Inorg. Chem.* **24**, 3502–3504 (1985).

40. Edwards, J.O. and Plimb, R.C. The chemistry of peroxonitrites. *Prog. Inorg. Chem.* **41**, 599–635 (1994).

41. Beckman, J.S., Beckman, T.W., Chen, J., Marshall, P.A., and Freeman, B.A. Apparent hydroxyl radical production by peroxynitrite: Implications for endothelial injury

from nitric oxide and superoxide. *Proc. Natl. Acad. Sci.* (USA), **87,** 1620–1624 (1990).

42. Pryor, W.A. and Squadrito, G.L. The chemistry of peroxynitrite: a product from the reaction of nitric oxide with superoxide. *Am. J. Physiol.* **268,** L699–L722 (1995).

43. Radi, R., Beckman, J.S., Bush, K.M., and Freeman, B.A. Peroxynitrite oxidation of sulfhydryls. *J. Biol. Chem.* **266,** 4244–4250 (1991).

44. Beckman, J.S., Ischiropoulos, H., Zhu, L., van der Woerd, M., Smith, C., Chen, J., Harrison, J., Martin, J.C., and Tsai, M. Kinetics of superoxide dismutase- and iron-catalyzed nitration of phenolics by peroxynitrite. *Arch. Biochem. Biophys.* **298,** 438–445 (1992).

45. Beckman, J.S., Ye, Y.Z., Anderson, P.G., Chen, J., Accavitti, M.A., Tarpey, M.M., and White C.E. Extensive nitration of protein tyrosines in human atherosclerosis detected by immunohistochemistry. *Biol. Chem. Hoppe-Seyler* **375,** 81–88 (1994).

46. Pryor, W.A., Jin, X., and Squadrito, G.L. One- and two-electron oxidations of methionine by peroxynitrite. *Proc. Natl. Acad. Sci. U.S.A.* **91,** 11173–11177 (1994).

47. Hogg, H., Joseph, J., and Kalyanaraman, B. The oxidation of α-tocopherol and trolox by peroxynitrite. *Arch. Biochem. Biophys.* **314,** 153–158 (1994).

48. Bartlett, D., Church, D.F., Bounds, P.L., and Koppenol, W.H. The kinetics of the oxidation of L-ascorbate by peroxynitrite. *Free Rad. Biol. Med.* **18,** 85–92 (1995).

49. Augusto, O., Gatti, R.M., and Radi, R. Spin-trapping studies of peroxynitrite decomposition and of 3-morpholinsydnonimine *N*-ethylcarbamide autooxidation: Direct evidence for metal-independent formation of free radical intermediates. *Arch. Biochem. Biophys.* **310,** 118–125 (1994).

50. Karoui, H., Hogg, N., Fréjaville, C., Tordo, P., and Kalyanaraman, B. Characterization of sulfur-centered radical intermediates formed during the oxidation of thiols and sulfite by peroxynitrite. *J. Biol. Chem.* **271,** 6000–6009 (1996).

51. Koppenol, W.H., Pryor, W.A., Moreno, J.J., Ischiropoulos, H., and Beckman, J.S. Peroxynitrite, a cloaked oxidant formed by nitric oxide and superoxide. *Chem. Res. Toxicol.* **5,** 834–842 (1992).

52. Hogg, N., Darley-Usmar, V.M., Wilson, M.T., and Moncada, S. Production of hydroxyl radicals from the simultaneous generation of superoxide and nitric oxide. *Biochem. J.* **281,** 419–424 (1992).

53. Darley-Usmar, V.M., Hogg, N., O'Leary, V.J., Wilson, M.T., and Moncada, S. The simultaneous generation of superoxide and nitric oxide can initiate lipid peroxidation in human low-density lipoprotein. *Free Rad. Res. Commun.* **17,** 9–20 (1992).

54. Hogg, N., Darley-Usmar, V.M., Wilson, M.T., and Moncada, S. The oxidation of α-tocopherol in human low-density lipoprotein by the simultaneous generation of superoxide and nitric oxide. *FEBS Lett.* **326,** 199–203 (1993).

55. Feelisch, M., Ostrowski, J., and Noack, E. On the mechanism of NO release from sydnonimines. *J. Cardiovasc. Pharm.* **14** (Suppl. 11), S13–S22 (1989).

56. Rubbo, H., Radi, R., Trujillo, M., Telleri, R., Kalyanaraman, B., Barnes, S., Kirk, M., and Freeman B. Nitric oxide regulation of superoxide and peroxynitrite-dependent lipid peroxidation. *J. Biol. Chem.* **42,** 26066–26075 (1994).

57. Wink, D.A., Hanbauer, I., Krishna, M.C., DeGraff, W., Gamson, J., and Mitchell, J.B. Nitric oxide protects against cellular damage and cytotoxicity from reactive oxygen species. *Proc. Natl. Acad. Sci. USA* **90**, 9813–9817 (1993).

58. Gergel, D., Misik, V., Ondrias, K., and Cederbaum, A.I. Increased cytotoxicity of 3-morpholinnsydnonimine to HepG2 cells in the presence of superoxide dismutase. *J. Biol. Chem.* **270**, 20922–20929 (1995).

59. Hecker, M., Priess, C., Schinikerth, V.B., and Busse R. Antioxidants differentially affect nuclear factor kappa-B-mediated nitric oxide synthase expression in vascular smooth muscle cells. *FEBS Lett,* **385**, 124 (1996).

60. Wu, M., Pritchard Jr., K.A., Kaminski, P.M., Fayngersh, R.P., Hintze, T.H., and Wolin, M.S. Involvement of nitric oxide and nitosothiols in relaxation of pulmonary arteries to peroxynitrite. *Am. J. Physiol.* **266**, H2108–H2113 (1984).

62. Moro, M.A., Darley-Usmar, V.M., Lizasoain, I., Su, Y., Knowles, R.G., Radomski, M.W., and Moncada, S. The formation of nitric oxide donors from peroxynitrite. *Br. J. Pharmacol.* **116**, 1999–2004 (1995).

62. Ravichandran, R.V., Johns, R.A. and Rengasamy, A. Direct and reversible inhibition of endothelial nitric oxide synthase by nitric oxide. *Am. J. Physiol.* **268**, H2216–H2223 (1995).

63. Cleeter, M.W.J., Cooper, J.M., Darley-Usmar, V.M., Moncada, S. and Schapira, A.H.V. Reversible inhibition of cytochrome *c* oxidase, the terminal enzyme of the mitochondrial electron transport chain, by nitric oxide. *FEBS Lett.* **345**, 50–54 (1994).

64. Borutaite, V. and Brown, G.C. Rapid reduction of nitric oxide by mitochondria, and reversible inhibition of mitochondrial respiration by nitric oxide. *Biochem. J.* **315**, 295–299 (1996).

65. Hermann, L. Ueber die Wirkungen des Stickoxydulgases auf das Blut. *Arch. Anat. Physiol. Lpz.* 469–481 (1865).

66. Kelm, M. and Schrader, J. Control of coronary vascular tone by nitric oxide. *Circ. Res.* **66**, 1561–1575 (1990).

67. Feelisch, M. and Noack, E.A. Correlation between nitric oxide formation during degradation of organic nitrates and activation of guanylate cyclase. *Eur. J. Pharmacol.* **139**, 19–30 (1987).

68. Gibson, Q.H. and Roughton, F.J.W. The kinetics and equilibria of the reactions of nitric oxide with sheep haemoglobin. *J. Physiol.* **136**, 507–526 (1957).

69. Westenberger, U., Thanner, S., Ruf, H.H., Gersonde, K., Sutter, G., and Trenz, O. Formation of free radicals and nitric oxide derivative of hemoglobin in rats during shock syndrome. *Free Rad. Res. Commun.* **11**, 167–178 (1990).

70. Konorev, E.A., Joseph, J., and Kalyanaraman, B. *S*-Nitrosoglutathione induces formation of nitrosylmyoglobin in isolated hearts during cardioplegic ischemia— an electron spin resonance study. *FEBS Lett.* **378**, 111–114 (1996).

71. Addison, A.W. and Stephanos, J.J. Nitrosyliron (III) hemoglobin: Autoreduction and spectroscopy. *Biochemistry* **25**, 4104–4113 (1986).

72. Griffith, O.W. and Meister A., Glutathione: Interorgan translocation, turnover and metabolism. *Proc. Natl. Acad. Sci. USA* **76**, 5606–5610 (1979).

73. Stamler, J.S., Singel, D.J., and Loscalzo, J. Biochemistry of nitric oxide and its redox-activated forms. *Science* **258**, 1898–1902 (1992).

74. Hogg, N., Singh, R.J., and Kalyanaraman, B. The role of glutathione in the transport and catabolism of nitric oxide. *FEBS Lett,* **382**, 223–228 (1996).

75. DeMaster, E.G., Quast, B.J., Redfern, B., and Nagasawa, H.T. Reaction of nitric oxide with free sulfhydryl group of human serum albumin yields a sulfenic acid and nitrous oxide. *Biochemistry* **34**, 11494–11499 (1995).

76. Davidson, C.A., Kaminski, P.M., Wu, M., and Wolin, M.S. Nitrogen dioxide causes pulmonary arterial relaxation via thiol nitrosation and •NO formation. *Am. J. Physiol.* **270**, H1038–H1043 (1996).

77. Myers, P.R., Minor Jr., R.L., Guerra Jr., R., Bates, J.N., and Harrison, D.G. Vasorelaxant properties of the endothelium-derived relaxing factor more closely resemble *S*-nitrosocysteine than nitric oxide. *Nature* **345**, 161–163 (1990).

78. Feelisch, M., te Poel, M., Zamora, R., Deussen, A., and Moncada, S. Understanding the controversy over the identity of EDRF. *Nature* **368**, 62–65 (1994).

79. Stamler, J.S., Jarkari, O., Osborne, J., Simon, D.I., Keaney, J., Vita, J., Singel, D., Valeri, C.R., and Loscalzo, J. Nitric oxide circulates in mammalian plasma primarily as an S-nitroso adduct of serum albumin. *Proc. Natl. Acad. Sci. USA* **89**, 7674–7677 (1992).

80. Jia, L., Bonaventura, C., Bonaventura, J., and Stamler, J.S. *S*-Nitrosohaemoglobin: A dynamic activity of blood involved in vascular control. *Nature* **380**, 221–226 (1996).

81. Stamler, J.S., Simon, D.I., Osborne, J.A., Mullins, M.E., Jaraki, O., Michel, T., Singel, D.J., and Loscalzo, J. *S*-Nitrosylation of proteins with nitric oxide: Synthesis and characterization of biologically active compounds. *Proc. Natl. Acad. Sci. USA.* **89**, 222–228 (1992).

82. Mayer, D., Kramer, H., Özer, N., Coles, B., and Ketterer, B. Kinetics and equilibria of *S*-nitrosothiol–thiol exchange between glutathione, cysteine, penicillamine and serum albumin. *FEBS Lett.* **345**, 177–180 (1994).

83. Barnett, D.J., Rios, A., and Williams, D.L.H. NO-group transfer (transnitrosation) between *S*-nitrosothiols and thiols. Part 2. *J. Chem. Soc. Perkin Trans.* **2**, 1279–1282 (1995).

84. Singh, R.J., Hogg, N., Joseph, J., and Kalyanaraman, B. Mechanism of nitric oxide release from S-nitrosothiols. *J. Biol. Chem.* **271**, 18596–18603 (1996).

85. Park, J-W., Billman, G.E., and Means, G.E. Transnitrosation as a predominant mechanism in the hypotensive effect of *S*-nitrosoglutathione. *Biochem. Mol. Biol. Int.* **30**, 885–891 (1993).

86. De Groote, M.A., Granger, D., Xu, Y., Campbell, G., and Prince, R. Genetic and redox determinants of nitric oxide cytotoxicity in a *Salmonella typhimurium* model. *Proc. Natl. Acad. Sci. USA* **92**, 6399–6403 (1995).

87. Askew, S.C., Barnett, D.J., McAninly, J., and Williams, D.L.H. Catalysis by Cu^{2+} of nitric oxide release from *S*-nitrosothiols (RSNO). *J. Chem. Soc. Perkin Trans.* **2**, 741–745 (1995).

88. McAninly, J., Williams, D.L.H., Askew, S.C., Butler, A.R., and Russell, C. Metal ion catalysis in nitrosothiol (RSNO) decomposition. *J. Chem. Soc., Chem. Commun.*, 1758–1759 (1993).

89. Singh, R.J., Hogg, N., Joseph, J., and Kalyanaraman, B. Photosensitized decomposition of *S*-nitrosothiols and 2-methyl-2-nitrosopropane. Possible use for site-directed nitric oxide production. *FEBS Lett.* **360**, 47–61 (1995).

90. Hogg, N., Kalyanaraman, B., Joseph, J., Struck, A., and Parthasarathy, S. Inhibition of low-density lipoprotein oxidation by nitric oxide. Potential role in atherogenesis. *FEBS Lett* **334**, 170–174 (1993).

91. Hayashi, K., Noguchi, N., and Niki, E. Action of nitric oxide as an antioxidant against oxidation of soybean phosphatidylcholine liposomal membranes. *FEBS Lett.* **370**, 37–40 (1995).

92. Padamaja, S. and Huie R.E. The reaction of nitric oxide with organic peroxyl radicals. *Biochem. Biophys. Res. Commun.* **195**, 539–544 (1993).

93. Rubbo, H., Parthasarathy, S., Barnes, S., Kirk, M., Kalyanaraman, B., and Freeman, B.A. Nitric oxide inhibition of lipoxygenase-dependent liposome and low-density oxidation: Termination of radical chain propagation reactions and formation of nitrogen-containing oxidized lipid derivatives. *Arch. Biochem. Biophys.* **324**, 15–25 (1995).

94. Goss, S.P.A., Hogg, N., and Kalyanaraman, B. The antioxidant effect of spermine NONOate in human low-density lipoprotein. *Chem. Res. Toxicol.* **8**, 800–806 (1995).

95. Boese, M., Mordvintcev, P.I., Vanin, A.F., Busse, R., and Mulsch,. A. *S*-Nitrosation of serum albumin by dinitrosyl-iron complex. *J. Biol. Chem.* **270**, 29244–29249 (1995).

96. Stadler, J., Bergonia, H.A., Di Silvio, M., Sweetland, M.A., Billiar, R., Simmons, R.L., and Lancaster, J.R. Nonheme iron–nitrosyl complex formation in rat hepatocytes: Detection by electron paramagnetic resonance spectroscopy. *Arch. Biochem. Biophys.* **302**, 4–11 (1993).

97. Bienert, H. and Kennedy, M.C. Aconitase, a two face protein: Enzyme and iron regulatory factor. *FASEB J.* **7**, 1442–1449 (1993).

98. Pantopoulos, K. and Hentze, M.W. Nitric oxide signaling to oron-regulatory protein: Direct control of ferritin mRNA translation and transferrin receptor mRNA stability in transfected fibroblasts. *Proc. Natl. Acad. Sci. USA* **92**, 1267–1271 (1995).

99. Lepoivre, M., Fieschi, F., Coves, J., Thelander, L., and Fontecave, M. Inactivation of ribonucleotide reductase by nitric oxide. *Biochem. Biophys. Res. Commun.* **179**, 442–448 (1991).

100. Moriguchi, M., Manning, L.R., and Manning, J.M. Nitric oxide can modify amino acid residues in proteins. *Biochem. Biophys. Res. Commun.* **183**, 598–604 (1992).

101. Gurbunov, N.V., Osipov, A.N., Sweetland, M.A., Day, B.W., Elsayed, N.M., and Kagen V.E. *Biochem. Biophys. Res. Commun.* **219**, 835–841 (1996).

102. Janzen, E.G., Wilcox, A.L., and Manoharan, V. *J. Org. Chem.* **58**, 3597–3599 (1993).

103. de Groot, H., Hegi, U. and Sies, H. (1993). *FEBS Lett,* **315,** 139–142.

104. Hogg, N., Singh, R.J., Goss, S.P.A., and Kalyanaraman, B. The reaction between nitric oxide and α-tocopherol: A reappraisal. *Biochem. Biophys. Res. Commun.* **224,** 696–702 (1996).

105. Farias-Eisner, R., Chaudhuri, G., Aeberhard, E., and Fukuto, J.M. The chemistry and tumoricidal activity of nitric oxide/hydrogen peroxide and the implications to cell resistance/susceptibility. *J. Biol. Chem.* **271,** 6144–6151 (1996).

2

Role of NO and Nitrogen Intermediates in Regulation of Cell Functions

*Young-Myeong Kim, Edith Tzeng, and
Timothy R. Billiar*

1. Introduction

Prior to the 1980s, nitric oxide (NO) was best known as a toxic reactive free radical found in atmospheric pollutants and carcinogens [1], a by-product of microbial nitrogen metabolism [2,3] and a potent activator of the mammalian enzyme heme-containing soluble guanylate cyclase [4]. As early as 1916, evidence for a mammalian nitrogen metabolic pathway was reported with the finding that several types of animals, including humans, excreted more urinary nitrate than could be accounted for by dietary intake [5]. Further evidence was reported by Tannenbaum and co-workers [6,7] who showed that in vivo mammalian nitrate formation was substantially enhanced by administration of the immunostimulant lipopolysaccharide (LPS) [8]. Stuehr and Marletta [9] first demonstrated that murine macrophages stimulated in vitro with LPS expressed nitrogen oxide synthetic activity and produced nitrite (NO_2^-) and nitrate (NO_3^-). Simultaneously, other investigators were trying to elucidate the identity of a short-lived, diffusable endothelium-derived relaxing factor (EDRF) produced by acetylcholine-treated endothelial cells that was responsible for mediating smooth muscle cell relaxation [10]. Based on the similarities in the pharmacological properties of EDRF and NO generated from acidified nitrite, Furchgott suggested that EDRF may be NO in 1986 [11]. At the same time, Ignarro et al. also proposed that EDRF may be NO or a closely related species [12]. The following year, two independent groups of investigators [13,14] demonstrated EDRF was indeed nitric oxide (NO).

Following the first report describing nitrite and nitrate biosynthesis by mammalian cells (macrophages) stimulated with LPS in 1985 [9], macrophage-generated NO was shown to cause inhibition of DNA synthesis and inhibition of mitochondrial respiration and aconitase activity in tumor target cells [15]. Subsequently, cytosols of cytokine-activated macrophages were shown to have an enzymatic

Table 2-1. Beneficial and Detrimental Effects of Nitric Oxide

Beneficial Effects	Detrimental Effects
Maintenance of tissue perfusion	Hypotension/vascular collapse
Scavenging of oxygen radicals	Formation of oxidizing agents (i.e., ONOO⁻)
Inhibition of Fenton-type reaction of hemoproteins	Iron release from ferritin (Fenton reaction)
Inhibition of microvascular thrombosis	DNA modification/damage
Inhibition of leukocyte adhesion	Inhibition of cellular protein synthesis
Inhibition of TNFα production	Apoptosis of host cells
Antimicrobial activity	
Tumoricidal activity (apoptosis)	
Induction of cytoprotective genes	

activity catalyzing an NADPH-dependent conversion of the amino acid L-arginine to NO [16], which is selectively inhibited by the L-arginine analog, N^ω-methyl-L-arginine [15]. NO synthase requires Flavin Adenine Dinucleotide (FAD), Flavin Mononucleotide (FMN), Nicotinamide Adenine Dinucleotide Phosphate (NADPH), tetrahydrobiopterin (BH_4), and heme as enzyme-bounding cofactors, and two substrates, L-arginine and molecular oxygen. It has been shown that many cell types have the capacity to generate NO from L-arginine. However, the level of NO production and the functional role of NO vary from cell to cell.

In biological systems, NO is a diffusible, highly reactive, free radical that interacts with many different types of biomolecules. It is now well accepted that NO is a mammalian biological messenger molecule involved in numerous homeostatic processes. Examples include the regulation of vasomotor tone [11–14], the cytotoxicity modulated by activated macrophages [15–18], regulation of cell proliferation [19], regulation of cell migration [20], activation of transcription factors [21], induction of cytoprotective genes [22], modulation of toxic cytokine production [23,24], and a neurotransmission in the central and peripheral nervous system [25]. NO acts like a double-edged sword, having both deleterious and beneficial effects in biological systems (Table 2-1). The ability of NO to mediate these disparate biological functions depends not only on its unique chemical properties but also on the site, local concentration, duration, or quantity produced. In this chapter, we review the function of NO in cytotoxicity, cytoprotection, and other cellular functions.

2. Chemistry of Nitric Oxide

Nitric oxide is capable of reacting with molecular oxygen, reactive oxygen species and radicals, transition metals, and some biological molecules. The most common and biologically significant target for NO is molecular oxygen. Although by Lewis' electron structure for O_2 indicates that it is not a radical, O_2 is indeed a diradical with two unpaired electrons in the Π^* antibonding orbital. NO gas can readily react with O_2 to form NO_2 gas, which can, in turn, dimerize to yield N_2O_4. N_2O_4 dismutates spontaneously in water to yield equimolecular amounts of NO_2^- and NO_3^-. However, NO in oxygenated aqueous solution does not yield NO_3^- [26]. NO synthesized by purified NOS breaks down predominantly into NO_2^- with only trace amounts of NO_3^- in the absence of contaminating hemoproteins. Similarly, macrophages (containing a small amount of hemoprotein) activated with immunostimulants generate mostly NO_2^- as the oxidized end product of NO and a small amount of NO_3^- is formed from the reaction of NO with O_2^- [27]. Hepatocytes, in contrast, contain large amounts of hemoproteins. Because of this, hepatocytes stimulated to express NOS-2 and synthesize NO generate equal amounts of NO_2^- and NO_3^- [28]. NO reacts with O_2^- and transition metals to produce peroxynitrite ($ONOO^-$) [29] and nitrosonium (NO_2^+) [30], respectively. These products can support additional nitrosative reactions at nucleophilic centers located within tyrosine residues and thiol groups, forming nitrotyrosine and *S*-nitrothiols. In biological systems, the most powerful scavengers for NO are hemoproteins such as hemoglobin and myoglobin. The reaction between hemoglobin (Hb) and NO is shown in Fig. 2-1. NO is capable of binding heme iron in either the ferrous (Fe^{2+}) or ferric (Fe^{3+}) oxidation state, but NO has a lower affinity for ferric heme. The reaction between NO and ferric heme reduces the iron moiety to the ferrous state and simultaneously generates nitrosonium ion (NO^+) [Reaction (1)], which reacts with thiols to yield *S*-nitrothiol [Reaction (2)]:

$$\text{Heme } Fe^{3+} + NO \longrightarrow \text{Heme } Fe^{3+} - NO \longleftrightarrow \text{Heme } Fe^{2+} - NO^+ \quad (1)$$
$$\text{Heme } Fe^{2+} - NO^+ + GS^- \longrightarrow Fe^{2+} + GSNO \quad (2)$$

NO also directly interacts with most reactive oxygen intermediates, which are the products of activated phagocytes or in several pathological conditions (Table 2-2) and forms a variety of reaction products (Table 2-3).

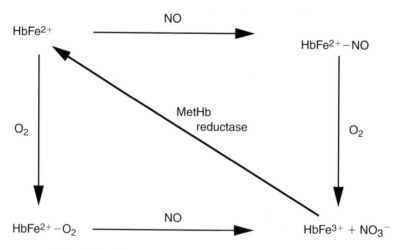

Figure 2-1. Hemoglobin-mediated scavenging of nitric oxide.

Table 2-2. Reactive Oxygen Species Generated in Biological Systems

Radicals	Nonradicals
Superoxide anion—$O_2^-\cdot$	Oxygen—O_2
Hydroxyl radical—$HO\cdot$	Hydrogen peroxide—H_2O_2
Peroxyl radical—$LOO\cdot$	Hypochlorous acid—$HOCl$
Alkoxyl radical—$LO\cdot$	
Hydroperoxyl radical—$HOO\cdot$	

Table 2-3. Possible Reactions and Products of Reactive Nitrogen and Oxygen Species in Biological Systems

Reactants	Products
$\cdot NO + O_2$	$\cdot ONOO$
$\cdot NO + O_2^-\cdot$	$ONOO^- + H^+ \longrightarrow ONOOH$
$\cdot NO + \cdot OH$	$HONO \longrightarrow NO_2^- + H^+$
$\cdot NO + LO\cdot$	$LONO$
$\cdot NO + LOO\cdot$	$LOONO$
$\cdot NO + HOO\cdot$	$ONOOH$
$\cdot NO + HOCl$	$\cdot NO_2 + Cl^- + H^+$
$\cdot NO + H_2O_2$	$\cdot NO_2 + H_2O$
$ONOOH + HOCl$	$NOCl + O_2 + H_2O$
$NO^+ + Cl^-$	$NOCl$

3.1. Biological Targets of NO

The biological activity of NO as an EDRF has been shown to be inhibited by Hb, thiol compounds, and superoxide anion, indicating that NO is highly reactive with heme-containing proteins as well as other biological molecules [31]. The toxicity of NO in macrophage-mediated immune reactions was attributed to the direct interaction of NO with iron–sulfur-containing mitochondrial enzymes as well as the inhibition of DNA synthesis through the inactivation of nonheme–iron-containing ribonucleotide reductase. Although NO reacts with a variety of biological molecules (Table 2-4), the reactivity of NO with these compounds is dependent on the local NO concentrations and the makeup of the microenvironment. For example, heme proteins such as guanylate cyclase are activated by low NO concentrations, whereas cytochrome P-450 and other nonheme proteins, such as ribonucleotide reductase and cyclooxygenase, can be inhibited by much greater concentrations of NO which are not typically biologically achievable. Further-more, it should be noted that the reaction of NO with thiol to form *S*-nitrosothiol does not directly occur; rather, "redox activation" of either the thiol to thioyl radical or further oxidation of NO to a species (NO^+ or N_2O_3) capable of nitrosation is required.

Table 2-4. Biologic Targets for Nitric Oxide and Nitrogen Intermediates

Reactive Sites	Protein Targets	Action (Refs.)
Heme	NOS	Inhibition (44–47)
	Guanylate cyclase	Activation (34)
	Cyclooxygenase	Activation (41)/inactivation (43)
	Indoleamine 2,3-dioxygenase	Inhibition (50)
	Hemoglobin	Oxidation/scavenger (51,52)
	Myoglobin	Inhibition/scavenger (53)
	Cytochrome P-450	Inhibition (48,49)
	Ferryl hemoprotein	Reduction (127)
Iron–Sulfur	Aconitase/IRE-BP	Inhibition (54,55)/activation (61,62)
	Mitochondrial complex I & II	Inhibition (55)
	SoxRS	Activation (174,175)
	Ferrochelatase	Inhibition (39,176)
	Metallothionein	Scavenger (56,57)
Nonheme iron	Ribonucleotide reductase	Inhibition (58)
	Lipoxygenase	Inhibition (43,177)
	Ferritin	Scavenger (64)/Fe release (69)
Sulfhydryl	Adenylate cyclase (type I)	Inhibition (76)
	GAPDH	Inhibition (72,73)
	Plasminogen activator	Activation (68,69)
	NMDA receptor	Inhibition (neuroprotection) (70)
	GSH	Redox signaling/NO carrier (78,98)
	p21/ras	Redox signaling (75)
	Protein kinase C	Inactivation (77)
	G-protein	Inhibition (178,179)
	Albumin	NO carrier (82–85)
	Hemoglobin	NO carrier (86)
	OMDM transferase	Inhibition (180)
	GSH peroxidase	Inhibition (145)
	GSH reductase	Inhibition (146)
Zn–sulfur	Metallothionein	Zn release (94)
	Fpg protein	Inhibition (95)
	Alcohol dehydrogenase	Inactivation (96)
	Transcription factor	Inhibition (94)
Copper	Cytochrome c oxidase	Inactivation (181)
	Caeruloplasmin	Cu release (182)
Others	Tyrosine	Nitration (30)/inhibition of signal transduction (113)
	DNA	Base modification (100,101)
	SOD	Tyrosine nitration/inhibition (30)
	Lipid	Peroxidation (106,125)
	Lipid radicals	Termination of lipid peroxidation (125)
	Tyrosine radical	Neutralizing (59)
	Ascorbic acid	Oxidative injury (183)
	NADPH oxidase	Inhibition (184)

3.1. Iron-Containing Proteins

Nitric oxide participates in cellular metabolism, signal transduction, and cellular protection by interacting with a variety of cellular components. A primary intracellular target for NO is intracellular iron [32,33]. NO nitrosylates the heme moiety in oxygen-carrying hemoproteins, perhaps the best example of which is the ferrous iron in hemoglobin and myoglobin. Another important heme-containing target for NO is soluble guanylyl cyclase [34,35]. NO alters the conformation of heme moiety of the enzyme, causing activation and cGMP synthesis. cGMP, in turn, activates cGMP-dependent protein kinases, which may then phosphorylate synaptic vesicle proteins associated with neurotransmitter release [36,37] and peripheral cGMP-gate ion channels to increase the intracellular level of Ca^{2+}. cGMP may also stimulate or inhibit cyclic nucleotide phosphodiesterase to control intracellular cAMP levels [38].

Other heme proteins are inhibited by NO. Catalase is a tetrameric ferric hemoprotein which is also sensitive to inhibition by NO in vitro [39,40]. Another heme-containing enzyme that is influenced by NO is cyclooxygenase, the rate-limiting enzyme in the biosynthesis of prostaglandins thromboxane A2 and prostacyclins. There is evidence that indicate that NO interacts with the heme prosthetic group of cyclooxygenase to either activate the enzyme to increase prostaglandin biosynthesis [41,42] or to inhibit enzyme activity [43]. NO can also regulate its own biosynthesis by interacting with the heme iron in NOSs [44–47]. This feedback inhibition may serve to control neurotransmitter release, endothelium-dependent relaxation, and NO-mediated cytotoxicity. Additionally, NO regulates several other heme-containing enzymes, including cytochrome P-450 [48,49] and indoleamine 2,3-dioxygenase [50]. NO also rapidly reacts with hemoglobin and myoglobin to form iron–nitrosyl hemoprotein or nitrate [51–53] and may represent a mechanism to attenuate NO-mediated biological functions.

Nitric oxide may also mediate cellular effects through interactions with enzymes containing nonheme iron. Such enzymes include the iron–sulfur cluster enzymes, mitochondrial *cis*-aconitase, complexes I and II of the mitochondrial electron-transfer chain [32,54,55], metallothionein [56,57] and ribonucleotide reductase [58–60]. Through its interactions with these enzymes, NO may inhibit cellular ATP synthesis and DNA synthesis. NO has recently been shown to influence intracellular iron metabolism at the posttranscriptional level by interacting with cytosolic aconitase (referred to as the iron-regulatory factor; IRF). NO disrupts the conformation of the iron–sulfur cluster at the catalytic site of the enzyme and increases the accessibility and binding of IRF to a specific mRNA structure (iron-responsive element; IRE) [61–63] contained in ferritin, transferrin receptor, erythroid 5-aminoelvulinate synthase, and mitochondrial aconitase transcripts. These proteins regulate iron homeostasis [64]. NO-induced binding of IRF to the IRE located in the 5' untranslated region of ferritin mRNA inhibits ribosomal binding and represses translation. In contrast, binding of IRF to the

3' IRE of the transferrin receptor mRNA protects the transcript from RNAse digestion and increases the mRNA half-life, resulting in increased protein synthesis. NO also directly interacts with the iron in ferritin [33], an iron-storage protein, to liberate iron [65]. Thus, the interaction between NO and iron may regulate cellular iron homeostasis or may mediate cellular toxicity [54,55,66].

3.2. Thiol-Containing Proteins

S-Nitrosylated proteins have been identified in biological tissues, including salivary glands of the blood-sucking insect, *Rhodnius prolixus,* and human plasma, airway-lining fluid, and neutrophils [67]. In order for NO to take part in S-nitrosylation, NO must first react with a transition metal, molecular oxygen, or superoxide anion to undergo a one-electron oxidation to form products (i.e., NO^+) that can support additional nitrosative reactions. S-Nitrosylation can regulate protein function by several different mechanisms. The activation of tissue-specific plasminogen activator and N-methyl-D-aspartic acid (NMDA) receptors occur through posttranslational modification of a single thiol [68–70], whereas the activation of calcium-dependent potassium channels is mediated by a conformational change following S-nitrosylation in vascular smooth muscle [71]. Also, the cytosolic glycolytic enzyme glyceraldehyde-3-phosphate dehydrogenate (GAPDH) is inhibited by modification of thiol groups located within the active site [72,73]. In the central nervous system, S-nitrosylation of the NMDA subtype of glutamate receptors by NO inactivates the receptor and may physiologically regulate glutaminergic neurotransmission [70], protecting against excessive excitation-induced neuronal cell death. NO also inhibits thioester-linked long-chain fatty acylation of neuronal proteins. Through direct modification of cysteine thiols in the neuronal growth cone, NO may reversibly inhibit the growth of dorsal root ganglion neurites [74]. S-Nitrosylation activates pertussis toxin-sensitive G proteins and p21[ras] [75], whereas this modification reduces enzymatic activity of adenylate cyclase (type I) and protein kinase C [76,77]. NO can deplete intracellular reduced glutathione levels through the formation of S-nitroglutathione and oxidation of glutathione which can lead to rapid activation of the hexose-mono-phosphate pathway [78] and induction of stress proteins such as heme oxygenase [79] and heat shock protein 70 [80]. S-Nitrosylated albumin has been identified in serum, but a biological function for this species of albumin has not yet been identified [81]. One possible function of S-nitrosothiols is to serve as a reservoir for NO, acting to reduce the functional availability of NO or as a carrier molecule for NO transport [82–86].

Nitric oxide also stimulates the apparent ADP-ribosylation of GAPDH in the presence of NAD^+ or NADH. NO-induced ADP-ribosylation is an inductive automodification, in contrast to the activity of an endogenous ADP-ribosyltransferase activated by bacterial toxins [87], and involves an active-site cysteine, probably Cys-149 [88]. ADP-ribosylation of this cysteine inhibits catalytic activity and depresses glycolysis. ADP-ribosylation is stimulated by RS-NO (usually liberating NO^+ + RS^- by heterolytic cleavage) and related nitrosating agents (i.e., NO^+) rather than by NO itself [89]. *S*-Nitrosylation of an active-site thiol subsequently promotes modification by NAD. It is theoretically possible to have transnitrosation from active-site RS-NO (NO^+) to NADH by either increasing the electrophilicity of the nicotinamide ring (i.e., attack at C6) or by making nicotinamide a better leaving group (i.e., attack ribose C1), thereby causing irreversible inhibition of the enzymatic activity [67]. ADP-ribosylation of GAPDH may represent a pathophysiological event associated with the inhibition of gluconeogenesis. The active-site thiol of GAPDH is readily oxidized by H_2O_2 or other oxidative stress conditions leading to the generation of disulfides or oxidant-specific *S*-thiolation [90] and results in irreversible inhibition. Therefore, *S*-nitrosylation of GAPDH may protect the enzyme from irreversible inactivation of the active-site thiol, perhaps regulating glucose metabolism [91].

3.3. Other Biomolecules

Zinc is the second most abundant metal in biologic tissues. It is usually complexed to thiol ligands of cysteine and/or imidazole nitrogen atoms of histidine-forming zinc-finger domains which are essential for specific DNA binding. Many redox-sensitive transcription factors including Sp1 and Egr-1 [64] bind cognate DNA via three zinc fingers of the Cys_2His_2 type [92]. Zinc is a relatively labile component of this structure and is most susceptible to removal by redox reactions or intracellular thiol compounds such as apo-metallothionein [93]. NO interacts with Zn–sulfur clusters in transcriptional factors [LAC9-containing $Cys_6Zn(II)$-type cluster] [94] and Zn-containing proteins such as metallothionein [94], DNA repair enzymes (Fpg protein) [95], and alcohol dehydrogenase [96], resulting in the destruction of the Zn–sulfur bond. This finding implies that NO may alter the structural integrity of DNA binding proteins which contain zinc finger motifs and suggests that NO-mediated cytotoxicity can be associated with the destruction of proteins containing Zn–sulfur clusters as well as through the interaction of NO with proteins containing Fe–S clusters. Another possible consequence of NO interactions is seen with metallothionein wherein Zn is released and can inhibit the catalytic activity of endonucleases involved in DNA fragmentation during apoptosis [97–99].

Nitric oxide gas and NO-generating compounds have been shown to be carcinogenic because of evidence indicating that NO or other reactive nitrogen intermediates can modify deoxyribonucleotides or induce DNA strand breaks. Wink et al. [100] reported that NO causes genomic alterations by deamination of DNA in vitro and in vivo by showing that cytosines are changed to thymidine in DNA sequences. Guanine can also be modified to form 8-nitrosoguanosine when intact DNA is exposed to peroxynitrite [101] or in NO-generating epithelial cells [102]. Furthermore, NO and/or peroxynitrite by inducing DNA strand breaks [103] activates the nuclear enzyme poly (ADP-ribose) polymerase (PARP). PARP activation has been proposed to decrease ATP synthesis and induce cell death by depletion of cellular NAD^+ during DNA repair.

4. Interaction with Reactive Oxygen Intermediates (ROIs)

Endogenously synthesized NO has a biological half-life of less than 5 s [52] as compared to a half-life of 500 s or longer of NO in pure aqueous solution [104]. It is possible that NO is rapidly converted into a less active or inactive product in biological tissues. As detailed earlier, NO can simply react with molecular oxygen; however, this is a very slow reaction at physiological concentrations of O_2. The reaction of NO with biological components such as metals, thiols, O_2, and O_2^- produces a variety of secondary products ranging from the innocuous oxidized compounds (NO_2^- and NO_3^-) and nitroxyl (NO^-) to the reactive intermediates such as nitrosonium (NO^+), peroxynitrite ($ONOO^-$), and nitrogen dioxide ($\bullet NO_2$). The predominant end products of all these intermediates are NO_2^- and NO_3^-, both easily quantified in biological systems.

In many pathological conditions such as inflammation, endotoxemia, and ischemia/reperfusion, NO and O_2^- are simultaneously produced. This creates an environment that favors the formation of peroxynitrite which is a much stronger oxidant than either NO or O_2^-. Peroxynitrite is an important biological bactericidal agent that is produced by activated macrophages to combat infection [105]. Although peroxynitrite has a half-life of under 1 s in phosphate buffer at pH 7.0, it is sufficiently stable under physiological conditions to diffuse some distance from its site of formation before reacting with target molecules such as membrane lipid [106], protein sulfhydryl groups [107], DNA [100,108], and antioxidants [109]. Peroxynitrite generated by endothelial cells has been identified as a potent mediator of endothelial injury through oxidative modification of low-density lipoprotein within the arterial wall [110]. This may then contribute to the formation of the fatty streak and plaque that are characteristic of the athereosclerotic lesion [29]. Peroxynitrite also reacts with Cu,Zn superoxide dismutase as well as low-molecular-weight transition metals like Fe^{3+}–EDTA. Peroxynitrite catalyzes nitration of tyrosine residues in many proteins including histone, lysozyme, and superoxide dismutase [30]. This nitration process may be responsible for the neuronal damage in amyotrophic lateral sclerosis [111]. Although peroxynitrite-mediated tyrosine nitration has been identified in a variety of pathological conditions, a precise role for this process has yet to be defined. There are several possible consequences of tyrosine nitration. It may lead to (i) inhibition of tyrosine phosphorylation [112] and inhibition of phosphorylation-mediated signal transduction or, conversely, mimic phosphorylation [113], (ii) alteration of protein conformation and function [114,115], (iii) "tagging" of proteins for proteolysis, (iv) initiation of autoimmune processes because these altered tyrosines structurally resemble dinitrophenol (a strongly antigenic compound), and (v) alteration of the dynamics of assembly and disassembly processes critical to motor neuron survival [116]. Peroxynitrite has also been shown to react with iron–sulfur clusters in aconitase and zinc–thiolate centers of numerous transcription factors and DNA repair enzymes. Through these interactions, peroxynitrite may inactivate metabolic pathways, induce DNA mutations, and modify gene expression. It may, in fact, account for the effects associated with NO.

5. Relevance to Cellular Toxicity and Disease

The role of NO in cellular toxicity is quite complex because it involves a variety of cellular targets within a wide range of cell types. The toxicity of NO depends on the cell types producing NO, as well as the local environment into which it is released, such as oxygen tension and the concentration of transition metals, reactive oxygen species, and NO scavenging molecules (i.e., hemoglobin).

5.1. Consequences of NO and Superoxide Anion Interaction

The direct interaction of NO with O_2^- to form $ONOO^-$ occurs 3.5 times faster than the dismutation of O_2^- by superoxide dismutase. Thus, the interaction between NO and O_2^- represents a major potentiating pathway for NO toxicity, as well as an inactivating route for endothelium-derived NO. Peroxynitrite is a potent biological oxidant capable of directly oxidizing amino acids, lipids, antioxidants, nucleic acids, and thiols in both one- and two-electron transfer reactions [30,107,108,117,118]. At near biological pH ($pK_a = 6.8$), peroxynitrite becomes protonated to form peroxynitrous acid (ONOOH). Peroxynitrous acid undergoes a unique •OH-like oxidative reaction with membrane lipid via a metal-independent mechanism [29]. Peroxynitrite can react with transition metal ions, such as Fe and Cu, to yield a species similar in reactivity to that of nitronium cation (NO_2^+) [119]. Accumulating data suggest that peroxynitrite is a major cytotoxic agent in many pathophysiologic processes such as ischemia/reperfusion injury, immune complex-mediated pulmonary edema, cytokine-induced oxidant lung injury, inflammatory cell-mediated pathogen killing/host injury, macrophage-mediated bacterial killing, and immune-mediated joint disease.

Alternatively, the reaction of NO with O_2^- may also serve to protect cells or tissues from reactive oxygen intermediate-mediated cytotoxicity. This diversionary reaction with NO (kinetically outcompeting SOD) forces O_2^- to form peroxynitrite and into a decomposition pathway that generates the nontoxic end-product nitrate. This pathway limits the accumulation of H_2O_2 and decreases the formation of more toxic secondary intermediate species such as •OH which would be derived from the metal-catalyzed Haber–Wiess reaction (Fenton reaction). NO also interacts with hemoproteins capable of catalyzing the Fenton-like reaction with H_2O_2 and so inhibits production of highly oxidized iron species [ferryl ion,

Fe(IV)=0] characteristic of •OH-like chemical reactivity. In animals exposed to LPS, inhibition of NO synthesis with a competitive inhibitor of NO synthase, N^G-monomethyl-L-arginine (L-NMA) enhances superoxide release [120] and simultaneously increases liver injury [121,122], suggesting that nitric oxide synthesis during endotoxemia prevents hepatic damage by reducing oxygen radical-mediated cellular injury.

Vascular endothelial cells and phagocytes produce both NO and O_2^-. In the microenvironment of these cells, interaction between these two radicals leads not only to the formation of peroxynitrite but also the inhibition of the biological function of NO as well as the scavenging of O_2^-. In advanced atherosclerotic lesions, the consumption of NO by O_2^- released by the lesion may contribute to increased incidence of thrombosis by promoting platelet aggregation [123]. This interaction of NO with O_2^- not only inhibits the homeostatic function of NO (e.g., inhibition of platelet aggregation and phagocyte adhesion) but also forms peroxynitrite, which is a strong oxidant and proaggregatory agent for platelets and may further contribute to cellular injury and the formation of thrombus in atherosclerosis [124].

Therefore, two biological consequences of NO interaction with O_2^-, either toxic peroxynitrite formation or scavenging of O_2^-, will depend upon the rate of formation of both radicals, local redox potential, and transition metal availability (Table 2-5).

5.2. Interaction of NO with Other Biological Free Radicals

Nitric Oxide can react with a number of other free radicals generated in biological systems. During inflammation or oxidative stress, hydroxyl radical (•OH) and lipid radicals, such as peroxyl (LOO•) and alkoxyl (LO•) radicals, are generated by the Fenton reaction and lipid peroxidation. NO reacts with these radicals at near-diffusion-limited rates (e.g., $1.3 \times 10^9 \ M^{-1} \ s^{-1}$ for LOO• and $10 \times 10^{10} \ M^{-1} \ s^{-1}$ for •OH). Therefore, NO can be a potential scavenger of •OH [125] and an inhibitor of the radical-induced chain propagation reaction [126] in lipid peroxidation via radical–radical interaction. The rate constant of the reaction of NO with

Table 2-5. *Metal-Dependent Cytoxicity Is Dependent on the Ratio of NO to O_2^-*

Ratio	Relative Cytotoxicity	
	Presence of Iron	Absence of Iron
NO only	–/+	–/+
NO > O_2^-	–	+++
NO + O_2^-	+	+++
NO < O_2^-	+++	+++
O_2^- only	+	+

LOO• is ~500 times greater than that for the reaction between α-tocopherol and LOO• radicals (2.5×10^6 M^{-1} s^{-1}). Because NO is a lipophilic molecule (lipid : water partition coefficient ~6.5 : 1) and concentrates in the cellular membrane compartment, NO can act as a vitamin E-like antioxidant in oxidant-mediated lipid peroxidation.

During bacterial infection, phagocytes first produce reactive oxygen species (ROS) to kill bacteria, and 6–9 h later, NO is generated by the inducible NO synthase. ROS has two consequences: the desirable one being bacterial killing and the undesirable one being host cell injury via lipid peroxidation. The delayed production of NO, therefore, may serve to prevent host cell injury by terminating lipid radical-induced chain propagation reaction induced by ROS production aimed at killing pathogens. Similarly, NO also reacts with highly oxidized iron, FE(IV)=0, a •OH-like potent oxidizing species generated from the reaction of hemoproteins, and inhibits lipid peroxidation of LDL [127,128]. Although the rate of reaction is significantly lower (2×10^7 M^{-1} s^{-1}), NO can reduce a ferric ion to the ferrous form, which directly enters the Haber–Weiss reaction in the presence of H_2O_2 to produce a highly toxic •OH. However, at a site of bacterial infection where ROS and NO are sequentially generated, NO may exert a protective role by termination of free radical-dependent chain propagation reaction in cell membranes [125,126] and reduction of toxic ferrylhemoproteins produced by the reaction of heme with H_2O_2 [127].

5.3. Direct Cytotoxicity of NO

As stated earlier, NO reacts rapidly with iron–sulfur-containing proteins and hemoproteins. Although these iron-containing proteins are the primary targets for NO, the reactivity of each of these proteins with NO may vary. EPR (electon paramagnetic resonance) spectra of iron–nitrosyl complexes have shown that NO has a much higher affinity for ferrous heme (probably cytosolic guanylate cyclase) than for iron–sulfur proteins in vascular smooth cells [129]. It is likely that the major role of NO in these cells is to activate guanylate cyclase to produce cGMP. In some cell types, the NO/cGMP signaling pathway may be associated with antiapoptotic functions (see below). Macrophage-derived NO has been shown to be a cytotoxic effector against tumor cells by mediating intracellular iron loss [32,130,131], inhibition of iron–sulfur-containing enzymes such as aconitase [130], and inhibition of NADPH–CoQ reductase (complex I) and succinate–CoQ (complex II) in the mitochondrial electron transport chain [55,131]. NO-induced inhibition of these enzymes blocks the metabolic pathway of cellular energy biosynthesis and depletes cellular ATP stores with consequent cytotoxicity. The inhibition of these enzymes is associated with the formation of the iron–nitrosyl complex in a catalytic active site which can be detected at **g** = 2.04 by EPR spectroscopy [32,108,132]. Macrophages and pancreatic β cells, when stimulated with cytokines and/or LPS, express iNOS and produce a large amount of NO. The overproduction of NO by iNOS inhibits mitochondrial function by inactivating iron–sulfur-containing enzymes that is believed to ultimately contribute to cell death. The cytotoxicity and $g = 2.04$ EPR signal were completely prevented by addition of L-NMA as well as by red blood cells, a biological scavenger of NO [28]. It is noteworthy that the EPR signal at $g = 2.04$ and accumulation of cGMP were both nearly inhibited in the coculture system of cytokine-activated hepatocytes and red blood cells, but NO production was not inhibited. This observation demonstrates that NO-induced cytotoxicity and cGMP signaling are mediated in a paracrine and/or autocrine manner. Interestingly, the large amounts of NO produced by hepatocytes do not inhibit mitochondrial function [133] and do not result in cellular injury [22]. This resistance of hepatocytes to NO cytotoxicity may be due to higher amounts of NO scavengers such as glutathione and cytosolic iron present in hepatocytes as compared to other cells. Thus, cellular resistance or susceptibility to NO depends on the presence or absence of cellular and microenvironmental factors such as thiols, metal ions, and the rate of cellular metabolism.

5.4. Cytoprotective Actions of NO

Nitric oxide can directly react with transcriptional factors, such as SoxR, a redox-sensitive transcriptional regulator found in the bacteria-containing Fe–S cluster. This factor is involved in the transcriptional regulation of approximately 80 inducible proteins responsible for protection against oxidative damage [134]. Mammalian cells also respond to unfavorable stimuli such as oxidative stress, heat, and heavy metals by expressing many different stress proteins. NO induces a number of stress proteins including heme oxygenase [22] and heat shock protein 70 [80] via interaction with intracellular glutathione or hemoproteins. We have shown that hepatocytes pretreated with NO were protected against subsequent exposure to high doses of NO and H_2O_2, and this protection was abrogated by Sn-protoporphyrin, a specific inhibitor of heme oxygenase. NO-pretreated hepatocytes were also resistant to TNFα-induced cytotoxicity. Thus, NO protects against cellular injury by inducing cytoprotective proteins through alterations of cellular redox potential [80,135] or heme mobilization [136].

Nitric oxide also modulates vascular function and platelet aggregation by altering the expression of adhesion molecules such as selectins and ICAM. The genes for these adhesion molecules possess NF-κB binding sites in their promoter regions [137] which are transcriptionally inhibited by NO. In animal studies, it has been demonstrated that pretreatment with L-NMA enhanced the expression of P-selectin in a concentration- and time-dependent manner and significantly increased leukocyte rolling and adhesion [138]. Furthermore, the coadministration of L-NMA with immunostimuli such as interleukin-1β or TNF increased platelet adhesion to endothelial cells and platelet accumulation in the lungs [139]. Inhibition of NOS activity during endotoxemia also has detrimental effects on tissue blood flow and oxygen delivery [140,141].

In a murine model of LPS-induced hepatic damage, nonselective inhibition of NO synthesis increased hepatic injury as measured by enzyme release as well as by histologic examination [142]. A subsequent study demonstrated that this hepatic injury was in part mediated by ROS such as O_2^-, H_2O_2, and •OH, supporting the cytoprotective function of NO in LPS-induced hepatic damage associated with ROS scavenging and by the inhibition of microvascular thrombosis [121]. Additional studies showed that NO acts synergistically with prostaglandins in ameliorating hepatic injury during endotoxemia [143]. During endotoxemia, macrophages including Kupffer cells produce TNFα, one of the major soluble cytotoxic cytokines in hepatic injury. NO-mediated cGMP production inhibited LPS-induced TNFα production by Kupffer cells in vitro [24]. Furthermore, administration of L-NMA increased TNFα expression in a murine model of endotoxemia [23]. These reports suggest that NO may protect from hepatic injury by direct scavenging of ROS and by suppression of TNFα expression. More recent studies from our group and others [144] suggest that the source of NO may be important in determining the mechanism of protection in the liver. Treatment with selective iNOS inhibitors does not result in increased hepatic necrosis in endotoxemia, whereas inhibitors that block cNOS do cause increased liver injury. We have shown that although iNOS inhibitors do not promote necrosis, they do induce apoptosis. Thus, NO generated by cNOS may block necrosis, whereas NO derived from iNOS may function to suppress apoptosis in acute hepatic injury.

5.5. Inhibition of Antioxidant Enzymes

Cellular antioxidant enzymes such as catalase, SOD, GSH reductase, and GSH peroxidase are important in host defense against oxidative injury. Superoxide is produced by activated phagocytes during infection, and xanthine oxidase is produced in ischemia/reperfusion as a by-product of mitochondrial respiration. Superoxide is dismutated to H_2O_2 either spontaneously or by superoxide dismutase. H_2O_2 is then dismutated to water and molecular oxygen by catalase or converted to water by GSH-peroxidase and glutathione. These activated species can also react with transition metals to produce another highly reactive oxygen radical, •OH, which has been implicated as the predominant cause of oxygen radical-induced cytotoxicity.

Oxygen radical-mediated damage is enhanced by inhibiting the antioxidant enzymes that normally function to inactivate these radicals. NO can interfere with the catalytic activity of many antioxidant enzymes though the transition metals or sulfhydryl groups found in these enzymes. NO can directly interact with heme iron in the catalytic site of catalase, inhibiting the enzymatic activity [40]. NO can also inactivate catalase activity indirectly via the induction of heme oxygenase which degrades heme in hemoproteins [39]. GSH peroxidase is also inactivated by both endogenous and exogenous NO [145]. The mechanism by which NO inhibits GSH peroxidase may be through an interaction with a thiol group in the active site of the enzyme. Peroxynitrite interacts with the transition metal in the catalytic site of SOD and generates a highly reactive nitronium ion (NO_2^+), which can then nitrate tyrosine residues in superoxide dismutase as well as other proteins. Self-nitration of tyrosine residues inactivates Mn- and Fe–superoxide dismutase but not Cu/Zn–superoxide dismutase [30]. GSH reductase is a dimeric flavoprotein and is the key enzyme of the GSH redox metabolism involved in the reduction of GSSG to GSH. NO can inhibit GSH reductase by nitrosylating Cys63 and/or Cys58 located in the catalytic site of the enzyme [146]. The ability of NO and peroxynitrite to inhibit major antioxidants may, at least in part, explain the cytotoxicity of NO in some cells or tissues.

5.6. Effects of NO on Apoptosis

Cell death occurs by either apoptosis or necrosis. Apoptosis is a morphologically and biochemically distinct form of cell death that occurs in many different cell types. It occurs in response to a variety of stimuli, including DNA damage, growth factor deficiency, ligation of cell surface receptors, heat shock, microorganism infection, and oxidative stress [147,148]. Although activated macrophages utilize NO to destroy microorganisms and tumor cell targets, they themselves are not immune to the NO they synthesize and have been shown to undergo apoptosis in a NO-dependent manner [149]. NO-mediated apoptosis has been associated with the overexpression of the cell death gene p53 [150,151] or the activation of proteases (including interleukin-1β-converting enzyme; ICE) [152] in macrophages. ICE-like proteases such as ICE and CPP32 inactivate the nuclear DNA repair enzyme, poly (ADP-ribose) polymerase by proteolysis [153,154], resulting in p53 induction or enhanced DNA fragmentation. NO generated from NO donors or synthesized by NOS induces cell death via apoptosis in a variety of different cell types [155–157].

Nitric oxide, nonenzymatically synthesized from azide and hydroxyamine, has been suggested to prevent apoptosis of human eosinophils during cytokine deprivation [158]. This protective effect was enhanced by the phosphodiesterase inhibitor 3-isobutyl-1-methylxanthine and mimicked by membrane-permeable cGMP analogs. Moreover, a soluble guanylate cyclase inhibitor reversed the effects of the NO donors in a dose-dependent manner. NO synthesized by LPS-stimulated hepatocytes protected the cells from apoptosis induced by treatment with transforming growth factor [159], which has been known to activate ICE-like protease [160]. This evidence suggests that NO may inhibit ICE-like protease-induced apoptosis. This protection can also be elicited by NO donor at concentrations between 2 and 100 μM, whereas higher concentrations of NO promoted both apoptosis and necrotic cell death. Recently, another study showed that cytokine-induced NO production and a cGMP analog protected rat ovarian follicles from apoptotic cell death [161]. Also, production of NO delays programmed cell death (apoptosis) in splenic B lymphocytes as determined by DNA fragmentation and flow cytometric analysis of DNA [162]. This effect was reproduced with membrane-permeable analogs of cGMP. The antiapoptotic effect of NO/cGMP may be associated with increased expression of the protooncogene bcl-2 at both the mRNA and protein levels. The protective role of NO against apoptosis in B cells contrasts with the NO-dependent induction of apoptosis observed in other cell types such as macrophages [157], indicating the existence of different cell-specific pathways in the response to NO. Similarly, cGMP can increase the sensitivity of tumor cells to TNFα-induced cytotoxicity but protects normal cells from TNFα toxicity [163].

5.7. Effect of NO on the Nervous System

Nitric oxide is an important neuromodulator in both central and peripheral neurons [164]. NO stimulates the release of neurotransmitters by stimulating guanylate cyclase to increase intracellular cGMP, which activates cGMP-dependent protein kinases to augment the phosphorylation of synaptic vesicle proteins associated with neurotransmitter release. In contrast, excessive NO is neurotoxic [165] via the massive release of glutamate. Glutamate activates the NMDA receptor and causes excessive exitation, thereby resulting in neuronal death [166]. NO-induced toxicity, like NMDA neurotoxicity, may play a role in neurodegenerative diseases such as Alzheimer's and Huntington's diseases. Furthermore, NO might exert a dual function, exhibiting either neurotoxic or neuroprotective effects, depending on its oxidation-reduction status [70]. The interaction of NO with O_2^- to form peroxynitrite may be the final neurotoxic species. On the other hand, NO^+ generated from the reaction of NO with transition metals nitrosylates the thiol group of the NMDA receptor and blocks glutamate neurotransmission [167].

5.8. Other Cytotoxic Effects

In contrast to the several beneficial or cytoprotective roles of NO in sepsis, neurodegeneration, and apoptosis, NO can be toxic to cultured pancreatic β cells exposed to interleukin-1β. This is a proposed mechanism for β-cell disruption that leads to diabetes [168]. The overproduction of NO damages the β-cell mitochondria by inactivating iron–sulfur-containing enzymes and also inhibits DNA synthesis which is believed to ultimately contribute to β-cell death. This cytotoxicity is inhibited by aminoguanidine, an inhibitor of iNOS. Other studies show, however, that low concentrations of NO increase insulin secretion from pancreatic islets by inducing Ca^{2+} release from mitochondria [168]. After organ transplantation, NO production can be detected during rejection of allogeneic heart grafts, but not in syngeneic grafts, by the EPR technique [169]. NO inhibition prolongs cardiac allograft survival, suggesting that NO may mediate organ injury in transplant rejection. Based on results of studies using the NOS inhibitor, NO may also mediate tissue injury in graft-versus-host disease [170,171], rheumatoid arthritis [172], and systemic lupus erythematosus [173].

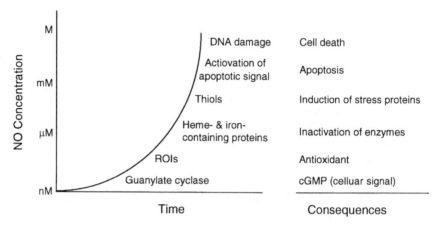

Figure 2-2. Intracellular targets for NO and pathophysiological consequences.

6. Summary

Nitric oxide is involved in signaling, cytotoxicity, and cytoprotection by interacting with different types of intracellular molecules. Figure 2-2 shows the simple correlation between NO concentration and pathophysiological events. At low concentrations, NO activates ferrous heme-containing guanylate cyclase to generate cGMP, which may be involved in vascular homeostasis and antiapoptotic effects. Higher doses of NO, as produced by iNOS, interact with thiol compounds and transition metals (e.g., Fe–S-containing proteins) and can induce cytoprotective proteins such as stress proteins. Even higher concentrations of NO may cause nonspecific DNA damage and activate PARP or p53, resulting in nonspecific cell death. Other factors, such as the presence of other radicals, antioxidant system, and NO scavengers, will have profound influences on NO actions.

References

1. The Committee on Nitrite and Alternative Curing Agents in Food, Assembly of Life Sciences. *The Health Effects of Nitrate, Nitrite,* and *N-Nitroso Compounds.* National Academy Press, Washington, DC, 1981.

2. Papen, H., von Berg, R., Hinkle, I., Thoene, B., and Rennenberg, H. Heterotrophic nitrification by Alicaligenes faecalis: NO_2^-, NO_3^-, and NO production in exponentially growing cultures. *Appl. Environ. Microbiol.* **55,** 2068–2072 (1989).

3. Sprent, J.I. *The Ecology of the Nitrogen Cycle.* Cambridge University Press, Cambridge, 1987.

4. Murad, F., Mittal, C.K., Arnold, W.P., Katsuki, S., and Kimura, H. Guanylate cyclase: Activation by azide, nitro compounds, nitric oxide, and hydroxyl radical and inhibition by hemoglobin and myoglobin. *Adv. Cyclic Nucl. Res.* **9,** 145–158 (1978).

5. Mitchell, H.H., Schonle, H.A., and Grindly, H.S. The origin of the nitrates in the urine. *J. Biol. Chem.* **24**, 461–490 (1916).

6. Green, L.C., De Luzuriaga, K.R., Wagner, D.A., Rand, W., Istfan, N., Young, V.R., and Tannenbaum, S.R. Nitrate biosynthesis in man. Proc Natl Acad Sci USA 1981; **78**, 7764–7768.

7. Green, L.C., Tannenbaum, S.R., and Goldman, P. Nitrate synthesis and reduction in the germ-free and conventional rat. *Science* **212**, 56–58 (1981).

8. Wagner, D.A., Young, V.R., and Tannenbaum, S.R. Mammalian nitrate biosynthesis: Incorporation of [^{15}N] ammonia into nitrate in enhanced by endotoxin treatment. *Proc. Natl. Acad. Sci. USA* **80**, 4519–4521 (1983).

9. Stuehr, D.J. and Marletta, M.A. Mammalian nitrate biosynthesis: Mouse macrophages produce nitrite and nitrate in response to Escherichia coli lipopolysaccharide. *Proc. Natl. Acad. Sci. USA* **82**, 7738–7742 (1985).

10. Furchgott, R.F. and Zawadzki, J.W. The obligatory role of endothelial cells in the relaxation of arterial smooth muscle by acetylcholine. *Nature* **288**, 373–376 (1980).

11. Furchgott, R.F. Studies on relaxation of rabbit aorta by sodium nitrite: The basis for the proposal that the acid activatible inhibitory factor from bovine retractor penis is inorganic nitrite and the endothelium-derived relaxing factor is nitric oxide. In: *Mechanisms of Vasodilatation.* Ed. Vanhoutte, P.M. Raven Press, New York, 1988, pp. 401–414.

12. Ignarro, L.J., Byrns, R.E., and Woods, K.S. Biochemical and pharmacological properties of endotheliaum-derived relaxing factor and its similarity to nitric oxide radical. In: *Mechanisms of Vasodilatation.* Ed. Vanhoutte P.M. Raven Press, New York, 1988, pp. 427–436.

13. Palmer, R.M., Ferrige, A.G., and Moncada, S. Nitric oxide release accounts for the biological activity of endothelium-derived relaxing factor. *Nature* **327**, 524–525 (1987).

14. Ignarro, L.J., Buga, G.M., Wood, R.E., Byrns, R.E., and Chaudhuri, G. Endothelium-derived relaxing factor produced and released from artery and vein is nitric oxide. *Proc. Natl. Acad. Sci. USA* **84**, 9265–9269 (1987).

15. Hibbs, Jr., J.B., Taintor, R.R., and Vavrin, Z. Macrophage cytotoxicity: Role for L-arginine deiminase and imino nitrogen oxidation to nitrite. *Science* **235**, 473–476 (1987).

16. Marletta, M.A., Yoon, P.S., Iyengar, R., Leaf, C.D., and Wishnok, J.S. Macrophage oxidation of L-arginine to nitrite and nitrate: Nitric oxide is an intermediate. *Biochemistry* **27**, 8706–8711 (1988).

17. Iyengar, R., Stuehr, D.J., and Marletta, M.A. Macrophage synthesis of nitrite, nitrate, and *N*-nitrosamine: Precursors and role of the respiratory burst. *Proc. Natl. Acad. Sci. USA* **84**, 6369–6373 (1987).

18. Hibbs, Jr., J.B., Vavrin, Z., and Taintor, R.R. L-arginine is required for the expression of the activated macrophage effector mechanism causing selective metabolic inhibition in target cells. *J. Immunol.* **138**, 550–565 (1987).

19. Morbidelli, L., Chang, C.H., Douglas, J.G., Granger, H.J., Ledda, F., and Ziche, M. Nitric oxide mediates mitogenic effect of VEGF on coronary venular endothelium. *Am. J. Physiol.* **39**, H411–H415 (1996).

20. Noiri, E., Peresleni, T., Strivastava, N., Webster, P., Bahou, W.F., Peunova, N., and Goligorsky, M.S. Nitric oxide is necessary for a switch from stationary to locomoting phenotype in epithelial cells. *Am. J. Physiol.* **270**, C794–C802 (1996).

21. Colasanti, M., Persichini, T., Menegazzi, M., Mariotto, S., Giordano, E., Caldarera, C. M., Sogos, V., Lauro, G. M., and Suzuki, H. Induction of nitric oxide synthase mRNA expression: Suppression by exogenous nitric oxide. *J. Biol. Chem.* **270**, 26731–26733 (1995).

22. Kim, Y.M., Bergonia, H., and Lancaster, Jr., J.R. Nitrogen oxide-induced autoprotection in isolated rat hepatocytes. *FEBS Lett.* **374**, 228–232 (1995).

23. Tiao, G., Rafferty, J., Ogle, C., Fischer, J.E., and Hasselgren, P.O. Detrimental effect of nitric oxide synthase inhibition during endotoxemia may be caused by high levels of tumor necrosis factor and interleukin-6. *Surgery* **116**, 332–337 (1994).

24. Harbrecht, B.G., Wang, S.C., Simmons, R.L., and Billiar, T.R. Cyclic GMP and guanylate cyclase mediate lipopolysaccharide-induced Kupffer cell tumor necrosis factor-alpha synthesis. *J. Leuk. Biol.* **57**, 297–302 (1995).

25. Bredt, D.S., Hwang, P.M., and Snyder, S.H. Localization of nitric oxide synthase indicating a neural role of nitric oxide. *Nature* **347**, 768–770 (1990).

26. Ignarro, L.J., Fukuto, J.M., Griscavage, J.M., Rogers, N.E., and Byrns, R.E. Oxidation of nitric oxide in aqueous solution to nitrite but not nitrate: Comparison with enzymatically formed nitric oxide from L-arginine. *Proc. Natl. Acad. Sci. USA* **90**, 8103–8107 (1993).

27. Lewis, R.S., Tamir, S., Tannenbaum, S.R., and Deen, W.M. Kinetic analysis of the fate of nitric oxide synthesized by macrophages in vitro. *J. Biol. Chem.* **270**, 29350–29355 (1955).

28. Stadler, J., Bergonia, H. A., Di Silvio, M., Sweetland, M.A., Billiar, T.R., Simmons, R.L., and Lancaster, Jr., J.R. Nonheme iron-nitrosyl complex formation in rat hepatocytes: Detection by electron paramagnetic resonance spectroscopy. *Arch. Biochem. Biophys.* **302**, 4–11 (1993).

29. Beckman, J.S., Beckman, T.W., Chen, J., Marshall, P.A., and Freeman, B.A. Apparent hydroxyl radical production by peroxynitrite: Implications for endothelial injury from nitric oxide and superoxide. *Proc. Natl. Acad. Sci. USA* **87**, 1620–1624 (1990).

30. Ischiropoulos, H., Zhu, L., Chen, J., Tsai, M., Martin, J.C., Smith, C.D., and Beckman, J.S. Peroxynitrite-mediated tyrosine nitration catalyzed by superoxide dismutase. *Arch. Biochem. Biophys.* **298**, 431–437 (1992).

31. Moncada, S., Palmer, R.M., and Higgs, E.A. Nitric oxide: Physiology, pathophysiology, and pharmacology. *Pharmacol. Rev.* **43**, 109–142 (1991).

32. Hibbs, Jr., J.B., Taintor, R.R., Vavrin, Z., Granger, D.L., Drapier, J.C., Amber, I.J., and Lancaster, Jr., J.R. Synthesis of nitric oxide from a terminal guanidino group nitrogen atom of L-arginine: A molecular mechanism regulating cellular proliferation that targets intracellular iron. In: *Nitric Oxide from L-Arginine: A*

Bioregulatory System. Eds. Moncada, S. and Higgs, A. Elsevier, Amsterdam, 1990, pp. 189–233.

33. Henry, Y., Lepoivre, M., Drapier, J.C., Ducrocq, C., Boucher, J.L., and Guissani, A. EPR characterization of molecular targets for NO in mammalian cells and organelles. *FASEB J.* **7**, 1124–1134 (1993).

34. Ignarro, L.J. Heme-dependent activation of soluble guanylate cyclase by nitric oxide: Regulation of enzyme activity by porphyrins and metalloporphyrins. *Sem. Hematol.* **26**, 63–76 (1989).

35. Ignarro, L.J. Biosynthesis and metabolism of endothelium-derived relaxing factor. *Annu. Rev. Pharmacol. Toxicol.* **30**, 535–560 (1990).

36. Montague, P.R., Gancayco, C.D., Winn, M.J., Marchase, R.B., and Friedlander, M.J. Role of nitric oxide production in NMDA receptor-mediated neurotransmitter release in cerebral cortex. *Science* **263**, 973–977 (1994).

37. Hirsch, D.B., Stener, J.P., Dawson, T.M., Mammen, A., Hayek, E., and Snyder, S.H. Neurotransmitter release regulated by nitric oxide in PC-12 cells and brain synaptosomes. *Curr. Biol.* **3**, 749–754 (1993).

38. Schmidt, H.H.H., Lohmann, S.M., and Walter, U. The nitric oxide and cGMP signal transduction system: Regulation and mechanism of action. *Biochim. Biophys. Acta* **1178**, 153–175 (1993).

39. Kim, Y.M., Bergonia, H.A., Muller, C., Pitt, B.R., Watkins, S.D., and Lancaster, Jr., J.R. Loss and degradation of enzyme-bound heme induced by cellular nitric oxide synthesis. *J. Biol. Chem.* **270**, 5710–5713 (1995).

40. Brown, G.C. Reversible binding and inhibition of catalase by nitric oxide. *Eur. J. Biochem.* **232**, 188–191 (1995).

41. Corbett, J.A., Kwon, G., Turk, J., and McDaniel, M.L. IL-1 beta induces the coexpression of both nitric oxide synthase and cyclooxygenase by islets of Langerhans: Activation of cyclooxygenase by nitric oxide. *Biochemistry* **32**, 13767–13770 (1993).

42. Salvemini, D., Misko, T.P., Masferrer, J.L., Seibert, K., Currie, M.G., and Needleman, P. Nitric oxide activates cyclooxygenase enzymes. *Proc. Natl. Acad. Sci. USA* **90**, 7240–7244 (1993).

43. Kanner, J., Harel, S., and Granit, R. Nitric oxide, an inhibitor of lipid oxidation by lipooxygenase, cyclooxygenase and hemoglobin. *Lipids* **27**, 46–59 (1992).

44. Abu-Soud, H.M., Wang, J., Rousseau, D.L., Fukuto, J.M., Ignarro, L.J., and Stuehr, D.J. Neuronal nitric oxide synthase self-inactivates by forming a ferrous-nitrosyl complex during aerobic catalysis. *J. Biol. Chem.* **270**, 22997–23006 (1995).

45. Griscavage, J.M., Rogers, N.E., Sherman, M.P., and Ignarro, L.J. Inducible nitric oxide synthase from a rat alveolar macrophage cell line is inhibited by nitric oxide. *J. Immunol.* **151**, 6329–6337 (1993).

46. Buga, G.M., Griscavage, L.M., Rogers, N.E., and Ignarro, L.J. Negative feedback regulation of endothelial cell function by nitric oxide. *Circ. Res.* **73**, 808–812 (1993).

47. Kim, Y.M., Nüssler, A., Sweetland, M.A., Billiar, T.R., Simmons, R.L., and Lancaster, Jr., J.R. Apparent differential sensitivity of nitric oxide synthase to feedback

inhibition (by nitric oxide) in cell versus extracts. In: *Biology of Nitric Oxide*. Eds. Moncada, S., Marletta, M.A., Hibbs, Jr., J.B., and Higgs, A. Portland Press, London, 1994, pp. 77–80.

48. Khatsenko, O.G., Gross, S.S., Rifkind, A.B., and Vane, J.R. Nitric oxide is a mediator of the decrease in cytochrome P450-dependent metabolism caused by immunostimulants. *Proc. Natl. Acad. Sci. USA* **90**, 11147–11151 (1993).

49. Wink, D.A., Osawa, Y., Darbyshire, J.F., Jones, C.R., Eshenaur, S.C., and Nims, R.W. Inhibition of cytochrome P450 by nitric oxide and a nitric oxide-releasing agent. *Arch. Biochem. Biophys.* **300**, 115–123 (1993).

50. Thomas, S.R., Mohr, D., and Stocker, R. Nitric oxide inhibits indoleamine 2,3-dioxygenase activity in interferon-gamma primed mononuclear phagocytes. *J. Biol. Chem.* **269**, 14457–14464 (1994).

51. Meyer, M., Schuster, K.D., Schulz, H., Mohr, M., and Piiper, J. Pulmonary diffusing capacities for nitric oxide and carbon monoxide determined by rebreathing in dogs. *J. Appl. Physiol.* **68**, 2344–2357 (1990).

52. Lancaster, Jr., J.R. Simulation of the diffusion and reaction of endogenously produced nitric oxide. *Proc. Natl. Acad. Sci. USA* **91**, 8137–8141 (1994).

53. LoBrutto, R., Wei, Y.H., Yoshida, S., Van Camp, H.L., Scholes, C.P., and King, T.E. Electron paramagnetic resonance- (EPR-) resolved kinetics of cryogenic nitric oxide recombination to cytochrome oxidase and myoglobin. *Biophys. J.* **45**, 473–479 (1984).

54. Hibbs, Jr., J.B., Taintor, R.R., Vavrin, Z., and Rachlin, E.M. Nitric oxide: A cytotoxic activated macrophage effector molecule. *Biochem. Biophys. Res. Commun.* **157**, 87–94 (1988).

55. Drapier, J.C., and Hibbs, Jr., J.B. Differentiation of murine macrophages to express nonspecific cytotoxicity for tumor cells results in L-arginine-dependent inhibition of mitochondrial iron-sulfur enzymes in the macrophage effector cells. *J. Immunol.* **140**, 2829–2838 (1988).

56. Schwarz, M.A., Lazo, J.S., Yalowich, J.C., Allen, W.P., Whitmore, M., Bergonia, H.A., Tzeng, E., Billiar, T.R., Robbins, P.D., Lancaster, Jr., J.R., and Pitt, B.R. Metallothionein protects against the cytotoxic and DNA-damaging effects of nitric oxide. *Proc. Natl. Acad. Sci. USA* **92**, 4452–4456 (1995).

57. Kennedy, M.C., Gan, T., Antholine, W.E., and Petering, D.H. Metallothionein reacts with Fe^{2+} and NO to form products with a g = 2.039 ESR signal. *Biochem. Biophys. Res. Commun.* **196**, 632–635 (1993).

58. Kwon, N.S., Stuehr, D.J., and Nathan, C.F. Inhibition of tumor cell ribonucleotide reductase by macrophage-derived nitric oxide. *J. Exp. Med.* **174**, 761–767 (1991).

59. Lepoivre, M., Flaman, J.M., Bobe, P., Lemaire, G., and Henry, Y. Quenching of the tyrosyl free radical of ribonucleotide reductase by nitric oxide: Relationship to cytostasis induced in tumor cells by cytotoxic macrophages. *J. Biol. Chem.* **269**, 21891–21897 (1994).

60. Lepoivre, M., Fischi, F., Coves, J., Thelander, L., and Fontecave, M. Inactivation of ribonucleotide reductase by nitric oxide. *Biochem. Biophys. Res. Commun.* **179**, 442–448 (1991).

61. Drapier, J.C., Hirling, H., Wiestzerbin, J., Kald, P., and Kuhn, L.C. Biosynthesis of nitric oxide activates iron-regulatory factor in macrophages. *EMBO J.* **12**, 3643–3649 (1993).

62. Weiss, G., Goossen, B., Doppler, W., Fuchs, D., Pantopolulos, K., Werner-Felmayer, G., Wachter, H., and Hentze, M.W. Translational regulation via iron-responsive elements by the nitric oxide/NO-synthase pathway. *EMBO J.* **12**, 3651–3657 (1993).

63. Pantopoulos, K. and Hentze, M.W. Nitric oxide signaling to iron-regulatory protein: Direct control of ferritin mRNA translation and transferrin receptor mRNA stability it transfected fibroblasts. *Proc. Natl. Acad. Sci. USA* **92**, 1267–1271 (1995).

64. O'Halloran, T.V. Transition metals in control of gene expression. *Science* **261**, 715–730 (1993).

65. Reif D.W. and Simmons, R.D. Nitric oxide mediates iron release from ferritin. *Arch. Biochem. Biophys.* **283**, 537–541 (1990).

66. Jaffrey, S.R., Cohen, N.A., Rouault, T.A., Klausner, R.D., and Snyder, S.H. The iron-responsive element binding protein: A novel target for synaptic actions of nitric oxide. *Proc. Natl. Acad. Sci. USA* **91**, 12994–12998 (1994).

67. Stamler, J.S. Redox signaling: Nitrosylation and related target interaction of nitric oxide. *Cell* **78**, 931–936 (1994).

68. Stamler, J.S., Simon, D.I., Osborne, J.A., Mullins, M.E., Jaraki, O., Michel, T., Singel, D.J., and Loscalzo, J. *S*-Nitrosylation of proteins with nitric oxide: Synthesis and characterization of biologically active compounds. *Proc. Natl. Acad. Sci. USA* **89**, 444–448 (1992).

69. Stamler, J.S., Simon, D.I., Jaraki, O., Osborne, J.A., Francis, S., Mullins, M., Singel, D.I., and Loscalzo, J. *S*-Nitrosylation of tissue-type plasminogen activator confers vasodilatory and antiplatelet properties on the enzyme. *Proc. Natl. Acad. Sci. USA* **89**, 8087–8091 (1992).

70. Lipton, S.A., Choi, Y.B., Pan, Z.H., Lei, S.Z., Chen, H.S., Sucher, N.J., Loscalzo, J., Singel, D.J., and Stamler, J.S. A redox-based mechanism for the neuroprotective and neurodestructive effects of nitric oxide and related nitroso-compounds. *Nature* **364**, 624–632 (1993).

71. Bolotina, V.M., Najibi, S., Palacino, J.J., Pagano, P.J., and Cohen, R.A. Nitric oxide directly activates calcium-dependent potassium channels in vascular smooth muscle. *Nature* **368**, 850–853 (1994).

72. Zhang, J. and Snyder, S.H. Nitric oxide stimulates auto-ADP-ribosylation of glyceraldehyde-3-phosphate dehydrogenase. *Proc. Natl. Acad. Sci. USA* **89**, 9382–9385 (1992).

73. Dimmeler, S., Lottspeich, F., and Brune, B. Nitric oxide causes ADP-ribosylation and inhibition of glyceraldehyde-3-phosphate dehydrogenase. *J. Biol. Chem.* **267**, 16771–16774 (1992).

74. Hess, D.T., Patterson, S.I., Smith, D.S., and Skene, J.H. Neuronal growth cone collapse and inhibition of protein fatty acylation by nitric oxide. *Nature* **366**, 562–565 (1993).

75. Lander, H.M., Ogiste, J.S., Teng, K.K., and Novogrodsky, A. p21[ras] as a common signaling target of reactive free radicals and cellular redox stress. *J. Biol. Chem.* **270**, 21195–21198 (1995).

76. Duhe, R.J., Nielsen, M.D., Dittman, A.H., Villacres, E.C., Choi, E.J., and Strom, J.R. Oxidation of critical cysteine residues of type I adenylyl cyclase by o-iodosobenzoate or nitric oxide reversibily inhibits stimulation by calcium and calmodulin. *J. Biol. Chem.* **269**, 7290–7293 (1994).

77. Gopalakrishna, R., Chen, Z.H., and Gundimeda, U. Nitric oxide and nitric oxide-generating agents induce a reversible inactivation of protein kinase C activity and phorbol ester binding. *J. Biol. Chem.* **268**, 27180–27185 (1993).

78. Clancy, R.M., Levartovsky, D., Leszczynska-Piziak, J., Yegudin, J., and Abramson, S.B. Nitric oxide reacts with intracellular glutathione and activates the hexose monophosphate shunt in human neutrophils: Evidence for *S*-nitrosoglutathione as a bioactive intermediary. *Proc. Natl. Acad. Sci. USA* **91**, 3680–3684 (1994).

79. Clancy, R.M. and Abramson, S.B. Novel synthesis of *S*-nitrosoglutathione and degradation by human neutrophils. *Anal. Biochem.* **204**, 365–371.

80. Kim, Y.M., de Vera, M.E., Wakins, S.C., and Billiar, T.R. Nitric oxide protects cultured rat hepatocytes from TNFα-induced apoptosis by induced heat shock protein 70 expression. Submitted.

81. Butler, A.R., Flintney, F.W., and Williams, D.L.H. NO, nitrosonium ions, nitroxide ion, nitrosothiols, and iron-nitrosyls in biology: A chemist's prospective. *Trends Pharmacol. Sci.* **16**, 8–22 (1995).

82. Pietraforte, D., Mallozzi, C., Scorza, G., and Minetti, M. Role of thiols in the targeting of *S*-Nitrosothiols to red blood cells. *Biochemistry* **34**, 7177–7185 (1995).

83. Scharfstein, J.S., Keaney, Jr., J.F., Slivka, A., Welch, G.N., Vita, J.A., Stamler, J.S., and Loscalzo, J. In vivo transfer of nitric oxide between a plasma protein-bound reservoir and low molecular weight thiols. *J. Clin. Invest.* **94**, 1432–1439 (1994).

84. Keaney, Jr., J.F., Simon, D.I., Stamler, J.S., Jaraki, O., Scharfstein, J., Vita, J.A., and Loscalzo, J. NO forms an adduct with serum albumin that has endothelium-derived relaxing factor-like properties. *J. Clin. Invest.* **91**, 1582–1589 (1993).

85. Stamler, J.S., Jaraki, O., Osborne, J., Simon, D.I., Keaney, J., Vita, J., Singel, D., Valeri, C.R., and Loscalzo, J. Nitric oxide circulates in mammalian plasma primary as an *S*-nitroso adduct of serum albumin. *Proc. Natl. Acad. Sci. USA* **89**, 7674–7677 (1992).

86. Jia, L., Bonaventura, C., Bonaventura, J., and Stamler, J.S. *S*-Nitrosylhemoglobin: A dynamic activity of blood involved in vascular control. *Nature* **380**, 221–226 (1996).

87. McDonald, L.J. and Moss, J. Stimulation by nitric oxide of an NAD linkage to glyceraldehyde-3-phosphate dehydrogenase. *Proc. Natl. Acad. Sci. USA* **90**, 6238–6241 (1993).

88. Dimmeler, S. and Brüne, B. Characterization of a nitric oxide-catalyzed ADP-ribosylation of glyceraldehyde-3-phosphate dehydrogenase. *Eur. J. Biochem.* **210**, 305–310 (1993).

89. Mohr, S., Stamler, J.S., and Brüne, B. Mechanism of covalent modification of glyceraldehyde-3-phosphate dehydrogenase at its active site thiol by nitric oxide, peroxynitrite, and related nitrosating agents. *FEBS Lett.* **348**, 223–227 (1994).

90. Schuppe-Koistinen, I., Moldeus, P., Bergman, T., and Cotgreave, L.A. *S*-Thiolation of human endothelial cell of glyceraldehyde-3-phosphate dehydrogenase after hydrogen peroxide treatment. *Eur. J. Biochem.* **221**, 1033–1037 (1994).

91. Mohr, S., Stamler, J.S., and Brüne, B. Posttranslational modification of glyceraldehyde-3-phosphate dehydrogenase by S-nitrosylation and subsequent NADH attachment. *J. Biol. Chem.* **271**, 4209–4214 (1996).

92. Kadonaga, J.T., Carner, K.R., Masiarz, F.R., and Tjian, R. Isolation of cDNA encoding transcription factor Sp1 and functional analysis of the DNA binding domain. *Cell* **51**, 1079–1090 (1987).

93. Zeng, J., Heuchel, R., Schaffner, W., and Kagi, J.H. Thionein (apometallothionein) can modulate DNA binding and transcription activation by zinc finger containing factor Sp1. *FEBS Lett.* **279**, 310–312 (1991).

94. Kroncke, K.D., Fehsel, K., Schmidt, T., Zenke, F.T., Dasting, I., Wesener, J.R., Bettermann, H., Breunig, K.D., and Kolb-Bachofen, V. Nitric oxide destroys zinc-sulfur clusters inducing zinc release from metallothionein and inhibition of the zinc finger-type yeast transcription activator LAC9. *Biochem. Biophys. Res. Commun.* **200**, 1105–1110 (1994).

95. Wink, D.A. and Laval, J. The Fpg protein, a DNA repair enzyme, is inhibited by the biomediator nitric oxide in vitro and in vivo. *Carcinogen* **15**, 2125–2129 (1994).

96. Crow, J.P., Beckman, J.S., and McCord, J.M. Sensitivity of the essential zinc-thiolate moiety of yeast alcohol dehydrogenase to hypochlorite and peroxynitrite. *Biochemistry* **34**, 3544–3552 (1995).

97. Gaido, M.L. and Cidlowski, J.A. Identification, purification, and characterization of a calcium-dependent endonuclease (NUC18) from apoptotic rat thymocytes. NUC18 is not histone H2B. *J. Biol. Chem.* **266**, 18580–18585 (1991).

98. Zalewski, P.D., Forbes, I.J., and Giannakis, C. Physiological role for zinc in prevention of apoptosis (gene-directed death). *Biochem. Int.* **24**, 1093–1101 (1991).

99. Obeid, L.M., Linardic, C.M., Karolak, L.A., and Hannun, Y.A. Programmed cell death induced by ceramide. *Science* **259**, 1769–1771 (1993).

100. Wink, D.A., Kasprzak, K.S., Maragos, C.M., Elespuru, R.K., Misra, M., Dunams, T.M., Cebula, T.A., Koch, W.H., Andrews, A.W., Allen, J.S., et al. DNA deaminating ability and genotoxicity of nitric oxide and its progenitors. *Science* **254**, 1001–1003 (1991).

101. Yermilov, V., Rubio, J., Becchi, M., Friesen, M.D., Pignatelli, B., and Ohshima, H. Formation of 8-nitroguanine by the reaction of guanine with peroxynitrite in vitro. *Carcinogen* **16**, 2045–2050 (1995).

102. Chao, C.C., Park, S.H., and Aust, A.E. Participation of nitric oxide and iron in the oxidation of DNA in asbestos-treated human lung epithelial cells. *Arch. Biochem. Biophys.* **326**, 152–157 (1996).

103. Heller, B., Wang, Z.Q., Wagner, E.F., Radons, J., Burkle, A., Fehsel, K., Burkart, V., and Kolb, H. Inactivation of the poly(ADP-ribose) polymerase gene affects oxygen radical and nitric oxide toxicity in islet cells. *J. Biol. Chem.* **270**, 11176–11180 (1995).

104. Wink, D.A., Darbyshire, J.F., Nims, R.W., Saavedra, J.E., and Ford, P.C. Reactions of the bioregulatory agent nitric oxide in oxygenated aqueous media: Determination of the kinetics for oxidation and nitrosation by intermediates generated in the NO/O2 reaction. *Chem. Res. Toxicol.* **6**, 23–32 (1993).

105. Zhu, L., Gunn, C., and Beckman, J.S. Bactericidal activity of peroxynitrite. *Arch. Biochem. Biophys.* **298**, 452–457 (1992).

106. Radi, R., Beckman, J.S., Bush, K.M., and Freeman, B.A. Peroxynitrite-induced membrane lipid peroxidation: The cytotoxic potential of superoxide and nitric oxide. *Arch. Biochem. Biophys.* **288**, 481–487 (1991).

107. Radi, R., Beckman, J.S., Bush, K.M., and Freeman, B.A. Peroxynitrite oxidation of sulfhydryls: The cytotoxic potential of superoxide and nitric oxide. *J. Biol. Chem.* **266**, 4244–4250.

108. Inoue, S. and Kawanishi, S. Oxidative DNA damage induced by simultaneous generation of nitric oxide and superoxide. *FEBS Lett.* **371**, 86–88 (1995).

109. Hogg, N., Joseph, J., and Kalyanaraman, B. The oxidation of a-tocopherol and Torlox by peroxynitrite. *Arch. Biochem. Biophys.* **314**, 153–158 (1994).

110. White, C.R., Brock, T.A., Chang, L.Y., Crapo, J., Briscoe, P., Ku, D., Bradley, W.A., Gianturco, S.H., Gore, J., Freeman, B.A., and Tarpey, M.M. Superoxide and peroxynitrite in atherosclerosis. *Proc. Natl. Acad. Sci. USA* **91**, 1044–1048 (1994).

111. Beckman, J.S., Carson, M., Smith, C.D., and Koppenol, W.H. ALS, SOD and peroxynitrite. *Nature* **364**, 584 (1993).

112. Martin, B.L., We, D., Jakes, S., and Graves, D.J. Chemical influences on the specificity of tyrosine phosphorylation. *J. Biol. Chem.* **265**, 7108–7111 (1990).

113. Berlett, B.S., Friguet, B., Yim, M.B., Chock, P.B., and Stadtman, E.R. Peroxynitrite-mediated nitration of tyrosine residues in *Escherichia coli* glutamine synthetase mimics adenylation: Relevance to signal transduction. *Proc. Natl. Acad. Sci. USA* **93**, 1776–1780 (1996).

114. Lundblad, R.L., Noyes, C.M., Featherstone, G.L., Harrison, J.H., and Jenzano, J.W. The reaction of bovine alpha-thrombin with tetranitromethane: Characterization of the modified protein. *J. Biol. Chem.* **263**, 3729–3734 (1988).

115. Deckers-Hebestreit, G., Schmid, R., Kiltz, H.H., and Altendorf, K. FO portion of *Escherichia coli* ATP synthase: Orientation of subunit *c* in the membrane. *Biochemistry* **26**, 5487–5492 (1985).

116. Nixon, R.A. The regulation of neurofilament protein dynamics by phosphorylation: Clues to neurofibrillary photobiology. *Brain Pathol.* **3**, 29–38 (1993).

117. Rubbo, H., Denicola, A., and Radi, R. Peroxynitrite inactivates thiol-containing enzymes of *Trypanosoma cruzi* energetic metabolism and inhibits cell respiration. *Arch. Biochem. Biophys.* **308**, 96–102 (1994).

118. Squadrito, G.L., Jin, X., and Pryor, W.A. Stopped-flow kinetic study of the reaction of ascorbic acid with peroxynitrite. *Arch. Biochem. Biophys.* **322**, 53–59 (1995).

119. Beckman, J.S., Ischiropoulos, H., Zhu, L., van der Woerd, M., Smith, C., Chen, J., Harrison, J., Martin, J.C., and Tsai, M. Kinetics of superoxide dismutase- and iron-catalyzed nitration of phenolics by peroxynitrite. *Arch. Biochem. Biophys.* **298**, 438–445 (1992).

120. Bautista, A.P. and Spitzer, J.J. Inhibition of nitric oxide formation in vivo enhances superoxide release by the perfused liver. *Am. J. Physiol.* **266**, G783–G788 (1994).

121. Harbrecht, B.G., Billiar, T.R., Stadler, J., Demetris, A.J., Ochoa, J., Curran, R.D., and Simmons, R.L. Inhibition of nitric oxide synthesis during endotoxemia promotes intrahepatic thrombosis and an oxygen radical-mediated hepatic injury. *J. Leuk. Biol.* **52**, 390–394 (1992).

122. Harbrecht, B.G., Billiar, T.R., Stadler, J., Demetris, A.J., Ochoa, J.B., Curran, R.D., and Simmons, R.L. Nitric oxide synthesis serves to reduce hepatic damage during acute murine endotoxemia. *Crit. Care Med.* **20**, 1568–1574 (1992).

123. Moncada, S. and Higgs, A. The L-arginine-nitric oxide pathway. *N. Engl. J. Med.* **329**, 2002–2012 (1993).

124. Moro, M.A., Darley-Usmar, V.M., Goodwin, D.A., Read, N.G., Zamora-Pino, R., Feelisch, M., Radomski, M.W., and Moncada, S. Paradoxical fate and biological action of peroxynitrite on human platelets. *Proc. Natl. Acad. Sci. USA* **91**, 6702–6706 (1994).

125. Rubbo, H., Radi, R., Trujillo, M., Telleri, R., Kalyanaraman, B., Barnes, S., Kirk, M., and Freeman, B.A. Nitric oxide regulation of superoxide and peroxynitrite-dependent lipid peroxidation: Formation of novel nitrogen-containing oxidized lipid derivatives. *J. Biol. Chem.* **269**, 26066–26075 (1994).

126. Padmaja, S. and Huie, R.E. The reaction of nitric oxide with organic peroxyl radicals. *Biochem. Biophys. Res. Commun.* **195**, 539–544 (1993).

127. Kanner, J., Harel, S., and Granit, R. Nitric oxide as an antioxidant. *Arch. Biochem. Biophys.* **289**, 130–136 (1991).

128. Gorbunov, N.V., Osipov, A.N., Day, B.W., Zayas-Rivera, B., Kagan, V.E., and Elsayed, N.M. Reduction of ferrylmyoglobin and ferrylhemoglobin by nitric oxide: A protective mechanism against ferryl hemoprotein-induced oxidations. *Biochemistry* **34**, 6689–6699 (1995).

129. Geng, Y.J., Petersson, A.S., Wennmalm, A., and Hansson, G.K. Cytokine-induced expression of nitric oxide synthase results in nitrosylation of heme and nonheme iron proteins in vascular smooth muscle cells. *Exp. Cell Res.* **214**, 418–428 (1994).

130. Drapier, J.C. and Hibbs, Jr., J.B. Murine cytotoxic activated macrophages inhibit aconitase in tumor cells: Inhibition involves the iron-sulfur prosthetic group and is reversible. *J. Clin. Invest.* **78**, 790–797 (1986).

131. Wharton, M., Granger, D.L., and Durack, D.T. Mitochondrial iron loss from leukemia cells injured by macrophages: A possible mechanism for electron transport chain defects. *J. Immunol.* **141**, 1311–1317 (1988).

132. Drapier, J.C., Pellat, C., and Henry, Y. Generation of EPR-detectable nitrosyl-iron complexes in tumor target cells cocultured with activated macrophages. *J. Biol. Chem.* **266**, 10162–10167 (1991).

133. Stadler, J., Billiar, T.R., Curran, R.D., Stuehr, D.J., Ochoa, J.B., and Simmons, R.L. Effect of exogenous and endogenous nitric oxide on mitochondrial respiration of rat hepatocytes. *Am. J. Physiol.* **260**, C910–C916 (1991).

134. Nunoshiba, T., deRojas-Walker, T., Wishnok, J.S., Tannenbaum, S.R., and Demple, B. Activation by nitric oxide of an oxidative-stress response that defends *Escherichia coli* against activated macrophages. *Proc. Natl. Acad. Sci. USA* **90**, 9993–9997 (1993).

135. Yee, E.L., Pitt, B.R., Billiar, T.R., and Kim, Y.M. The effects of nitric oxide on heme metabolism in pulmonary artery endothelial cells. *Am. J. Physiol.* (in press).

136. Kim, Y.M., Bergonia, H.A., Muller, C., Pitt, B.R., Watkins, S.D., and Lancaster, J.R., Nitric oxide and intracellular heme. *Adv. Pharmacol.* **34**, 277–291 (1995).

137. Chen, C.C., Rosenbloom, C.L., Anderson, D.C., and Manning, A.M. Selective inhibition of E-selectin, vascular cell adhesion molecule-1, and intercellular adhesion molecule-1 expression by inhibitors of I kappa B-alpha phosphorylation. *J. Immunol.* **155**, 3538–3545 (1995).

138. Davenpeck, K.L., Gauthier, T.W., and Lefer, A.M. Inhibition of endothelial-derived nitric oxide promotes P-selectin expression and actions in the rat microcirculation. *Gastroenterology* **107**, 1050–1058 (1994).

139. Radomski, M.W., Vallance, P., Whitley, G., Foxwell, N., and Moncada, S. Platelet adhesion to human vascular endothelium is modulated by constitutive and cytokine induced nitric oxide. *Cardiovasc. Res.* **27**, 1380–1382 (1993).

140. Zhang, H., Rogiers, P., Preiser, J.C., Spapen, H., Manikis, P., Metz, G., and Vincent, J.L. Effects of methylene blue on oxygen availability and regional blood flow during endotoxic shock. *Crit. Care Med.* **23**, 1711–1721 (1995).

141. Petros, A., Lamb, G., Leone, A., Moncada, S., Bennett, D., and Vallance, P. Effects of a nitric oxide synthase inhibitor in humans with septic shock. *Cardiovasc. Res.* **28**, 34–39 (1994).

142. Billiar, T.R., Curran, R.D., Harbrecht, B.G., Stuehr, D.J., Demetris, A.J., and Simmons, R.L. Modulation of nitrogen oxide synthesis in vivo: N^G-monomethyl-L-arginine inhibits endotoxin-induced nitrite/nitrate biosynthesis while promoting hepatic damage. *J. Leuk. Biol.* **48**, 565–569 (1990).

143. Harbrecht, B.G., Stadler, J. Demetris, A.J., Simmons, R.L., and Billiar, T.R. Nitric oxide and prostaglandins interact to prevent hepatic damage during murine endotoxemia. *Am. J. Physiol.* **266**, G1004–G1010 (1994).

144. Szabo, C., Southan, G.J., and Thiemermann, C. Beneficial effects and improved survival in rodent models of septic shock with S-methylisothiourea sulfate, a potent and selective inhibitor of inducible nitric oxide synthase. *Proc. Natl. Acad. Sci. USA* **91**, 12472–12476 (1994).

145. Asahi, M., Fujii, J., Suzuki, K., Seo, H.G., Kuzuya, T., Hori, M., Tada, M., Fujii, S., and Taniguchi, N. Inactivation of glutathione peroxidase by nitric oxide: Implication for cytotoxicity. *J. Biol. Chem.* **270**, 21035–21039 (1995).

146. Becker, K., Gui, M., and Schirmer, R.H. Inhibition of human glutathione reductase by *S*-nitrosoglutathione. *Eur. J. Biochem.* **234**, 472–478 (1995).

147. Thompson, C.B. Apoptosis in the pathogenesis and treatment of disease. *Science* **267**, 1456–1462 (1995).

148. Steller, H. Mechanisms and genes of cellular suicide. *Science* **267**, 1445–1449 (1995).

149. Albina, J.E., Caldwell, M.D., Henry, Jr., W.L., and Mills, C.D. Regulation of macrophage functions by L-arginine. *J. Exp. Med.* **169**, 1021–1029 (1989).

150. Messmer, U.K., Ankarcrona, M., Nicotera, P., and Brune, B. p53 expression in nitric oxide-induced apoptosis. *FEBS Lett.* **355**, 23–26 (1994).

151. Fehsel, K., Kroncke, K.D., Meyer, K.L., Huber, H., Wahn, V., and Kolb-Bachofen, V. Nitric oxide induces apoptosis in mouse thymocytes. *J. Immunol.* **155**, 2858–2865 (1995).

152. Meβmer, U.K., Reimer, D.M., Reed, J.C., and Brüne, B. Nitric oxide induced poly (ADP-ribose) polymerase cleavage in RAW264.7 macrophage apoptosis is blocked by Bcl-2. *FEBS Lett.* **384**, 162–166 (1996).

153. Nicholson, D.W., Ali, A., Thornberry, N.A., Vaillancourt, J.P., Ding, C.K., Gallant, M., Gareau, Y., Griffin, P.R., Labelle, M., Lazebnik, Y.A., Munday, N.A., Raju, S.M., Smulson, M.E., Yamin, T.T., Yu, V.L., and Miller, D.K. Identification and inhibition of the ICE/CED-3 protease necessary for mammalian apoptosis. *Nature* **376**, 37–43 (1995).

154. Tewari, M., Quan, L.T., O'Rourke, K., Desnoyers, S., Zeng, Z., Beidler, D.R., Poirier, G.G., Salvesen, G.S., and Dixit, V.M. Yama/CPP32 beta, a mammalian homolog of CED-3 is a CrmA-inhibitable protease that cleavese the death substrate poly(ADP-ribose) polymerase. *Cell* **81**, 801–809 (1995).

155. Kaneto, H., Fujii, J, Seo, H.G., Suzuki, K., Matsuoka, T., Nakamura, M., Tatsumi, H., Yamasaki, Y., Kamada, T., and Taniguchi, N. Apoptotic cell death triggered by nitric oxide in pancreatic beta-cells. *Diabetes* **44**, 733–738 (1995).

156. Terenzi, F., Diaz-Guerra, M.J., Casado, M., Hortelano, S., Leoni, S., and Bosca, L. Bacterial lipopeptides induce nitric oxide synthase and promote apoptosis through nitric oxide-independent pathways in rat macrophages. *J. Biol. Chem.* **270**, 6017–6021 (1995).

157. Albina, J.C., Cui, S., Mateo, R.B., and Reichner, J.S. Nitric oxide-mediated apoptosis in murine peritoneal macrophages. *J. Immunol.* **150**, 5080–5085 (1993).

158. Beauvais, F., Michel, L., and Dubertret, L. The nitric oxide donors, azide and hydroxylamine, inhibit the programmed cell death of cytokine-deprived human eosinophils. *FEBS Lett.* **361**, 229–232 (1995).

159. Martin-Sanz, P., Dlaz-Guerra, M.J.M., Casado, M., and Bosca, L. Bacterial lipopolysaccharide antagonizes transforming growth factor β1-induced apoptosis in primary cultures of hepatocytes. *Hepatology* **23**, 1200–1207 (1996).

160. Cain, K., Inayat-Hussain, S.H., Couet, C., and Cohen, G.M. A cleavage-site directed inhibitor of interleukin-lb-converting enzyme-like proteases inhibits apoptosis in primary cultures of rat hepatocytes. *Biochem. J.* **314**, 27–32 (1996).

161. Chun, S.Y., Eisenhauer, K.M., Kubo, M., and Hsueh, A.J. Interleukin-1 beta suppresses apoptosis in rat ovarian follicles by increasing nitric oxide production. *Endocrinology* **136**, 3120–3127 (1995).

162. Genaro, A.M., Hortelano, S., Alvarez, A., Martinez, A.C., and Bosca, L. Splenic B lymphocyte programmed cell death is prevented by nitric oxide release through mechanisms involving sustained bcl-2 level. *J. Clin. Invest.* **95**, 1884–1890 (1995).

163. Higuchi, M., Higashi, N., Nishimura, Y., Toyoshima, S., and Osawa, T. TNF-mediated cytotoxicity: Importance of intracellular cGMP level for determining TNF-sensitivity. *Molec. Immunol.* **28**, 1039–1044 (1991).

164. Snyder, S.H. Nitric oxide: First in a new class of neurotransmitters. *Science* **257**, 494–496 (1992).

165. Dawson, T.M., Dawson, V.L., and Snyder, S.H. A novel neuronal messenger molecule in brain: The free radical, nitric oxide. *Ann. Neurol.* **32**, 297–311 (1992).

166. Choi, D.W. Bench to bedside: the glutamate connection. *Science* **258**, 241–243 (1992).

167. Lei, S.Z., Pan, Z.H., Aggarwal, S.K., Chen, H.S., Hartman, J., Sucher, N.J., and Lipton, S.A. Effect of nitric oxide production on the redox modulatory site of the NMDA receptor-channel complex. *Neuron* **8**, 1087–1099 (1992).

168. Corbett, J.A., Lancaster, Jr., J.R., Sweetland, M.A., and McDaniel, M.L. Interleukin-1 beta-induced formation of EPR-detectable iron-nitrosyl complexes in islets of Langerhans: Role of nitric oxide in interleukin-1 beta-induced inhibition of insulin secretion. *J. Biol. Chem.* **266**, 21351–21354 (1991).

169. Lancaster, Jr., Langrehr, J.M., Bergonia, H.A., Murase, N., Simmons, R.L., and Hoffman, R.A. EPR detection of heme and nonheme iron-containing protein nitrosylation by nitric oxide during rejection of rat heart allograft. *J. Biol. Chem.* **267**, 10994–10998 (1992).

170. Hoffman, R.A., J.R., Langrehr, J.M., Wren, S.M., Dull, K.E., Ildstad, S.T., McCarthy, S.A., and Simmons, R.L. Characterization of the immunosuppressive effects of nitric oxide in graft vs host disease. *J. Immunol.* **151**, 1508–1518 (1993).

171. Langrehr, J.M., Murase, N., Markus, P.M., Cai, X., Neuhaus, P., Schraut, W., Simmons, R.L., and Hoffman, R.A. Nitric oxide production in host-versus-graft and graft-versus-host reactions in the rat. *J. Clin. Invest.* **90**, 679–683 (1992).

172. McCartney-Francis, N., Allen, J.B., Mizel, D.E., Albina, J.E., Xie, Q.W., Nathan, C.F., and Wahl, S.M. Suppression of arthritis by an inhibitor of nitric oxide synthase. *J. Exp. Med.* **178**, 749–754 (1993).

173. Kallenberg, C.G. Overlapping syndromes, undifferentiated connective tissue disease, and other fibrosing conditions. *Curr. Opin. Rheumatol.* **7**, 568–573 (1995).

174. Nunoshiba, T., deRojas-Walker, T., Wishnok, J.S., Tannenbaum, S.R., and Demple, B. Activation by nitric oxide of an oxidative-stress response that defends *Escherichia coli* against activated macrophages. *Proc. Natl. Acad. Sci. USA* **90**, 9993–9997 (1993).

175. Hidalgo, E. and Demple, B. An iron-sulfur center essential for transcriptional activation by the redox-sensing SoxR protein. *EMBO J.* **13**, 138–146 (1994).

176. Furukawa, T., Kohno, H., Tokunaga, R., and Taketani, S. Nitric oxide-mediated inactivation of mammalian ferrochelatase in vivo and in vitro: possible involvement of the iron-sulphur cluster of the enzyme. *Biochem. J.* **310**, 533–538 (1995).

177. Nakatsuka, M., and Osawa, Y. Selective inhibition of the 12-lipooxygenase pathway of arachidonic acid metabolism by L-arginine or sodium nitroprusside in intact human platelets. *Biochem. Biophys. Res. Commun.* **200**, 1630–1634 (1994).

178. Hess, D.T., Lin, L.H., Freeman, J.A., and Norden, J.J. Modification of cysteine residues within G(o) and other neuronal proteins by exposure to nitric oxide. *Neuropharmacology* **33**, 1283–1292 (1994).

179. Pozdnyakov, N., Lloyd, A., Reddy, V.N., and Sitaramayya, A. Nitric oxide-regulated endogenous ADP-ribosylation of rod outer segment proteins. *Biochem. Biophys. Res. Commun.* **192**, 610–615 (1993).

180. Laval, F. and Wink, D.A. Inhibition by nitric oxide of the repair protein, O 6-methylguanine-DNA-methyltransferase. *Carcinogen* **15**, 443–447 (1994).

181. Cleeter, M.W., Cooper, J.M., Darley-Usmar, V.M., Moncada, S., and Schapira, A.H. Reversible inhibition of cytochrome *c* oxidase, the terminal enzyme of the mitochondrial respiratory chain, by nitric oxide: Implications for neurodegenerative diseases. *FEBS Lett.* **345**, 50–54 (1994).

182. Swain, J.A., Darley-Usmar, V., and Gutteridge, J.M. Peroxynitrite releases copper from ceruloplasmin: implications for atherosclerosis. *FEBS Lett.* **342**, 49–52 (1994).

183. Bell, J.A., Beglan, C.L., and London, E.D. Interaction of ascorbic acid with the neurotoxic effects of NMDA and sodium nitroprusside. *Life Sci.* **58**, 367–371 (1995).

184. Clancy, R.M., Leszczynska-Piziak, J., and Abramson, S.B. Nitric oxide, an endothelial cell relaxation factor, inhibits neutrophil superoxide anion production via a direct action on the NADPH oxidase. *J. Clin. Invest* **90**, 1116–1121 (1992).

3

Nitric Oxide Synthases and Their Cofactors

Steven S. Gross

1. Introduction

Production of the free radical NO via a five-electron oxidation of a guanidino nitrogen from arginine is no small enzymatic task. To this end, nature has invented a complex chemistry, not yet fully understood, that is catalyzed by proteins of the NO synthase (NOS) gene family. In mammals, three distinct NOS genes have been cloned and shown to encode highly related enzyme isoforms (see Chapter 4). Nomenclature to describe the NOS gene products derives from the tissues in which isoforms were first recognized: neuronal NOS (nNOS), endothelial NOS (eNOS), and inducible NOS (iNOS, from immunostimulant-activated macrophages). Although these gene products share only 50–60% identity with one another, they all possess a common genomic organization, are homodimeric, catalyze an identical chemistry, and display identical cofactor requirements [1]. Molecular, anatomical, and functional distinctions among these three NOS isoforms are discussed in this chapter and throughout the volume.

A fundamental difference between NOS isoforms lies in their mechanisms of regulation [2]; NO is either synthesized by a constitutively expressed enzyme that lies dormant and is activated on a moment-to-moment basis in response to transient increases in levels of intracellular Ca^{2+} (nNOS and eNOS) or NO is produced by a transcriptionally induced enzyme that possesses full activity at all levels of intracellular Ca^{2+} (iNOS). The latter isoform, iNOS, is responsible for the "high-output" NO pathway that can be induced in perhaps all nucleated cells following exposure to an appropriate immunostimulant. Once induced, iNOS may become limited by substrate availability (see Chapter 23) or subject to autoinactivation by the NO it produces [3–5]. iNOS is devoid of a 45 amino acid insert found in constitutively expressed NOSs, which we speculate to be autoinhibitory to activity at resting intracellular levels of Ca^{2+} [6] (see below).

It is notable that each of the three mammalian NOS genes are expressed within the kidney in specific cell types (Chapter 7); in the case of constitutive NOSs, subcellular localization is conferred by *N*-terminal protein binding motifs (nNOS [7,8]) or myristoylation (eNOS [9,10]). Prevailing views regarding the enzymatic function of NOS cofactors and how they regulate NOS activity in tissues are presented below.

2. An Overview of NOS Structure and Function

Nitric oxide synthase stereospecifically converts L-arginine to NO and L-citrulline; this involves two successive monooxygenation reactions which, in total, consume 1.5 moles of reduced pyridine nucleotide and 2 moles of molecular oxygen and yield N^{ω}-hydroxy-L-arginine as an isolatable intermediate [11] (Fig. 3-1). A balanced reaction would require unitary consumption of reduced pyridine nucleotide, thus, more properly, two molecules of NO are produced in coupled synthetic reactions which use electrons derived from three molecules of reduced pyridine nucleotide. To accomplish and regulate catalysis, NOSs rely on a variety of cofactors and prosthetic groups: reduced nicotine adenine dinucleotide phosphate (NADPH), flavin adinine dinucleotide (FAD), flavin mononucleotide (FMN), iron protoporphrin IX (heme), tetrahydrobiopterin (BH4), and calmodulin (CaM). Although each of these factors are individually utilized by other mammalian enzymes, NOSs are unique in requiring all simultaneously. The present understanding of NOS catalytic mechanisms was spurred by the recognition of requirements for specific cofactors and an appreciation of how these species function within other enzymes.

Analysis of the deduced amino acid sequence of nNOS cDNA, the first cloned member of the NOS gene family, immediately suggested a bidomain protein structure [12]. A canonical 25–35 residue CaM binding motif was observed to

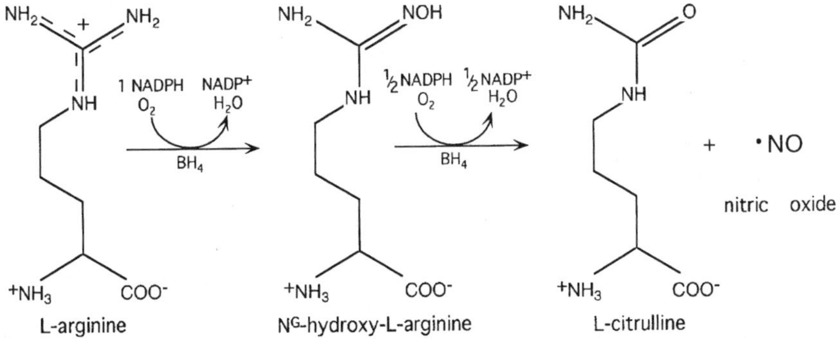

Figure 3-1. Reaction mechanism and stoichiometry for the two-step oxidation of L-arginine by the NOS protein family.

bridge a C-terminal reductase-related polypeptide with a presumed N-terminal substrate-binding (oxygenase) polypeptide. This pattern was preserved in the subsequently cloned iNOS [13] and eNOS [14,15] isoforms. Later, functionality of the isolated reductase and oxygenase domains from nNOS was established [16–18]. The bidomain structure of NOSs could be further dissected into a series of functional subdomains, discrete structural "modules" which are likely to have originated through a process of duplication, rearrangement, and fusion of ancestral genetic building blocks. The modular structure of NOS proteins is consistent with intron/exon boundaries identified within the cloned NOS genes [19–21]. Alignment of domains and subdomains within the three mammalian NOS isoforms and some related homologs are depicted in Fig. 3-2.

It is now understood that the reductase domain serves to funnel reducing equivalents from NADPH, through the flavins, to heme–iron within the oxygenase domain, where molecular oxygen is activated for catalysis. Binding of CaM to the "bridge" between reductase and oxygenase domains is essential for interdomain

MODULAR STRUCTURE OF NITRIC OXIDE SYNTHASES

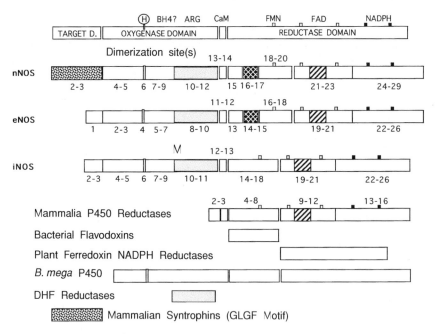

Figure 3-2. Linear schematics which depict the domain structure of the three mammalian NOS proteins and their relationship to protein homologs. Numbers beneath the modules in NOS and cytochrome P-450 reductase indicate their respective exon origins. Target D. refers to an N-terminal targeting domain identified in nNOS and eNOS, which has not yet been examined for function in iNOS.

electron flux [22]. The perpetual activation of iNOS is explained by its remarkably high affinity for CaM; CaM binding to iNOS is sustained at all levels of intracellular Ca^{2+} concentration and is even stable to boiling in SDS/PAGE sample loading buffer [23].

As isolated, NOSs are predominantly homodimeric proteins comprising subunits of 160 kD (nNOS [24–26], 130 kD (iNOS [27]), and 133 kD (eNOS [28]). In vitro dissociation of iNOS into subunits has been shown to result in a complete loss of activity, concomitant with the dissociation of heme and BH4 [29]. Although it is likely that a similar quaternary structure is also required for activity of nNOS and eNOS, this has not formally been established. Interestingly, the activity and the dimeric structure of iNOS can be reconstituted upon incubation of subunits with L-arginine, heme, and BH4 [29]. Because analytical ultracentrifugation studies reveal an equilibrium in the distribution of monomer/dimer (nNOS) and monomer/dimer/tetramer (eNOS) (P. Martasek and B.S.S. Masters et al., unpublished communication), subunit assembly could be a significant site for regulation of NOS activity in vivo. The posttranslational sequence of events culminating in subunit assembly, and factors which modulate this important process, await clarification.

Bacterial expression of NOS fragments and tryptic cleavage studies of holo-NOSs reveal the presence of dimerization sites within the oxygenase domains of nNOS, eNOS, and iNOS [17,18]. Although the isolated reductase domain of nNOS and iNOS appears incapable of dimerization, dimerization of the reductase domain of eNOS may be possible [30]. Thus, NOSs probably assemble in a head-to-head manner. Complementation studies, where one subunit of a chimeric iNOS heterodimer has its reductase domain deleted and the other subunit possesses a dysfunction in its oxygenase domain that precludes BH4 binding, raise the possibility that electron flux can crisscross between subunits [31]. If correct, this mechanism offers a clear basis for the observed reliance of activity on dimerization.

3. NOS Cofactors and Prosthetic Groups

3.1. NADPH and Flavins

The reductase domain of the NOSs comprises the C-terminal 600–700 amino acids, encoded by NOS gene exons 15–29 (nNOS), 14–26 (iNOS), and 13–26 (eNOS). Residing within this domain are identified consensus binding sites for NADPH (both ribose and adenine moieties), FAD, and FMN (both isoalloxazine ring and pyrophosphate moieties). In NOS-containing cytosol from immunostimulant-activated murine macrophages, even before NO was established as the primary product of the arginine-derived nitrite pathway, NADPH was appreciated as the sole reductant that could support this pathway. The presumptive NADPH binding site on NOS and mode for transmission of reducing equivalents from

NADPH was deduced after cloning of the neuronal NOS (nNOS) gene [12]. The amino acid sequence of the C-terminal half of nNOS was noted to be homologous with only one mammalian protein, cytochrome P-450 reductase (CPR; 58% identity). CPR is a FAD/FMN-containing enzyme which serves mammalian cells by providing NADPH-derived reducing equivalents to the evolutionarily diverse and extensive P-450 monooxygenase gene family. Whereas CPR and P-450 monooxygenases carry distinct reductase and oxygenase activities in their respective polypeptide sequences, the NOSs possess both activities in a single polypeptide, making them catalytically self-sufficient. Prior to the discovery of NOSs, CPR was the only mammalian enzyme known to contain both FAD and FMN within a single polypeptide, although precedent was found in several bacterial enzymes which additionally contain NOS-like oxygenase domains (i.e., nitrite reductase, sulfite reductase, and the unusual P-450 monooxygenase BM-3 from *Bacillus (B.) megatarium* [32]).

Cytochrome P-450 reductase itself has been dissected into two subdomains based on a remarkable degree of homology with FMN-containing electron transferases (e.g., bacterial flavodoxins) and FAD/NADPH-containing transhydrogenases (e.g., plant ferredoxin–NADP$^+$ reductase; FNR) [32,33]. Because high-resolution three-dimensional crystal structures have been obtained for flavins in complex with FNR and flavodoxins [34], binding sites and structure of these single flavin-containing enzymes have been established with rigor. Fusion of such ancestral flavoproteins within a single polypeptide, initially appreciated as the evolutionary origin of CPR [32], more recently NOS, is affirmed by identification of the distinct boundaries between FMN and FAD/NADPH subdomains. Moreover, homology with CPR and ancestral flavoproteins defines the specific amino acid residues within NOSs which are most certainly engaged in binding of pyrophoshate groups and isoalloxazine rings of NADPH, FAD, and FMN. It is notable that FMN ring-shielding aromatic residues are also 100% conserved among the flavodoxins, CPR (Tyr-140 and Tyr-178), and all NOSs; these serve to maintain a hydrophobic environment for effective interflavin electron tunneling and may participate in π-π stacking [35]. Also conserved between NOSs and CPR is a $\approx$120 amino acid insertion within the FAD subdomain that is not present in proteins containing a single flavin and which likely provides the needed orientation for efficient electron flux between flavin subdomains [32].

By analogy with CPR, whose enzymology has been studied in detail [36], the reductase domain of NOSs has evolved to accomplish at least two pivotal functions. One is to deliver electrons to the oxygenase domain's heme–iron, one-by-one, from an obligate two-electron donor, NADPH. This is made possible by the unusual ability of flavins to accept two electrons and donate them singly. Studies of CPR establish that electrons flux from NADPH through FAD to FMN, in that order [36], a sequence which is assuredly preserved in NOSs. Second, the capacity of the flavins to store up to four electrons (two per flavin) provides a reservoir for the odd electron which remains after an initial cycle of NO generation, for use in the subsequent cycle (appreciate that three NADPH molecules donate six electrons; however, only three electrons are used per mole of NO formed). It is not surprising therefore that, as isolated, NOS has been shown to contain a stable flavin radical, presumably FMNH• [37].

Oxidoreductases, to varying degrees, all engage in uncoupled electron transfer (i.e., delivery of electrons to acceptors other than the natural substrate). For nNOS, this "inefficiency" in utilization of reducing equivalents is exaggerated under conditions where L-arginine or BH4 are limiting [38], resulting in substantial production of superoxide anion from O_2 [39,40]. Notably, uncoupled generation of superoxide by iNOS is less pronounced than for nNOS (one-sixth [40]). Because superoxide reacts at a near-diffusion-limited rate with NO to form the reactive metabolite peroxynitrite, NOS uncoupling may contribute significantly to NOS-mediated pathophysiology. Uncoupled electron transfer also confers NOSs with the ability to reduce cytochrome c, ferricyanide, and dyes such as dichlorophenolindopenol (DCIP) and nitrobluetetraziolium (NBT), typically at rates far exceeding those observed for maximal NO formation. The selective ability of NOSs in aldehyde-fixed tissues to reduce NBT to an insoluble blue dye, formazan (i.e., NADPH diaphorase activity), has been used as an effective histochemical stain for NOS localization [41,42] (Chapter 7).

Potentially, NADPH-derived electrons may "leak" to alternate acceptors from either FAD, FMN, or heme. In actuality, however, heme appears to be the only significant site of superoxide generation as this can be blocked by arginine analogs [43,44] and iron-binding agents [45]. Just as coupled electron transfer to arginine requires CaM, the ability of nNOS to reduce O_2 has been shown to be accelerated 100-fold by CaM [40]. In contrast to O_2, several other electron acceptors appear to be directly reduced by the flavins, principally FMN, in a manner which is either CaM dependent (cytochrome c [16,44,46]) or CaM independent (DCPIP and NBT [47]).

It is unlikely that flavin availability ever serves to limit NOS activity in a physiological setting. Although not covalently bound, FAD and FMN are bound tightly to NOS and typically copurify in equimolar quantities with NOS subunits from mammalian cells. The site, timing, and binding constants for insertion of flavins into maturing NOS subunits remains to be elucidated.

3.2. Calmodulin

The NOSs are the first enzymes in which CaM has been shown to gate electron flux [40]. This presumably reflects the ability of CaM to align the reductase and oxygenase domains of NOSs for productive interdomain electron transfer, from FAD to heme. Nonetheless, CaM has also been shown to promote interflavin electron transit [22], which may be essential for heme reduction. CaM binding, initiated by increased levels of intracellular Ca^{2+}, serves as the principal means for activating eNOS and nNOS activity; in contrast, iNOS is perpetually active by virtue of its ability to bind CaM in a Ca^{2+}-independent manner. An appreciation that nNOS requires calmodulin for activity was enabling for the first purification of a NOS isoform by Bredt and Snyder [24].

Calmodulin-binding domains in NOSs are prototypic in their canonical content of basic and lipophillic amino acids. They span 25–35 amino acids and each arises from two exons, the second of which contributes the N-terminal aspect of the FMN subdomain. Thus, it is likely that CaM binding has evolved from an ancestral FMN-containing flavoprotein. Alignment of the CaM-binding domains in NOS isoforms suggests that eNOS has a 4 amino acid deletion near its N-terminus, conceivably resulting in reduced affinity for calmodulin relative to nNOS. Nonetheless, using a solution phase [^{125}I]-CaM-binding assay, performed in the presence of Ca^{2+}, we have found comparable equilibrium dissociation constants for equilibrium binding to purified rat nNOS and bovine eNOS (3–5 nM; Q. Liu and S. Gross, unpublished). Based on measurements of fluorescence quenching, nNOS was previously reported to bind CaM with a 1 nM Kd [16]. In another study, a 23 amino acid peptide from the binding site of nNOS was found to bind CaM in a 1 : 1 ratio and 2–3 nM Kd in the presence of calcium [48]. Two-dimensional NMR and circular dichroism measurements suggest that upon interaction with CaM, the nNOS CaM-binding site assumes an α-helical conformation, essentially identical to that previously shown for the CaM site of myosin light-chain kinase [48].

Calmodulin is a 17-kD protein composed of an eight-turn central α-helix, flanked by globular N- and C-terminal Ca^{2+}-binding lobes. Upon Ca^{2+} binding, these globular lobes compact, exposing sequestered hydrophobic and acidic amino acids that engage in binding to target sites [49]. It is notable that each lobe of CaM has been shown to independently bind to nNOS [50]. Although the C-lobe of CaM binds Ca^{2+} with ≈10-fold greater affinity than the N-lobe [51], and thus should bind first in response to rising levels of intracellular Ca^{2+}, binding of the N-lobe is essential for nNOS activation [52]. Based on kinetic studies of CaM binding and activation of purified nNOS and eNOS (Weissman, Jones, Liu, Persechini, and Gross, unpublished), the following sequential steps are proposed: (1) Ca^{2+} binds to CaM's C-lobe, exposing hydrophobic and acidic binding residues, (2) the C-lobe of CaM binds NOS, (3) Ca^{2+} binds to the N-lobe of NOS-bound CaM, exposing N-lobe NOS-binding residues, and (4) the N-terminal lobe of CaM binds NOS. Notably, NOS activation would occur only after Step 4. Deactivation of nNOS, upon intracellular Ca^{2+} sequestration, would proceed in reverse order, allowing for some CaM to be bound to inactive nNOS solely via the C-lobe. Whether occupancy of one or both of the CaM sites on NOS dimers is a minimum requirement for NOS activation remains to be ascertained.

Studies of CaM binding to chimeric eNOS and iNOS, in which isoform-specific CaM-binding sites have been exchanged, reveal that the Ca^{2+} dependence of binding is contributed largely by the cognate CaM-binding sequence; however, the protein backbone in which the CaM site resides is also a determinant [53].

A possible clue as to the role of the NOS backbone in CaM binding (i.e., sites external to the CaM-binding site itself) is provided by comparative alignment of amino acid sequences in NOSs and related flavoproteins. Unique to the Ca^{2+} dependent isoforms of NOS (both nNOS and eNOS), a ≈ 45 amino acid peptide insert is observed in the FMN-binding domain; this peptide is conspicuously absent from iNOS and all otherwise homologous flavoproteins. Because CaM-dependent enzymes protypically contain an autoinhibitory sequence which is disengaged upon CaM binding, we examined whether the unique peptide serve an analogous function for cNOSs [6].

Computer-assisted three-dimensional modeling of the FMN domain of nNOS and eNOS, based on homology to bacterial flavodoxins with crystal structures that have been solved at high resolution, suggests domain overlap between the peptide insert and NOS-bound CaM. This predicts that CaM binding to nNOS (and eNOS) would be sterically hindered by the putative autoinhibitory domain. Consistent with this view, synthetic peptides contained within the FMN domain polypeptide insert of both eNOS and nNOS were found to potently block [^{125}I]-CaM binding and activation of NOS. Binding of CaM is thus suggested to displace the autoinhibitory peptide to elicit NOS activation. This view is also consistent with studies of limited NOS proteolysis. Although tryptic cleavage of nNOS in the absence of CaM results almost exclusively in the bissection of NOS at its CaM-binding site, tryptic cleavage in the presence of CaM occurs predominantly within the putative autoinhibitory peptide. Some pressing questions await answers: Where does the autoinhibitory peptide bind within NOS? Is dysinhibition the sole function of calmodulin? Does the phosphorylation status of the autoinhibitory peptide regulate its ability to bind and, hence, restrict catalysis (as is the case for CaM kinases [54])?

3.3. Heme

Macmillan et al. were first to show a heme prosthetic group in NOSs, as revealed by a CO-induced 445-nm chromophore that is characteristic of the ferrous–carbonyl complex [55]. Several labs have shown that heme is present in purified nNOS and that stoichiometric quantities of heme are needed to obtain full catalytic activity [37,55,56]. Moreover, NOS activity can be abolished by exposure to heme-binding ligands such as CO in the presence of NADPH [56], imidazoles [57], and arginine analogs known to additionally bind heme [44], suggesting heme to be the site of oxygen activation for oxidation of substrate. The interaction of arginine with heme in nNOS is indicated by an arginine-induced "type I" difference spectrum [58], arising from a shift in the iron from low-spin to high-spin states. Resonance Raman spectra demonstrate that heme coordination is pentavalent and verify that thiolate serves as a fifth axial ligand [59]. Heme function in NOS catalysis of L-arginine to N^{ω}-hydroxy-L-arginine is considered to be mechanistically analogous to P-450$_{CAM}$, whereas the precise function of heme in the second oxygenation step is more speculative [1]. Mutation complementation studies raise the possibility that both hemes on dimeric NOS may function as a single active site [31].

Inasmuch as NO has extremely high avidity for heme, it is not surprising that NO has been reported to inactivate NOS by serving as a sixth axial ligand to heme [3–5]. The real dilemma is to explain how NOS largely manages to escape permanent inactivation by its own product. Speculations include that heme reactivity is prevented by water (as hydroxy-heme), prior to the diffusion of NO from the enzyme active site [1] and that nitroso-heme is formed, but reactivation rapidly ensues by a mechanism involving BH4 [3].

Tentative identification of the specific cysteine residues in NOS isoforms involved in heme coordination was based on the observed correspondence to the consensus motif FXXGXXXCXG found in P-450s and related enzymes [55]. Confirmation that Cys415 of rat nNOS serves as the axial heme-thiolate ligand was obtained by site-directed mutagenesis [60]. A similar mutational analysis supports the view that the homologous cysteine in bovine eNOS (Cys186) also provides the fifth ligand to heme [61]. This cysteine is conserved in all isoforms of NOS and in human iNOS and eNOS, corresponds to Cys200 and Cys184, respectively. Exons encoding the heme-thiolate-containing sequence of NOSs are also highly conserved; each encodes a 54 amino acid sequence corresponding to exon 6 (nNOS and iNOS) or exon 4 (eNOS) (see Fig. 3-2).

3.4. BH4

Prior to the discovery of NOS's heme cofactor, BH4 was identified as an absolute catalytic requirement [62,63]. By analogy with all known enzymes that use this pterin cofactor, phenylalanine hydroxylase being the best studied [64], the prediction was that BH4 would prove to be redox active in NOSs, serving to activate O_2 for catalysis. This view was challenged by a striking difference in BH4 function for NOSs and the aromatic amino acid hydroxylases: Whereas aromatic amino acid hydroxylation requires stoichiometric quantities of BH4, NO production can be sustained with catalytic quantities of BH4, in the absence of a dihydropteridine reductase for continued regeneration of reduced pterin cofactor [65]. Thus, either NOS possesses an inherent dihydropteridine reductase activity of its own or NOSs use a BH4-independent mechanism to activate O_2. Although the former possibility remains untested, the identification of heme's presence and function in NOSs support the latter view. The precise function for BH4 has been elusive and is perhaps the least understood aspect of NOS catalysis.

That the oxygenase domain of NOSs contains the BH4 binding site was first shown in studies of isolated domains. As discussed above, trypsin preferentially cleaves a site within the CaM site of nNOS [16], iNOS [17], and eNOS (Q. Liu and S. Gross, unpublished). Spectral characterization of the N-terminal heme-containing tryptic fragment of nNOS by Sheta et al. [16] (rat nNOS$_{1-728}$) and the homologous recombinant peptide expressed in bacteria [60], revealed a type 1 difference spectra upon addition of either arginine or BH4, confirming the presence of respective binding sites. Tryptic fragments of iNOS were similarly shown to bind arginine, CO, and imidazole [17].

One, but perhaps not the only, role for BH4 in NOS appears to be as an allosteric effector. Radioligand-binding studies with [^{3}H] N^ω-nitro-L-arginine (NNA), a selective NOS inhibitor, have been used to probe the arginine site of rat nNOS [66,67]. Because BH4 is largely retained in NOS after its purification, the importance of BH4 for modulation of the arginine-binding site was not fully appreciated in these initial reports. However, studies of NNA binding to BH4-depleted iNOS revealed that BH4 is an absolute requirement for specific NNA binding to occur [33]. It is notable that the potency of tetrahydropterin analogs for eliciting NNA binding correlates with their relative abilities to support catalysis by iNOS (S. Gross, unpublished). Similar results were obtained using a truncated oxygenase domain comprising nNOS$_{220-721}$, expressed in *Escherichia coli*. This NOS fragment bound NNA with a similar high affinity to that observed with native holo-nNOS (50 nM) and, again, BH4 was found to be essential for binding [18]. This confirms by direct measurement that binding sites for BH4 and arginine reside within nNOS$_{220-721}$ and homologous sequences of other NOS isoforms. Recognition of a 160 amino acid sequence within the oxygenase domain of nNOS with apparent homology to the pterin-binding enzyme dihydrofolate reductase [18] raised the possibility that this "DHFR module" (rat nNOS$_{558-721}$, see Fig. 3-2) was the site of BH4 binding. Such a localization of the BH4-binding site is also suggested by mutational studies of iNOS performed by Nathan and colleagues [68]. Nonetheless, although the putative DHFR module was observed to bind NNA, albeit at lower affinity than holo-nNOS, binding was observed to be BH4 independent [18]. These findings suggest that some additional N-terminal amino acid sequence is needed for high-affinity substrate binding and to confer modulation of the arginine site by BH4.

It is notable that *E. coli* cannot synthesize BH4; thus, in contrast to NOS purified from mammalian cells, bacterial-expressed NOSs are pterin-free. Consistent with this distinction, binding of NNA to bacterial expressed nNOS [69] and eNOS [70] have been shown to depend on BH4, at least in part. Additional evidence for an allosteric modulation of NOS by BH4 comes from studies by Stuehr and colleagues, who show that reassembly of iNOS subunits into dimers requires BH4 (in addition to heme and L-arginine). Although these studies clearly define BH4 as an allosteric modulator of NOSs, an additional redox role cannot be ruled out.

Induction of iNOS by cytokines and immunostimulants occurs concomitantly with induction of BH4 synthesis in many cell types. Indeed, studies in fibroblasts [71], endothelial cells [72], vascular smooth muscle [73], and various other cell types have shown that induced BH4 synthesis is essential for the activity of co-induced iNOS. This requirement for *de novo* BH4 synthesis to support the high-output NO pathway has similarly been observed in vivo. It is noteworthy that BH4 availability may be rate limiting to the high-output NO pathway. Accordingly, NO production by immunostimulant-activated cells can be further increased twofold to threefold upon supplementation with excess BH4 [73]. Induction of BH4 synthesis by immunostimulants occurs, at least in part, by transcriptional upregulation of the gene-encoding GTP cyclohydrolase I (GTPCH), the first of three enzymes involved in conversion of GTP to BH4 [74]. The molecular basis for coordinate induction of iNOS and GTPCH gene transcription remains to be explored.

4. Conclusions

Nitric oxide synthases represents perhaps the most complex class of enzymes yet identified. Their ability to catalyze a five-electron oxidation of a nitrogen, leading to a free radical, is unprecedented in biochemistry. A remarkable modular structure hints at their evolutionary heritage, and the cofactors that drive their function offer clues to catalytic mechanisms. It is only the immense importance of NOSs to biology which could have provoked such progress in understanding their structure/function, during the brief period since discovery. These insights will undoubtedly assist in disclosing roles for NO and its regulation in the kidney, during health and disease.

References

1. Griffith, O.W. and Stuehr, D.J. *Annu. Rev. Physiol.* **57,** 707–36 (1995).
2. Xie, Q. and Nathan, C. *J. Leuk. Biol.* **56,** 576–82 (1994).
3. Griscavage, J.M., Fukuto, J.M., Komori, Y., and Ignarro, L.J. *J. Biol. Chem.* **269,** 21644–21649 (1994).
4. Rogers, N.E. and Ignarro, L.J. *Biochem. Biophys. Res. Commun.* **189,** 242–259 (1992).
5. Klatt, P., Schmidt, K., and Mayer, B. *Biochem. J.* **288,** 15–17 (1992).
6. Gross, S.S., Liu, Q., Jones, C.L., Weissman, B.A., Martasek, P., Roman, L.J., Masters, B.S.S., Harris, D., Irazarry, K., Patel, B., Moraks, A.J., Smith, S.M.E. and Salerno, J.C., FASEB J., in press.
7. Brenman, J.E., Chao, D.S., Xia, H., Aldape, K., and Bredt, D.S. *Cell* **82,** 743–752 (1995).
8. Brenman, J.E., Chao, D.S., Gee, S.H, McGee, A.W., Craven, S.E., Santillano, D.R., Wu, Z., Huang, F., Xia, H., Peters, M.F., Frochner, S.C., and Bredt, D.S. *Cell* **84,** 757–767 (1996).
9. Busconi, L. and Michel, T. *J. Biol. Chem.* **268,** 8410–8413 (1993).
10. Sessa, W.C., Barber, C.M., and Lynch, K.R. *Circ. Res.* **72,** 921–924 (1993).
11. Stuehr, D.J., Kwon, N.S., Nathan, C.F., Griffith, O.W., Feldman, P.L. and Wiseman, J. *J. Biol. Chem.* **266,** 6259–6273 (1991).

12. Bredt, D.S., Hwang, P.M., Glatt, C.E., Lowenstein, C., Reed, R.R., and Snyder, S. *Nature* **351,** 714–718 (1991).

13. Xie, Q.W., Cho, H.J., Calay, J., Mumford, R.A., Swiderek, K.M., Lee, T.D., and Nathan, C. *Science* **256,** 225–228 (1992).

14. Sessa, W.C., Harrison, J.K., Barber, C.M., Zeng, D., Durieux, M.E., D'angelo, D.D., Lynch, K.R., and Peach, M.J. *J. Biol. Chem.* **267,** 15274–15276 (1992).

15. Lamas, S., Marsden, P.A., Li, G.K., Tempst, P., and Michel, T. *Proc. Natl. Acad. Sci. USA* **89,** 6348–6352 (1992).

16. Sheta, E.A., McMillan, K., and Masters, B.S. *J. Biol. Chem.* **269,** 15147–15153 (1994).

17. Ghosh, D.K. and Stuehr, D.J. *Biochemistry* **34,** 801–807 (1995).

18. Nishimura, J.S., Martasek, P., McMillan, K., Salerno, J., Liu, Q., Gross, S.S., and Masters, B.S.S. *Biochem. Biophys. Res. Commun.* **210,** 288–294 (1995).

19. Chartrain, N.A., Geller, D.A., Koty, P.P., Sitrin, N.F., Nussler, A.K., Hoffman, E.P., Billiar, T.R., Hutchinson, N.I., and Mudgett, J.S. *J. Biol. Chem.* **269,** 6765–6772 (1994).

20. Miyahara, K., Kawamoto, T., Sase, K., Yui, Y., Toda, K., Yang, L.X., Hattori, R., Aoyama, T., Yamamoto, Y., and Doi, Y. et al. *Eur. J. Biochem.* **223,** 719–726 (1994).

21. Hall, A.V., Antoniou, H., Wang, Y., Cheung, A.H., Arbus, A.M., Olson, S.L., Lu, W.C., Kan, C.L., and Marsden, P.A. *J. Biol. Chem.* **269,** 33082–90 (1994).

22. Abu-Soud, H.M., Yoho, L.L., and Stuehr, D.J. *J. Biol. Chem.* **269,** 32047–32050 (1994).

23. Cho, H.J., Xie, Q.W., Calaycay, J., Mumford, R.A., Swiderak, K.M., Lee, T.D., and Nathen, C. *J. Exp. Med.* **176,** 599–604 (1992).

24. Bredt, D.S., and Snyder, S.H. *Proc. Natl. Acad. Sci. USA* **87,** 682–685 (1990).

25. Mayer, B., John, M., and Bohme, E. *FEBS Lett.* **277,** 215–219 (1990).

26. Schmidt, H.H., Smith, R.M., Nakane, M., and Murad, F. *Biochemistry* **31,** 3243–3249 (1992).

27. Hevel, J.M., White, K.A., and Marletta, M.A. *J. Biol. Chem.* **266,** 22789–22791 (1991).

28. Pollock, J.S., Forstermann, U., Mitchell, J.A., Warner, T.D., Schmidt, H.H., Nakane, M., and Murad, F. *Proc. Natl. Acad. Sci. USA* **88,** 10480–10484 (1991).

29. Baek, K.J., Thiel, B.A., Lucas, S. and Stuehr, D.J. *J. Biol. Chem.* **268,** 21120–21129 (1993).

30. Lee, C.M., Robinson, L.J., and Michel, T. *J. Biol. Chem.* **270,** 27403–27406 (1995).

31. Xie, Q.-W., Leung, M., Fuortes, M., Sassa, S., and Nathan, C. *Proc. Natl. Acad. Sci. USA* **93,** 4891–4896 (1996).

32. Porter, T.D. *Trends Biochem. Sci.* **16,** 154–158 (1991).

33. Liu, Q., and Gross, S.S. In: *Methods in Enzymology.* Ed. Packer, L. Academic Press, San Diego, 1996, pp. 311–324.

34. Watenpaugh, K., Sieker, L., and Jensen, L. *Proc. Natl. Acad. Sci. USA* **70,** 3857–3860 (1973).

35. Porter, T.D. and Kasper, C.B. *Biochemistry* **25,** 1682–1687 (1986).

36. Vermilion, J.L., Ballou, D.P., Massey, V., and Coon, M.J. *J. Biol. Chem.* **256,** 266–277 (1981).

37. Stuehr, D.J. and Ikeda-Saito, M. *J. Biol. Chem.* **267,** 20547–20550 (1992).

38. Klatt, P., Schmidt, K., Uray, G., and Mayer, B. *J. Biol. Chem.* **268,** 14781–14787 (1993).

39. Pou, S., Pou, W.S., Bredt, D.S., Snyder, S.H., and Rosen, G.M. *J. Biol. Chem.* **267,** 24173–24176 (1992).

40. Abu-Soud, H.M. and Stuehr, D.J. *Proc. Natl. Acad. Sci. USA* **90,** 10769–10772 (1993).

41. Dawson, T.M., Bredt, D.S., Fotuhi, M., Hwang, P.M., and Snyder, S.H. *Proc. Natl. Acad. Sci. USA* **88,** 7797–7801 (1991).

42. Hope, B.T., Michael, G.J., Knigge, K.M., and Vincent, S.R. *Proc. Natl. Acad. Sci. USA* **88,** 2811–2814 (1991).

43. Heinzel, B., John, M., Klatt, P., Bohme, E., and Mayer, B. *Biochem. J.* **281,** 627–630 (1992).

44. Frey, C., Narayanan, K., McMillan, K., Spack, L., Gross, S.S., Masters, B.S.S., and Griffith, O.W. *J. Biol. Chem.* **269,** 26083–26091 (1994).

45. Wolff, D.J., Datto, G.A., Samatovicz, R.A., and Tempsick, R.A. *J. Biol. Chem.* **268,** 9425–9429 (1993).

46. Klatt, P., Heinzel, B., John, M., Kastner, M., Bohme, E., and Mayer, B. *J. Biol. Chem.* **267,** 11374–11378 (1992).

47. Schmidt, H.H.H.W., Lohman, S.M., and Walter, U. *Biochem. Biophys. Acta* **1178,** 153–175 (1993).

48. Zhang, M., and Vogel, H.J. *J. Biol. Chem.* **269,** 981–985 (1994).

49. Babu, Y.S., Bugg, C.E., and Cook, W.J. *J. Mol. Biol.* **204,** 191–204 (1988).

50. Persechini, A., McMillan, K., and Leakey, P. *J. Biol. Chem.* **269,** 16148–16154 (1994).

51. Linse, S., Helmersson, A., and Forsen, S. *J. Biol. Chem.* **266,** 8050–8054 (1991).

52. Persechini, A., Gansz, K.J., and Paresi, R.J. *Biochemistry* **35,** 224–228 (1996).

53. Venema, R.C., Sayegh, H.S., Kent, J.D., and Harrison, D.G. *J. Biol. Chem.* **271,** 6435–6440 (1996).

54. Braun, A.P. and Schulman, H. *Annu. Rev. Physiol.* **57,** 417–745 (1995).

55. McMillan, K., Bredt, D.S., Hirsch, D.J., Snyder, S.H., Clark, J.E., and Masters, B.S. *Proc. Natl. Acad. Sci. USA* **89,** 11141–11145 (1992).

56. White, K.A. and Marletta, M.A. *Biochemistry* **31,** 6627–6631 (1992).

57. Wolff, D.J., Datto, G.A., and Samatovicz, R.A. *J. Biol. Chem.* **268,** 9430–9436 (1993).

58. McMillan, K. and Masters, B.S. *Biochemistry* **32,** 9875–9880 (1993).

59. Wang, J., Stuehr, D.J., Ikeda-Saito, M., and Rousseau, D.L. *J. Biol. Chem.* **268,** 22255–22258 (1993).

60. McMillan, K. and Masters, B.S. *Biochemistry* **34,** 3686–3693 (1995).

61. Chen, P.F., Tsai, A.L., and Wu, K.K. *J. Biol. Chem.* **269,** 25062–25066 (1994).

62. Tayeh, M.A. and Marletta, M.A. *J. Biol. Chem.* **264,** 19654–19658 (1989).

63. Kwon, N.S., Nathan, C.F., and Stuehr, D.J. *J. Biol. Chem.* **264,** 20496–20501 (1989).

64. Kaufman, S. *Annu. Rev. Nutr.* **13,** 261–286 (1993).

65. Giovanelli, J., Campos, K.L., and Kaufman, S. *Proc. Natl. Acad. Sci. USA* **88,** 7091–7095 (1991).

66. Michel, A.D., Phul, R.K., Stewart, T.L., and Humphrey, P.P. *Br. J. Pharmacol.* **109,** 287–288 (1993).

67. Klatt, P., Schmidt, K., Brunner, F., and Mayer, B. *J. Biol. Chem.* **269,** 1674–1680 (1994).

68. Cho, H.J., Martin, E., Xie, Q.W., Sassa, S., and Nathan, C. *Proc. Natl. Acad. Sci. USA* **92,** 11514–11518 (1995).

69. Roman, L.J., Sheta, E.A., Martasek, P., Gross, S.S., Liu, Q., and Masters, B.S.S. *Proc. Natl. Acad. Sci. USA* **92,** 8428–8432 (1995).

70. Martasek, P., Liu, Q., Liu, J., Roman, L.J., Gross, S.S., Sessa, W.C., and Masters, B.S. *Biochem. Biophys. Res. Commun.* **219,** 359–365 (1996).

71. Werner-Felmayer, G., Werner, E.R., Fuchs, D., Hansen, A., Reibnegger, G., and Wachter, H. *J. Exp. Med.* **172,** 1599–1607 (1990).

72. Gross, S.S., Jaffe, E.A., Levi, R., and Kilbourn, R.G. *Biochem. Biophys. Res. Commun.* **178,** 823–829 (1991).

73. Gross, S.S. and Levi, R. *J. Biol. Chem.* **267,** 25722–25729 (1992).

74. Smith, J. and Gross, S.S. (submitted).

4

Regulation of the NOS Gene Family

Andreas Papapetropoulos and William C. Sessa

1. Introduction

The nitric oxide synthase (NOS) family of proteins catalyze the five-electron oxidation of L-arginine to generate nitric oxide (NO) and L-citrulline [1]. The three prototypical NOS isoforms, neuronal NOS (nNOS or NOS 1), inducible NOS (iNOS or NOS 2), and endothelial NOS (eNOS or NOS 3) are coded for by three distinct genes in the mammalian genome [2–4]. The recent discoveries of genes for new NOS isoforms in lower species [5] suggest that NO plays a role in basic cellular processes, in addition to the complex functions attributed to NO in mammalian cell biology. As there are several excellent reviews discussing the historical perspectives, pharmacology, and biochemistry of NOS and NO, this chapter will focus on recent advances in the molecular regulation of NOS expression [1,6–8].

2. General Features and Nomenclature of NOS Isoforms

In general, NOS 1 and NOS 3 are basally expressed and activated by elevations in cytoplasmic calcium in specific cells types, and NOS 2 is transcriptionally induced by immune activators such as cytokines and is not further activated by increases in cytoplasmic calcium [6]. This paradigm led researchers to group NOS 1 and NOS 3 as "constitutive and calcium-dependent NOSs" and NOS 2 as "inducible or calcium-independent NOS." However, because recent data show that NOS 2 in certain cell types is "constitutively" expressed whereas NOS 1 and NOS 3 can be induced transcriptionally (see below) and, under certain conditions, calcium-dependent NOS 3 can become activated independent of rises in cytoplasmic calcium [9–11], this old nomenclature is misleading and should be replaced either with historical nomenclature based on the cell type used for the original cDNA cloning of the representative NOS isoforms (nNOS, iNOS, and eNOS) or a less cell-based nomenclature, based on the genetic isoform expressed (NOS 1, 2, and 3). The latter nomenclature will be used throughout the text.

3. Cloning and Organization of NOS Genes

After the cloning of cDNAs for NOS isoforms from several species between 1991 and 1993 [12–23], it became evident that the NOS family of proteins would comprise a gene family. NOS proteins share approximately 50% amino acid sequence identity with NOS 1 and NOS 3 (isoforms dependent on exogenous calcium and calmodulin for activation), sharing the highest degree of identity (60%). Within each isoform group, there is a high degree of amino acid identity (80–94%) across species, demonstrating that the differences in each group are species related and not due to the existence of different enzymes or genes. Within cofactor-binding regions of NOSs (i.e. heme, calmodulin, FMN, FAD, and NADPH) domains, there is a very high degree of conservation among the proteins [8].

The similarity in primary protein structure also transcends to similar genomic organization. The cloning and mapping of the three representative NOS genes demonstrate distinct loci for NOS 1 (12q24.2), NOS 2 (17q11.2–q12), and NOS 3 (7q35–q36) in the human genome [4,24,25]. Although on separate genes, the intron/exon boundaries of the three genes and number of exons are very similar, suggesting a recent gene duplication event and subsequent translocation of NOS DNA to different chromosomes. The molecular reasons for such similarity in gene structure yet such diversity in chromosomal location of related proteins are not known. For readers interested in detailed organization of NOS genes, a comprehensive review of the molecular diversity of NOS gene structure has recently been provided [26].

4. NOS 1

The neuronal type of nitric oxide synthase (NOS 1) was the first of the three isoforms to be cloned [12]. Although its name implies restricted expression of this isoform to neuronal tissues, it is now well established that NOS 1 is expressed outside the central and peripheral nervous system. NO in the central nervous system has been implicated in long-term synaptic depression, long-term potentiation, and glutamate neurotoxicity and is implicated as a nonadrenergic noncholinergic neurotransmitter in the peripheral nervous system [27]. The expanding list of neurons that express NOS 1 include basket and granule cells in the cerebellum, interneurons in the striatum, cortex, and hippocampus, and neurosecretory neurons in the hypothalamus [27]. Extraneuronal tissues displaying NOS 1 immunoreactivity, mRNA transcripts, and/or activity are the skeletal muscle, pancreatic β cells, the pituitary gland, the adrenal medulla, epithelial cells, the macula densa, and distal nephron in the kidney and the male sex organ [26]. In addition, we have recently detected the presence of NOS 1 mRNA and protein in cultured rat macrovascular and microvascular endothelial cells (A. Papapetropoulos and W.C. Sessa, unpublished observations). As is a common theme for all NOS isoforms, the mRNA and protein is found in cells types where the functional role of NO is not yet discovered.

The human NOS 1 gene occupies more than 160 kb of DNA and consists of 29 exons and 28 introns [2]. Sequencing of the 5'-flanking region of the NOS 1 gene expressed in neurons reveals the presence of multiple potential cis-acting elements such as AP-2, TEF-1/MCBF, CREB/ATF/c-fos, NRF-1, Ets, NF-1 and NF-κB binding motifs. However, its still unclear which of these elements in the promoter/enhancer region, if any, contribute to the transcriptional regulation of nNOS expression. In the only published report on the regulation of promoter activity, phorbol esters increase luciferase activity in reporter constructs transfected into HeLa cells [28].

Few examples of the regulation of NOS 1 expression induced by pharmacologic agents, second messenger pathways, and physiologic or pathophysiologic states exist. To date, most of the studies have focused on the regulation of NOS 1 in intact animals without investigating the cellular and molecular basis for this regulation. In a study by Herdegen et al. [29], transection of the medial forebrain bundle and mamillothalamic tract of rats led to increases in NOS 1 mRNA levels; 80% of the NOS 1 immunoreactive neurons stained for the transcription factor c-jun. In a different study, occlusion of the middle cerebral artery in rats increases NADPH–diaphorase activity, nNOS immunoreactivity, and mRNA levels [30]. The mechanism of this increase is not known. Estrogen administration increases calcium-dependent NOS activity in the heart, kidney, skeletal muscle, and cerebellum of female guinea pigs as well as steady-state levels of NOS 1 mRNA levels in skeletal muscle [31]. Kadowaki et al. [32] reported that salt loading led to increases in NOS 1 mRNA levels and NOS activity in the supraoptic and paraventricular nuclei in the hypothalamus and in the posterior pituitary gland. Finally, exposure of rats to hypoxia enhances NOS 1 and NOS 3 gene expression in the lung [33]. All of the above-mentioned studies demonstrated that NOS 1 expression is dynamically regulated in the context of the intact organism. However, the molecular mechanisms of induction in all the studies need further elucudation, perhaps in vitro or in transgenic animals.

An alternative form of molecular diversity found in the NOS 1 gene is the appearance of multiple transcripts. Some of them result from the usage of different transcriptional start sites, whereas others are generated by alternative splicing of pre-mRNA. Heterogeneity in exon 1 is observed in NOS 1 transcripts isolated from human tissues [2,34]. Seven different first exons are spliced to a common second exon. As the translational start site is found within exon 2, all of these transcripts should give rise to the same protein. Each of the alternative forms of exon 1 is flanked by its unique 5′-upstream sequence; thus, there is the possibility of unique promoters driving the NOS 1 expression. Xie et al. identified two separable promoter elements that drive luciferase expression in transfected cells [35]. The advantage of such complexity is not obvious, but it may facilitate tissue-specific and/or developmental regulation of NOS 1 expression. Differences in the 5′-untranslated region (5′ UTR) in the NOS 1 mRNAs may regulate mRNA processing, localization, stability, or translational efficiency. It should be noted that NOS 1 expression appears to be developmentally regulated in the rat lung [36]. In addition to the differences in the 5′ UTR of NOS 1 mRNA, cassette deletions of exons 9/10 and 10 have been demonstrated in human and murine tissues and human cell lines [2,34,37]. Alternative splicing results in an mRNA species that is 315 bps (base pairs) shorter (exons 9/10). This form is expressed at lower levels in many areas of the nervous system and gives rise to a protein lacking 105 amino acids located at the amino terminal of the calmodulin-binding domain. This shorter NOS protein retains NADPH–diaphorase activity, although it is unable to convert arginine to citrulline [38]. Splice variants lacking exon 10 have also been described but are expected to yield an inactive protein because a premature stop codon is introduced. Whether these alternatively spliced NOS 1 mRNAs are translated in vivo remains to be elucidated. Recently, Silvagano et al. [39] reported a novel NOS 1 protein that they termed nNOSμ, which is expressed only in differentiated skeletal muscle. nNOSμ contains an extra 102 bps between exons 16 and 17. Although it carries 34 additional amino acids at the carboxy side of the calmodulin-binding domain, it has similar catalytic activity with that expressed in the rat cerebellum. The neuronal isoform of NOS has been previously localized in skeletal muscle, and under normal conditions is membrane associated via a PDZ domain found in the amino terminus [40,41]. The PDZ domain contains the signature amino acid motif, GLGF, and is found in a diverse series of proteins such as protein tyrosine phosphatases and junctional proteins like ZO-1. This motif facilitates NOS 1 interaction with α1-syntrophin and dystrophin and recruits the protein into the sarcolemma of skeletal muscle. The sarcolemmal membrane association of NOS 1 is lost in humans who have Duchenne muscular dystrophy and in a mouse model of the disease [42]. The functional relevance of such an interaction is not clear but may relate to the ability of NO to influence the mechanical properties of skeletal muscle.

5. NOS 2

Unlike NOS 1 and NOS 3, the expression of NOS 2 has been documented in almost all cells studied so far. NOS 2 is transcriptionally regulated and can be detected following cytokine or bacterial lipopolysaccharide (LPS) stimulation [7]. Cells capable of expressing NOS 2 include, but are not limited to, neutrophils, macrophages, vascular smooth muscle, endothelium, messangial cells, cardio-myocytes, hepatocytes, lympocytes, glia and neurons, chondrocytes, platelets, and epithelial cells, as well as tumor cells [43–45]. In general, cells de-rived from rodents are more easily stimulated to produce NO than human cells, at least in vitro. Recently, there is increasing evidence for "constitutive" expres-sion of NOS 2 in certain cells (airway and renal epithelium, microvascular endothelium, and cell lines such as PC-12, T84, AM) both in vivo and in vitro [46–48].

Proinflammatory cytokines that increase NOS 2 transcription and NO produc-tion are tumor necrosis factor-α (TNF-α), interleukin-1 (IL-1), IL-2, and inter-feron-γ (IFN-γ). It should be noted that cytokine potency and efficacy differ with the cell type studied, and frequently cytokine combinations, with or without LPS, exert a synergistic effect. Many different classes of molecules can either increase or decrease steady-state iNOS mRNA and/or protein levels. Among them are inhibitory cytokines (IL-4, IL-8, IL-10, IL-13), peptides, and growth factors platelet-derived growth factor, transforming growth factor-β; (TGF-β), angioten-sin-II], redox-based transcription factor inhibitors (pyrrolidine dithiocarbamate), tyrosine kinase inhibitors (genistein, tyrphostins, erbstatin A), cytoskeleton-modi-fying agents (colchicine, nocodazole, taxol), second messengers and second-messenger pathway activators (cAMP, cGMP, phorbol esters, and phosphatase inhibitors), hormones, glucocorticoids (dexamethasone), immunosuppressant (cyclosporin A), and anti-inflammatory drugs (aspirin). The above list is by no means complete and readers are referred to recent reviews [43,44,49,50]. Herein, we will discuss only agents for which the molecular mechanisms have been elucidated in some detail. Moreover, as many agents may act differently depend-ing on the species studied and/or cell type used, the regulation of NOS 2 expression will be discussed in a cell-specific context (for example, TGF-β inhibits LPS and cytokine-induced NOS 2 expression in murine macrophages and rat aortic smooth muscle cells but potentiates the induction in 3T3 fibroblasts and bovine retinal pigmented epithelial cells [51–53]). Regulation of NOS 2 expression will be discussed in murine macrophages, rat vascular smooth muscle cells, and human epithelial cells.

Most of the work on the regulation of NOS 2 expression has been performed using freshly isolated murine macrophages or macrophage cell lines. Expression of NOS 2 in these cells mediates their cytotoxic actions and plays an important role in the immune response [6]. Cloning of a 1.7-kb fragment flanking the transcriptional start site of the murine gene reveals several putative transcription factor binding sequences including ten IFN-γ response elements (IFN-RE), three γ-activated sites (GAS), two consensus sequences for nuclear factor-κB (NF-κB) binding, and four for NF-IL6, two TNF-α-response elements (TNFα-RE), two activating protein-1 binding motifs (AP-1), three interferon-α-stimulated response elements (ISRE), and a basal transcription recognition site (TATA box [54,55]). Many of these elements are also present in the human NOS 2 promoter [56]. Initial experiments using deletion constructs of the murine promoter placed upstream of a luciferase reporter gene revealed two regions (−48 to −209 and −1029 to −913) containing cis-acting elements important for regulation of NOS 2 transcription [55]. The −48 to −209 region is responsible for transcriptional regulation in response to LPS, whereas the −1029 to −913 region, in spite of not being able to increase luciferase activity by itself, potentiates luciferase activity by 10-fold in response to IFN-γ. These data help explain the synergy afforded by LPS and IFN-γ in the induction of NOS 2 at the molecular level. Additional studies, confirmed the finding that the proximal region of the promoter (−85 to −76) is essential for LPS-induced NOS 2 transcription and identified p50/c-rel and p50/RelA heterodimers as components of the NF-κB-trans acting factor [57]. A different series of experiments, confirmed that a cluster of four enhancer elements known to bind IFN-γ responsive transcription factors (−951 to −911) are important for NOS 2 transcription [58]. Mutations in two nucleotides in the interferon-regulatory binding factor site (IRF; −913 to −923) in the context of the full-length promoter ablated the synergistic action of IFN-γ in the transcription of NOS 2. Electrophoretic mobility gel shift assays reveal a binding complex recognized by an anti-IFN-γ regulatory factor-1 (IRF-1) antibody in the nuclei of IFN-γ-treated macrophages. In yet another study, macrophages from mice with targeted disruption of the IRF-1 gene fail to respond to stimulation by increasing NOS 2 mRNA and nitrite production [59]. In addition to the positive data obtained about the involvement of the above-mentioned binding sites and transcription factors in NOS 2 expression, experiments using promoter constructs with the AP-1 consensus sequence deleted show that this site is not necessary for NOS 2 induction. Melillo et al. recently identified a functionally important hypoxia responsive enhancer in the murine promoter located at −227 to −209 [60].

Posttranslational regulation of NOS 2 expression also has been described. The 3′ untranslated region of the NOS 2 mRNA contains a "AUUUA" motif which can potentially destabilize mRNAs. This is reflected by the short half-life of the NOS 2 mRNA in freshly isolated murine macrophages (approximately 3 h); however, direct demonstration that this motif affects mRNA destabilization is lacking. Weisz et al. reported that potentiation of IFN-γ-induced NOS 2 gene expression in RAW 264.7 cells by LPS is due to an increase the half-life of NOS 2 mRNA from 1.5 h in cells stimulated with IFN-γ to 6 h in cells treated with both LPS and IFN-γ [61]. In contrast to the stabilizing action of LPS on NOS 2 mRNA levels in IFN-γ-stimulated macrophages, TGF-β was shown to decrease NOS 2 expression by decreasing mRNA stability [52]. In addition, TGF-β decreases translation of mRNA without affecting the rate of transcription, and increases NOS protein degradation. These actions of TGF-β collectively led to a decrease in NOS activity even after the protein was expressed. Although the exact mechanism of action of IL-4 is not known, this cytokine decreases NOS 2 steady-state mRNA levels [62]. There are yet additional means of regulating NOS 2 expression posttranscriptionally. Expression of a maximally active NOS 2 requires an efficient arginine transport mechanism and availability of cofactors such as tetrahydrobiopterin (BH_4 [50]). In particular, BH_4 is required for the formation of an active NOS 2 dimer [63]. Most of the enzymes responsible for the synthesis of the substrate and cofactors are concomitantly induced with NOS 2 following cytokine treatment. Interference with the induction of either the y^+ transporter (increases the uptake of arginine) or GTP-cyclohydrolase I (responsible for the increased BH_4 synthesis after cytokine treatment) would decrease the amount of the NO formed. Although moderatively selective pharmacologic inhibitors for NOS 2 have been developed [64–66], specific inhibition of the NOS activity can be achieved by modulating the transcriptional and postranscriptional events responsible for NOS 2 expression.

Another cell type frequently used to study the regulation of NOS 2 expression is rat vascular smooth muscle (RVSM). The function of NO in this cell type is not as clear as with macrophages, and the interest of modulating NOS 2 induction lies primarily in the potential benefits of inhibiting its expression in pathological states such as septic shock [67]. Although a battery of substances have been shown to modulate NOS activity (as reflected by arginine to citrulline conversion and nitrite accumulation), the molecular mechanism of action has been studied for only a few of them. Analogous to NOS 2 in other cells, its expression in RVSM is under transcriptional control [68]. Treatment with proinflammatory cytokines or LPS induces NOS 2 mRNA. However, the exact signal transduction pathways by which extracellular signals lead to increased gene transcription are not clear. In RVSM cells, as well as in a variety of other cell types, tyrosine kinase inhibitors reduce NOS 2 activity, mRNA, and protein levels [69,70]. The proteins phosphorylated by these kinases in response to cytokines or LPS remain largely unidentified. A potential mechanism linking cytokine-induced NOS 2 activation to tyrosine phosphorylation has recently been elucidated in rat ventricular myocytes [71]. Incubation of ventricular myocytes with IL-1β or IFN-γ leads to activation of mitogen-activated protein kinases (ERK1/ERK2) and the transcription factor STAT-1a that binds to a cis-acting element in the promoter region. Blockade of ERK1/ERK2 activation inhibits NOS 2 induction by IFN-γ and IL-1β; however, phosphorylation of STAT-1a in the absence of ERK1/ERK2 activation is not sufficient for NOS 2 induction. A similar mechanism for activation for NOS 2 in macrophages and RVSM is likely. Other second-messenger pathways, such as cAMP and cGMP, can induce NOS 2 directly or potentiate the actions of cytokines [72–74]. Many of these pathways, at least in 3T3 fibroblasts, were shown to converge in the activation of NF-κB [75].

Modulators of protein kinase C (PKC) activity and microtubule-depolymerizing agents also influence NOS 2 expression [76,77]. The latter decrease steady-state NOS 2 mRNA and protein levels in RVSM, whereas the former act in a cell-specific manner. Phorbol esters induce NOS 2 in macrophages, fibroblasts, and hepatocytes, whereas in one report they reduced NOS 2 mRNA in RVSM [75,77–79]. The precise level of regulation of NOS 2 expression (transcriptional or posttranscriptional) by pharmacologic agents affecting the aforementioned signal transduction pathways is currently unknown.

Recently, Spink et al. [68] reported on the transcriptional control of NOS 2 expression in the RVSM cell line A7r5 using promoter/reporter gene constructs. A 1.7-kb 5' upstream fragment of the NOS 2 murine gene functions as a promoter within RVSM but not as efficiently as in murine macrophages. The combination of IL-1, TNF-α, and IFN-γ increase transcriptional activation of the reporter gene construct in RVSM by 8-fold as compared to the 44-fold increase afforded by LPS and IFN-γ in murine macrophages. In RVSM, a 112-bp region (−890 to −1002) containing NF-κB, GAS/ISRE, and IRF-1 sites is required for full promoter activity. A major difference in the behavior of the NOS 2 promoter between RVSM and murine macrophages was shown using promoter constructs that carried mutations in one or both NF-κB sites: the NF-κB site that was more important for NOS induction in RVSM was the upstream site (−971 to −962), whereas, as mentioned earlier, in murine macrophages, the site responsible for the LPS response was the downstream sequence (−85 to −76). Deletion of both sites practically abolished induction by the cytokine mixture in RVSM. It should be noted that the RVSM cell line used was unresponsive to LPS. In addition, the protein binding to the necessary NF-κB site in RVSM was identified as a p65 together with an unidentified 50-kDa protein. Postranscriptional regulation of NOS 2 expression has also been demonstrated in RVSM cells. Imai et al. [74] showed that the NOS mRNA half-life is 2–3 h in RVSM (comparable to murine macrophages and endothelial cells) and that treatment with cycloheximide causes a superinduction of NOS mRNA levels by prolonging its half-life to more than 12 h. Caution should, however, be used when extrapolating these results in vivo, as neither the synergy between LPS and IFN-γ nor the superinduction by cycloheximide was found in rat aortic strips [80]. TGF-β, similarly to murine macrophages, inhibits NOS induction in RVSM [81]. However, the only proven action of TGF-β in RVSM is through a reduction in NOS 2 transcription, a mechanism different from that observed in the murine macrophages (see above).

Considerably less work on the regulation of NOS 2 expression has been done using epithelial cells, but there are some important observations made in the human cell line A549 that are noteworthy. Glucocorticoids are known to inhibit NOS 2 induction in almost all cell types. However, the mechanism through which dexamethasone inhibits NOS was only recently described. Dexamethasone inhibits the binding of the transcription factor NF-κB, but not of AP-1, to the NOS 2 promoter, thereby inhibiting its transcription [82]. It should be noted that dexamethasone did not increase the mRNA for the NF-κB inhibitor, I-κB, and does not reduce the nuclear content of the NF-κB proteins p65 and p50, suggesting that dexamethasone inhibits NOS 2 transcription through a protein–protein interaction between the activated glucocorticoid receptor and the active NF-κB complex. As mentioned earlier, although the human and murine NOS 2 promoters contain similar putative, consensus sequences for the binding of transcription factors, it is more difficult to induce NOS 2 in human cells. As recently described, transfection of a human kidney epithelial cell line (AKN) with a luciferase reporter constructs containing a 3.8-kb fragment upstream of the human NOS 2 gene did not exhibit activation in the presence of cytokines, whereas the production of nitrite/nitrate, products of endogenous NOS activation, are increased [83]. Unlike results obtained with the murine promoter, at least 5.8 kb of the human promoter is required for minimal cytokine induction (threefold) and a 16-kb fragment is necessary for a 10-fold induction. These data demonstrate that considerable differences exist in the regulation of the human and murine promoters.

Another important observation made in a human epithelial cell line and in primary cells is the presence of transcripts with considerable diversity in the 5′ untranslated region (UTR) of the mRNA [48]. Although the majority of NOS 2 mRNA transcripts originate downstream of the TATA box, approximately 6% of them originate upstream of the TATA box. This, in combination with the fact that some mRNAs lack exon 1 as a result of alternative splicing, increases the complexity of the 5′ UTR of human iNOS. The significance of this finding with respect to regulation of gene expression remains to be elucidated.

6. NOS 3

The NOS 3 protein was originally purified and the corresponding cDNA cloned from endothelial cells (EC) and is the NOS isoform responsible for producing the classic endothelium-derived relaxing factor as described by Furchgott et al. [84]. Direct evidence for the importance of NOS 3-derived NO stems from the recent generation of mice with targeted disruption of the NOS 3 gene locus [85]. F2-generation NOS 3 knockout mice (−/−) are hypertensive relative to wild-type littermate control mice (+/+) of the same generation. Importantly, the pressor effect of nitro-L-arginine, a NOS inhibitor, is attenuated in the −/− mice and endothelium-dependent relaxation in response to acetylcholine is abrogated in vessels isolated from the −/− mice. This fundamental finding is direct "proof of principal" for the major contribution of NO in vasomotor control.

Recent experimentation demonstrating that NOS 3 in EC can be regulated and that cells other than EC express NOS 3 including cardiac myocytes, hippocampal neurons, certain epithelial cells, and platelets suggests that NO may subserve other functions in addition to being an endothelium-derived vasodilator [26]. However, because the identification of other cell types that express NOS 3 is a relatively new finding, the functional significance of NOS 3 in nonendothelial cell types is less clear. The best characterized, nonendothelial source of NOS 3 is cardiac myocytes [86,87]. In myocytes, locally produced NO may modulate the inotropic and chronotropic state of the heart as NOS inhibitors influence the force of contraction and spontaneous rate of beating in isolated myocytes. Because the regulation of NOS 3 expression in this cell type has not been examined, the regulation of NOS 3 will be discussed in the context of EC.

The regulation of NOS 3 expression in EC is a relatively recent area of NO biology for the following reasons: (1) Basal expression of NOS 3 mRNA and protein is seen in most primary isolates of EC and was not suspected to be regulated; (2) the induction process that occurs in vivo and in vitro is usually never greater than twofold to fourfold, making detailed molecular characterization difficult and tedious. Activators that increase NOS 3 mRNA levels are mechanical forces (shear stress and cyclic strain [23,88,89]) and their in vivo correlate, exercise training [90], cell proliferation [91], lysophosphotidylcholine (LPC [92]), TGF-β [93], and basic fibroblast growth factor [94]. All these agents (except LPC, which increases NOS 3 expression more than fourfold in human umbilical vein EC) induce NOS 3 by twofold to fourfold. Although small in magnitude relative to the induction of NOS 2, the functional relevance of a twofold to fourfold induction of NOS 3 should not be discounted due to the steep dose-response relationship to NO in blood vessels (i.e., small changes in NO concentration have dramatic effects on vascular tone [95]). This is supported by data demonstrating that chronic exercise training in dogs increases endothelium-dependent relaxations in vivo, and NO release and NOS 3 expression in vitro [90]. Conversely, decompensated heart failure in dogs decreases endothelium-dependent relaxations in vivo, NO release, and NOS 3 expression [96] supporting the concept that changes in blood flow influence EC gene expression and that such changes have a functional consequence on EC control of vasomotor tone.

Cloning of the human and bovine NOS 3 genes revealed a 5′ regulatory region containing a "TATA-less" promoter and a variety of cis elements for the putative binding of transcription factors [4,97]. TATA-less promoters are commonly found in genes that are under tight transcriptional control and not cytokine activated. In these genes, other sequences in the absence of a TATA box can serve as binding sites for RNA polymerase II and the basal transcriptional machinery. Specific sites that may influence basal transcription found in the NOS 3 gene include Sp1 and GATA sites, whereas potential sites for stimulated transcription include a sterol regulatory element (SRE), activator proteins 1 and 2 elements (AP1, AP2), a nuclear factor-1 element (NF-1), partial estrogen responsive elements (ERE), a cAMP response element (CRE), and a shear-stress response element (SSRE). The presence of such sites gives the investigator a molecular "road map" to examine gene regulation by substances that activate known transcription factors that can bind to putative elements.

Initial experiments using reporter gene constructs of the NOS 3 promoter reveals that the proximal Sp1 is necessary for basal transcription [97–99]. Deletion or site-directed mutagenesis of this site reduces promoter activity by 90–95% when transfected into EC. Incubation of an oligonucleotide probe encompassing the Sp1 site with nuclear extracts from human or bovine EC demonstrates the specific binding of three nuclear protein complexes, one of which that can be "supershifted" with a Sp1 antibody. This suggests that Sp1 does bind to the Sp1 element in the NOS 3 gene and that other Sp1 family members may also participate in coordinating basal transcription. In addition to Sp1 regulating basal transcription, GATA-2, a transcription factor highly expressed in EC, also modulates basal activity [98]. Mutation of the GATA-binding site in the context of the full-length NOS 3 promoter reduces reporter gene activity by 25%. Recombinant GATA can bind to the NOS 3 element and can be depleted with GATA-2 antisera. Considering the relative hierarchy of these two basal transcription factors needed for constitutive NOS 3 activation, Sp1 is necessary for basal expression and GATA-2 regulates the level of expression.

Very little is known about the molecular mechanisms of NOS 3 induction by mechanical forces or growth factors. Shear stress and cyclic strain, in vitro, increase NOS 3 mRNA and protein levels. Such an effect is the most likely mechanism for increased NOS 3 expression with exercise training and high flow states. For cyclic strain, activation is due to increases in NOS 3 transcription, not to mRNA stabilization. Transcriptional activation is most likely occurring in response to shear stress via the SSRE; however, there are no reports yet demonstrating this. Interestingly, the mechanism of induction by cyclic strain is unrelated to the SSRE because deletion of this region does not influence the ability of NOS 3 reporter constructs to be activated by cyclic strain (B.E. Sumpio and W.C. Sessa, unpublished observations). Understanding the similarities and differences of NOS 3 gene expression via mechanotransduction (shear and strain) is a new, exciting area with far-reaching implications in the fields of vascular biology and medicine.

Molecular analysis for TGF-β activation of NOS 3 has recently been reported [93]. TGF-β induces NOS 3 mRNA, protein, and NOS activity in bovine aortic EC. Transient transfection assays using bovine NOS 3 promoter reporter constructs demonstrates that the TGF-β responsive element resides between −935 and −1269 nucleotides (nt) upstream of the transcriptional start site. Gel shift assays and point mutation analysis show that TGF-β increased the binding of the CCAAT transcription factor/nuclear factor-1 to the NF-1 site (−1014 to −1026 nt) in the NOS 3 promoter. However, this site is necessary but not sufficient for TGF-β activation of NOS 3, suggesting that additional factors are most likely required.

The most recent and unexpected activator of NOS 3 is LPC. LPC content of atherosclerotic blood vessels is higher than that of normal vessels and is a major phospholipid found in oxidized low-density lipoproteins. Treatment of human umbilical vein EC with LPC initiates a dose- and time-dependent induction of NOS 3 mRNA [92]. The magnitude of the mRNA induction is 11-fold with a consistent, but not equal, increase in the transcription rate of the NOS 3 gene, as measured in nuclear runoff studies. NOS protein levels, activity, and biologically active NO are all increased with LPC treatment. However, there is a clear discrepancy between the levels of NOS 3 mRNA (11-fold), protein (5-fold), and activity (2-fold), suggesting complex posttranscriptional and posttranslational regulation. These data are consistent with higher levels of NOS 3 protein found in atherosclerotic rabbits [100]. However, what is the relevance of greater NOS 3 expression in atherosclerosis? Perhaps, during the development of atherosclerosis prior to lesion formation, LPC induction of NOS 3 affords protection against proatherosclerotic mechanisms [recruitment of mononuclear cells and oxidation of low-density lipoproteins (LDL)] required for the lesion development.

In contrast to the stimulatory effects of LPC on NOS 3, Liao et al. reported that oxidized LDL decreases NOS 3 expression in human EC [101]. At early time points, the suppression appears to be mediated through a partial decrease in the rate of transcription and a destabilization of the NOS 3 mRNA. However, at later time points, the rate of transcription increases paradoxically relative to the reduction in steady-state mRNA levels. The ability of lipids derived from LDL and oxidized LDL, per se, to influence NOS 3 expression and activity clearly merits further evaluation before one can extrapolate these in vitro findings to observations made in human atheromas.

7. Summary and Future Directions

The recent cloning of the NOS family of genes will allow investigators to elucidate the plenitude of factors that have the capacity to influence NOS gene expression. Elucidation of the mechanisms of NOS 1 and NOS 3 expression will no doubt lead to novel strategies that will influence the basal expression of these physiologically relevant genes. Insights into the complexities of NOS 2 regulation will be gleaned from examining its expression in human tissues and comparing and contrasting its regulation to other genes activated by inflammatory stimuli. Finally, the potential identification of relevant polymorphisms in NOS loci and misense mutations that influence protein function in human diseases will shed light on the role of NO in health and disease.

Acknowledgments

We apologize to colleagues whose references were omitted for the sake of brevity or cited in reviews. This work is supported by grants from the National Institutes of Health, the American Heart Association, and the Patrick and Catherine Weldon Donaghue Medical Research Foundation. One of the authors (WCS) is an Established Investigator of the American Heart Association.

References

1. Moncada, S., Palmer, R.M., Higgs, and E.A. Nitric oxide: Physiology, pathophysiology, and pharmacology. *Pharmacol. Rev.* **43,** 109–142 (1991).
2. Hall, A.V., Antoniou, H., Wang, Y., Cheung, A.H., Arbus, A.M., Olson, S.L., Lu, W.C., Kau, C.L., and Marsden, P.A. Structural organization of the human neuronal nitric oxide synthase gene (NOS1). *J. Biol. Chem.* **269,** 33082–33090 (1994).

3. Chartrain, N.A., Geller, D.A., Koty, P.P., Sitrin, N.F., Nussler, A.K., Hoffman, E.P., Billiar, T.R., Hutchinson, N.I., and Mudgett, J.S. Molecular cloning, structure, and chromosomal localization of the human inducible nitric oxide synthase gene. *J. Biol. Chem.* **269**, 6765–6772 (1994).

4. Marsden, P.A., Heng, H.H., Scherer, S.W., Stewart, R.J., Hall, A.V., Shi, X.M., Tsui, L.C., and Schappert, K.T. Structure and chromosomal localization of the human constitutive endothelial nitric oxide synthase gene. *J. Biol. Chem.* **268**, 17478–17488 (1993).

5. Regulski, M. and Tully, T. Molecular and biochemical characterization of dNOS: a Drosophila Ca^{2+}/calmodulin-dependent nitric oxide synthase. *Proc. Natl. Acad. Sci. USA* **92**, 9072–9076 (1995).

6. Nathan, C. Nitric oxide as a secretory product of mammalian cells. *FASEB J.* **6**, 3051–3064 (1992).

7. Nathan, C. and Xie, Q.W. Nitric oxide synthases: Roles, tolls, and controls. *Cell* **78**, 915–918 (1994).

8. Sessa, W.C. The nitric oxide synthase family of proteins. *J. Vasc. Res.* **31**, 131–143 (1994).

9. Kuchan, M.J. and Frangos, J.A. Role of calcium and calmodulin in flow-induced nitric oxide production in endothelial cells. *Am. J. Physiol.* **266**, C628–C636 (1994).

10. Tsukahara, H., Gordienko, D.V., Tonshoff, B., Gelato, M.C., and Goligorsky, M.S. Direct demonstration of insulin-like growth factor-I-induced nitric oxide production by endothelial cells. *Kidney Int.* **45**, 598–604 (1994).

11. Ayajiki, K., Kindermann, M., Hecker, M., Fleming, I., and Busse, R. Intracellular pH and tyrosine phosphorylation but not calcium determine shear stress-induced nitric oxide production in native endothelial cells. *Circ. Res.* **78**, 750–758.

12. Bredt, D.S, Hwang, P.M., Glatt, C.E., Lowenstein, C., Reed, R.R., and Snyder, S.H. Cloned and expressed nitric oxide synthase structurally resembles cytochrome P-450 reductase. *Nature* **351**, 714–718 (1991).

13. Nakane, M., Schmidt, H.H., Pollock, J.S., Forstermann, U., and Murad, F. Cloned human brain nitric oxide synthase is highly expressed in skeletal muscle. *FEBS Lett.* **316**, 175–180 (1993).

14. Xie, Q.W., Cho, H.J., Calaycay, J., Mumford, R.A., Swiderek, K.M., Lee, T.D., Ding, A., Troso, T., and Nathan, C. Cloning and characterization of inducible nitric oxide synthase from mouse macrophages. *Science* **256**, 225–228 (1992).

15. Lyons, C.R., Orloff, G.J., and Cunningham, J.M. Molecular cloning and functional expression of an inducible nitric oxide synthase from a murine macrophage cell line. *J. Biol. Chem.* **267**, 6370–6374 (1992).

16. Lowenstein, C.J., Glatt, C.S., Bredt, D.S., and Snyder, S.H. Cloned and expressed macrophage nitric oxide synthase contrasts with the brain enzyme. *Proc. Natl. Acad. Sci. USA* **89**, 6711–6715 (1992).

17. Wood, E.R., Berger, H.J., Sherman, P.A., and Lapetina, E.G. Hepatocytes and macrophages express an identical cytokine inducible nitric oxide synthase gene. *Biochem. Biophys. Res. Commun.* **191**, 767–774 (1993).

18. Geller, D.A., Lowenstein, C.J., Shapiro, R.A., Nussler, A.K., Di, S.M., Wang, S.C., Nakayama, D.K., Simmons, R.L., Snyder, S.H., and Billiar, T.R. Molecular cloning and expression of inducible nitric oxide synthase from human hepatocytes. *Proc. Natl. Acad. Sci. USA* **90**, 3491–3495 (1993).

19. Lamas, S., Marsden, P.A., Li, G.K., Tempst, P., and Michel, T. Endothelial nitric oxide synthase: Molecular cloning and characterization of a distinct constitutive enzyme isoform. *Proc. Natl. Acad. Sci. USA* **89**, 6348–6352 (1992).

20. Marsden, P.A., Schappert, K.T., Chen, H.S., Flowers, M., Sundell, C.L., Wilcox, J.N., Lamas, S., and Michel, T. Molecular cloning and characterization of human endothelial nitric oxide synthase. *FEBS Lett.* **307**, 287–293 (1992).

21. Sessa, W.C., Harrison, J.K., Barber, C.M., Zeng, D., Durieux, M.E., D'Angelo, D.D., Lynch, K.R., and Peach, M.J. Molecular cloning and expression of a cDNA encoding endothelial cell nitric oxide synthase. *J. Biol. Chem.* **267**, 15274–15276 (1992).

22. Janssens, S.P., Shimouchi, A., Quertermous, T., Bloch, D.B., and Bloch, K.D. Cloning and expression of a cDNA encoding human endothelium-derived relaxing factor/nitric oxide synthase. *J. Biol. Chem.* **267**, 14519–14522 (1992). [Erratum: *J. Biol. Chem.* **267**(31); 22694 (1992).]

23. Nishida, K., Harrison, D.G., Navas, J.P., Fisher, A.A., Dockery, S.P., Uematsu, M., Nerem, R.M., Alexander, R.W., and Murphy, T.J. Molecular cloning and characterization of the constitutive bovine aortic endothelial cell nitric oxide synthase. *J. Clin. Invest.* **90**, 2092–2096 (1992).

24. Xu, W., Charles, I.G., Moncada, S., Gorman, P., Sheer, D., Liu, L., and Emson, P. Mapping of the genes encoding human inducible and endothelial nitric oxide synthase (NOS2 and NOS3) to the pericentric region of chromosome 17 and to chromosome 7, respectively. *Genomics* **21**, 419–422 (1994).

25. Marsden, P.A., Heng, H.H., Duff, C.L., Shi, X.M., Tsui, L.C., and Hall. A.V. Localization of the human gene for inducible nitric oxide synthase (NOS2) to chromosome 17q11.2–q12. *Genomics* **19**, 183–185 (1994).

26. Wang, Y. and Marsden, P.A. Nitric oxide synthases: gene structure and regulation. [Review]. *Adv. Pharmacol.* **34**, 71–90 (1995).

27. Vincent, S.R. and Hope, B.T. Neurons that say NO. [Review]. *Trends Neurosci.* **15**, 108–113 (1992).

28. Young, A.P., Murad, F., Vaessin, H., Xie, J., and Rife, T.K. Transcription of the human neuronal nitric oxide synthase gene in the central nervous system is mediated by multiple promoters. [Review]. *Adv. Pharmacol.* **34**, 91–112 (1995).

29. Herdegen, T., Brecht, S., Mayer, B., Leah, J., Kummer, W., Bravo, R., and Zimmermann, M. Long-lasting expression of JUN and KROX transcription factors and nitric oxide synthase in intrinsic neurons of the rat brain following axotomy. *J. Neurosci.* **13**, 4130–4145 (1993).

30. Zhang, Z.G., Chopp, M., Gautam, S., Zaloga, C., Zhang, R.L., Schmidt, H.H., Pollock, J.S., and Forstermann, U. Upregulation of neuronal nitric oxide synthase and mRNA, and selective sparing of nitric oxide synthase-containing neurons after focal cerebral ischemia in rat. *Brain Res.* **654**, 85–95 (1994).

31. Weiner, C.P., Lizasoain, I., Baylis, S.A., Knowles, R.G., Charles, I.G., and Moncada, S. Induction of calcium-dependent nitric oxide synthases by sex hormones. *Proc. Natl. Acad. Sci. USA* **91,** 5212–5216 (1994).

32. Kadowaki, K., Kishimoto, J., Leng, G., and Emson, P.C. Up-regulation of nitric oxide synthase (NOS) gene expression together with NOS activity in the rat hypothalamo-hypophysial system after chronic salt loading: evidence of a neuromodulatory role of nitric oxide in arginine vasopressin and oxytocin secretion. *Endocrinology* **134,** 1011–1017 (1994).

33. North, A.J., Star, R.A., Brannon, T.S., Ujiie, K., Wells, L.B., Lowenstein, C.J., Snyder, S.H., and Shaul, P.W. Nitric oxide synthase type I and type III gene expression are developmentally regulated in rat lung. *Am. J. Physiol.* **266,** L635–41 (1994).

34. Fujisawa, H., Ogura, T., Kurashima, Y., Yokoyama, T., Yamashita, J., and Esumi, H. Expression of two types of nitric oxide synthase mRNA in human neuroblastoma cell lines. *J. Neurochem.* **63,** 140–145 (1994).

35. Xie, J., Roddy, P., Rife, T.K., Murad, F., and Young, A.P. Two closely linked but separable promoters for human neuronal nitric oxide synthase gene transcription. *Proc. Natl. Acad. Sci. USA* **92,** 1242–1246 (1995).

36. Shaul, P.W., North, A.J., Brannon, T.S., Ujiie, K., Wells, L.B., Nisen, P.A., Lowenstein, C.J., Snyder, S.H., and Star, R.A. Prolonged in vivo hypoxia enhances nitric oxide synthase type I and type III gene expression in adult rat lung. *Am. J. Respir. Cell Mol. Biol.* **13,** 167–174 (1995).

37. Ogura, T., Yokoyama, T., Fujisawa, H., Kurashima, Y., and Esumi, H. Structural diversity of neuronal nitric oxide synthase mRNA in the nervous system. *Biochem. Biophys. Res. Commun.* **193,** 1014–1022 (1993).

38. Nathan, C. and Xie, Q.W. Regulation of biosynthesis of nitric oxide. [Review]. *J. Biol. Chem.* **269,** 13725–12728 (1994).

39. Silvagno, F., Xia, H., and Bredt, D.S. Neuronal nitric-oxide synthase-m, an alternatively spliced isoform expressed in differentiated skeletal muscle. *J. Biol. Chem.* **271,** 11204–11208 (1996).

40. Brenman, J.E., Chao, D.S., Gee, S.H., McGee, A.W., Craven, S.E., Santillano, D.R., Wu, Z., Huang, F., Xia, H., Peters, M.F., Froehner, S.C., and Bredt, D.S. Interaction of nitric oxide synthase with the postsynaptic density protein PSD-95 and a1-syntrophin mediated by PDZ domains. *Cell* **84,** 757–767 (1996).

41. Kobzik, L., Reid, M.B., Bredt, D.S., and Stamler, J.S. Nitric oxide in skeletal muscle. *Nature* **372,** 546–548 (1994).

42. Brenman, J.E., Chao, D.S., Xia, H., Aldape, K., and Bredt, D.S. Nitric oxide synthase complexed with dystrophin and absent from skeletal muscle sarcolemma in Duchenne muscular dystrophy. *Cell* **82,** 743–752 (1995).

43. Xie, Q. and Nathan, C. The high-output nitric oxide pathway: role and regulation. *J. Leuk. Biol.* **56,** 576–782 (1994).

44. Forstermann, U., Kleinert, H., Gath, I., Schwarz, P., Closs, E.I., and Dun, N.J. Expression and expressional control of nitric oxide synthases in various cell types. *Adv. Pharmacol.* **34,** 171–186 (1995).

45. Nussler, A.K. and Billiar, T.R. Inflammation, immunoregulation, and inducible nitric oxide synthase. *J. Leuk. Biol.* **54,** 171–178 (1993).

46. Guo, F.H., De, R.H., Rice, T.W., Stuehr, D.J., Thunnissen, F.B., and Erzurum, S.C. Continuous nitric oxide synthesis by inducible nitric oxide synthase in normal human airway epithelium in vivo. *Proc. Natl. Acad. Sci. USA* **92,** 7809–7813 (1995).

47. Kumar, M., Liu, G.J., Floyd, R.A., and Grammas, P. Anoxic injury of endothelial cells increases production of nitric oxide and hydroxyl radicals. *Biochem. Biophys. Res. Commun.* **219,** 497–501 (1996).

48. Chu, S.C., Wu, H.P., Banks, T.C., Eissa, N.T., and Moss, J. Structural diversity in the 5′-untranslated region of cytokine-stimulated human inducible nitric oxide synthase mRNA. *J. Biol. Chem.* **270,** 10625–10630 (1995).

49. Szabo, C. and Thiemermann, C. Regulation of the expression of the inducible isoform of nitric oxide synthase. *Adv. Pharmacol.* **34,** 113–153 (1995).

50. Morris, S.J. and Billiar, T.R. New insights into the regulation of inducible nitric oxide synthesis. *Am. J. Physiol.* **266,** E829–39 (1994).

51. Vodovotz, Y. and Bogdan, C. Control of nitric oxide synthase expression by transforming growth factor-beta: implications for homeostasis. *Prog. Growth Factor Res.* **5,** 341–351 (1994).

52. Vodovotz, Y., Bogdan, C., Paik, J., Xie, Q.W., and Nathan, C. Mechanisms of suppression of macrophage nitric oxide release by transforming growth factor beta. *J. Exp. Med.* **178,** 605–613 (1993).

53. Gilbert, R.S. and Herschman, H.R. Transforming growth factor beta differentially modulates the inducible nitric oxide synthase gene in distinct cell types. *Biochem. Biophys. Res. Commun.* **195,** 380–384 (1993).

54. Xie, Q.W., Whisnant, R., and Nathan, C. Promoter of the mouse gene encoding calcium-independent nitric oxide synthase confers inducibility by interferon gamma and bacterial lipopolysaccharide. *J. Exp. Med.* **177,** 1779–1784 (1993).

55. Lowenstein, C.J., Alley, E.W., Raval, P., Snowman, A.M., Snyder, S.H., Russell, S.W., and Murphy, W.J. Macrophage nitric oxide synthase gene: two upstream regions mediate induction by interferon gamma and lipopolysaccharide. *Proc. Natl. Acad. Sci. USA* **90,** 9730–9734 (1993).

56. Nunokawa, Y., Ishida, N., and Tanaka, S. Promoter analysis of human inducible nitric oxide synthase gene associated with cardiovascular homeostasis. *Biochem. Biophys. Res. Commun.* **200,** 802–807 (1994).

57. Xie, Q.W., Kashiwabara, Y., and Nathan, C. Role of transcription factor NF-kappa B/Rel in induction of nitric oxide synthase. *J. Biol. Chem.* **269,** 4705–4708 (1994).

58. Martin, E., Nathan, C., and Xie, Q.W. Role of interferon regulatory factor 1 in induction of nitric oxide synthase. *J. Exp. Med.* **180,** 977–984 (1994).

59. Kamijo, R., Harada, H., Matsuyama, T., Bosland, M., Gerecitano, J., Shapiro, D., Le, J., Koh, S.I., Kimura, T., Green, S.J. et al. Requirements for transcription factor IRF-1 in NO synthase induction in macrophages. *Science* **263,** 1612–1615 (1994).

60. Melillo, G., Musso, T., Sica, A., Taylor, L.S., Cox, G.W., and Varesio, L. A hypoxia-responsive element mediates a novel pathway of activation of the inducible nitric oxide synthase promoter. *J. Exp. Med.* **182,** 1683–1693 (1995).

61. Weisz, A., Oguchi, S., Cicatiello, L., and Esumi, H. Dual mechanism for the control of inducible-type NO synthase gene expression in macrophages during activation by interferon-gamma and bacterial lipopolysaccharide. Transcriptional and post-transcriptional regulation. *J. Biol. Chem.* **269**, 8324–8333 (1994).

62. Sands, W.A., Bulut, V., Severn, A., Xu, D., and Liew, F.Y. Inhibition of nitric oxide synthesis by interleukin-4 may involve inhibiting the activation of protein kinase C epsilon. *Eur. J. Immunol.* **24**, 2345–2350 (1994).

63. Albakri, Q.A. and Stuehr, D.J. Intracellular assembly of inducible NO synthase is limited by nitric oxide-mediated changes in heme insertion and availability. *J. Biol. Chem.* **271**, 5414–5421 (1996).

64. Garvey, E.P., Oplinger, J.A., Tanoury, G.J., Sherman, P.A., Fowler, M., Marshall, S., Harmon, M.F., Paith, J.E., and Furfine, E.S. Potent and selective inhibition of human nitric oxide synthases. Inhibition by non-amino acid isothioureas. *J. Biol. Chem.* **269**, 26669–26676 (1994).

65. Furfine, E.S., Harmon, M.F., Paith, J.E., Knowles, R.G., Salter, M., Kiff, R.J., Duffy, C., Hazelwood, R., Oplinger, J.A., and Garvey, E.P. Potent and selective inhibition of human nitric oxide synthases. Selective inhibition of neuronal nitric oxide synthase by *S*-methyl-L-thiocitrulline and *S*-ethyl-L-thiocitrulline. *J. Biol. Chem.* **269**, 26677–26683 (1994).

66. Szabo, C., Southan, G.J., and Thiemermann, C. Beneficial effects and improved survival in rodent models of septic shock with *S*-methylisothiourea sulfate, a potent and selective inhibitor of inducible nitric oxide synthase. *Proc. Natl. Acad. Sci. USA* **91**, 12472–12476 (1994).

67. Nava, E., Palmer, R.M., and Moncada, S. Inhibition of nitric oxide synthesis in septic shock: how much is beneficial? *Lancet* **338**, 1555–1557 (1991).

68. Spink, J., Cohen, J., and Evans, T.J. The cytokine responsive vascular smooth muscle cell enhancer of inducible nitric oxide synthase. Activation by nuclear factor-kappa B. *J. Biol. Chem.* **270**, 29541–29547 (1995).

69. Kanno, K., Hirata, Y., Imai, T., Iwashina, M., and Marumo, F. Regulation of inducible nitric oxide synthase gene by interleukin-1 beta in rat vascular endothelial cells. *Am. J. Physiol.* **267**, H2318–24 (1994).

70. Marczin, N., Papapetropoulos, A., and Catravas, J.D. Tyrosine kinase inhibitors suppress endotoxin- and IL-1 beta-induced NO synthesis in aortic smooth muscle cells. *Am. J. Physiol.* (1993).

71. Singh, K., Balligand, J.L., Fischer, T.A., Smith, T.W., and Kelly, R.A. Regulation of cytokine-inducible nitric oxide synthase in cardiac myocytes and microvascular endothelial cells. Role of extracellular signal-regulated kinases 1 and 2 (ERK1/ERK2) and STAT1 alpha. *J. Biol. Chem.* **271**, 1111–1117 (1996).

72. Hirokawa, K., O'Shaughnessy, K., Moore, K., Ramrakha, P., and Wilkins, M.R. Induction of nitric oxide synthase in cultured vascular smooth muscle cells: the role of cyclic AMP. *Br. J. Pharmacol.* **112**, 396–402 (1994).

73. Inoue, T., Fukuo, K., Nakahashi, T., Hata, S., Morimoto, S., and Ogihara, T. cGMP upregulates nitric oxide synthase expression in vascular smooth muscle cells. *Hypertension* **25**, 744–747 (1995).

74. Imai, T., Hirata, Y., Kanno, K., and Marumo, F. Induction of nitric oxide synthase by cyclic AMP in rat vascular smooth muscle cells. *J. Clin. Invest.* **93**, 543–549 (1994).

75. Kleinert, H., Euchenhofer, C., Ihrig-Biedert, I., and Förstermann, U. In murine 3T3 fibroblasts, different second messenger pathways resulting in the induction of NO synthase II (iNOS) converge in the activation of transcription factor NF-κB. *J. Biol. Chem.* **271**, 6039–6044 (1996).

76. Marczin, N., Papapetropoulos, A., Jilling, T., and Catravas, J.D. Prevention of nitric oxide synthase induction in vascular smooth muscle cells by microtubule depolymerizing agents. *Br. J. Pharmacol.* **109**, 603–605 (1993).

77. Hortelano, S., Genaro, A.M., and Bosca, L. Phorbol esters induce nitric oxide synthase and increase arginine influx in cultured peritoneal macrophages. *FEBS Lett.* **320**, 135–139 (1993).

78. Hortelano, S., Genaro, A.M., and Bosca, L. Phorbol esters induce nitric oxide synthase activity in rat hepatocytes. Antagonism with the induction elicited by lipopolysaccharide. *J. Biol. Chem.* **267**, 24937–24940 (1992).

79. Geng, Y.J., Wu, Q., and Hansson, G.K. Protein kinase C activation inhibits cytokine-induced nitric oxide synthesis in vascular smooth muscle cells. *Biochim. Biophys. Acta* **1223**, 125–132 (1994).

80. Sirsjo, A., Soderkvist, P., Sundqvist, T., Carlsson, M., Ost, M., and Gidlof, A. Different induction mechanisms of mRNA for inducible nitric oxide synthase in rat smooth muscle cells in culture and in aortic strips. *FEBS Lett.* **338**, 191–196 (1994).

81. Perrella, M.A., Patterson, C., Tan, L., Yet, S., Hsieh, C., Yoshizumi, M., and Lee, M. Suppression of interleukin-1b-induced nitric-oxide synthase promoter/enhancer activity by transforming growth factor-β1 in vascular smooth muscle cells. *J. Biol. Chem.* **271**, 13776–13780 (1996).

82. Kleinert, H., Euchenhofer, C., Ihrig-Biedert, I., and Forstermann, U. Glucocorticoids inhibit the induction of nitric oxide synthase II by down-regulating cytokine-induced activity of transcription factor nuclear factor-kappa B. *Mol. Pharmacol.* **49**, 15–21 (1996).

83. de Vera, M., Shapiro, R.A., Nussler, A.K., Mudgett, J.S., Simmons, R.L., Morris, S.J., Billiar, T.R., and Geller, D.A. Transcriptional regulation of human inducible nitric oxide synthase (NOS2) gene by cytokines: initial analysis of the human NOS2 promoter. *Proc. Natl. Acad. Sci. USA* **93**, 1054–1059 (1996).

84. Furchgott, R.F. and Zawadzki, J.V. The obligatory role of endothelial cells in the relaxation of arterial smooth muscle by acetylcholine. *Nature* **288**, 373–376 (1980).

85. Huang, P.L., Huang, Z., Mashimo, H., Bloch, K.D., Moskowitz, M.A., Bevan, J.A., and Fishman, M.C. Hypertension in mice lacking the gene for endothelial nitric oxide synthase. *Nature* **377**, 239–242 (1995).

86. Balligrand, J.L., Kelly, R.A., Marsden, P.A., Smith, T.W., and Michel, T. Control of cardiac muscle cell function by an endogenous nitric oxide signaling system. *Proc. Natl. Acad. Sci. USA* **90**, 347–351 (1993).

87. Balligand, J.L., Kobzik, L., Han, X., Kaye, D.M., Belhassen, L., O'Hara, D.S., Kelly, R.A., Smith, T.W., and Michel, T. Nitric oxide-dependent parasympathetic

signaling is due to activation of constitutive endothelial (type III) nitric oxide synthase in cardiac myocytes. *J. Biol. Chem.* **270**, 14582–14586 (1995).

88. Uematsu, M., Ohara, Y., Navas, J.P., Nishida, K., Murphy, T.J. Alexander, R.W., Nerem, R.M., and Harrison, D.G. Regulation of endothelial cell nitric oxide synthase mRNA expression by shear stress. *Am. J. Physiol.* **269**, C1371–C1378 (1995).

89. Awolesi, M.A., Sessa, W.C., and Sumpio, B.E. Cyclic strain upregulates nitric oxide synthase in cultured bovine aortic endothelial cells. *J. Clin. Invest.* **96**, 1449–1454 (1995).

90. Sessa, W.C., Pritchard, K., Seyedi, N., Wang, J., and Hintze, T.H. Chronic exercise in dogs increases coronary vascular nitric oxide production and endothelial cell nitric oxide synthase gene expression. *Circ. Res.* **74**, 349–353 (1994).

91. Arnal, J.F., Yamin, J., Dockery, S., and Harrison, D.G. Regulation of endothelial nitric oxide synthase mRNA, protein, and activity during cell growth. *Am. J. Physiol.* **267**, C1381–C1388 (1994).

92. Zembowicz, A., Tang, J.L., and Wu, K.K. Transcriptional induction of endothelial nitric oxide synthase type III by lysophosphatidylcholine. *J. Biol. Chem.* **270**, 17006–170010 (1995).

93. Inoue, N., Venema, R.C., Sayegh, H.S., Ohara, Y., Murphy, T.J., and Harrison, D.G. Molecular regulation of the bovine endothelial cell nitric oxide synthase by transforming growth factor-beta 1. *Arterioscler. Thromb. Vasc. Biol.* **15**, 1255–1261 (1995).

94. Kostyk, S.K., Kourembanas, S., Wheeler, E.L., Medeiros, D., McQuillan, L.P., D'Amore, P.A., and Braunhut, S.J. Basic fibroblast growth factor increases nitric oxide synthase production in bovine endothelial cells. *Am. J. Physiol.* **269**, H1583–H1589 (1995).

95. Myers, P.R., Minor, Jr., R.L., Guerra, Jr., R., Bates, J.N., and Harrison, D.G. Vasorelaxant properties of the endothelium-derived relaxing factor more closely resemble *S*-nitrosocysteine than nitric oxide. *Nature* **345**, 161–163 (1990).

96. Smith, C.J., Sun, D., Hoegler, C., Roth, B.S., Zhang, X., Zhao, G., Xu, X.B., Kobari, Y., Pritchard, Jr., K., Sessa, W.C., and Hintze, T.H. Reduced gene expression of vascular endothelial NO synthase and cyclooxygenase-1 in heart failure. *Circ. Res.* **78**, 58–64 (1996).

97. Venema, R.C., Nishida, K., Alexander, R.W., Harrison, D.G., and Murphy, T.J. Organization of the bovine gene encoding the endothelial nitric oxide synthase. *Biochim. Biophys. Acta* **1218**, 413–420 (1994).

98. Zhang, R., Min, W., and Sessa, W.C. Functional analysis of the human endothelial nitric oxide synthase promoter. Sp1 and GATA factors are necessary for basal transcription in endothelial cells. *J. Biol. Chem.* **270**, 15320–15326 (1995).

99. Tang, J.L., Zembowicz, A., Xu, X.M., and Wu, K.K. Role of Sp1 in transcriptional activation of human nitric oxide synthase type III gene. *Biochem. Biophys. Res. Commun.* **213**, 673–680 (1995).

100. Kanazawa, K., Kawashima, S., Mikami, S., Miwa, Y., Hirata, K., Suematsu, M., Hayashi, Y., Itoh, H., and Yokoyama, M. Endothelial constitutive nitric oxide

synthase protein and mRNA increased in rabbit atherosclerotic aorta despite impaired endothelium-dependent vascular relaxation. *Am. J. Pathol.* **148,** 1949–1956 (1996).

101. Liao, J.K., Shin, W.S., Lee, W.Y., and Clark, S.L. Oxidized low-density lipoprotein decreases the expression of endothelial nitric oxide synthase. *J. Biol. Chem.* **270,** 319–324 (1995).

5

Role of NO in Cell Locomotion

Michael S. Goligorsky and Eisei Noiri

1. Introduction

Among various pleiotropic actions of nitric oxide discussed in the preceding chapters, its effects on different aspects of cell motility have been by and large neglected. Undoubtedly, this is due to the subtlety of these phenomena [especially when compared to the overtness of other NO-dependent processes (e.g., smooth muscle relaxation)], rather than to the lack of interest or insight into the biological significance of locomotion.

Cell migration is one of the hallmarks of morphogenesis; this function is maintained throughout the development and during the life cycle of organisms, thereby forming and protecting the structural integrity and functional plasticity of various tissues. Ontogenetic programs such as compartmentalization of bodily fluids through development of epithelial barriers or functionally guided meander of vascularization are two classical examples of processes governed by cell migration. Perpetual occurrence of breaks in the integrity of epithelial or endothelial layers by exfoliated cells does not seriously compromise barrier functions because of a rapid reestablishment of integrity by migrating cells. Many pathological situations are accompanied by a more challenging loss of epithelial integrity or requirement for angiogenesis. Initial stages of processes like these strongly depend on cell migration, whereas cell proliferation lags behind this process [1,2]. Release of various growth factors and proinflammatory mediators under these conditions may affect the rate of epithelial wound healing or growth of new vessels. Not surprisingly, therefore, the process of cell migration is thoroughly regulated. The role of growth factors [3–6], as well as matrix proteins, cytokines, cytoskeleton, and integrin receptors, has been comprehensively studied and reviewed [7–11].

The descriptive term "movement" or "locomotion" is comprised of several mechanistic elements, and some of them, under appropriate conditions, may become a target for NO regulation or modulation, as we shall show later. The unitary mechanics of changing position of a cell relative to its substratum consists of (1) local cytoskeletal rearrangements (e.g., gel–sol cytoplasmic transition and extension of lamellipodia, (2) nondirectional micromotions (or podokinesis) which result in accordionlike fluctuations in the distance between the ventral cell surface and the substratum (see below), (3) establishment, dissociation, and reestablishment of focal adhesions, and (4) generation of chemical gradients within a moving cell and the acquisition of directional movement. Although these components of locomotion are spatially and temporally integrated, each of them has autonomic functions. The most obvious example of effects of NO on one such function is presented by platelet aggregation via the αIIbβ3 integrins, as discussed later. In this chapter, we shall examine the mechanisms of locomotion, provide available evidence on NO regulation of each, speculate on some other possible targets of NO action, and illustrate them with pertinent physiological and pathophysiological phenomena.

2. A Panoramic View on the Participation of NO in Cell Adhesion and Locomotion

The spectrum of known effects of NO on various functions involving cell adhesion and migration is broadening. Figure 5-1 illustrates some of these actions. The best studied among them are the effects of NO on transendothelial migration

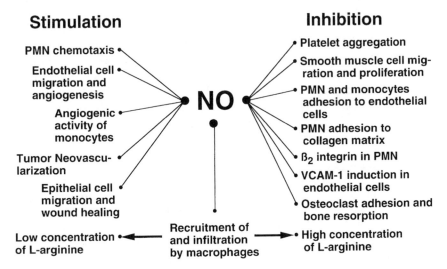

Figure 5.1 Nitric oxide elicits pleiotropic effects by acting on basic mechanisms of cell adhesion and migration.

of neutrophils, platelet aggregation, angiogenesis, neointimal formation after endothelial cell injury, osteoclastic bone resorption, and epithelial wound healing. Despite apparent distinctions between these processes, the common denominator in each and every case is the effect of NO on cell adhesion and motility.

2.1. Neutrophil–Endothelial Cell Adhesion

Studies of inflammation using intravital videomicroscopy and superfusion of cat mesenteric preparation with inhibitors of nitric oxide synthase (NOS), L-NMMA, or L-NAME, revealed that the number of adherent and immigrated leukocytes increased 15-fold [12]. This process was partially reversed by L-arginine but not by D-arginine. The investigators argued that the effects of NOS inhibitors were confined mostly to the venular endothelium, resulting in the increased adhesiveness through the CD11/CD18 leukocyte adhesion molecule. In vitro studies utilizing primary cultures of human umbilical vein endothelial cells reinforced these findings. L-NAME-treated endothelial cells exhibited increased leukocyte adhesiveness [13]. This was due to the enhanced neutrophil binding through the interaction of the β2 integrin CD18 with ICAM expressed on the endothelium. The similar inhibition of neutrophil β2 integrin function is elicited by platelet-released NO [14]. Furthermore, leukocyte adhesion to type I collagen is also inhibited by NO in a cyclic GMP-independent manner [15]. In addition, NO inhibits monocyte adhesion to the endothelial cells [16]. These effects of NO can be due in part to its effect on adhesion molecules in endothelial cells. De Caterina et al. [16] have demonstrated decreased cytokine-induced expression of vascular cell adhesion molecule-1 (VCAM-1), E-selectin, and intercellular adhesion molecule-1 (ICAM-1) on endothelium pretreated with the NO donor, whereas inhibition of NO production resulted per se in the induction of VCAM-1 expression. NO-induced repression of VCAM-1 occurs at the level of VCAM-1 gene transcription, in part due to the inhibition of NF-κB activation. On the other hand, NO serves as a chemotactic signal for neutrophils and may be involved in their recruitment to the sites of inflammation [17]. In addition, although the chemotaxis of peritoneal macrophages induced by the zymosan-activated serum was inhibited by 1–10 mM L-arginine, it was also stimulated by lower concentrations of the amino acid [18].

2.2. Angiogenesis

Formation of new blood vessels, driven by the morphogenetic program and/or by the functional demand for increased blood supply, is initiated by the budding of endothelial cells off the microcirculatory bed. The possible role of NO in the formation of capillary circuits has recently been considered by several investigative teams. In the in vivo model of the chick embryo chorioallantoic membrane, both sodium nitroprusside and L-arginine inhibited, whereas L-NMMA stimulated angiogenesis [19]. In contrast, Leibovich et al. [20] have provided solid evidence that the production of angiogenic activity by activated monocytes (assayed by chemotaxis of endothelial cells and corneal angiogenesis) is absolutely dependent on L-arginine and NO synthase. These observations are in concert with findings reported by Ziche et al. [21] who detected the potentiation by sodium nitroprusside of angiogenic effect of substance P in the rabbit cornea. These investigators hypothesized that endogenous NO production induced by vasoactive agents serves as an autocrine mediator of angiogenesis. Our own observations expand this function to the classical angiogenic signal, the vascular endothelial growth factor. We demonstrated that endogenous NO production by the endothelial cells is a prerequisite for the motogenic and angiogenic effects of this factor [22]. These findings gain in significance when applied to the pathophysiological situations (see below).

2.3. Neointimal Formation After Endothelial Denudation

After endothelial denudation, migration and proliferation of vascular smooth muscle cells, exposed to the vessel lumen, initiate neointimal formation which represents an early event in atherogenesis. NO effects on vascular smooth muscle cells seem to have certain peculiarities, distinguishing them from the endothelial cells. Dubey et al. [23] studied migration of rat aortic smooth muscle cells in a modified Boyden chamber. When angiotensin II was used to promote transwell migration, the addition of sodium nitroprusside or *S*-nitroso-*N*-acetylpenicilla-mine inhibited this process. The effect of NO donors was mimicked by a cell-permeant form of cyclic GMP and counteracted by the pretreatment of cells with LY83583 (an inhibitor of soluble guanylyl cyclase) and by KT5823 (an inhibitor of cGMP-dependent protein kinase). This, as well as the known antiproliferative effect of NO [24], may have a prominent function in preventing neointimal formation. In fact, in vivo transfer of Sendai virus/liposome-encapsulated cDNA encoding endothelial NOS after denudation by balloon injury of rat carotid artery resulted in 70% inhibition of neointimal formation by day 14 after the procedure [25]. Similar results were obtained after balloon angioplasty of femoral artery in rabbits with the application of long-lived NO adducts [26].

2.4. Epithelial Wound Healing

The integrity of epithelial barriers is continuously challenged during their physio-logic function and, especially, in the course of pathophysiologic events. Migration of epithelial cells is the major mechanism of wound healing. Studies of wound healing in epithelial cells or barriers provided a set of observations consistent with the promoting effects of NO on cell migration and restitution of barrier integrity. Our findings in cultured epithelial cells, BSC-1 cells, showed that several growth factors (EGF, IGF-I, HGF, b-FGF) exerted a motogenic effect which was abrogated by pretreatment with L-NAME [106]. L-Arginine or NO donor, S-nitroso-N-acetylpenicillamine (SNAP), exerted motogenic effect in epithelial cells. Inhibition of NOS with L-NAME or a selective knockout of inducible NOS (iNOS) with antisense oligodeoxynucleotides reduced the rate of spontaneous or EGF-induced BSC-1 cell migration. Although the constitutive endothelial NO synthase (ecNOS) did not show any detectable spatial or temporal changes associated with wounding, the iNOS became expressed 3 h after wounding and showed higher abundance at the edges of epithelial wounds (Fig. 5-2). Consistent with this dynamics of NOS expression, NO release from migrating epithelial BSC-1 cells displayed a biphasic response to the inflicted wounds: An initial transient release of nitric oxide is followed by a delayed sustained elevation. Based on above observations, we hypothesize that NO serves as a switch from stationary to locomoting epithelial phenotype.

2.5. Platelet Aggregation

Platelet aggregation is governed by the fibrinogen receptor belonging to the integrin family, activation-dependent $\alpha IIb\beta 3$, whereas adhesion of platelets to the basement membrane of deendothelialized vascular wall is mediated through other members of the same family which are constitutively active. Radomski et al. [27,28] have provided evidence that platelets produce NO by a calcium-dependent mechanism which requires L-arginine. Direct monitoring of NO release with a porphyrinic microsensor showed that resting platelets do not release it, whereas platelet activation results in a rapid NO production [29]. Platelet aggregation induced by collagen is inhibited by L-arginine and potentiated by L-NMMA, whereas the release of NO follows the inverse relationship with platelet aggregation. Different aggregating agents (collagen, adenosine diphosphate, thrombin, epinephrine) act on specific receptors and trigger the elevation of intraplatelet $[Ca^{2+}]$. This event is critical for the activation of the fibrinogen receptor, usually dormant, and initiation of aggregation cascade. The process has a built-in mechanism of self-termination: this is achieved via the simultaneous stimulation of NO production which inhibits intraplatelet $[Ca^{2+}]$ transients and counteracts the activation of the $\alpha IIb\beta 3$ integrin receptor and platelet aggregation. This process can be pharmacologically regulated by organic nitrates [30].

2.6. Osteoclastic Bone Resorption

Another example of NO action on cell adhesion resulting in profound functional changes is seen in osteoclasts. These bone-resorbing cells of monocytic lineage form tight focal adhesions with the matrix proteins. Upon stimulation, H^+ pumps are recruited to the basal cell surface which permits the generation of a low-pH subcompartment underneath the cell, leading to bone resorption [31]. MacIntyre et al. [32] demonstrated that NO donors caused retraction of osteoclasts and reduced their adhesion to the matrix and motility. The physiological consequence of this NO effect is the inhibition of bone resorption, as judged by the extent of

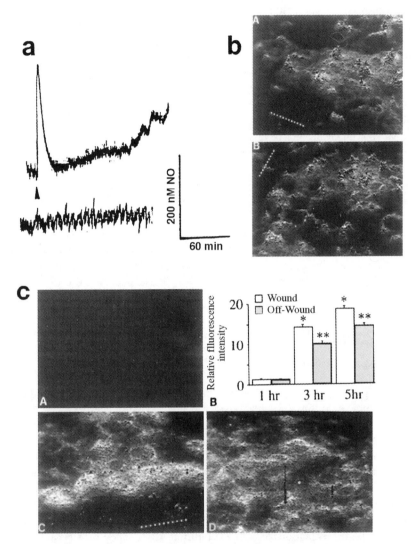

Figure 5.2 Nitric oxide and epithelial BSC-1 cell wound healing. (a) Nitric oxide release from wounded epithelial cells (upper tracing) shows two phases—the first is a rapid and variable in its amplitude phase and the second is a delayed and highly reproducible. Lower tracing represents a recording of no release by time-matched intact monolayer. (b) Immunocytochemical detection of eNOS in epithelial cells shows no spatial or temporal changes after disruption of monolayer integrity. (c) Immunocytochemical staining of iNOS in epithelial cells shows its appearance by 3 h after wounding, and prevalence of iNOS at wound edges. Panel D shows that cells peripheral to the wound are stained less intensely for iNOS. (Compiled from Ref 106 with permission of the American Physiological Society)

excavations (resorption pits) on the bone surface. These findings were reconfirmed by Kasten et al. [33] who also showed in vitro that inhibitors of NOS dramatically increased both the number of pits and the resorption area per pit. In a model of osteoporosis, NOS inhibitors accelerated the loss of bone. Because osteoclasts express a constitutive and an inducible forms of NOS, and the NO produced modulates cell function [34], this autocrine mechanism is currently considered as an important regulator of bone remodeling and mineral homeostasis. An additional level of regulatory complexity can be provided by osteoblasts and osteocytes releasing NO in response to mechanical stress [35]. Thus, acting in an autocrine or paracrine mode, NO inhibition of osteoclast adhesion to the bone matrix results in both the regulation of local bone accretion–resorption as well as profound systemic effects on mineral metabolism.

In conclusion, the data presented in this section convincingly demonstrate that NO participates in the regulation of distinct physiological processes based on its ability to interfere with cognate mechanisms of cell adhesion and migration. To gain further insights into the mode(s) of NO action, we shall next focus on the general mechanisms governing cell attachment and detachment from the substrate in the process of locomotion, defining along these lines some established or hypothetical targets for NO.

3. Components of Cell Locomotion and Their Regulation by NO

3.1. Unitary Elements: Extension of a Lamellipodium and Micromotion

The mechanics of cell migration has been described in a series of studies by Abercrombie et al. [36–38] and considerably expanded in later years [7,10]. The key element of cell movement is the extension of a lamellipodium. The force required for the protrusion of a certain portion of a cell is derived from the reversible cytochalasin B-inhibitable polymerization of G- to F-actin. According to the Brownian ratchet model [39], a thermally vibrating plasma membrane is displaced a distance sufficient to accommodate the addition of an actin monomer to the barbed end of F-actin, thus preventing the elastic recoil of the membrane. Determination of flicker spectra of erythrocytes, indeed, showed that the membrane oscillates over a distance of ~0.1 μm at a frequency of 5 Hz [40]. Thus formed, membrane protrusions represent a unitary element of motility with the vector of generated protrusive force directed toward the lamellipodium and tangential to the substratum.

When a directional signal is absent, the ostensibly stationary confluent cells nevertheless undergo yet another type of spontaneous shape change referred to as micromotions [41], probably distinct from the above protrusion of lamellipodia. Originally, these micromotions were detected with a sensitive impedance registration and analysis in cells growing on the surface of a miniature gold electrode in tissue culture. These experiments showed the existence of spontaneous microoscillations in cell monolayer resistance which had fractal characteristics. Giaever and Keese proposed a mathematical model ascribing changes in the electrical resistance to either of two parameters: Rb—resistance of the paracellular pathway; α—resistance of the slit space between the ventral cell surface and the substratum. We have recently analyzed the changes in endothelial cell impedance, and again detected spontaneous fluctuations (unpublished observations). Exogenous NO significantly enhanced the amplitude of oscillations (Fig. 5-3). Analysis of components responsible for this effect of NO showed that it was due primarily to the broadening of the distance between the ventral cell surface and the substratum, hence suggesting a second type of micromotional activity, already with unitary displacement occurring in the direction perpendicular to the substratum, which we termed podokinesis. Important for this discussion is the fact that this type of micromotions was significantly accentuated by NO. The relation between the ratchet movements and podokinesis, though possible, remains unexplored.

3.2. Function of Focal Adhesions

The above unitary elements of cell movement are functioning autonomously but may become a part of a cooperative, spatially and temporally well-coordinated process of directional cell migration. The process is initiated by extending lamellipodia in the direction of a void or a chemoattractant signal and eventual "pulling" of the cell body. There is a considerable amount of data showing the importance of establishing new focal contacts between the leading edge and the matrix in cell migration. Understanding these phenomena will require an excursus into structure and function of focal adhesions and their key elements, integrins.

The integrins are noncovalently bound heterodimeric glycoproteins composed of α and β subunits. Members of this large family of receptors share several common features. Both subunits have a single hydrophobic transmembrane domain, relatively short cytoplasmic tails, and massive extracellular domains. The extracellular domains are compactly folded by virtue of disulfide bonding, associated together, and both chains contribute to the formation of the binding domain. All α subunits contain a sevenfold repeat of a homologous segment, the last three or four repeats of which are likely to contribute to divalent cations binding. The β subunits contain four cysteine-rich repeats responsible for the folding via internal disulfide groups. Cytoplasmic domains of the β subunits are indispensable to connecting the receptors to the cytoskeleton via talin and α-actinin [42], as detailed below.

Extracellular Domain: Ligand Recognition and Interaction

The arginine-glycine-aspartic acid (RGD) motif in matrix proteins turned to be one of several key integrin recognition sequences. Three contact sites on the integrin are needed to bind ligand: one on the β subunit and two on the α subunit

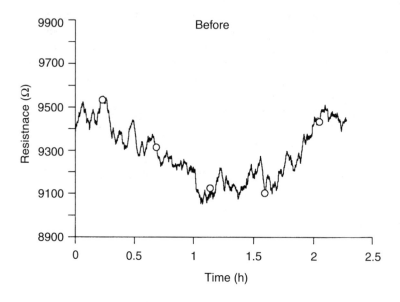

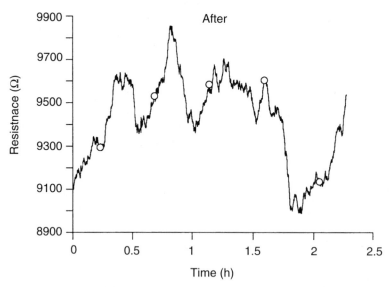

Figure 5.3 Spontaneous and NO-induced oscillations in the resistance of endothelial monolayers. The resistance was measured with a sensitive miniature electrode technique, as detailed in the text. Note the existence of spontaneous oscillations in endothelial resistance (upper panel). After the application of 100 μ*M* SNAP (lower panel), the amplitude of oscillations has increased.

[43]. As indicated earlier, all α subunits contain sequences homologous to the EF-hand of Ca-binding proteins, with the common distinction that integrins lack the essential aspartic acid residue which is present in position 12 of EF-hand Ca-binding site. It has been hypothesized, therefore, that the aspartic acid on integrin ligands, by providing the missing cation-coordinating residue, forms a ternary complex with the receptor-bound divalent ion [44]. The cation displacement model emphasizes the role of initial receptor–ligand binding via coordinating aspartic residue, followed by destabilization of a divalent ion bond to the EF-hand, extrusion of a divalent ion, and completion of ligand–receptor interaction. Recent demonstration of cation binding to the β3 (118–131) fragment, assessed by terbium luminescence and mass spectrometry, and its displacement by RGD-containing ligands supports this hypothesis [45]. This mechanism, in addition to the cytoplasmic domain-triggered conformational changes of the extracellular integrin domain (see below) may be responsible for the affinity modulation and the conversion from dormancy to active state of integrin receptors.

Intracellular Domain: Cytoskeletal Interactions

In contrast to the extended extracellular domains, the cytoplasmic tails of integrins are short (except for the β4 subunit). Numerous studies with truncated cytoplasmic domains of integrins have demonstrated the essential role of the β subunit in establishing integrin–cytoskeletal interactions [46–48]. Using the β1 subunit as a prototype, three potentially important cytoplasmic regions have been identified. The sequence HDRREFAKFEKE, denoted as the cyto-1 region, appears to be essential for the targeting of the β1 subunit to focal adhesions [49]. Four amino acid sequences NPIY and NPKY (cyto-2 and cyto-3, respectively) not only participate in localizing the integrin to focal adhesions but also, quite unexpectedly, represent a concensus signal for clathrin-coated pit-mediated internalization of membrane proteins [50]; the significance of this finding awaits elucidation. Above all, the NPxY motif has been found to represent an alternative binding site for Shc (Margoulis, personal communication). Interestingly, these three regions are conserved in many integrin's β subunits. LaFlamme et al. [51] constructed chimeric receptors consisting of the extracellular and transmembrane domains of the human interleukin-2 (IL-2) receptor connected to the intracellular domain of either β1, β3, β3B, or β5 subunit. Although the β3B-containing chimera was expressed diffusely on the cell surface, other chimeric receptors were localized at focal adhesions of transfected human fibroblasts. Expressed at higher levels, β1 and β3 chimeras functioned as dominant negative mutants and inhibited endogenous integrins function in cell spreading and migration. These elegant studies convincingly demonstrate the role of cytoplasmic domains of β subunits in regulating integrin clustering at focal adhesions and mediating cell spreading and locomotion.

The functions of α subunits cytoplasmic domains have been extensively studied using genetic engineering approaches. Using the α2β1 collagen/laminin receptor, Chan et al. [52] constructed chimeras with cytoplasmic domains of the α4 and α5 subunits. These studies revealed a different role of individual cytoplasmic domains in cell migration and contraction of collagen gel.

Focal Adhesions and Involvement of Integrins in Signal Transduction

Focal adhesions are highly specialized domains of the plasma membrane which, on the extracellular side, form the closest contact between integrin receptors and matrix proteins, and, on the cytoplasmic face, represent the sites where converging actin filaments terminate and interact with integrins. In these complex structures, several cytoskeletal proteins are participating in anchoring actin to integrins: vinculin, talin, α-actinin, fimbrin, tensin, paxillin, and zyxin [53], whereas several other focal adhesion-associated components express enzymatic or yet unidentified activities. Several proteins are substrates for PKC phosphorylation (vinculin, talin, tensin, filamin) and tyrosine kinase phosphorylation (vinculin, talin, tensin, paxillin). Using shearing and quick-freezing procedures, Samuelsson et al. [53] evaluated cytoplasmic surface and the three-dimentional organization of focal contacts and associated actin bundles and observed that type I aggregates contained β1 integrins, vinculin, talin, and anchoring actin filaments, whereas type II aggregates did not contain vinculin and talin and were not associated with the actin cytoskeleton. These findings suggest that type I aggregates are relatively stable and represent the classical focal adhesions, whereas type II aggregates containing the β1 integrin subunits are unanchored, significantly more mobile, and represent the pool of integrins that can be recruited to form new focal adhesions. Engagement of these integrins with the particular epitopes on the extracellular matrix triggers distinct cell signaling events and remodeling of cell shape.

The establishment of focal adhesions via integrin–extracellular ligand binding initiates a cascade of signaling events (Fig. 5-4). The formation of cell–matrix and cell–cell contacts triggers cellular responses as diverse as the rapid activation of the Na^+/H^+ exchanger and cell alkalinization [54,55], the elevation of cytosolic calcium concentration (possibly, due to the activation of a 50-kDa β3 integrin-associated protein, presumed to represent an integrin-regulated calcium channel in endothelial cells and neutrophils [56,57]), the delayed stimulation of a K^+ current [58], and a series of tyrosine phosphorylation reactions mediated via activation of focal adhesion kinase (FAK) [59,60]. Conversely, the reduction of cell–substrate adhesion and eventual detachment of cells from their matrix is associated with the activation of a phosphotyrosine phosphatase and decreased tyrosine phosphorylation of focal adhesions [61]. The diversity of signaling mechanisms triggered by the establishment or dissolution of focal adhesions can provide a means for the conformational changes of proteins comprising the focal adhesions (e.g., the β1 integrin, paxillin, and tensin, all containing phosphotyro-

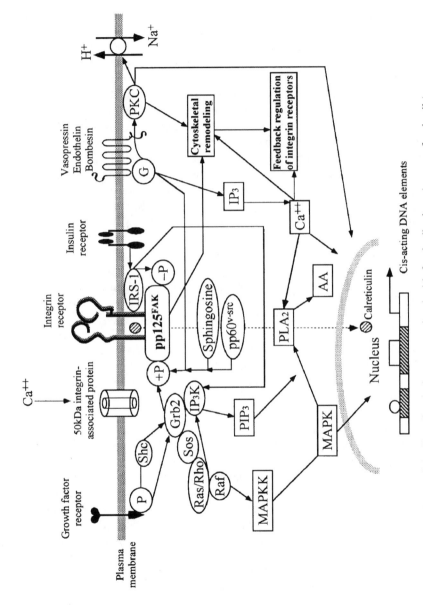

Figure 5.4 Signaling pathways associated with focal adhesions (see text for details).
Reprinted from Kidney International 48: 1375–1385, 1995 with permission by Blackwell Science Inc.

sine) and explain the phenomenon of affinity modulation of integrins toward the extracellular matrix proteins, outside the cell, and toward the cytoskeletal elements, inside the cell [47].

The list of the substrates for tyrosine kinase phosphorylation is growing and presently includes vinculin, talin, tensin, paxillin, integrins, pp125^{FAK}, cadherins, and catenins. Pp125^{FAK} represents, so far, the best studied tyrosine kinase associated with focal adhesions. This protein is autophosphorylated during cell adhesion (the cytoplasmic domain of β subunits is necessary for this reaction), but it also undergoes autophosphorylation in response to pp60^{v-src}, bradykinin, endothelin, and sphingosine, resulting in the increase in its tyrosine kinase activity. Dephosphorylation of pp125^{FAK} may be accomplished in part via activation of the SH2 domain-containing protein tyrosine phosphatase Syp (SHPTP2) which is activated by the insulin receptor substrate-1 (IRS-1). In turn, Grb2 and PI(3)-kinase are two known substrates of pp125^{FAK} [62,63]. Collectively, these studies interconnect integrins with receptors for hormones and growth factors and implicate the state of pp125^{FAK} phosphorylation in modulation of the Ras/MAPK and PIP$_2$ signal transduction pathways, which are all indispensable for the process of cytokinesis.

3.3. Chemical Gradients Within a Moving Cell

When cells receive a directional motogenic signal, either in the form of an impaired integrity of a monolayer or of a chemotactic gradient, vectorial movement ensues. Thus targeted, protrusion of lamellipodia at a leading edge coincides with the formation of new focal adhesions; simultaneously, the developing traction of a trailing edge is accompanied by the dissolution of previously formed focal contacts [64,65]. These coordinated activities ensure attachment of the leading edge to the substratum and "pulling" of the cell body in the direction of the movement. For these reciprocal processes to occur in synchrony, the cytoplasmic microenvironment at the leading and trailing edges should be efficiently compartmentalized to accommodate the simultaneous action of opposing signaling cascades. Indeed, several investigators observed the development of chemical gradients within a moving cell, as schematically depicted in Fig. 5-5. Singer and Kupfer [8] have provided evidence that the microtubule-organizing center and the Golgi apparatus rapidly reorient themselves in the direction of movement in cultured endothelial cells or fibroblasts rendered motile by inflicting wound to the monolayers. This polarization of organelles within cells destined to migrate occurs with a half-time of several minutes and long before a leading edge has been formed. In many cells, application of microtubule-depolymerizing agents inhibits migration, in agreement with a role for the microtubule-organizing center and microtubules in cytokinesis. The leading edge is characterized by actin polymerization and gelation of the cytoplasm, whereas F-actin depolymerization and solation of the cytoplasm occur predominantly at the trailing end. Moreover, myosin is excluded from the lamellipodium and actin is concentrated within this

Chemical gradients within a moving cell

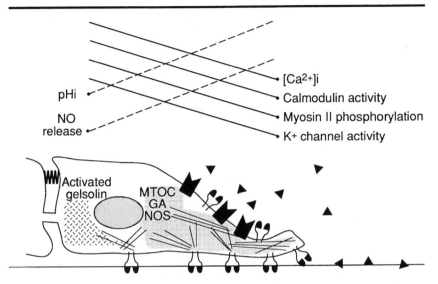

Figure 5.5 Established and hypothetical chemical gradients within a moving cell. Solid lines represent the established gradients; dotted lines show hypothetical gradients. MTOC = microtubule-organising center; GA = Golgi apparatus. Reprinted from Experimental Nephrology 4: 314–321, 1996 with permission by S. Karger AG.

structure [8]. Uncapping the barbed ends of actin filaments, paralleled by the release of actin monomers from the tymosin $\beta 4$ and increased availability of actin, result in the nucleated growth of actin filaments at the leading edge [10]. Newly growing actin filaments undergo bundling and cross-linking, leading to cytoplasmic gelation and volume expansion by osmotic gel force, thus providing mechanical thrust to form and extend cell protrusions.

In view of the role of actin filaments in cytoskeletal structure and function, several investigators addressed the question of NO effects on F-actin. In subcellular fractions of neutrophils, NO caused ADP ribosylation of a 43-kDa cytosolic protein identified as actin but not the $G\alpha i$ protein which is the substrate for pertussis toxin [66]. The functional significance of these in vitro observations is unclear. In intact cultured endothelial cells, however, neither L-arginine nor L-NAME elicited any changes in actin cytoskeleton [67]. Pretreatment of endothelial cells with a phorbol ester revealed that L-arginine elicits the increased phalloidin staining of the peripheral band and prevents the increase in permeability to 40-kDa dextran, whereas L-NAME produces the opposite effect. These data are consistent with an NO-induced increase in F-actin. In contrast, the polymerization of actin was decreased after application of a long-acting NO donor *S*-nitroso-*N*-acetylpenicillamine [15]. Hence, these discrepant observations demonstrate that the elucidation of the role played by NO in actin cytoskeleton is far from being resolved.

It has been observed that migrating neutrophils or newt eosinophils are characterized by a gradient in cytosolic free calcium concentration ($[Ca^{2+}]_i$): it is elevated at the trailing end and decreased in the front of a cell [68,69]. When a cell changes its direction of movement, in response to a repositioned chemotactic target, the $[Ca^{2+}]_i$ gradient reestablishes with enhanced levels at the newly formed trailing end [68]. This gradient may be responsible for the uneven activation of calcium-dependent potassium channels, as it was observed in transformed MDCK cells [70], although the exact mechanistic role of these channels in locomotion is unknown. It appears that this $[Ca^{2+}]_i$ gradient is important for cell locomotion. Chelation of cytosolic calcium or pretreatment of neutrophils with an inhibitor of calcium–calmodulin-sensitive protein phosphatase 2B, calcineurin, abolished cell migration by interfering with the release of cells from sites of attachment while preserving the ability to form lamellipodia [69]. Given the role of integrins in the formation of focal adhesions, the participation of these molecules in the above reciprocal processes can be envisaged. Indeed, the activity of calmodulin and the phosphorylation of myosin II at the trailing end are increased compared to the leading edge, thus providing a motor force in the tail [71]. It is highly possible that several Ca-dependent enzymes are also activated at the rare portion of the cell; among them, the activation of actin-severing proteins, like gelsolin, would be of great pertinence. Activation of these proteases will lead to actin depolymerization, liberation of focal adhesions from anchorage to actin filaments, and the loss of attachment to a substrate at the trailing end. This hypothetical mechanism can explain the predominant solation of the trailing portion of the migrating cell and the localized destabilization of focal adhesions. In fact, when elevations of cytosolic calcium are inhibited, neutrophils become stuck to fibronectin or vitronectin [72].

The puzzling question is: What keeps the low $[Ca^{2+}]_i$ at the leading edge? Although the answer to this puzzle is unknown, some observations may suggest a possible mechanism. Recent findings in migrating epithelial cells showed that iNOS becomes preferentially polarized to the front portion, where the Golgi apparatus is located [106]. We postulated, therefore, that NO production follows this iNOS gradient from the leading edge, where it is highest, to the trailing end of the cell. If this inference is correct, the NO cyclic GMP–GMP kinase signaling cascade, activated in the cell front, should result in the inhibition of calcium release from the endoplasmic reticulum. This inference is based on the previously reported action of protein kinase G on G proteins, inhibiting the inositol 1,4,5-trisphosphate production [73] (although it may also act by accelerating the extrusion or compartmentalization of the released calcium). In CHO cells transfected with cGMP-dependent protein kinase, thrombin-induced formation of inositol trisphosphate and calcium release were blunted by pretreatment with 8-bromo-cGMP. Such a mechanism would explain the observed paradoxical generation and maintenance of a low-$[Ca^{2+}]_i$ microenvironment at the leading edge, against the principle of the propagating calcium-induced calcium release from the intracellular $[Ca^{2+}]$ stores. Obviously, these speculations are awaiting experimental proof.

3.4. Cell Adhesion and Motility: Potential Targets of NO

Based on the previous discussion, it is clear that cell migration proceeds by virtue of coordinated establishment of focal adhesions at the leading edge, thus preventing its slippage, and dissociation of focal adhesions at the cell's trailing end, thus permitting forward movement. It is not surprising, therefore, that both the excessive cell adhesion, as well as prevention of cell attachment should cause inhibitory effects on cell motility, thus conforming with the classical *meden agan*—nothing in excess—principle. Indeed, experimental evidence shows that agents which only mildly destabilize focal adhesions (e.g. hyaluronate or low concentrations of RGD peptides [65,69]), enhance cell migration, whereas excessive deposition of matrix proteins inhibits it. Several integrin receptors contribute to locomotion. Receptors responsible for cell motility are occasionally distinct from those responsible for cell adhesion and spreading, as it was convincingly demonstrated for smooth muscle cells expressing β1 receptors involved in adhesion and β3 integrins governing migration [74]. In endothelial cells, the αVβ3 receptor is responsible for adhesion, spreading and migration on vitronectin; but when cells are cultured on collagen, all these functions are handled by the α2β1 integrin [75]. Monoclonal antibodies directed to the α chain of collagen receptor, α2β1 integrin, in endothelial cells cultured in collagen gels converted them from a proliferative phenotype toward locomotive phenotype, resulting in enhanced capillary tube formation [76]. When endothelial cells were cultured in collagen gels of increased density, the capillary tube formation, however, was decreased. In in vivo studies, Friedlander et al. [77] demonstrated the involvement of predominantly αVβ3 in bFGF-induced angiogenesis, and αVβ5 in the VEGF-induced process. In view of the role of protein tyrosine kinase FAK in the formation of focal adhesions, as discussed above, the recent studies in FAK-deficient mice (FAK gene was mutated by inserting the neomycin phosphotransferase gene in the beginning of its kinase domain) revealed reduced cell motility and enhanced focal adhesion formation in primary-cultured fibroblasts [78]. These data implicate FAK in the turnover of focal adhesions during cell locomotion, thus reinforcing the above notion on the relationship between adhesion and migration.

In this vein, the observed effects of NO on cell migration should be gauged. As shown in Fig. 5-6, endothelial cell migration is enhanced by stimulation of NO production [106]. We have next determined the effects of NO on endothelial cell attachment and detachment. When suspended cells were allowed to attach to fibronectin in the presence of SNAP, cell adhesion was decreased. Conversely, when SNAP was added to confluent endothelial cell monolayers, cell attachment to fibronectin was decreased. These findings suggest that the observed enhance-

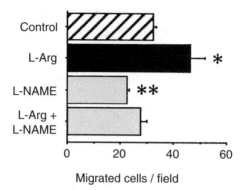

Figure 5.6 Transwell migration of microvascular endothelial cells. Endothelial cell migration was initiated by the addition of 10 ng/ml EGF (control). In the presence of L-arginine, the number of migrated cells was increased, whereas L-NAME blocked migration. Experiments were performed in arginine-free culture medium. * denotes $p<.05$ versus control; ** denotes $p<.05$ versus L-arginine + L-NAME. (Modified from Ref. 106.)

ment of endothelial cell migration by NO is due in part to its effect on integrins, most probably $\alpha V\beta 3$ integrins, which are responsible for endothelial cell adhesion and locomotion. The exact mechanism whereby NO acts on integrins and/or formation of focal adhesions remains elusive. The potential sites of NO action are summarized in Fig. 5-7. As discussed above, extracellular domains of integrins are folded by disulfide bonding and β subunits contain four cysteine-rich repeats. These SH– and S–S groups may represent targets for nitrosylation [79], resulting in conformational changes of heterodimeric receptors. Although this mechanism is plausible, its existence awaits experimental proof.

The second potential target for NO is located on the cytoplasmic domain of focal adhesions and is related to tyrosine phosphorylation of components of focal contacts. Tyrosine phosphorylation is critical for cell motility [80]. It is important to emphasize, therefore, that integrins and growth factor receptors converge within the focal adhesion complex [81], thus providing a spatial compartmentalization for tyrosine kinase and its substrates. As mentioned earlier, formation of focal contacts is accompanied by tyrosine phosphorylation of several constituents of focal adhesions [82], and the dissociation of focal contacts results in activation of phosphotyrosine phosphatases [61]. Because phosphotyrosine protein phosphatases contain two highly reactive thiol groups within the active site, Caselli et al. [83] explored the hypothesis that NO may regulate the activity of the enzyme. Using a low-molecular-weight phosphotyrosine protein phosphatase as a model representing a common reaction mechanism and active site motif, these investigators demonstrated its inactivation by NO donors. In view of the possible head-to-tail NO gradient in migrating cells, the above NO-induced mechanism of

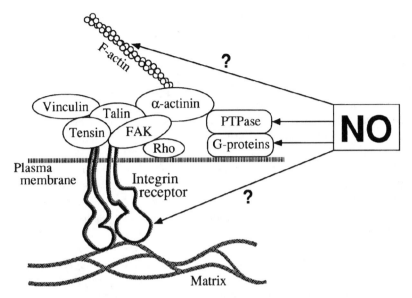

Figure 5.7 Potential sites of NO action within the focal adhesion complex (see text for the details). Reprinted from Experimental Nephrology 4: 314–321, 1996 with permission by S. Karger AG.

phosphotyrosine phosphatase inactivation has a potential to facilitate formation of focal adhesion at the leading edge and promote locomotion.

4. Pathophysiological Implications of Disturbances in NO Regulation of Adhesion and Migration

4.1. NO and Atherosclerosis

Studies by DeCaterina et al. have demonstrated an increase in monocyte adhesion to endothelial cells treated with NOS inhibitors [16]. Neointimal formation is NO dependent and is enhanced after endothelial injury, but NO prevents it [16]. It has been shown that NO exerts antiproliferative effect on vascular smooth muscle cells [24]. Based on these findings, it was suggested that NO exerts an antiatherogenic effect [84].

Endothelial production of NO is inhibited in hypercholesterolemia and athero-sclerosis, in some forms of hypertension, and after tissue ischemia and reperfusion [85–90]. Indeed, it has been recently found that oxidized low-density lipoprotein decreases the expression of endothelial NO synthase [91]. Collectively, these observations shed light on the additional beneficial mechanisms of therapy with organic nitrates and justify the search for means of targeted NOS gene delivery [25] and/or development of long-lasting NO adducts [92].

4.2. NO and Platelet Thrombi

As detailed earlier, platelet activation by collagen, thrombin, or ADP elicits an increase in platelet [Ca^{2+}] which, in turn, transforms the fibrinogen receptor from a dormant to the active conformation [93], thus triggering platelet aggregation. The process is self-limited by virtue of concomitant stimulation of NOS and production of NO. It has been shown that NO-induced inhibition of platelet aggregation is due to a decrease in affinity for fibrinogen binding to its receptor, explaining the decrease in number of fibrinogen molecules bound to the platelet surface [94]. Exogenous NO donors have been used as antithrombotic agents for a long time. Probably, in pathophysiological situations accompanied by increased platelet aggregability, the endogenous NO generation is insufficient, advocating the use of NO or S-nitrosothiol donors [30]. It has been observed by Loscalzo [92] that both nitroglycerin and nitroprusside inhibit platelet aggregation, and this process was accompanied by the elevation of intraplatelet cGMP levels. Conversely, inhibition of NO production is complicated by excessive thrombogenesis and vascular occlusion, as has been demonstrated by Westberg et al. [95]. Most recently, an NO-releasing aspirin derivative, acetylsalicylic acid 4-(nitroxy)-butylester, with vastly enhanced antiplatelet activity has been synthesized and tested in vitro and in vivo [96]. Thus, the deficient generation of endogenous NO can be overcome with various NO donors, resulting in the therapeutic inhibition of platelet aggregation.

4.3. Angiogenesis and Its Targeted Stimulation or Inhibition

Angiogenic stimuli fall into one of two categories: They are either beneficial, providing supplemental circulation to functioning organs, or they are pathological, emanating from neoplastic tissues. In the first case, ischemic myocardium or other tissue is the source of angiogenic growth factors (e.g., vascular endothelial growth factor) [97]. Despite the local presence of angiogenic stimuli, their efficacy may be diminished due to an inadequate generation of NO by the endothelial cells. As demonstrated earlier, production of NO is a prerequisite for endothelial cell migration induced by vascular endothelial growth factor [22] or substance P [21]. This scenario would call for a means to enhance NO generation. Conversely, inhibition of tumor angiogenesis and growth [98,99] or enhanced vascularization in diabetic retinopathy may require local suppression of NO production.

Occasionally, the need to therapeutically enhance and inhibit angiogenesis exists simultaneously, as in atherosclerosis. On the one hand, stimulation of vascularization of ischemic tissues is in demand, whereas, on the other, prevention of developing *vasa vasorum* within the atherosclerotic vascular wall (the frequent site of hemorrhage and thrombosis) represents a therapeutic goal. Because it is rather difficult to accomplish both, strategies for the targeted gene therapy, like Sendai virus transfer of the eNOS gene to denuded arteries [25], are under intense investigation [100].

4.4. NO and Wound Healing

Studies of L-arginine metabolism in wound fluid revealed an early pattern of NOS activation, as judged by the increase in NO and L-citrulline generation, followed by activation of arginase and a significant decrease in the local L-arginine availability, accompanied by an increase in L-ornithine, a substrate for synthesis of polyamines [101–103]. It was therefore surmised that the latter arginase-L-ornithine-polyamines pathway is essential for wound healing, and the stimulatory effect of polyamines on migration has been demonstrated in two intestinal epithelial cell lines [104]. Similarly, in a randomized double-blind study, L-arginine stimulated wound healing in elderly patients [105]. The above-mentioned observations from our laboratory [106] suggest that, in addition to the stimulation of polyamines pathway in the process of wound healing, L-arginine is utilized for the synthesis of NO which is required for the epithelial wound healing.

5. Conclusions

We have summarized herein available information on adhesion and migration of diverse cells as they are affected by NO, discussed the mechanics of cell adhesion and locomotion, delineated some established or possible targets of NO action, and outlined the pathophysiological consequences of inappropriate NO regulation of adhesion and migration. The gestalt of this chapter implies that archaic functions such as adhesion and migration utilize an archaic regulatory mechanism, NO production, and that disregulation of these fundamental properties of a living matter brings about ominous sequelae.

Several potential sites for NO regulation of adhesion and migration have been proposed. It appears that NO affects integrin receptor affinity toward its ligands. This mode of NO action is crucial for the effect on platelet aggregation, osteoclastic bone resorption, and neutrophil or monocyte adhesion to the endothelium. Directional migration of many cells consists of a turnover of focal adhesions creating a nonvectorial "tap dancing," which we termed podokinesis, and vectorial protrusion of lamellipodia accompanied by the traction of cell body along a gradient in the concentration of a chemoattractant. Recent findings show that NO increases the turnover of focal adhesions, thus agitating cellular "tap dancing." Furthermore, in some moving cells, a head-to-tail gradient in NO production may be responsible for the decreased cytosolic calcium concentration at the leading edge. This intracellular calcium gradient, in turn, seems to be critical for the development of gel–sol anisotropy and coordinated attachment–detachment of focal adhesions in the locomoting cell. Finally, NO may act on protein tyrosine phosphatases, G-proteins, and F-actin, thus modulating cell–matrix adhesion and migration. The typical examples of NO participation in cell migration are represented by its effects on angiogenesis and epithelial wound healing. Disregulation of these NO-linked functions is associated with either a deficient angiogenesis in atherosclerosis or excessive neovascularization of some tumors, as well as a defective wound healing.

A more profound future understanding of the role played by NO in processes of cell adhesion and migration should open new vistas in therapeutic use of its donors or inhibitors. The seeds of such approaches (e.g., NO adducts or selective inhibitors of NOS) have already been planted. However, the rational use of this universal cell messenger or inhibitors of its synthesis will require additional efforts in designing strategies for their effective targeting to a selected cell population.

Note Added in Proof

Studies from authors laboratory were supported by NIH grants DK45695 and DK45462. Permissive role of NO in endothelin-induced migration of endothelial cells has been described most recently by Noiri et al. in J. Biol. Chem. 272: 1747–1752, 1997.

References

1. Kartha, S. and Toback, F.G. *J. Clin. Invest.* **90**, 288–292 (1992).

2. Nusrat, A., Delp, C., and Madara, J.L. *J. Clin. Invest.* **80**, 1501–1511 (1992).

3. Barrandon, Y. and Green, H. *Cell* **50**, 1131–1137 (1987).

4. Blay, J. and Brown, K.D. *J. Cell Physiol.* **124**, 107–112 (1985).

5. Mignatti, P., Morimoto, T., and Rifkin, D.B. *Proc. Natl. Acad. Sci. USA* **88**, 11007–11011 (1991).

6. Nakao-Hayashi, J., Ito, H., Kanayasu, T., Morita, I., and Mirota, S. *Atherosclerosis* **92**, 141–149 (1992).

7. Bretcher, M.S. *Science* **224**, 681–686 (1984).

8. Singer, S.J. and Kupfer, A. *Annu. Rev. Cell Biol.* **2**, 337–365 (1986).

9. Theriot, J.A. and Mitchison, T.J. *Nature* **352**, 126–131 (1991).

10. Condeelis, J. *Annu. Rev. Cell Biol.* **9**, 411–444 (1993).

11. Cooper, J. *Annu. Rev. Physiol.* **53**, 585–605 (1991).

12. Kubes, P., Suzuki, M., and Granger, D.N. *Proc. Natl. Acad. Sci. USA* **88**, 4651–4655 (1991).

13. Niu, X., Smith, C.W., and Kubes, P. *Circ. Res.* **74**, 1133–1140 (1994).

14. Thom, S., Ohnishi, T., and Ischiropoulos, H. *Tox. Appl. Pharm.* **128**, 105–110 (1994).

15. Sundquist, T., Forslund, T., Bengtsson, T., and Axelsson, K. *Inflammation* **18**, 625–631 (1994).

16. De Caterina, R., Libby, P., Peng, H., Thannickal, V., Rajavashisth, T., Gimbrone, M., Shin, W., and Liao, J. *J. Clin. Invest.* **96**, 60–68 (1995).

17. Beauvais, F., Michel, L., and Dubertret, L. *J. Cell Physiol.* **165**, 610–614 (1995).

18. Reyes, A., Porras, B., Chasalow, F., and Klahr, S. *Kidney Int.* **45**, 1346–1354 (1994).

19. Pipili-Synetos, E., Sakkoula, E., and Maragoudakis, M.E. *Br. J. Pharmacol.* **108**, 855–857 (1993).

20. Leibovich, S.J., Polverini, P.J., Fong, T.W., Harlow, L.A., and Koch, A.E. *Proc. Natl. Acad. Sci. USA* **91**, 4190–4194 (1994).

21. Ziche, M., Morbidelli, L., Masini, E., Amerini, S., Granger, H., Maggi, C., Geppetti, P., and Ledda, F. *J. Clin. Invest.* **94**, 2036–2044 (1994).

22. Noiri, E., Testa, J., Quigley, J., Colflesh, D., Keese, C., Giaever, I., and Goligorsky, M.S. (submitted).

23. Dubey, R., Jackson, E., and Luscher, T.F. *J. Clin. Invest.* **96**, 141–149 (1995).

24. Garg, U.C. and Hassid, A. *J. Clin. Invest.* **83**, 1774–1777 (1989).

25. Von der Leyen, H., Gibbons, G., Morishita, R., Lewis, N., Zhang, L., Nakajima, M., Kaneda, Y., Cooke, J., and Dzau, V.J. *Proc. Natl. Acad. Sci. USA* **92**, 1137–1141 (1995).

26. Marks, D., Vita, J., Folts, J., Keaney, J., Welch, G., and Loscalso, J. *J. Clin. Invest.* **96**, 2630–2638 (1995).

27. Radomski, M., Palmer, R., and Moncada, S. *Proc. Natl. Acad. Sci. USA* **87**, 5193–5197 (1990).

28. Radomski, M., Palmer, R., and Moncada, S. *Trends Pharmacol. Sci.* **12**, 87–88 (1991).

29. Malinski, T., Radomski, M., Taha, Z., and Moncada, S. *Biochem. Biophys. Res. Commun.* **194**, 960–965 (1993).

30. Loscalzo, J. *Am. J. Cardiol.* **70**, 18B–22B (1992).

31. Blair, H., Teitelbaum, S., Ghiselli, R., and Gluck, S. *Science* **245**, 855–857 (1989).

32. MacIntyre, I., Zaidi, M., Alam, A., Datta, H., Moonga, B., Lidbury, P., Hecker, M., and Vane, J.R. *Proc. Natl. Acad. Sci. USA* **88**, 2936–2940 (1991).

33. Kasten, T., Collin-Osdoby, P., Patel, N., Osdoby, P., Krukowski, M., Misko, T., Settle, S., Currie, M., and Nickols, G.A. *Proc. Natl. Acad. Sci USA* **91**, 3569–3573 (1994).

34. Brandi, M., Hukkanen, M., Umeda, T., Morandi-Bidhendi, N., Bianchi, S., Gross, S.S., Polak, J.M., and MacIntyre, I. *Proc. Natl. Acad. Sci. USA* **92**, 2954–2958 (1995).

35. Pitsillides, A., Rawlinson, S., Suswillo, R., Bourrin, S., Zaman, G., and Lanyon, L. *FASEB J.* **9**, 1614–1622 (1995).

36. Abercrombie, M., Heaysman, J.E.M., and Pegrum, S.M. *Exp. Cell Res.* **62**, 389–398 (1970).

37. Abercrombie, M., Heaysman, J.E.M., and Pegrum, S.M. *Exp. Cell Res.* **67,** 359–367 (1971).

38. Abercrombie, M., Heaysman, J.E.M., and Pegrum, S.M. *Exp. Cell Res.* **73,** 536–539 (1972).

39. Peskin, C., Odell, G., and G Oster. *Biophys J.* **65,** 316–324 (1993).

40. Fricke, K. and Sackmann, E. *Biochem. Biophys. Acta* **803,** 145–152 (1984).

41. Giaever, I. and Keese, C.R. *Proc. Natl. Acad. Sci. USA* **88,** 7896–7900 (1991).

42. Otey, C.A., Vasquez, G.B., Burridge, K., and Erickson, B.W. *J. Biol. Chem.* **268,** 21193–21197 (1993).

43. Segel, I. *Enzyme Kinetics.* John Wiley and Sons, New York, 1975.

44. Smith, J.W. and Cheresh, D.A. *J. Biol. Chem.* **266,** 11429–11432 (1991).

45. D'Souza, S.E., Haas, T., Piotrowicz, R., Byers-Ward, V., McGrath, D., Soule, H., Cierniewski, C., Plow, E., and Smith, J.W. *Cell* **79,** 659–667 (1994).

46. Solowska, J., Guan, J.L., Marcantonio, E.E., Trevithick, J.E., Buck, C.A., and Hynes, R.O. *J. Cell Biol.* **109,** 853–861 (1989).

47. LaFlamme, S.E., Akiyama, S.K., and Yamada, K.M. *J. Cell Biol.* **117,** 437–447 (1992).

48. Geiger, B., Salomon, D., Takeichi, M., and Hynes, R.O. *J. Cell Sci.* **103,** 943–951 (1992).

49. Reszka, A.A., Hayashi, Y., and Horwitz, A.F. *J. Cell Biol.* **117,** 1321–1330 (1992).

50. Chen, W.J., Goldstein, J.L., and Brown, M.S. *J. Biol. Chem.* **265,** 3116–3123 (1990).

51. LaFlamme, S., Thomas, L., Yamada, S., and Yamada, K.M. *J. Cell Biol.* **126,** 1287–1298 (1994).

52. Chan, B.M., Kassner, P.D., Schiro, J.A., Byers, H.R., Kupper, T.S., and Hemler, M.E. *Cell* **68,** 1051–1060 (1992).

53. Samuelsson, S.J., Luther, P.W., Pumplin, D.W., and Bloch, R.J. *J. Cell Biol.* **122,** 485–496 (1993).

54. Schwartz, M.A., Cragoe, E.J., and Lechene, C. *J. Biol. Chem.* **265,** 1327–1332 (1990).

55. Galkina, S.I., Sudina, G.F., and Margolis, L.B. *Exp. Cell Res.* **200,** 211–214 (1992).

56. Jaconi, M.E., Theler, J.M., Schlegel, W., Appel, R.D., Wright, S.D., and Lew, P.D. *J. Cell Biol.* **112,** 1249–1257 (1991).

57. Schwartz, M.A., Brown, E.J., and Fazeli, B. *J. Biol. Chem.* **268,** 19931–19934 (1993).

58. Becchetti, A., Arcangeli, A., Riccarda del Bene, M., Olivotto, M., and Wanke, E. *Proc. R. Soc. Lond. B. Biol. Sci.* **248,** 235–240 (1992).

59. Schaller, M.D., Borgman, C.A., Cobb, B.S., Vines, R.R., Reynolds, A.B., and Parsons, J.T. *Proc. Natl. Acad. Sci. USA* **89,** 5192–5196 (1992).

60. Kornberg, L., Earp, H.S., Parsons, J.T., Schaller, M., and Juliano, R.L. *J. Biol. Chem.* **267,** 23439–23442 (1992).

61. Maher, P.A. *Proc. Natl. Acad. Sci. USA* **90**, 11177–11181 (1993).

62. Schlaepfer, D., Hanks, S., Hunter, T., and van der Geer, P. *Nature* **372**, 786–791 (1994).

63. Chen, H.-C. and Guan, J.-L. *Proc. Natl. Acad. Sci. USA* **91**, 10148–10152 (1994).

64. Schmidt, C.E., Horwitz, A.F., Lauffenburger, D.A., and Sheetz, M.P. *J. Cell Biol.* **123**, 977–991 (1993).

65. Hall, C.L., Wang, C., Lange, L.A., and Turley, E.A. *J. Cell Biol.* **126**, 575–588 (1994).

66. Clancy, R., Leszczynska-Piziak, J., and Abramson, S. *Biochem. Biophys. Res. Commun.* **191**, 847–852 (1993).

67. Liu, S. and Sundqvist, T. *Exp. Cell Res.* **221**, 289–293 (1995).

68. Brundage, R., Fogarty, K., Tuft, R., and Fay, F.S. *Am. J. Physiol.* **265**, C1527–C1543 (1993).

69. Hendey, B., Klee, C.B., and Maxfield, F.R. *Science* **258**, 296–299 (1992).

70. Schwab, A., Wojnowski, L., Kerstin, G., and Oberleithner, H. *J. Clin. Invest.* **93**, 1631–1636 (1994).

71. Post, P., DeBiasio, R., and Taylor, D.L. *Mol. Biol. Cell* **6**, 1755–1768 (1995).

72. Marks, P., Hendey, B., and Maxfield, F.R. *J. Cell Biol.* **112**, 149–158 (1991).

73. Ruth, P., Wang, G., Boekhoff, I., May, B., Pfeifer, A., Penner, R., Korth, M., Breer, H., and Hofmann, F. *Proc. Natl. Acad. Sci. USA* **90**, 2623–2627 (1993).

74. Clyman, R., Mauray, F., and Kramer, R. *Exp. Cell Res.* **200**, 272–284 (1992).

75. Leavesley, D., Schwartz, M.A., Rosenfeld, M., and Cheresh, D.A. *J. Cell Biol.* **121**, 163–170 (1993).

76. Gamble, J., Matthias, L., Meyer, G., Kaur, P., Russ, G., Faull, R., Berndt, M., and Vadas, M.A. *J. Cell Biol.* **121**, 931–943 (1993).

77. Friedlander, M., Brooks, P., Shaffer, R., Kincaid, C., Varner, J., and Cheresh, D.A. *Science* **270**, 1500–1502 (1995).

78. Ilic, D., Furuta, Y., Kanazawa, S., Takeda, N., Sobue, K., Nakatsuji, N., Nomura, S., Fujimoto, J., Okada, M., Yamamoto, T., and Aizawa, S. *Nature* **377**, 539–544 (1995).

79. Stamler, J.S. *Cell* **78**, 931–936 (1994).

80. Klemke, R., Yebra, M., Bayna, E., and Cheresh, D.A. *J. Cell Biol.* **127**, 859–866 (1994).

81. Plopper, G., McNamee, H., Dike, L., Bojanowski, K., and Ingber, D.E. *Mol. Biol. Cell* **6**, 1349–1365 (1995).

82. Miyamoto, S., Teramoto, H., Coso, O., Gutkind, S., Burbelo, P., Akiyama, S., and Yamada, K.M. *J. Cell Biol.* **131**, 791–805 (1995).

83. Caselli, A., Camici, G., Manao, G., Moneti, G., Pazzagli, L., Cappugi, G., and Ramponi, G. *J. Biol. Chem.* **269**, 24878–24882 (1994).

84. Cooke, J. and Tsao, P. *Atheroscler. Thromb.* **14,** 653–655 (1994).

85. Lefer, A.M. and Ma, X. *Arterioscler. Thromb.* **13,** 771–776 (1993).

86. Flavahan, N.A. *Circultion* **85,** 1927–1938 (1992).

87. Lefer, A.M., Tsao, P., Lefer, D.J., and Ma, X. *FASEB J.* **5,** 2029–2034 (1991).

88. Radomski, M.W., Palmer, R., and Moncada, S. *Lancet* **2,** 1057–1058 (1987).

89. Ma, X., Weyrich, A., Lefer, D.J., and Lefer, A.M. *Circ. Res.* **72,** 403–412 (1993).

90. May, G.R., Crook, P., Moore, P., and Page, C.P. *Br. J. Pharmacol.* **102,** 759–763 (1991).

91. Liao, J., Shin, W., Lee, W., and Clark, S. *J. Biol. Chem.* **270,** 319–324 (1995).

92. Loscalzo, J. *J. Clin. Invest.* **76,** 703–708 (1985).

93. Shattil, S., Ginsburg, M., and Brugge, J.S. *Curr. Opin. Cell Biol.* **6,** 695–704 (1994).

94. Mendelsohn, M., O'Neill, S., George, D., and Loscalzo, J. *J. Biol. Chem.* **265,** 19028–19034 (1990).

95. Westberg, G., Shultz, P., and Raij, L. *Kidney Int.* **46,** 711–716 (1994).

96. Wallace, J., McKnight, W., Del Soldato, P., Baydoun, A., and Cirino, G. *J. Clin. Invest.* **96,** 2711–2718 (1995).

97. Dvorak, H., Brown, L., Detmar, M., and Dvorak, A.M. *Am. J. Pathol.* **146,** 1029–1039 (1995).

98. Folkman J. *Ann. Surg.* **175,** 409–416 (1972).

99. Folkman J. *Nature Med.* **1,** 27–31 (1995).

100. Fan, T.-P., Jaggar, R., and Bicknell, R. *Trends Pharm. Sci.* **16,** 57–66 (1995).

101. Albina, J.E., Mills, C.D., Barbul, A., Thurkill, C.E., Henry, W.L., Mastrofrancesco, B., and Caldwell, M.D. *Am. J. Physiol.* **254,** E459–E467 (1988).

102. Albina, J.E., Mills, C.D., Henry, W.L., and Caldwell, M.D. *J. Immunol.* **144,** 3877–3880 (1990).

103. Seifert, E., Rettura, G., and Levenson, S.M. *Surgery* **84,** 224–230 (1978).

104. McCormack, S.A., Wang, J.-Y., Viar, M.J., Tague, L., Davies, P.J.A., and Johnson, L.R. *Am. J. Physiol.* **267,** C706–C714 (1994).

105. Kirk, S.J., Hurson, M., Regan, M.C., Holt, D.R., Wasserkrug, H.L., and Barbul, A. *Surgery* **114,** 155–160 (1993).

106. Noiri, E., Peresleni, T., Srivastava, N., Weber, P., Bahou, W.F., Peunova, N., and Goligorsky, M.S. *Am. J. Physiol. (Cell Physiol.)* **270,** C794–C802 (1996).

PART II

Renal Expression of NOSs and Production of NO

6

Metabolic Pathways and Cycles of L-Arginine Synthesis and Utilization

Saulo Klahr

In addition to its role in the synthesis of proteins, the amino acid L-arginine is essential for the synthesis of urea, creatine, nitric oxide, agmatine, and polyamines and influences the release of hormones and the synthesis of pyrimidine bases [1]. Arginine was identified and isolated from proteins more than a century ago [2,3]. It was not until the 1930s that its prominent role in normal metabolism began to unfold. This chapter summarizes available information on the synthesis of L-arginine and various pathways of its utilization: arginase-catalyzed synthesis of urea and ornithine, production of agmatine through L-arginine metabolism by arginine decarboxylase, and nitric oxide synthase-catalyzed conversion of L-arginine to citrulline and NO.

1. Synthesis of L-Arginine

It has been known for more than 50 years that the kidney has a significant capacity for converting citrulline to arginine [4,5]. The small intestine is the principal source of circulating citrulline in adult mammals [6]. The synthesis of citrulline in the intestine is catalyzed by carbamoyl phosphate synthetase and ornithine transcarbamoylase, which are located in mucosal epithelial cells [7,8]. The regulation of carbamoyl phosphate synthetase and ornithine transcarbamoylase in the intestine differs from that in the liver. The activity of these two enzymes in the small intestine is either unaffected or decreased by increases in dietary protein, whereas, in the liver, the activity of most of the enzymes of the urea cycle is increased in response to high protein diets. [9] The roles of the small intestine and the kidney with regard to arginine biosynthesis change during development. Whereas levels of argininosuccinate lyase are relatively high in the liver and kidney, they are very low in the small intestine of adults. Activities of carabamoyl phosphate synthase and arginosuccinate synthase and lyase are relatively high in small intestine but low in the kidney for the first week or two after birth [10]. This finding is consistent with the virtual absence of arginase activity in small intestine during the first two weeks after birth, followed by a rapid increase in the activity of this enzyme to adult levels [10–12]. Thus, the small intestine appears to be the principal biosynthetic organ for arginine at and shortly after birth, whereas this function is divided between the small intestine and kidney in adults.

In 1941, the synthesis of L-arginine from L-citrulline and a nitrogen donor, usually L-aspartic acid, in the kidney was reported. Subsequent studies by Windmueller and Spaeth [6] demonstrated that the intestine is the major, almost exclusive source of L-citrulline for the renal synthesis of L-arginine. As shown in Fig. 6-1 approximately 28% of the glutamine, an end product of amino acid catabolism in muscle, is removed by the intestine from the circulation and released into the bloodstream as L-citrulline [13]. Smaller amounts, less than 0.025%, of the L-citrulline is taken up by the liver and 83% is converted to L-arginine in the kidney [6]. In vitro studies revealed that endothelial cells incubated with glutamine concentrations about one-third [14] to one-half [15] those found in normal plasma, significantly inhibit the recycling of L-citrulline into L-arginine and thus may affect the production of nitric oxide [14].

It has also been reported that the renal proximal tubule is the major site of synthesis of L-arginine in rat [16], mouse, and rabbit [17]. Although arginase is present in other segments of the nephron, this metabolic pathway accounts for only 6% of the metabolism of L-arginine in the proximal tubule [18]. Arginase activity, as described above, has also been observed in glomeruli obtained from rats with nephrotoxic nephritis [19], in fibroblasts [20], and in macrophages [21]. Most of the L-arginine synthesized in the proximal tubule enters the systemic circulation and only a small portion is utilized and metabolized to urea in the kidney. Thus, the kidney has a key role in maintaining a constant supply of L-arginine for utilization by other organs, as demonstrated by Featherson et al. [22]. These investigators administered labeled L-citrulline intravenously to rats and studied its conversion to L-arginine in muscle, liver, brain, and kidney. They found significantly lower levels of labeled L-arginine in animals with bilateral nephrectomy performed 1 h prior to the administration of the labeled citrulline than in similar animals with intact kidneys with or without hepatectomy [22]. In the rat [23] and in humans [24], the conversion of L-citrulline to L-arginine in the kidney is constant and independent of the intake of L-arginine or protein. As a result, L-arginine homeostasis is achieved by a balance between dietary intake and L-arginine degradation. When L-arginine degradation and/or utilization is increased (such as in growth, wound healing, trauma, injury, and sepsis), L-arginine becomes an essential amino acid [24] and its dietary intake should be increased. Under physiological conditions, the plasma levels of L-arginine and the utilization of L-arginine by extrarenal organs depend on the synthesis of this amino acid in the kidney by a mechanism that appears to be independent of the level of the

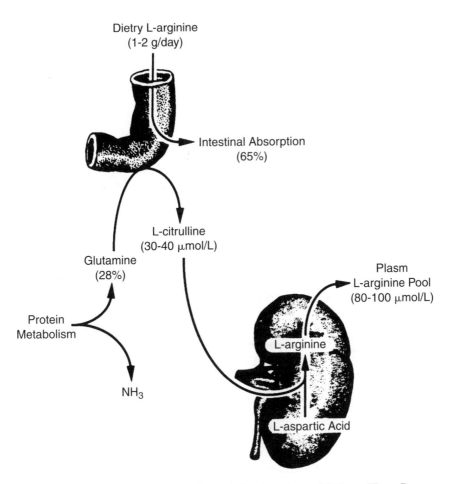

Figure 6-1. Sources and synthesis of L-arginine: intestine and kidney. [From Reyes, A.A., Karl, I.E., and Klahr, S. Role of arginine in health and in renal disease. *Am. J. Physiol.* **267** (Renal Fluid Electrolyte Physiol. 36), F331–F346 (1994). Used with permission.]

dietary intake of L-arginine or the dietary intake of protein. However, homeostasis of L-arginine is related to the net difference between the metabolic turnover of L-arginine and its supply in the diet. The normal plasma levels of arginine are approximately 80–100 μ*M* and intracellular concentrations are even greater (up to 1 m*M*).

The role of the kidney in maintaining plasma levels of L-arginine has been examined in normal humans and in patients with chronic renal failure. Tizianello et al. [25] found an arteriovenous renal gradient for L-citrulline and for L-arginine in normal humans indicating renal uptake of L-citrulline and renal release of L-arginine. Both processes (i.e., net uptake of L-citrulline and net release of

Table 6-1. Plasma Levels of Free L-Citrulline and L-Arginine in Healthy Subjects and in Patients with Varying Degrees of Renal Insufficiency

| | Controls | Creatinine Clearance | | | Hemodialysis | |
		25–60 ml/min (n = 33)	10–25 ml/min (n = 46)	<10 ml/min (n = 59)	(n = 52)	(n = 32)
L-Cit, μmol/L		30 ± 14	57 ± 20	73 ± 27	85 ± 41	106 ± 52
L-Arg, μmol/L		81 ± 21	67 ± 24	67 ± 21	76 ± 30	77 ± 35
L-Arg/L-Cit		2.70	1.18	0.92	0.89	0.73

Note: Values are means ± SE.

Source: Adapted from Jungers, P., Chauveau, P., Ceballos, I., Bardet, J., Parvy, P., Hannedouche, T., and Kamoun, P. Plasma free amino acid alterations from early to end-stage chronic renal failure. J. Nephrol. 7, 48–54 (1994).

L-arginine) were of similar magnitude, suggesting that the L-citrulline entering the kidney was converted to arginine, which was then released into the renal vein. The same group [26] reported that L-citrulline uptake and L-arginine released by the kidneys of patients with chronic renal insufficiency decreased to approximately 40% of values found in normal controls. Jungers et al. [27] found an inverse relationship between renal function and plasma levels of citrulline (Table 6-1). A progressive decrease in renal function (creatinine clearance) was associated with higher plasma levels of citrulline. Other investigators have reported similar findings [28,29].

2. Metabolism of L-Arginine

2.1. Urea Cycle

Studies on the synthesis of urea in liver slices incubated with arginine, citrulline, or ornithine, which allowed Krebs and Henseleit to postulate the existence of the urea cycle in 1932 [30], also established the key role of arginine in a major metabolic pathway. The urea cycle is an essential metabolic pathway for disposal of the toxic metabolite ammonia in most terrestrial vertebrates. On the other hand, in marine elasmobranchs, the urea synthesized by this pathway is used for osmoregulation [31]. The reactions and intermediates in the biosynthesis of 1 mole of urea from 1 mole each of ammonia and carbon dioxide and of the alpha amino nitrogen of aspartate are shown in Fig. 6-2. The overall process requires 3 moles of ATP and the successive participation of five enzymes catalyzing the numbered reactions shown in Fig. 6-2. Of the six amino acids involved in urea synthesis, one, n-acetyl-glutamate, functions as an enzyme activator rather than as an intermediate. The remaining five amino acids, aspartate, arginine, ornithine, citrulline, and arginosuccinate, all function as carriers of atoms which ultimately become urea. Two of these amino acids (aspartate and arginine) occur in proteins;

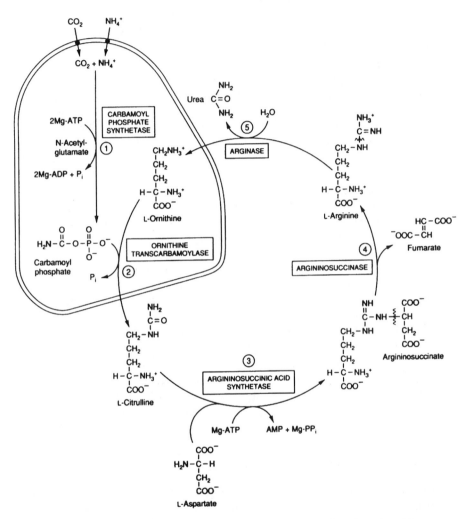

Figure 6-2. Reactions and intermediates of urea biosynthesis. Reactions 1 and 2 occur in the matrix of liver mitochondria and reactions 3–5 in liver cytosol. CO_2 (as bicarbonate), ammonium ion, and ornithine and citrulline traverse the mitochondrial matrix via specific carriers (•) present in the inner membrane of liver mitochondria. (From Rodwell, V.W. Catabolism of proteins and of amino acid nitrogen. In: *Harper's Biochemistry,* 23rd ed. Eds. Murray, R.K., Granner, D.K., Mayes, P.A., and Rodwell, V.W. Appleton & Lange, Norwalk, CT, 1993. Used with permission.)

the remaining three (ornithine, citrulline, and arginosuccinate) do not. The major metabolic role of these latter three amino acids in mammals is urea synthesis. As noted in the Fig. 6-2, urea formation is in part a cyclical process. The ornithine used in reaction 2 is regenerated in reaction 5. Thus, there is no net loss or gain of ornithine, citrulline, arginosuccinate, or arginine during urea synthesis; however, ammonia, CO_2, ATP, and aspartate are consumed. Both enzymes, carbamoyl phosphate synthetase (reaction 1) and L-ornithine transcarbamoylase (reaction 2), are located in liver mitochondria. Reaction 5, the cleavage of arginine to ornithine and urea, completes the urea cycle and regenerates ornithine as a substrate needed for reaction 2. Hydrolytic cleavage of the guanidino group of arginine is catalyzed by a cobalt- or manganese-activated enzyme, arginase, which is present in the livers of all ureotelic organisms. Smaller amounts of arginase are also present in renal tissue, brain, mammary gland, testicular tissue, and skin. Arginase is also present in macrophages. Ornithine and lysine are potent competitive inhibitors of arginase. Although the full complement of enzymes of the urea cycle is expressed only in the liver, some of the enzymes of this metabolic pathway are expressed also in the kidney and small intestine, thereby constituting an independent arginine biosynthetic pathway.

The urea, formed mainly in the liver, is not further metabolized; it is distributed in total body water and is excreted by the kidney. Since the beginning of this century, it has been known that urea production in adult humans varies as a function of dietary protein intake [32]. Changes in liver arginase activity related to changes in dietary protein intake were reported in 1939 [33]. This study represents one of the earliest examples of metabolic adaptation at the level of enzyme activity in mammals. In the 1960s, it was demonstrated that activities of all five urea cycle enzymes in rat liver varied as a function of dietary protein intake [34–36]. Activities of the urea cycle enzymes are highest in response to starvation and high-protein diets and are reduced in response to low-protein or protein-free diets [37–43].

2.2. Synthesis of Creatine and Creatinine

L-Arginine is also involved in the synthesis of creatine and creatinine (Fig. 6-3). Creatine is present in muscle, brain, and blood, both in the free state and as phosphocreatine. Creatinine is the anhydride of creatine [44]. Three amino acids— arginine, glycine, and methionine—are directly involved in the synthesis of creatine. L-Arginine is initially converted to guanidino acetic acid by the addition of L-glycine and, subsequently, guanidino acetic acid undergoes methylation in the liver to form creatine, which is then transported to muscle and actively taken up by this organ. Creatine undergoes a nonenzymatic and irreversible dehydration to form creatinine at a constant rate of 1–2% per day. The amount of creatinine formed is dependent on muscle mass [45]. Creatinine is not retained in muscle and may be further metabolized [46] and even recycled to creatine [47]. Creatinine

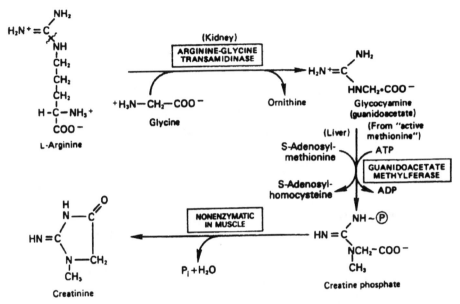

Figure 6-3. Biosynthesis of creatine and creatinine. (From Rodwell, V.W. Conversion of amino acids to specialized products. In: *Harper's Biochemistry,* 23rd ed. Eds. Murray, R.K., Granner, D.K., Mayes, P.A., and Rodwell, V.W. Appleton & Lange, Norwalk, CT, 1993. Used with permission.)

is distributed in total body water and eliminated by the kidney by filtration and tubular secretion.

2.3. Synthesis of Nitric Oxide

A metabolic pathway that utilizes L-arginine as an exclusive precursor is the synthesis of nitric oxide with the release of L-citrulline (Fig. 6-4) by cells containing the enzyme(s) nitric oxide synthase [48–50]. The formation of nitric oxide, one of the smallest (30 Da) and simplest biosynthetic products, is catalyzed by enzymes that are among the largest (300 kDa) and most complicated. Nitric oxide synthases are homodimers whose monomers are themselves two enzymes fused, a cytochrome reductase and a cytochrome, that require three cosubstrates [L-arginine, nicotinamide adenine dinucleotide phosphate (NADPH), and oxygen] and five cofactors of prosthetic groups [flavin adenine dinucleotide (FAD), flavin mononucleotide (FMN), calmodulin, (6R)-tetrahydrobiopterin, and heme). The Michaelis-Menton constant (Km) for arginine use as a substrate for NO synthase is on the order of $1-10 \mu M$. Thus, there would appear to be a vast surplus of substrate (plasma concentration of L-arginine is 80–100 μM). Nitric oxide has a half-life of only a few seconds, but its biological activity may last 1–2 min because of the in vivo formation of complexes of nitric oxide with S-nitroso adducts [51].

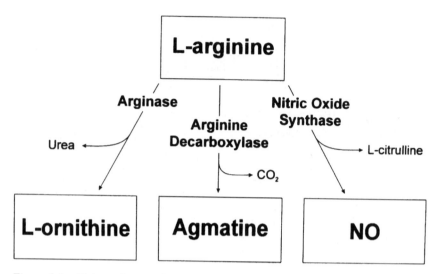

Figure 6-4. Major pathways of L-arginine metabolism. L-Arginine may be metabolized by the urea cycle enzyme arginase to L-ornithine and urea, by arginine decarboxylase to agmatine and CO_2, or by nitric oxide synthase to nitric oxide (NO) and L-citrulline.

As described elsewhere in this book, nitric oxide has a key role in the regulation of vascular tone [52], immune system function [53], neurotransmission [54,55], and platelet aggregation and adhesion [56], among other processes (Fig. 6-5 and Chapters 2, 5, and 8). Most of the effects of NO are mediated by second messengers, mainly cyclic GMP and protein kinases. Nitric oxide is synthesized from L-arginine in a reaction catalyzed by one of a family of nitric oxide synthase enzymes [57–59]. There are three major isoforms of nitric oxide synthase, which are encoded by three separate genes (see chapters 3 and 4). It is now known that all nitric oxide synthase (NOS) isoforms require L-arginine, oxygen, and NADPH as cosubstrates and FAD, FMN, heme, and (6R)-tetrahydro-L-biopterin as cofactors [60–62 and Chapter 3). Constitutively expressed NOS depends on an additional cofactor, calmodulin [63–65]. Elevated calcium levels lead to the binding of calmodulin to NOS and subsequent activation of the enzyme. A calmodulin consensus sequence also has been found in the inducible form of NOS (iNOS), thus pointing to a calcium–calmodulin dependence in the macrophage type iNOS, which requires much less calcium than constitutive NOS (cNOS). Once generated, NO is not very reactive under physiologic pH, but its paramagnetic properties (odd number of electrons) account for its strong binding affinity for the heme iron and thereby its inactivation by hemoglobin and other hemoproteins [66]. The same mechanism applies for the activation of soluble guanylate cyclase, because NO binds to the heme group of this enzyme. Reduced iron (Fe^{2+}) in the form of heme is required for the activation of soluble granylate cyclase, which represents the main molecular target of NO, although NO also interacts with a

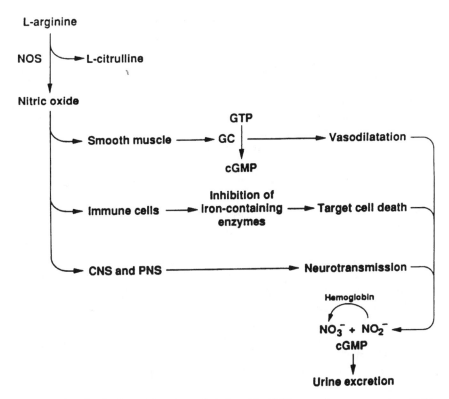

Figure 6-5. Synthesis and functions of nitric oxide. CNS, central nervous system; PNS, peripheral nervous system; NOS, nitric oxide synthase; GC, guanylate cyclase. [From Reyes, A.A., Karl, I.E., and Klahr, S. Role of arginine in health and in renal disease. *Am. J. Physiol.* **267** (Renal Fluid Electrolyte Physiol. 36) F331–F346 (1994). Used with permission.]

variety of other iron-sulfur enzymes [50] and promotes adenosine diphosphosphate ribosyl transfer [67].

2.4. Nitric Oxide Synthases

The expanding family of NOS isoforms can be divided into inducible and constitutive isoforms. The first isoform purified and cloned was discovered in neurons but is also present in skeletal muscle, neutrophils, pancreatic islets, endometrium, and respiratory and gastrointestinal epithelium. The isoform purified and cloned from endothelial cells is also expressed in neurons [68]. The isoform first purified and cloned from macrophages is inducible in cell types from all branches of the histogenetic tree, among them neurons and endothelial cells [69]. Thus, cell-type designations can be exceedingly confusing. Ambiguity is minimized by a one-to-one correspondence between names and genes, achieved by simplifying an earlier numerical nomenclature. However, most workers in the field use descriptive terms. A useful classification discriminates NO synthases on the basis of a physiological critical biochemical feature, dependence on a calcium transient (greater than 100 nM) in the whole cell to sustain the binding of calmodulin, signified by the letter "c," versus independence of elevated calcium to bind calmodulin [70], denoted by the letter "i." These terms have led to some confusion when chelation of virtually all the calcium in cell lysate has partially inactivated iNOS in some species; such results do not imply that the activity of the enzyme is dependent on calcium above the concentration present in the resting cell. The two cNOS are generally constitutive in the developed mammalian organism, whereas iNOS is generally inducible.

Nitric oxide synthases can be characterized also as low versus high output. A determinant of these phenotypes is the duration of NOS activity in natural host cells under physiological conditions. Because activity of the constitutive NOS is triggered by agonists that increase calcium, it is transient (seconds to minutes). Effective agonists are many, ranging from bradykinin, thrombin, and sheer stress in endothelial cells, to glutamate, human immunodeficiency gp virus (HIVGP) 120 [71], and a β-amyloid peptide in neurons [72]. In addition, the V_{max} of endothelial cNOS is reportedly far less than the V_{max} of neuronal cNOS and iNOS. Hence, constitutive NOSs comprise the low-output pathway involved in homeostatic processes such as regulation of blood pressure, neurotransmission, and peristalsis. Nonetheless, in some circumstances, such as stimulation of neurons in vitro, neuronal cNOS can contribute to cellular toxicity. In contrast, iNOS lies on the high-output path. Some normal tissue may express iNOS antigens, such as the uterus of the pregnant rabbit [73] and large airways in humans [74]. However, expression of iNOS is more often related to infection or inflammation and involved in host defense. Inductive signals include a wide range of microbes and microbial products, some tumor cells, and numerous cytokines, often in synergistic combinations [50]. The stimuli which lead to the upregulation of iNOS vary with cell type and species and in some cases an array of cytokines is necessary for induction. In rodent macrophages, interferon gamma is a potent inducer and this effect is potentiated by tumor necrosis factor α (TNFα), interleukin-1 (IL-1), and lipopolysaccharide [50,75–77]. There are a number of agents that are also capable of suppressing iNOS induction, including glucocorticoids, macrophage deactivating factor, isoforms of transforming growth factor beta (TGF-β), platelet-derived growth factor (PDGF), epidermal growth factor, IL-4 and IL-10 [50,75,76]. For reasons which are not understood, inducible NOS appears to be particularly sensitive to changes in extracellular arginine [78].

2.5. Regulation of Nitric Oxide Synthases

Different physiologic agents may upregulate or downregulate the activity of NOS. Agonists may increase the synthesis of NO within seconds or a few minutes by acting on constitutive NOS (cNOS) without affecting transcription or translation. This effect requires calmodulin binding, an event controlled by the level of intracellular calcium. Agonists in this category include bradykinin, acetylcholine, leukotrienes, platelet-activating factor, excitatory amino acids, and calcium ionophores. "Stretch" and electric stimulation have an important role in endothelium-dependent vasodilatation [50,79]. Other cells which respond with a rapid and transitory release of NO in response to such agonists include peripheral blood neutrophils, mast cells, and some neurons [50,66,80]. No evidence for downregulation of cNOS, once these enzymes are activated, is available. However, at the mRNA level, tumor necrosis factor α (TNFα) may downregulate cNOS protein expression activity in endothelial cells by shortening the mRNA half-life [81,82]. It has been shown that NO *per se* can downregulate NOS in the cerebellum [83]. Agents that increase NO synthesis through inducible nitric oxide synthases (iNOS) do so within hours. Their action may be prevented by inhibiting transcription and translation using drugs such as actinomycin D and cycloheximide.

2.6. Mechanism of Action of Nitric Oxide

Nitric oxide is a highly diffusable gas within and between cells and exerts its actions by binding to a diversity of target molecules. The major target sites are metal- and thiol-containing proteins. Nitric oxide has a higher affinity for iron, both heme and nonheme iron in the prosthetic groups of protein. It is the affinity for heme groups which leads to the activation of soluble guanylyl cyclase and the synthesis of cyclic GMP. Other heme enzymes to which nitric oxide binds include cyclooxygenase, the activity of which is increased [84], and nitric oxide synthase itself, which may be inhibited. The binding of NO to iron-containing enzymes is also a major mechanism of cytotoxicity; it binds to iron–sulfur clusters and enzymes of the mitochondrial electron transport chain, inhibiting mitochondrial respiration and *cis*-aconitase. A reaction of NO with thiol groups may also regulate protein functions. Some thiol-containing proteins, such as plasminogen activator, are upregulated by nitrosylation, whereas several enzymes, such as protein kinase C, cathepsin B, aldolase, and glyceraldehyde 3-phosphate dehydrogenase, are inhibited. Formation of S-nitroso thiols may also be a mechanism by which the activity of NO, as, for example, a vasodilator, is maintained for longer periods than would be expected for free NO [85].

2.7. Synthesis of Agmatine

Agmatine [4-amino butyl guanidine] has long been known to be a biochemical component in bacteria, but it was not recognized as a product in mammalian systems until recently. Agmatine and arginine decarboxylase, which catalyzes the formation of agmatine from arginine (Fig. 6-6), were recently shown to be present in mammalian brain [86]. Previous studies had identified a component

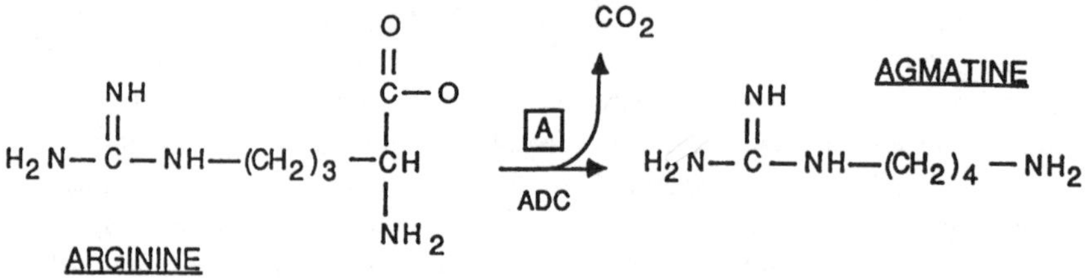

Figure 6-6. Synthesis of agmatine from arginine catalyzed by the enzyme arginine decarboxylase (ADC).

of calf brain that competed with clonidine for binding to α_2-adrenergic receptors [87] and which also bound to imidazoline receptors [88,89]. The latter class of receptors is found in the central nervous system, but an endogenous ligand for these receptors had not previously been identified. This "clonidine-displacing substance" was a candidate molecule for the endogenous imidazoline receptor ligand. Li et al. [86] purified a derivatized form of clonidine-displacing substance which on high-performance liquid chromatographic (HPLC) analysis coeluted with the derivatized form of agmatine. Agmatine and the clonidine-displacing substance exhibited closely similar properties of binding to imidazoline and α_2-adrenoreceptors in vitro and closely similar abilities to stimulate epinephrine or norepinephrine release in a dose-dependent manner in a α_2-adreno receptor activity assay. This evidence suggests that agmatine may be the endogenous imidazoline receptor ligand and that it is also a noncatecholamine ligand for α_2-adrenergic receptors.

Using homology-based polymerase chain-reaction amplification, we have demonstrated the presence of arginine decarboxylase mRNA (Table 6-2) in tissues involved in arginine metabolism (brain, kidney, gut, adrenal gland, and liver of the rat) but not in organs (lung, heart) in which arginine metabolism is low or absent [90]. The polymerase chain-reaction product from the kidney had a nucleotide sequence 61% identical to that of the *Escherichia coli* biosynthetic arginine decarboxylase. On a whole-tissue basis, kidney homogenates were three times more active than brain homogenates at decarboxylating [^{14}C]-labeled arginine. Subcellular fractionation localized the arginine decarboxylase activity of the kidney to the mitochondrial fraction. L-Agmatine, the product of arginine decarboxylation, was found to inhibit nitric oxide formation by postmitochondrial supernatants of the brain or kidney. Thus, it appears that arginine is metabolized to two structurally different signaling molecules, nitric oxide and agmatine. Furthermore, preliminary evidence indicates that agmatine can influence the nitric oxide synthase pathway. Measurements of the levels of agmatine using stable isotope dilution gas chromatography and negative-ion chemical ionization mass spectrometry have revealed that the highest concentration of agmatine per unit protein mass is present in the kidney. As mentioned above, the mRNA for

Table 6-2. Arginine Decarboxylase Activity of Rat Brain and Kidney

Tissue	Activity (pmol CO_2/h/mg protein)
Whole brain	74 ± 3
Whole kidney	245 ± 21, $p < .004$
Kidney cortex	199 ± 38
Glomeruli	41 ± 2
Tubules	165 ± 24, $p < .02$
Kidney outer medulla	
Outer stripe	240 ± 70
Inner stripe	184 ± 18
Kidney inner medulla	Not greater than blank values

Note: Values shown are the mean ± SD of three separate determinations. A heat-inactivated blank "activity" was routinely subtracted from each value. This blank usually amounted to 10–15 pmol/h/mg protein.

Source: From Morrissey, J., McCracken, R., Ishidoya, S., and Klahr, S. Partial cloning and characterization of an arginine decarboxylase in the kidney. *Kidney Int.* **47,** 1458–1461 (1995); used with permission.

mammalian arginine decarboxylase was characterized and found to be abundant in rat kidney compared to other rat tissues [90]. Endogenous production of agmatine may contribute to the regulation of kidney function through binding to imidazoline and α_2 adrenergic receptors. Indeed, imidazoline receptor agonists increased urinary flow and sodium excretion in rats [91]. The effects on sodium excretion may result from inhibition of a renal tubular sodium-proton exchanger [92].

Early work in plants demonstrated that agmatine was avidly degraded by the enzyme diamine oxidase, which converts agmatine to guanidino butylaldehyde [93]. Novotny et al. isolated diamine oxidase and found that the amino acid sequence was identical to that of amiloride-binding protein [94]. Amiloride and amiloride analogs significantly inhibit diamine oxidase by binding at the active site of the enzyme. It is of interest that aminoguanidine, a structurally simplified version of agmatine which is known to inhibit inducible NOS [95,96], is also a potent inhibitor of diamine oxidase. Agmatine is a biologically active substance, but the mode and site of its action have not been fully defined. Recent studies suggest that this alternate pathway of arginine metabolism is of physiological importance to renal function [97]. The perfusion of agmatine into the renal interstitium and into the urinary space of surface glomeruli of Wistar-Fromter rats produced reversible increases in nephron filtration rate and in absolute proximal reabsorption [97]. Renal denervation did not alter the effects on nephron filtration rate but prevented the changes in absolute proximal reabsorption.

2.8 The Metabolism of L-Arginine Through the Arginase Pathway

In addition to the L-arginine–nitric oxide pathway and the L-arginine decarboxyl-ation-agmatine pathway, L-arginine is also metabolized to L-ornithine and urea (Fig. 6-4) by the enzyme arginase [98,99]. L-Ornithine is further metabolized to polyamines—putrescein, spermin, spermadine—and L-proline (see Fig. 6-7) [99]. Polyamines play an important role in the regulation of cellular proliferation and DNA and RNA synthesis, and L-proline is a substrate for collagen synthesis; both pathways may be involved in tissue repair [100–103]. However, little is known about the influence of cytokines on these pathways. The rate-limiting enzyme for the polyamine synthesis, ornithine decarboxylase, and the first enzyme of L-proline synthesis, ornithine aminotransferase, are partially controlled by hypophyseal hormones and their activity follows a circadian rhythm. Thus, it is possible that insulinlike growth factor 1, which is a mediator of growth hormone effects in tissues, might play a role in the activation of these enzymes [104]. Recently, it was shown that cultured mesangial cells expressed ornithine decar-boxylase activity following stimulation with the platelet-derived growth factor and that the increased ornithine decarboxylase activity caused mesangial cells to

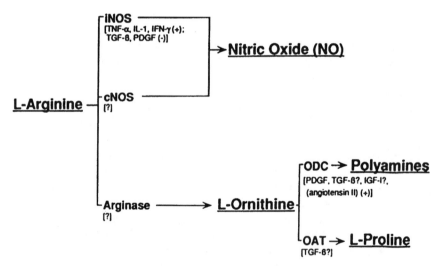

Figure 6-7. Schematic overview of pathways of L-arginine metabolism. Cytokines in-volved in enzyme regulation are shown in brackets. iNOS, inducible NO synthase; cNOS, constitutive, calmodulin-dependent NO synthase; ODC, ornithine decarboxylase; OAT, ornithine aminotransferase; TNF-α, tumor necrosis factor-α; IL-1, interleukin-1; IFN-γ, interferon-γ; TGF-β, transforming growth factor-β; PDGF, platelet-derived growth factor; IGF-1, insulinlike growth factor 1. +, stimulation; –, inhibition. [From Ketteler, M., Border, W.A., and Noble, N.A. Cytokines and L-arginine in renal injury and repair. *Am. J. Physiol.* **267** (Renal Fluid Electrolyte Physiol. 36), F197–F207 (1994). Used with permission.]

proliferate [105]. Little is known about the effects of cytokines on arginase expression or activity [106].

2.9. Synthesis of Orotic Acid

L-Arginine levels may also influence the synthesis of pyrimidines necessary for cell and tissue growth by modulating the generation of orotic acid, the first precursor in this metabolic pathway [107–109]. The nitrogen generated from protein metabolism is converted to urea in the liver, a metabolic pathway in which L-arginine has a key role as a donor of the two nitrogen atoms of urea [107]. A relative deficiency of L-arginine may impair the disposal of nitrogen through urea generation. When the amount of protein metabolized generates nitrogen in excess of the L-arginine available in the liver, this may lead to retention of ammonia in plasma and accumulation of carbamoyl phosphate, the mitochondrial precursor of urea. Carbamoyl phosphate may diffuse into the cytoplasm and eventually leak out of the hepatocyte to promote the synthesis of orotic acid [107–110], the initial step in the synthesis of pyrimidine nitrogen bases and RNA [111–114]. Greater availability of orotic acid stimulates the synthesis of nucleic acid pyrimidines and their sugar-containing derivatives [108], all of which may promote growth [111]. Increased excretion of orotic acid in the urine suggests a deficiency of L-arginine. Increased production and excretion of orotic acid in the urine may also occur as a result of enzymatic defects in the urea cycle. Interestingly, plasma levels of orotic acid are increased sevenfold in patients with end-stage renal disease, compared with normal subjects, and decreased to only 60% of normal levels after hemodialysis [115]. The pathophysiological implications of the increased levels of orotic acid in this setting remain to be explored.

3. Effect of L-Arginine on the Release of Hormones

L-Arginine has a marked effect on the release of hormones. Administration of L-arginine stimulates the secretion of insulin, glucagon, growth hormone, and endogenous steroids [116–120], all of which can, alone or in combination, affect a vast array of biological processes. Thus, the mechanisms of a given response observed after L-arginine administration may reflect interactions between the different pathways in which L-arginine has a key role and/or the multiple effects these amino acids have on hormonal release. Glucocorticoids exert their antiinflammatory actions through a variety of effects including inhibition of cytokine release. In addition, glucocorticoids have been shown to inhibit directly the induction of iNOS [121] via downregulation of protein expression [122]. Thus, the action of glucocorticoide can be similar to that achieved with substrate inhibition of cNOS, with the difference that glucocorticoids prevent the induction of the enzyme, whereas cNOS inhibitors inhibit activity [123]. In vivo, the metabolism of L-arginine may be assessed by determination of plasma levels of L-arginine, L-citrulline, ammonia, urea, and creatinine, and by the determination of the excretion in the urine of cyclic GMP, nitrate plus nitrite, orotic acid, urea, and creatinine.

References

1. Reyes, A.A., Karl, I.E., and Klahr, S. *Am. J. Physiol.* **267** (Renal Fluid Electrolyte Physiol. 36), F331–F346 (1994).

2. Schulze, E. and Steiger, E. *Z. Physiol. Chem.* **11**, 43–65 (1886).

3. Schulze, E. and Steiger, E. *Ber. Dtsch. Chem. Ges.* **19**, 1177–1180 (1886).

4. Borsook, H. and Dubnoff, J.W. *J. Biol. Chem.* **141**, 717–738 (1941).

5. Cohen, P.P. and Hayano, M. *J. Biol. Chem.* **166**, 239–250 (1946).

6. Windmueller, H.G. and Spaeth, A.E. *Am. J. Physiol.* **241**, E473–E480 (1981).

7. Hamano, Y., Kodama, H., Yanagisawa, M., Haraguchi, Y., Mori, M. et al. *J. Histochem. Cytochem.* **36**, 29–35 (1988).

8. Ryall, J., Nguyen, M., Bendayan, M., Shore, G.C. *Eur. J. Biochem.* **152**, 287–292 (1985).

9. Morris, S.M. *Annu. Rev. Nutr.* **12**, 81–101 (1992).

10. Hurwitz, R. and Kretchmer, N. *Am. J. Physiol.* **251**, G103–G110 (1986).

11. Herzfeld, A. and Raper, S.M. *Biochim. Biophys. Acta* **428**, 600–610 (1976).

12. Malo, C., Qureshi, I.A., and Letarte, J. *Am. J. Physiol.* **250**, G177–G184 (1986).

13. Windmueller, H.G. and Spaeth, A.E. *J. Biol. Chem.* **255**, 107–112 (1980).

14. Sessa, W.C., Hecker, M., Mitchell, J.A., and Vane, J.R. *Proc. Natl. Acad. Sci. USA* **87**, 8607–8611 (1990).

15. Wu, G. and Meininger, C.J. *Am. J. Physiol.* **265** (Heart Circ. Physiol. 34), H1965–H1971 (1993).

16. Levillain, O., Hus-Citharel, A., Morel, F., and Bankir, L. *Am. J. Physiol.* **259** (Renal Fluid Electrolyte Physiol. 28), F916–F923 (1990).

17. Levillain, O., Hus-Citharel, A., Morel, F., and Bankir, L. *Am. J. Physiol.* **264** (Renal Fluid Electrolyte Physiol. 33), F1038–F1045 (1993).

18. Dhanakoti, S.N., Brosnan, J.T., Brosnan, M.E., and Herzberg, G.E. *J. Nutr.* **122**, 1127–1164 (1992).

19. Jansen, A., Lewis, S., Cattell, V., and Cook, H.T. *Kidney Int.* **42**, 1107–1112 (1992).

20. Konarska, L., Wiessmann, U., and Colombo, J.P. *Clin. Chim. Acta* **757**, 191–195 (1983).

21. Chen, P.C. and Broome, J.D. *Proc. Soc. Exp. Biol. Med.* **163**, 354–359 (1980).

22. Featherson, W.R., Rogers, Q.R., and Freedland, R.A. *Am. J. Physiol.* **224**, 127–129 (1973).

23. Dhanakoti, S.N., Brosnan, J.T., Herzberg, G.R., and Brosnan, M.E. *Am. J. Physiol.* **259** (Endocrinol. Metab. 22), E437–E442 (1990).

24. Castillo, L., Chapman, T.E., Yu, Y.M., Ajami, A., Burke, J.F., and Young, V.R. *Am. J. Physiol.* **265** (Endocrinol. Metab. 28), E532–E539 (1993).

25. Tizianello, A., De Ferrari, G., Garibotto, G., Gurrei, G., Robaudo, C., Acquarone, N., and Ghiggeri, G.M. *J. Clin. Invest.* **69**, 240–250 (1982).

26. Tizianello, A., De Ferrari, G., Garibotto, G., Gurrei, G., and Robaudo, C. *J. Clin. Invest.* **65**, 1162–1173 (1980).

27. Jungers, P., Chauveau, P., Ceballos, I., Bardet, J., Parvy, P., Hannedouche, T., and Kamoun, P. *J. Nephrol.* **7**, 48–54 (1994).

28. Garibotto, G., Deferrari, G., Robaudo, C., Saffioti, S., Paoletti, E., Pontremoli, R., and Tizianello, A. *Nephron* **64**, 216–225 (1993).

29. Laidlaw, S.A., Berg, R.L., Kopple, J.D., Naito, H., Walker, W.G., and Walser, M. *Am. J. Kidney Dis.* **23**, 504–513 (1994).

30. Krebs, H.A. and Henseleit, H. *Hoppe-Seyler's Z. Physiol. Chem.* **210**, 33–66 (1932).

31. Anderson, P.M. *Biochem. Cell Biol.* **69**, 317–319 (1991).

32. Folin, O. *Am. J. Physiol.* **13**, 66–113 (1905).

33. Lightbody, H.D. and Kleinman, A. *J. Biol. Chem.* **129**, 71–78 (1939).

34. Schimke, R.T. *J. Biol. Chem.* **237**, 459–468 (1962).

35. Schimke, R.T. *J. Biol. Chem.* **237**, 1921–1924 (1962).

36. Schimke, R.T. *J. Biol. Chem.* **238**, 1012–1018 (1963).

37. Brebnor, L.D., Grimm, J., and Balinsky, J.B. *Cancer Res.* **41**, 2692–2699 (1981).

38. Brown, C.L., Houghton, B.J., Souhami, R.L., and Richards, P. *Clin. Sci.* **43**, 371–376 (1972).

39. Christowitz, D., Mattheyse, F.J., and Balinsky, J.B. *Enzyme* **26**, 113–121 (1981).

40. Das, T.K. and Waterlow, J.C. *Br. J. Nutr.* **32**, 353–373 (1974).

41. Felipo, V., Minana, M.-D., and Grisolia, S. *Arch. Biochem. Biophys.* **285**, 351–356 (1991).

42. McIntyre, P., DeMartinis, M.L., and Hoogenraad, N. *Biochem. Int.* **6**, 365–373 (1983).

43. Nuzum, C.T., and Snodgras, P.J. *Science* **172**, 1042–1043 (1971).

44. Munro, H.N. and Crim, M.C. The proteins and amino acids. In: *Modern Nutrition in Health and Disease.* Eds. Shils, M.E. and Young, V.R. Lea & Febiger, Philadelphia, 1988, pp. 8–9.

45. Levey, A.S. *Kidney Int.* **38**, 167–184 (1990).

46. Jones, J.D. and Burnett, P.C. *Kidney Int.* **7**, S294–S298 (1975).

47. Mitch, W.E., Collier, V.U., and Walser, M. *Clin. Sci. Lond.* **58**, 327–335 (1980).

48. Langrehr, J.M., Hoffman, R.A., Lancaster, J.R., and Simmons, R. *Transplantation* **55**, 1205–1212 (1993).

49. Moncada, S. and Higgs, A. *N. Engl. J. Med.* **329**, 2002–2012 (1993).

50. Nathan, C. *FASEB J.* **6**, 3051–3064 (1992).

51. Stamler, J.S., Jaraki, O., Osborne, J., Simon, D.L., Keaney, J., Vita, J., Singel, D., Valeri, C.R., and Loscalzo, J. *Proc. Natl. Acad. Sci. USA* **89**, 7674–7677 (1992).

52. Palmer, R.M.J., Ashton, D.S., and Moncada, S. *Nature* **333**, 664–666 (1988).

53. Albina, J.E., Caldwell, M.D., Henry, W.L., and Mills, C.D. *J. Exp. Med.* **169**, 1021–1029 (1979).

54. Bredt, D.S., Hwang, P.M., Glatt, C.E., Lowenstein, C., Reed, R.R., and Snyder, S.H. *Nature* **351**, 714–718 (1991).

55. Liu, L., Liu, G., and Barajas, L. *J. Am. Soc. Nephrol.* **4**, 558 (1993).

56. Lowenstein, C.J., Dinerman, J.L., and Snyder, S.H. *Ann. Intern. Med.* **120**, 227–237 (1994).

57. Knowles, R.G. and Moncada, S. *Biochem. J.* **298**, 249–258 (1994).

58. Nathan, C. and Xie, Q. *Cell* **78**, 915–918 (1994).

59. Marletta, M.A. *Cell* **78**, 927–930 (1994).

60. Tayeh, M.A. and Marletta, M.A. *J. Biol. Chem.* **264**, 19654–19658 (1989).

61. Kwon, N.S., Nathan, C.F., and Stuehr, D.J. *J. Biol. Chem.* **264**, 20496–20501 (1989).

62. Stuehr, D.J., Cho, H.J., Kwon, N.S., Weise, M., and Nathan, C.F. *Proc. Natl. Acad. Sci. USA* **88**, 7773–7777 (1991).

63. Klatt, P., Heinzel, B., John, M., Kastner, M., Boehme, E., and Mayer, B. *J. Biol. Chem.* **267**, 11374–11378 (1992).

64. Schmidt, H.H.H.W., Pollock, J.S., Nakane, M., Gorsky, L.D., Forstermann, U., and Murad, F. *Proc. Natl. Acad. Sci. USA* **88**, 365–369 (1991).

65. Yui, Y., Hatori, R., Eizawa, H., Hiki, K., Okhawa, S., Ohnishi, K., Terao, S., and Kawai, C. *J. Biol. Chem.* **266**, 3369–3371 (1991).

66. Marletta, M.A. *J. Biol. Chem.* **268**, 12231–12234 (1993).

67. Dimmeler, S. and Brune, B. *Eur. J. Biochem.* **210**, 305–310 (1992).

68. Dinerman, J.L., Dawson, T.M., Schell, M.J., Snowman, A., and Snyder, S.H. *Proc. Natl. Acad. Sci. USA* **91**, 4214–4218 (1994).

69. Oswald, I.P., Eltoum, I., Wynn, T.A., Schwartz, B., Caspar, P., Paulin, D., Sher, A., and James, S.L. *Proc. Natl. Acad. Sci. USA* **91**, 999–1003 (1994).

70. Cho, H.J., Xie, Q.W., Calaycay, J., Mumford, R.A., Swiderek, K.M., Lee, T.D., and Nathan, C. *J. Exp. Med.* **176**, 599–604 (1992).

71. Dawson, V.L., Dawson, T.M., Uhl, G.F., and Snyder, S.H. *Proc. Natl. Acad. Sci. USA* **90**, 3256–3259 (1993).

72. Hu, J. and El-Fakahany, E.E. *Neurochemistry* **4**, 760–762 (1993).

73. Sladek, S.M., Regenstein, A.C., Lykins, D., and Roberts, J.M. *Am. J. Obstet. Gynecol.* **169**, 1285–1291 (1993).

74. Nathan, C. and Xie, Q.-W. *J. Biol. Chem.* **269**, 13725–13728 (1994).

75. Pfeilschifter, J., Kunz, D., and Mhhl, H. *Nephron* **64**, 518–525 (1993).

76. Cattell, V. and Cook, H.T. *Exp. Nephrol.* **3**, 265–280 (1993).

77. Lamas, S., Michel, T., Tucker, C., Brenner, B.M., and Marsden, P.A. *J. Clin. Invest.* **90**, 879–887 (1992).

78. Baydoun, A.R., Bogle, R.G., Pearson, J.D., and Mann, G.E. *Br. J. Pharmacol.* **110**, 1401–1406 (1993).

79. Forstermann, U., Pollock, J.S., and Nakane, M. *Trends Cardiovasc. Med.* **3**, 104–110 (1993).

80. Snyder, S.H. *Science* **257**, 494–496 (1992).

81. Nishida, K., Harrison, D.G., Navas, J.P., Fisher, A.A., Dockery, S.P., Uematsu, M., Nerem, R.M., Alexander, R.W., and Murphy, T.J. *J. Clin. Invest.* **90**, 2092–2096 (1992).

82. Yoshizumi, M., Perrella, M.A., Burnett, J.C., and Lee, M.E. *Circ. Res.* **73**, 205–209 (1993).

83. Rengasamy, A. and Johns, R.A. *Mol. Pharmacol.* **44**, 124–128 (1993).

84. Salvemini, D., Misko, T.P., Masferrer, J.L., Seibert, K., Currie, M.G., and Needleman, P. *Proc. Natl. Acad. Sci. USA* **90**, 7240–7244 (1993).

85. Scharfstein, J.S., Keaney, Jr., J.F., Slivka, A., Welch, G.N., Vita, J.A., Stamler, J.S., and Loscalzo, J. *J. Clin. Invest.* **94**, 1432–1439 (1994).

86. Li, G., Regunathan, S., Barrow, C.J., Eshraghi, J., Cooper, R., and Reis, D.J. *Science* **263**, 966–969 (1994).

87. Atlas, D. and Burstein, Y. *Eur. J. Pharmacol.* **144**, 287–293 (1984).

88. Ernsberger, P., Meeley, M.P., Mann, J.J., and Reis, D.J. *Eur. J. Pharmacol.* **134**, 1–13 (1987).

89. Ernsberger, P., Meeley, M.P., and Reis, D.J. *Brain Res.* **441**, 309–318 (1988).

90. Morrissey, J., McCracken, R., Ishidoya, S., and Klahr, S. *Kidney Int.* **47**, 1458–1461 (1995).

91. Li, P., Penner, S.B., and Smyth, D.D. *Br. J. Pharmacol.* **112**, 200–206 (1994).

92. Bidet, M., Poujeol, P., and Parini, A. *Biochim. Biophys. Acta* **1024**, 173–178 (1990).

93. Smith, T. *Anal. Biochem.* **92**, 331–337 (1979).

94. Novotny, W.F., Chassande, O., Baker, M., Lazdunski, M., and Barbry, P. *J. Biol. Chem.* **269**, 9921–9925 (1994).

95. Yagihashi, S., Kamijo, M., Baba, M., Yagihashi, N., and Nagai, K. *Diabetes* **41**, 47–52 (1992).

96. Edelstein, D. and Brownlee, M. *Diabetes* **41**, 26–29 (1992).

97. Lortie, M.J., Novotny, W.F., Peterson, O.W., Vallon, V., Malvery, K., Mendonci, M., Satriano, J., Insel, P., Thomson, S.C., and Blantz, R.C. *J. Clin. Invest.* **97**, 413–420 (1996).

98. Barbul, A. *J. Parenter. Enteral Nutr.* **10,** 227–238 (1986).

99. Shih, V.E. *Enzyme* **26,** 254–258 (1981).

100. Davis, R.H., Morris, D.R., and Coffino, PP. *Microbiol. Rev.* **56,** 280–290 (1992).

101. Pegg, A.E. *Cancer Res.* **48,** 759–774 (1988).

102. Rojkind, M. and DeLeon, L.D. *Biochim. Biophys. Acta* **217,** 512–522 (1970).

103. Smith, R.J. and Phang, J.M. *Metabolism* **27,** 685–694 (1978).

104. Hammerman, M.R. and Miller, S.B. *Am. J. Physiol.* **265** (Renal Fluid Electrolyte Physiol. 34), F1–F14 (1993).

105. Schulze-Lohoff, E., Brand, K., Fees, H., Netzker, R., and Sterzel, R.B. *Kidney Int.* **40,** 684–690 (1991).

106. Ketteler, M., Border, W.A., and Noble, N.A. *Am. J. Physiol.* **267,** F197–F207 (1994).

107. Brusilow, S.W. and Horwich, A.L. Urea cycle enzymes. In: *The Metabolic Basis of Inherited Disease,* 6th ed. Eds. Scriver, A.L., Beaudet, A.L., Sly, W.S., and Valle, D. McGraw-Hill, St. Louis, MO, 1989, pp. 629–663.

108. Milner, J.A. and Visek, W.J. *Nature* **245,** 211–213 (1973).

109. Visek, W.J. *Cancer Res.* **52,** 2082S–2084S (1992).

110. Shoemaker, J.D. and Visek, W.J. *Exp. Mol. Pathol.* **50,** 371–384 (1989).

111. Cortes, P., Dumler, F., Paielli, D.L., and Levin, N.W. *Am. J. Physiol.* **255** (Renal Fluid Electrolyte Physiol. 24), F647–F655 (1988).

112. Donohoe, J.A., Rosenfeldt, F.L., Munsch, C.M., and Williams, J.F. *Int. J. Biochem.* **25,** 163–182 (1993).

113. Rogers, Q.R. and Visek, W.J. *J. Nutr.* **115,** 505–508 (1985).

114. Rudolph, F.B. *J. Nutr.* **124,** 124S–127S (1994).

115. Daniewska-Michalska, D., Motyl, T., Gellert, R., Kukulska, W., Podgurniak, M., Opechowska-Pacocha, E., and Ostrowski, K. *Nephron.* **64,** 193–197 (1993).

116. Alba-Roth, J., Muller, A., Schopohl, J., and von Werder, K.K. *J. Clin. Endocrinol Metab.* **67,** 1186–1189 (1988).

117. Barbul, A., Rettura, G., Levenson, S.M., and Seifter, E. *Am. J. Clin. Nutr.* **37,** 786–794 (1993).

118. Castellino, P., Hunt, W., and DeFronzo, R.A. *Kidney Int.* **32,** 15S–20S (1987).

119. Mulloy, A.L., Kari, F.W., and Visek, W.J. *Horm. Metab. Res.* **14,** 471–475 (1982).

120. Reyes, A.A., Purkerson, M.L., Karl, I., and Klahr, S. *Am. J. Kidney Dis.* **20,** 168–176 (1992).

121. Radomski, M.W., Palmer, R.M.J., and Moncada, S. *Proc. Natl. Acad. Sci. USA* **87,** 10043–10047 (1990).

122. Geller, D.A., Nussler, A.K., DiSilvio, M., Lowenstein, C.J., Shapiro, R.A., Wang, A.C., Simmons, R.L., and Bilias, T.R. *Proc. Natl. Acad. Sci. USA* **90,** 522–526 (1993).

123. Moncada, S. and Higgs, E.A. *FASEB J.* **9,** 1319–1330 (1995).

7

Distribution of NOSs in the Kidney

Sebastian Bachmann

1. Nitric Oxide Synthase Isoforms in the Kidney

A variety of renal cell types are capable of synthesizing nitric oxide (NO). Three isoforms have been identified to date. There are the constitutive isoforms NOS I [also termed neuronal (n) or brain (b) type NOS based on its first identification in the brain of various species] and NOS III [or endothelial (ec) NOS, although this isoform is not solely expressed in endothelia but has now also been detected in epithelial cells, see below], and the inducible isoform NOS II (or iNOS) which appears to occur in the kidney as a macrophage-type and a vascular smooth muscle cell-type NOS II [1]. NOS I and NOS III, although constitutively expressed, are quiescent until activated by increased Ca^{2+}_i levels that sustain calmodulin binding [2–6]. In contrast, NOS II, the inducible form, is present after transcriptional activation by cytokines or lipopolysaccharide (LPS), the principal component of bacterial endotoxin [7–11]. This NOS isoform remains active for longer periods and does not require Ca^{2+}_i levels for its activation; exceptions to this definition were reported for the kidney as detailed below.

2. Techniques to Localize NOS Isoforms

Localization of NOS isoforms in the kidney may be performed with a variety of techniques. NOS protein may be measured simply in kidney homogenates with isoform-specific antibodies using dot blots, Western blots, or ELISA. With the help of specific molecular probes, RNAase protection assay, Northern blots, or dot blot techniques have been applied to determine isoform-specific NOS mRNA expression. By separating renal zones, the regional distribution of NOS isoforms can be studied with these techniques [12]. The same techniques may be used for the analysis of renal cell lines or primary cell cultures [13,14]. In a highly sensitive approach, relative levels of specific NOS mRNA have been analyzed in glomeruli and single microdissected segments of the renal tubule using polymerase chain reaction (PCR) coupled to reverse transcriptase (RT-PCR) [15,16]. Histochemical demonstration of NOS isoforms can be performed on renal tissue sections by the NADPH-diaphorase (NADPH-d) reaction, immunohistochemistry, and in situ hybridization [17]. The NADPH-d reaction detects an enzyme associated with the NOS molecule but does not distinguish between the isoforms of NOS [18,19]. Of note, the intensity of the NADPH-d reaction varies with the NADPH-d activity of NOS, thereby indicating NOS enzyme activity at a given time point of fixation or freezing of the tissue. The reaction with NADPH is specific to NOS only if no other NADPH-requiring enzymes are present in a given cell. Immunocytochemical methods have been applied using isoform-specific monoclonal and polyclonal antibodies that were raised against purified NOSs and biochemically characterized [3,20–26]. Molecular cloning of NOS isoforms has provided specific sequences used for biochemical and histochemical probing. The cytosolic detection of specific mRNAs coding for NOS isoforms has been achieved using nonisotopically labeled riboprobes for in situ hybridization [25,27,28].

3. NOS in the Renal Vasculature

3.1. General Aspects of Vascular NOS

The formation of nitric oxide from L-arginine in blood vessels is catalyzed predominantly by the endothelial NOS III isoform [3,29–31]. In addition, the formation of NO can be elicited in most vascular cell types when mediators of vascular injury or inflammation such as interleukin-1β (IL-1β) and tumor necrosis factor α (TNFα), or bacterial endotoxin are present [31–33]. Most often, the vascular smooth muscle cells were reported to produce NOS II on the effect of inducing stimuli, and NOS II has also been demonstrated in tumor vessels [35–37]. Likewise, evidence has been presented that induction of NOS II also occurs in the endothelium and that such an induction is dependent on the same factors that trigger the high-output pathway for NO in vascular smooth muscle cells. The constitutively expressed endothelial NOS III isoform normally acts in a Ca^{2+}/calmodulin-dependent manner, and regulation is thought to be governed mainly by changes in the intracellular concentration of calcium [3,38,39]; major factors that increase endothelial NOS activity in a calcium-dependent manner are physical stimuli (shear stress) or endocrine/paracrine substances such as bradykinin and acetylcholine [2,40,41]. In addition, recent studies examining the expression of NOS III under chronic stimuli such as exercise [2] and flow exposure [38,42,43] have provided evidence for a sustained elevation of NOS III gene expression in a cytokine-independent manner, and putative sheer stress-responsive elements have now been described in the NOS III promoter sequence [44–47]. Similarly, lysophosphatidylcholine, a component of atherogenic lipoproteins and atherosclerotic lesions, induced an elevation of NOS III mRNA levels and a concomitant increase of NOS III protein [48].

Subcellular localization of NOS III in endothelial cells has been the focus of several investigations. Biochemically, 90–95% of NOS III has been found in the particulate fraction of endothelial cells [2,3,44,49], but upon stimulation, phosphorylation and translocation may occur, leading to a shift toward the cytosolic fraction. High-resolution localization of endothelial NADPH-d activity has demonstrated reaction product in the endoplasmic reticulum and/or Golgi apparatus [25,50,51]. Ultrastructural analysis of NOS III immunostaining showed a clear association of both NADPH-d and NOS III immunoreactivity with perinuclear Golgi stacks and associated vesicles; NOS III was suggested to be exteriorized by fusion of these vesicles with the plasma membrane [52]. Others have shown the enzyme to be associated with the plasma membrane and membranes of cytoplasmic vesicles [53] or to be present within the cytoplasm [51,53]. A recent study described the presence of NOS III in renal mitochondria, where the enzyme was suggested to control oxidative phosphorylation; however, the particular origin of the mitochondria has not been given [54]. As an unexpected observation, also NOS I immunoreactive cells were described in endothelia [25]; in rabbit aorta, a small subpopulation of endothelial cells revealed NOS I staining in the cytoplasm and in association with ribosomes [24].

3.2. Renal Vasculature and Glomerular Tuft—Morphology

A brief morphological overview of the renal vasculature is given to cover the topographical background for NOS localization in renal blood vessels. The renal artery is dividing near the hilum and establishes interlobar arteries which then enter the renal parenchyma at the border between the cortex and medulla, where they are called arcuate arteries [55]. The latter give rise to the interlobular (cortical radial) arteries. Interlobar, arcuate, and interlobular veins accompany the corresponding arteries. From the interlobular arteries, the afferent arterioles arise to the glomerular tufts of the renal corpuscles (Fig. 7.1).

Within the glomerular tuft, three cell types may be encountered—endothelial cells, mesangial cells, and visceral epithelial cells. Their numerical ratio has been estimated to be 3 : 2 : 1 [56]. Endothelial cells consist of a body and large fenestrated peripheral parts; the fenestrations lack a diaphragm. The mesangium is established by mesangial cells and the mesangial matrix.

Mesangial cells are thought to be contractile cells; processes contain microfilaments that attach to the glomerular basement membrane by the interposition of extracellular microfibrils [55,57]. The mesangial matrix contains a variety of collagen types (type III–VI [55,58]). The mesangium is thought to form a supporting framework which maintains the structural integrity of the glomerular tuft. The visceral epithelium, which principally belongs to the nephron (see below),

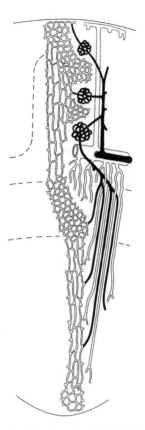

Figure 7-1. Distribution of endothelial constitutive NOS (indicated in black) in the renal vasculature. Vessels with positively stained endothelia are the arcuate interlobular and afferent arterioles, the glomerular capillaries, the afferent arterioles, and, originating from juxtamedullary efferent arterioles, the descending vasa recta. The remaining vessels and capillaries are negative. The broken lines are delineating cortex, medullary rays, outer and inner stripe of the outer medulla, and inner medulla. Results from Ref. 25. (Modified from Ref. 22.)

is constituted by the highly differentiated podocytes which give rise to long primary cell processes. The latter possess numerous foot processes forming the filtration slits. Podocytes contain the cytoskeletal components, vimentin, and bundles of microfilaments [55,59]. Both mesangium and podocytes are regarded as structure-stabilizing systems that maintain the folded pattern of the glomerular basement membrane and counteract the expansion of glomerular capillaries [58]. Failure of one system will lead to an involvement of the other so that these are considered the key events in glomerular pathology.

The efferent arterioles drain the glomerular capillary tufts, and those of superficial, midcortical, and juxtamedullary glomeruli must be distinguished. Whereas superficial and midcortical efferent arterioles supply the cortical peritubular capillaries, the efferent arterioles of juxtamedullary glomeruli turn toward the renal medulla where they divide into the descending vasa recta, which, together with the ascending vasa recta, form the vascular bundles of the medulla. The descending vasa recta traverse the inner stripe of the outer medula in cone-shaped vascular bundles from which single vasa recta branch off successively to supply the interstitium with capillaries. Ascending venous vessels of the medulla empty into the arcuate and interlobular veins.

The intrarenal arteries and proximal portions of the afferent arterioles are similar to vessels of equal size from elsewhere in the body [55]. Terminal portions of the afferent arterioles contain the granular (renin producing) cells which, together with the adjacent macula densa and extraglomerular mesangium, constitute the juxtaglomerular apparatus (JGA). The granular cells are transformed vascular smooth muscle cells, but normally contain relatively few myofilaments. Like the afferent arterioles, efferent arterioles possess a proper media of smooth muscle cells. Compared to the afferent arteriole, endothelial cells of the efferent arteriole show more profiles per cross section and are protruding into the lumen [55]; this pattern is particularly well developed in juxtamedullary efferent arterioles. In the descending vasa recta, the smooth muscle cells are gradually replaced by pericytes forming an incomplete layer of contractile cells. Ascending vasa recta have a fenestrated capillary wall structure; also, the walls of the large interlobular and arcuate veins are constituted accordingly and can therefore be classified as wide capillaries [55,60]

3.2. Constitutive NOS in the Renal Vasculature, Glomerular Tuft, and Interstitium

A comprehensive overview of vascular localization of NOS is given in Table 7-1. Despite the critical importance of NO in the regulation of renal microcirculation [40,61,62], relatively few studies are making reference to the particular localization and local activity of NOS isoforms in the kidney. At the mRNA level, constitutive NOS expression has been localized to the interlobular arteries [63] and arcuate arteries [15] after microdissection using RT-PCR. In the glomerulus, the presence of a constitutive NOS has been reported [15], and glomerular NOS III mRNA abundance in vivo was shown to be modulated by L-NAME, suggesting that not only the enzyme activity but also the amount of enzyme was downregulated by the NOS inhibitor [63]. At the histochemical level, NOS III localization in the renal vasculature has been performed with the aid of NADPH-d reaction and specific antibody staining [25,26,53]. In several mammalian species, including man, renal interlobar, arcuate, and interlobular arteries show significant NADPH-d staining, which, apart from a subpopulation of accompanying vascular nerves (see below), has been located exclusively in the endothelium [26], and

Table 7-1. *Distribution of Constitutive (c) NOS (Evidence for NOS I or NOS III or Both) and Inducible (i) NOS II in the Renal Vasculature*

Vascular Localization	cNOS Histochemical[a]	Ref.	cNOS Biochemical[a]	Ref.	iNOS Histochemical[a]	Ref.	iNOS Biochemical[a]	Ref.
Interlobular artery	+	25			–	28		
Arcuate artery	+	25	+	15	–	28	+	1
Interlobular artery	+	25, 53	+	63	–	28	+	1
Afferent arteriole	+	25, 26			+	26		
					–	28		
Efferent arteriole	+	25			–	28		
Mesangium	–	25			+	87	+	1, 82–84
					(+)	85		
Glomerular capillaries/glomerulus	+	25	(+)	15, 63	–	28		
Vasa recta	+	25	+	15	–	28	+	78
Peritubular capillaries	–	25			–	28		

[a]Refers to the assays used for detection of NOS.

staining was colocalized with specific NOS III immunostaining [25,53] (Figs. 7-2 and 7-3). Both signals were present also in the glomerular afferent and efferent arterioles [25,26], with preferential NADPH-d staining found in the efferent arteriole [25] (Fig. 7-4). Surprisingly, selective NOS I immunoreactivity was detected additionally in the efferent arteriolar endothelium; this finding was particularly clear in guinea pig kidney [25]. Regarding the particular microanatomy of the efferent arteriolar endothelium (see above), a shear stress-mediated activation of NOS could possibly be well accomplished in the efferent arteriole, so that high levels of NOS might be related with a well-developed susceptibility for flow changes in this segment. In rat and in other species, endothelial NADPH-d and NOS I and III signals are present already in the intraglomerular initial part of the efferent arteriole where NO may have a regulatory function in glomerular blood flow as well [64]. The preferential occurrence of NOS in the efferent arteriole is, however, somewhat controversial to functional studies which have indicated a preferential role for the afferent, and not the efferent arteriolar NO synthesis [65–67], but it is in agreement with another study on the microvasculature in the hydronephrotic kidney, where a preferential efferent reaction in response to NOS inhibition was registered [68].

In the glomerular capillaries, significant NADPH-d staining was obvious, and moderate antibody staining for NOS III was detected as well [25]. Together with

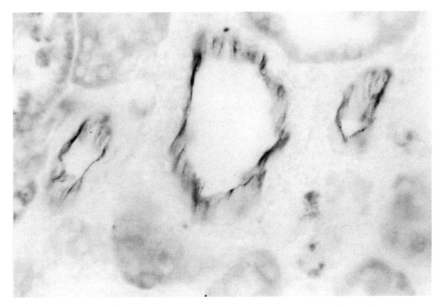

Figure 7-2. Vascular localization of NOS activity in human kidney. NADPH-diaphorase staining is present in the endothelia of one interlobular artery (center) and two glomerular afferent arterioles. Magnification: 740 ×.

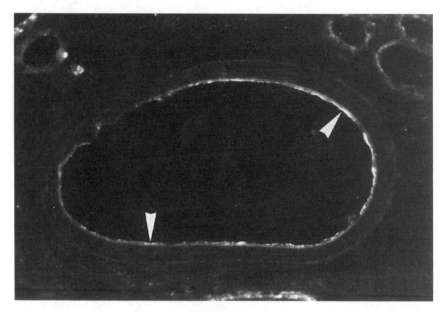

Figure 7-3. Vascular localization of NOS III immunoreactivity in the endothelium of an interlobular artery of rat kidney (arrowheads). Specific antibody against NOS III was applied (Ref. 53). Note that the smooth muscle layer is unreactive. Bound antibody was detected by Texas red epifluorescence. Magnification: 420 ×.

the results on glomerular constitutive NOS expression [15,63], a role for NOS with NO serving as a local regulator of intraglomerular capillary pressure appears possible; NO-induced effects could be mediated either by the mesangium or the podocytes [69]. In cell processes of the latter, significant immunoreactivity for the α_1 subunit of the NO "receptor," soluble guanylate cyclase, was detected [70]. Apart from glomerular capillaries, no other renal cortical capillaries have been shown to contain a signal indicating synthesis of NO; in the cortical interstitium, histochemical staining ends with the terminal branching of the efferent arteriole.

Regulation of endothelial NOS in the cortical renal vasculature could also be established with histochemical methods [71]. In two-kidney, one-clip Goldblatt hypertension in rats, a marked increase in NADPH-d signal was observed in interlobular and glomerular afferent arterioles of the contralateral, nonclipped kidney, whereas a decrease in signal was seen in the stenotic kidney [71]. The increase in NADPH-d signal was paralleled by an enhanced immunoreactivity for nitrotyrosine in the arteriolar wall, indicating a footprint for an enhanced release of NO from the endothelium into the vascular media [71,72]. It is currently thought that NO-dependent vasodilation in the Goldblatt condition serves to maintain contralateral renal perfusion despite elevated angiotensin II levels [73].

In the renal medulla, indirect evidence for NO-dependent effects has been

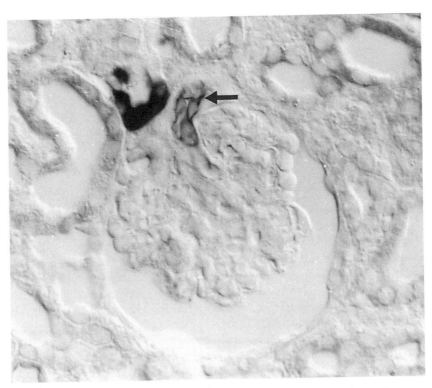

Figure 7-4. Localization of NADPH-diaphorase activity in the rat juxtaglomerular appa-ratus. NADPH-d positive macula densa is shown next to an efferent arteriole (arrow) in which NADPH-d label is apparent in the endothelium but absent from the smooth muscle layer. The efferent arteriole is identified by the large number of endothelial cells. Interfer-ence contrast microscopy. Magnification: 340 ×.

documented showing increased cGMP production in medullary tissue slices fol-lowing incubation with endothelium-dependent vasodilators [74], and stimulation in the inner medulla was more pronounced than in the outer medulla. The presence of endothelial-type NOS has been detected in rat renal medulla using the Western blot technique, and an upregulation of inner medullary endothelial NOS by 145% was encountered when rats maintained on a high compared to a low NaCl diet were compared [12]. It has been suggested that a relaxation of the vasa recta could account for the increase in medullary blood flow and interstitial pressure that induces natriuresis [40]. In more detail, microdissection of vasa recta bundles and evaluation by RT-PCR has identified the presence of constitutive NOS in these vessels [15]. Histochemically, descending vasa recta were shown to produce endothelial NADPH-d signal from the beginning at the branching of juxtamedul-lary efferent arterioles extending through the outer medulla down to the inner medulla [25]; Fig. 7-1); contrary to the dominant cGMP levels described for the inner medulla [40], the NADPH-d signal was more pronounced in the outer medulla so that it is possible that sources other than the vascular endothelium may contribute to medullary NO-dependent effects. NADPH-d staining has also been seen in medullary interstitial cells which may add to the local release of NO (Mundel and Bachmann, unpublished observations). In situ hybridization has recently identified NOS expression in these cells in normal rat; yet, these results were obtained with a NOS II probe (see below) and results were interpreted as to be based on a "steady-state" expression of inducible NOS.

3.3. Inducible NOS in the Renal Vasculature, Glomerular Tuft, and Interstitium

Contrary to the constitutive activity of NOS which serves to maintain basal vascular tone, to mediate a portion of the vasodilation secondary to hormones and amino acids, and possibly also to prevent contraction and proliferation of glomerular mesangial cells, inducible NOS II in the kidney is mostly activated in pathological conditions [75–78]. Thus, in the sepsis syndrome, high levels of NO synthesis have been reported in glomeruli [61]. In immune-mediated glomerular inflammation, increased NO production has been measured in rat [76,77]. Although, principally, renal NOS II should be expected to be exclusively present upon adequate induction, there are results on a "steady-state" presence of iNOS also under normal conditions [1,26,28,78].

The presence of NOS II in the arterioles of the renal cortex has not evidently been reported at the histochemical level. In the heart, NOS II expression in endotoxemic rats was most convincingly demonstrated in the media of myocardial arterioles [79], whereas no signal for NOS II mRNA was detected histochemically in renal vasculature irrespective of the presence or absence of endotoxemia [28], although the terminal afferent arteriolar portion had previously been reported to show iNOS immunoreactivity under normal conditions [26]; treatment with LPS had caused little difference in the latter result [26]. Biochemically, vascular smooth muscle-type NOS II has been identified in arcuate and interlobular arteries [1].

In the glomerulus, an important issue regarding NO synthesis during inflammation is the identity of the glomerular cell type that constitutes the source for NO. Cattell et al. [80] have shown that in glomerular injury, blood-borne macrophages are a dominant source for glomerular NO, but the concomitant participation of intrinsic glomerular cells has been considered as well, as macrophages may initially provide a source for the local release of cytokines which in the mesangium may then induce the release of the same cytokines, and in turn these could induce NOS by paracrine or autocrine mechanisms [61,81–83]. Mohaupt et al. have reported a macrophage-type NOS II to be tonically expressed in cultured mesangial cells, contrary to a vascular smooth muscle form of NOS II that was expressed after cytokine-dependent induction in mesangium but not in vascular media cells in situ [1,28]. However, histochemical analysis using NADPH-d reaction failed to show mesangial staining under physiologic conditions, except for single macrophages or blood monocytes that were occasionally detected in human kidney [25]. In rat specific antibody to NOS II failed to stain the unstimulated mesangium so (in mice, constitutive NOS II is immuno-detectable, Bachmann and Pfeilschifter, unpublished observations) [84,85], it has to be questioned whether functionally relevant amounts of NOS II are translated from mRNA detected under basal conditions. In unilateral immune complex glomerulonephritis induced by cationized IgG, immunohistochemistry and in situ hybridization probing for NOS II have mostly labeled glomerular macrophages, whereas clear evidence for mesangial labeling was absent; it was concluded that mesangial cells either did not synthesize NO in vivo in this model, or that the level of enzyme expression was very low [85]. By contrast, in anti-Thy-1 antibody glomerulonephritis, NOS II expressing cells were mostly polymorphonuclear leukocytes [86]. In endotoxin-treated rats, however, positive signal in the mesangial stalk has been shown by immunohistochemistry; positive cells were putatively identified as mesangial cells [87], and this result has been confirmed by in situ hybridization using NOS II antisense probe by Ahn et al. [28].

In the renal medulla, NOS II has been found after stimulation with LPS in the vasa recta bundles using RT-PCR [78]. Apart from NO synthesis by the renal tubule, as discussed below, medullary interstitial cells have recently been shown to contain vascular smooth muscle-type NOS II and to react upon stimulation by cytokines [88] and LPS [28]. Lau et al. concluded from their findings that NO may act as an autocrine activator of soluble guanylate cyclase in rat medullary interstitial cells [88].

4. NOS in the Renal Tubule

4.1. Glomerular Epithelia and Renal Tubule—Morphology

The specific structural unit of the kidney is the nephron. The nephron consists of the renal corpuscle which is connected to the renal tubule. Ontogenetically defined onset of the nephron epithelium is the highly specialized visceral epithelium of the glomerulus, which is described in Section 3.2. At the vascular pole, the visceral epithelium is reflected to become the squamous parietal epithelium of Bowman's capsule, which, at the urinary pole, transforms abruptly into the proximal tubule. In some species and particularly in some but not all rat strains, proximal tubule epithelium is lining significant portions of Bowman's capsule. Based on the location of the renal corpuscle in the cortex, superficial, midcortical, and juxtamedullary nephrons may be distinguished. The course of the renal tubule is arranged differently according to this classification [55]. The proximal tubule is divided into a convoluted part (PCT) located in the cortical labyrinth, and a straight part (PST) lying in the medullary rays (superficial and midcortical nephrons) and the outer medulla. Proximal straight parts of the juxtamedullary nephrons enter the outer medulla directly. Three subsegments, S1, S2, and S3, may be divided, and all are characterized by cellular interdigitation and a brush border, both of which vary in complexity from one subsegment to the next. The loop of Henle is composed of the straight part of the proximal tubule, a thin-walled intermediate tubule (thin limbs) and a thick ascending limb (TAL), or distal straight tubule with a medullary (mTAL) and a cortical (cTAL) portion. Long loops originating from juxtamedullary glomeruli may give rise to long descending thin limbs which descend to the inner medulla, and long ascending thin limbs which transform into the TAL at the border between inner and outer medulla. Short-looped nephrons have only short descending thin limbs, which are located in the inner stripe. Thin limb epithelia are flat and mostly lack significant amplifi-cations [89]. The epithelium of the mTAL consists of complex, interdigitating cells in the inner stripe, whereas the cTAL has a simpler structure. The terminal portions of the TAL contain the macula densa (MD), a cell plaque of roughly 20 polygonal cells which establish a contact to the extraglomerular mesangium at the vascular pole of its parent renal corpuscle. Shortly beyond the site of contact, the distal convoluted tubule (DCT) starts with an epithelium that is morphologically similar to that of the mTAL. The DCT and the cortical collecting duct (CCD) are connected by the connecting tubule (CNT). CNT and CCD cells are polygonal cells with mostly basally located membrane amplification. They intermingle with intercalated cells which exist in different forms and extend to the outer medullary collecting duct (OMCD) and the beginning of the inner medullary collecting duct (IMCD). Along the course of the collecting duct from the cortex to the inner medulla, the epithelium undergoes gradual changes; CCD cells are rather flat and reveal elaborate basal membrane folding, whereas IMCD cells are large and columnar with few membrane specializations, and, finally, the papillary collecting duct cells at the papillary tip bear some similarities with the papillary surface epithelium.

4.2. NOS in the Glomerular Epithelia

In the glomerular visceral epithelium, no histochemical evidence for the presence of NOS I has been reported [25]. Under inflammatory condition, only weak focal staining for NOS II has been encountered in the podocytes [85], but final evidence for a significant involvement of podocytes in NOS induction could not be deduced from that study.

The parietal epithelium of the renal corpuscle shows groups of NOS-I-positive cells [20,25] which have been identified by double-labeling histochemistry colo-calizing NOS I-antibody staining with NADPH-d reactivity. In the nephritic glomerulus, some of these cells were staining for NOS II as well [85]. A functional significance of NOS in the parietal epithelium is presently not evident.

4.3. NOS in the Proximal Tubule

Initially, it should be stated that a group of researchers from the University of Gainesville, Florida [1,26,28,102,113,116] have presented the majority of studies on NOS II distribution in the kidney. In these studies, apart from other issues they report "steady-state" expression of inducible NOS, which they found in almost every nephron segment of the normal rat kidney in the absence of inducing stimuli using RT-PCR, in situ hybridization, and immunohistochemistry. Few other groups have so far addressed this particular issue. A comprehensive overview of epithelial NOS localization is given in Table 7-2.

NOS I

A number of functional studies have emphasized a role for NO in renal volume control via a tubular effect, possibly by directly influencing tubular reabsorption [40,62]. NOS I mRNA has been found by Northern blot analysis in human proximal tubule primary cultures of different developmental stages [90] with elevated expression in old kidneys; this finding has been discussed in the context of ischemia/hypoxia-induced renal cell injury and the decreasing capacity to express osteopontin, which has putative effects protecting cells from NO by reducing its rate of synthesis [90]. In the porcine LLC-PK1 kidney epithelial cell line, which shares some properties with the proximal tubule, Ca^{2+}-dependent NOS activity was encountered, and the presence of NOS III was confirmed in the particulate fraction of these cells by Western blot [13,91]. In this context, it should be noted that the kidney is the principal source of arginine with the proximal tubule as the predominant site in the nephron [92]; local arginine production may, thus, potentially serve as a substrate for NO synthesis. In addition, soluble guanylate cyclase has been localized at the mRNA level in PCT and PST [15]. However, from a biochemical preparation using RT-PCR analysis of constitutive NOS mRNA in microdissected PST of rat kidney [15], negative results were reported, and likewise, although histochemically proximal tubules show intensive NADPH-d staining after longer exposure to the reaction, neither NOS I- nor NOS III-specific antibody or in situ hybridization probes have provided evidence that the NADPH-d staining in the proximal tubule is specific for constitutive NOS activity [20,25]. Therefore, concepts that have suggested NO-based effects on proximal tubular sodium excretion and pressure natriuresis [93–95] still await further confirmation.

NOS II

Expression of the inducible NOS II isoform has been reported from cultured rat proximal tubule cells by RT-PCR following stimulation with TNF-α and IFN-γ, and release of nitrite was considered to be comparatively high [14]. In another

Table 7-2. Distribution of Constitutive NOS (Evidence for NOS I or NOS III or Both) and Inducible (i) NOS II in the Renal Epithelia

Epithelial Localization	cNOS Histochemical[a]	Ref.	cNOS Biochemical[a]	Ref.	iNOS Histochemical[a]	Ref.	iNOS Biochemical[a]	Ref.
Visceral glomerular epithelium	–	25			(+)	85		
Parietal glomerular epithelium	+	20,25			(+)	85		
Proximal tubule	–	25	+	13, 99	+	28	+	1, 14, 78, 96–97
			–	15			–	78
Intermediate tubule			+	15	+	28		
Thick ascending limb	–/+	25	–	15	+	28, 66	+	1, 78
Macula densa	+	20, 22, 25, 26, 103, 104	+	105	–	28		
Distal convoluted tubule	–	25			+	26, 28		
Connecting tubule					+	113		
Cortical collecting duct			(+)	15	+	28		
Medullary collecting duct			+	15	+	28		
Intercalated cells					+	28, 113	+	1, 14, 116
Papillary surface epithelium					+	28		

[a]Refers to the assays used for detection of NOS.

study, LPS-stimulated NOS activity in a similar preparation was reported to be Ca^{2+}/calmodulin dependent, but no isoform-specific identification has been performed [96]. Likewise, activation of iNOS by LPS and IFN-γ in cultured SV 40-transfected mouse proximal tubule cells led to enhanced levels of NO production [97]. This increase was suggested to be partially mediated by activation of the transcription factor NF-κB [98,99]. Induction of NOS II by LPS alone, however, is not as efficient in epithelia, as in macrophages, and it was speculated whether in the proximal tubule, activation of another transcription factor, c-rel, was required for maximal tissue-specific activation of NOS II [98]. Another study reported the tonical expression of a macrophage-type NOS II mRNA in microdissected rat proximal tubule [1], but levels were reported to be low compared to the distal tubule; however, Morrissey et al. [78] could not detect NOS II mRNA in microdissected outer medullary PST segments under normal unstimulated conditions. The effect of LPS stimulation on renal epithelial NOS I expression has also been studied by in situ hybridization and, contrary to the study by Morrissey et al., the S3 segment of the proximal tubule was found to express iNOS transcripts which were not significantly different under basal conditions and after stimulation [28]. It is thus conceivable that the weak NADPH-d staining observed in the proximal tubule of normal rats could be related to the presence of NOS II, but, so far, enzyme protein has not been detected in this condition. Likewise, a possible functional relevance of NOS II expression in the proximal tubule is presently not transparent. NO production does not seem to be sufficient to account for direct toxic effects in endotoxemia; however, in terms of tubular transport function, it has been recorded that Na/K ATPase activity after 24 h of LPS/IFN γ treatment was significantly decreased, indicating that an NO-dependent effect, possibly via formation of peroxynitrite, may lead to alterations in tubular sodium handling [97].

4.4. NOS in the Intermediate Tubule

So far, only one report has described the presence of constitutive NOS mRNA by RT-PCR in microdissected thin limbs from the rat inner medulla [15], and the same authors were able to localize soluble guanylate cyclase mRNA in the thin limbs as well; however, only moderate NOS mRNA levels were detected in these nephron segments. These results have so far not been confirmed by histochemistry.

NOS II has been localized, as well, in the thin limbs of normal rats using in situ hybridization; stimulation of NOS II was reported to cause a rather moderate decrease in signal [28].

4.5. NOS in the Distal Tubule

NOS I

In the thick ascending limb, absence of constitutive NOS has been reported following biochemical [15] and histochemical evaluation [25]. In rat and rabbit, however, single scattered cTAL cells in the vicinity of the macula densa displayed a NOS I signal of similar intensity as the MD itself [25]. Such a cellular heterogeneity of the terminal cTAL has also been found when mRNA expression for inducible cyclooxygenase-2 was probed; constitutive expression of this enzyme has been detected in the vicinity of the MD and in premacula segments (Bachmann, unpublished observations, [100]).

NOS II

The inducible form of NOS appears to be dominant in the rat renal medulla [12], although much of the increase of NOS mRNA after LPS treatment seems to be due to an increase in the vascular smooth muscle form of NOS II [1]. Basal macrophage-type NOS II gene expression measured by RT-PCR was high in the mTAL [1,78]. By in situ hybridization, mTAL and cTAL were reported to show a significant signal, the same as the cTAL by immunohistochemistry using antibody to a vascular smooth muscle-type in rat under basal conditions [26]. The capacity of the TAL to react upon stimulation of NOS II was considered to be weak compared to the collecting duct [14,78], and it has been speculated whether Tamm Horsfall glycoprotein, the major secretory product of the TAL with cytokine-binding capacity [101], prevents significant NOS II induction in this tubular segment [36,78]. By contrast, NOS II expression in a mouse mTAL cell line was shown to be effectively enhanced by combined endotoxin and cytokine treatment [102].

4.6 NOS in the Macula Densa

Several groups have identified the significant presence of NOS I mRNA (Fig. 7-5), immunoreactivity, and the concomitant localization of strong NADPH-d activity in the MD of several species, including man [20,22,25,26,103,104] (Fig. 7-6), and, using isolated glomeruli from rat kidney, NOS I mRNA has also been identified by RT-PCR in glomerulus-associated cells [16]. Ultrastructural analysis showed a general cytosolic localization of the enzyme [25], and some reactivity was associated also with cytoplasmic vesicles [26,105]. Of all renal epithelia, the NOS I signal in MD cells is by far the strongest, suggesting high amounts of NO to be released at the glomerular vascular pole. Synthesis of NO from the MD may be involved in the mediation or modulation of tubulo-vascular signal transfer directed from the MD (i.e., tubulo-glomerular feedback response and/or MD-dependent regulation of afferent arteriolar renin production and release) [16,27,103,107,108]. NOS I activity and gene expression were inversely related to chronic changes in renal perfusion, salt balance, and salt transport at the distal tubule, and these changes were paralleled by changes in renin expression

Figure 7-5. Localization of NOS I expression at the juxtaglomerular apparatus: NOS I mRNA in rat kidney macula densa is shown by nonisotopic in situ hybridization using digoxigenin-labeled antisense NOS riboprobe [25] and an alkaline–phosphatase detection system. Interference contrast microscopy. Magnification: 340 ×.

[16,27,106]; this relation is particularly intriguing when the intimate topographical relation between MD and the renin-producing cells (Fig. 7-7) is considered.

From microperfusion experiments, it was suggested that juxtaglomerular NO exerts an upward pressure on single nephron glomerular filtration rate and reduces the efficiency of the tubuloglomerular feedback system in stabilizing proximal tubular flow rate [109]. Synthesis of NO and of NOS seem to be governed by NaCl transport at the MD [16,110], which, in turn, is controlled by a Na,K,2Cl-cotransporter that has recently been localized in MD and TAL cells [111]. In two-kidney, one-clip Goldblatt hypertension, upregulation of NOS in MD of the stenotic kidney was so strong that tyrosine nitration product in the adjacent extraglomerular mesangium could be detected; this was taken as evidence for the release of NO from the MD toward the extraglomerular mesangium, and it was speculated that from the extraglomerular mesangium, a secondary stimulus could be transmitted to the glomerular vasculature [71]. During nephrogenesis, the early presence of NOS I in future MD cells has been detected in rat with peak expression at postnatal day 6 [112].

NOS II was absent from MD, as revealed by in situ hybridization both with and without LPS stimulation [28].

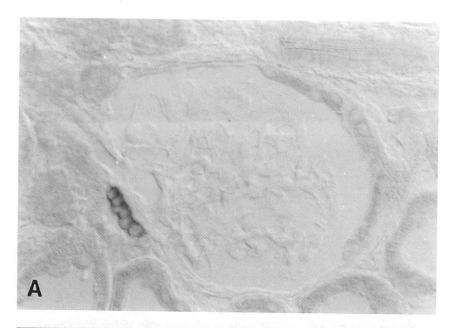

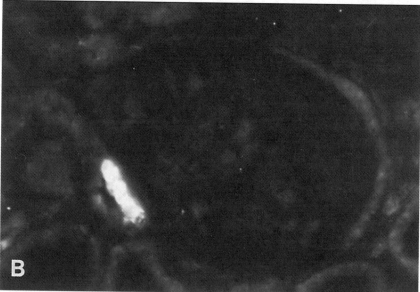

Figure 7-6. NOS I localization at the juxtaglomerular apparatus. Double-labeling by NADPH-diaphorase staining (A) and immunoreactivity for NOS I (B) shows positive reaction of macula densa in rat kidney. Interference contrast microscopy (A); in (B), specific antibody against NOS I was applied [20] and detected by Texas red epifluorescence. Magnification: 280 ×.

Figure 7-7. Renin mRNA expression by the juxtaglomerular apparatus. Renin mRNA is present in the wall of the preglomerular afferent arteriole in rat kidney. Nonisotopic in situ hybridization using digoxigenin-labeled specific riboprobe. Magnification: 320 ×.

4.7. NOS in the Distal Convoluted Tubule

NOS I was not detected in the DCT, whereas reports on NOS II expression [28] and immunoreactivity [26] have demonstrated signals in the DCT under basal conditions with no significant increase in mRNA signal after LPS stimulation [28].

4.8. NOS in the Connecting Tubule and Cortical Collecting Duct

In the connecting tubule, no evidence for the presence of NOS I has been reported so far [25], and a study on NOS II immunoreactivity mentions merely a faint staining in this segment [113].

The cortical collecting duct cells possess only weak levels of NOS I, as revealed by RT-PCR [15]. The CCD principal cells have not been found to produce NO, but they may be target to NO effects, because NO has been shown to inhibit ion transport by affecting apical membrane channels in CCD cells in culture [114]; similar results were obtained in microperfused CCD segments [115].

Intercalated cells may show significant NADPH-d activity which is particularly well developed in guinea pig kidney (Bachmann, unpublished observation). NADPH-d reactivity in these cells may be based on the presence of NOS II, as intercalated cells in rat CCD have been shown to contain NOS II by immunohisto-chemistry using antibody against a macrophage-type NOS II [113]. Strong staining was observed in the apical region of type A intercalated cells in the normal rat, whereas type B intercalated cells were reported to be weakly labeled [113]. NO from these cells is thought to be involved in the regulation of proton transport, possibly via activation of guanylyl cyclase [70,113].

4.9. NOS in the Medullary Collecting Duct

NOS I has been detected in microdissected outer medullary collecting duct of rat kidney and—to higher extent—in the initial as well as terminal inner medullary collecting duct [15]. The presence of guanylate cyclase transcripts were detected in these segments as well [15].

NOS II was shown by in situ hybridization to be localized in OMCD and IMCD under steady-state conditions [28]; the OMCD signal was found to be weak, whereas the IMCD segments showed an increase in signal toward the papillary tip, and prominent labeling was also noted in the medullary intercalated cells, which revealed an enhanced staining after endotoxemic stimulation [28]. These results have been confirmed using microdissected IMCD and RT-PCR analysis under basal conditions [1]. However, the same group has recently reported that under basal conditions, NO release from a murine IMCD cell line in culture was negligible unless cytokine stimuli were applied, leading to a significant increase in NOS II and nitrite formation [116], and it was speculated that NOS induction in IMCD under pyelonephritic conditions could serve for a defense against invading pathogens. Likewise, a previous study on primary cultures of rat IMCD cells has reported a significant increase in NOS II mRNA after cytokine stimulation; this was paralleled by increases in nitrite formation, whereas basal NOS levels were negligible [14].

4.10. NOS in Papillary Surface Epithelium

No evidence for NOS I expression in the renal papillary surface epithelium has been reported so far. NOS II expression in the papillary surface epithelium was reported to be weak, as revealed by in situ hybridization in normal rat, and the signal was found to be increased after endotoxemic stimulation [28].

4.11. NOS in Renal Nerves

NADPH-d positive nerve fibers in rat and human kidney have been observed to occur in association with renal arteries and arterioles and in the connective tissue beneath the renal pelvic epithelium. These fibers were also immunoreactive for NOS I-specific antibody [25]. The nerve fibers were occasionally seen to extend toward the JGA; however, significant innervation of the JGA by NOS-containing fibers has not been found [25].

NOS I-containing neuronal somata have been identified in association with nerve bundles and ganglionated plexus at the hilus of rat kidney, near the interlobular arteries, and on the wall of the renal pelvis [117]. Some of these ganglia were later shown to coexpress dopamine β-hydroxylase [118]. It is not known whether these neuronal somata belong to the sympathetic or the parasympathetic system [118]. Functionally, renal arterial tone appears to be dependent on the antagonistic effects of NO-mediated vasodilator and adrenergic vasoconstrictor fibers [119]. Proximal tubular reabsorption has been suggested to be under the influence of NOS-containing nerves as well [120]. It is notable that rich amounts of NOS-containing nerves have also been described in the genitourinary tract; this issue has been subject to a recent review [121].

References

1. Mohaupt, M.G., Elzie, J.L., Ahn, K.Y., Clapp, W.L., and Wilcox, C.S. *Kidney Int.* **46,** 653–665 (1994).

2. Sessa, W.C. *J. Vasc. Res.* **31,** 131–143 (1994).

3. Förstermann, U., Closs, E.I., Pollock, J.S., et al. *Hypertension* **23,** 1121–1131 (1994).

4. Nathan, C. *FASEB J.* **6,** 3051–3064 (1992).

5. Schmidt, H.H.H.W. and Walter, U. *Cell* **78,** 919–925 (1994).

6. Marletta, M.A. *Cell* **78,** 927–930 (1994).

7. Xie, Q., Whisnant, R., and Nathan, C. *J. Exp. Med.* **177,** 1779–1784 (1993).

8. De Vera, M., Shapiro, R.A., Nussler, A.K., et al. *Proc. Natl. Acad. Sci. USA* **93,** 1054–1059 (1996).

9. Wei, X., Charles, I.G., Smith, A., et al. *Nature* **375,** 408–411 (1995).

10. Nathan, C. *Cell* **82,** 873–876 (1995).

11. Xie, Q.W. and Nathan, C. *J. Leuk. Biol.* **56,** 576–582 (1994).

12. Mattson, D.L. and Higgins, D.J. *Hypertension* **27,** 688–692 (1996).

13. Tracey, W.R., Pollock, J.S., Murad, F., Nakane, M., and Förstermann, U. *Am. J. Physiol.* **266,** C22–C28 (1994).

14. Markewitz, B.A., Michael, J.R., and Kohan, D.E. *J. Clin. Invest.* **91,** 2138–2143 (1993).

15. Terada, Y., Tomita, K., Nonoguchi, H., and Mammo, F. *J. Clin. Invest.* **90,** 659–665 (1992).

16. Singh, I.J., Graham, M., Smart, A., Schnermann, J., and Briggs, J.P. *Am. J. Physiol.* **270,** F1027–F1037 (1996).

17. Beesley, J.E. *Histochem. J.* **27,** 757–769 (1995).

18. Pearse, A.G.E. *J. Histochem. Cytochem.* **5,** 515–527 (1957).

19. Matsumoto, T., Nakane, M., Pollock, J.M., Kuk, J.E., and Förstermann, U. *Neurosci. Lett.* **155,** 61–64 (1993).

20. Mundel, P., Bachmann, S., Bader, M., et al. *Kidney Int.* **42,** 1017–1019 (1992).

21. Mayer, B., John, M., and Boehme, E. *FEBS Lett.* **277,** 215–219 (1990).

22. Schmidt, H.H.H.W., Gagne, G.D., Nakane, M., Pollock, J.M., Miller, M.F., and Murad, F. *J. Histochem. Cytochem.* **40,** 1439–1456 (1992).

23. Springall, D.R., Riveros-Moreno, V., Buttery, L., et al. *Histochemistry* **98**, 259–266 (1992).

24. Loesch, A. and Burnstock, G. *Endothelium* **1**, 23–29 (1993).

25. Bachmann, S., Bosse, H.M., and Mundel, P. *Am. J. Physiol.* **268**, F885–F898 (1995).

26. Tojo, A., Gross, S.S., Zhang, L., et al. *J. Am. Soc. Nephrol.* **4**, 1438–1447 (1994).

27. Bosse, H.M., Boehm, R., Resch, S., and Bachmann, S. *Am. J. Physiol.* **26**, F793–F805 (1995).

28. Ahn, K.Y., Mohaupt, M.G., Madsen, K.M., and Kone, B.C. *Am. J. Physiol.* **267**, F748–F757 (1994).

29. Moncada, S. and Higgs, E.A. *FASEB J.* **9**, 1319–1330 (1995).

30. Palmer, R.M.J., Ashton, D.S., and Moncada, S. *Nature* **333**, 664–666 (1988).

31. Schini, V.B., Busse, R., and Vanhoutte, P.M. *Exp. Nephrol.* **2**, 139–144 (1994).

32. Schini, V.B., Catovsky, S., Schray-Utz, B., Busse, R., and Vanhoutte, P.M. *Circ. Res.* **1**, 24–32 (1994).

33. Schneider, F., Bucher, B., Schott, C., André, A., Julou-Schaeffer, G., and Stoclet, J.-P. *Am. J. Physiol.* **266**, H191–H198 (1994).

34. Nathan, C. and Xie, Q.-W. *Cell* **78**, 915–918 (1994).

35. Buttery, L.D.K., Springall, D.R., Andrade, S.P., et al. *J. Pathol.* **171**, 311–319 (1993).

36. Beasley, D. and Eldridge, M. *Am. J. Physiol.* **266**, R1197–R1203 (1994).

37. Imai, T., Hirata, Y., Kanno, K., and Marumo, F. *J. Clin. Invest.* **93**, 543–549 (1994).

38. Ranjan, V., Xiao, Z., and Diamond, S.L. *Am. J. Physiol.* **269**, H550–H555 (1995).

39. Busse, R. and Mülsch, A. *FEBS Lett.* **265**, 133–136 (1990).

40. Romero, J.C., Lahera, V., Salom, M.G., and Biondi M.L. *J. Am. Soc. Nephrol.* **2**, 1371–1387 (1992).

41. MacAllister, R. and Vallance, P. *J. Am. Soc. Nephrol.* **5**, 1057–1065 (1994).

42. Berk, B.C., Corson, M.A., Peterson, T.E., and Tseng, H. *J. Biomech.* **28**, 1439–1450 (1995).

43. Uematsu, M., Ohara, Y., Navas, J.P., et al. *Am. J. Physiol.* **269**, C1371–C1378 (1995).

44. Förstermann, U., Kleinert, H., Gath, I., Schwarz, P., Closs, E.I., and Dun, N.J. *Adv. Pharmacol.* **34**, 171–185 (1995).

45. Resnick, N., Collins, T., Atkinson, W., Bonthron, D.T., Dewey, C.F., Jr., and Gimbrone, M.A., Jr. *Proc. Natl. Acad. Sci. USA* **90**, 4591–4595 (1993).

46. Nadaud, S., Bonnardeaux, A., Lathrop, M., and Soubrier, F. *Biochem. Biophys. Res. Commun.* **198**, 1027–1033 (1994).

47. Robinson, L.J., Weremowicz, S., Morton, C.C., and Michel, T. *Genomics* **19**, 350–357 (1994).

48. Zembowicz, A., Tang, J.-I., and Wu, K.K. *J. Biol. Chem.* **270**, 17006–17010 (1995).

49. Hecker, M., Mülsch, A., Bassenge, E., Förstermann, U., and Busse, R. *Biochem. J.* **299**, 247–252 (1994).

50. Morin, A.M. and Stanboli, A. *J. Neurosci. Res.* **36**, 272–279 (1993).

51. Loesch, A., Belai, A., and Burnstock, G. *Cell Tissue Res.* **274**, 539–545 (1993).

52. O'Brien, A.J., Young, H.M., Povey, J.M., and Furness, J.B. *Histochemistry* **103**, 221–225 (1995).

53. Pollock, J.S., Nakane, M., Buttery, L.D.K., et al. *Am. J. Physiol.* **265**, C1379–C1387 (1993).

54. Bates, T.E., Loesch, A., Burnstock, G., and Clark, J.B. *Biochem. Biophys. Res. Commun.* **218**, 40–44 (1996).

55. Kriz, W. and Kaissling, B. In: *The Kidney: Physiology and Pathophysiology.* Eds. Seldin, D.W. and Giebisch G. Raven, New York, 1992, pp. 707–777.

56. Helmchen, U.E. Inaugural dissertation, Tuebingen, 1980.

57. Kreisberg, J.F., Venkatachalam, M., and Troyer, D. *Am. J. Physiol.* **249**, F457–F469 (1985).

58. Bachmann, S. and Kriz, W. In: *ILSI Monographs on Pathology of Laboratory Animals, Urinary System.* Ed. Jones, D.V.M. Springer-Verlag, New York, 1996, pp. 5–25.

59. Drenckhahn, D. and Franke, R.P. *Lab. Invest.* **59**, 673–682 (1988).

60. Dieterich, H.J. *Norm. Pathol. Anat.* **35**, 1–127 (1978).

61. Raij, L. and Baylis, C. *Kidney Int.* **48**, 20–32 (1995).

62. Bachmann, S. and Mundel, P. *Am. J. Kidney Dis.* **24**, 112–129 (1994).

63. Ujiie, K., Yuen, J., Hogart, L., Danziger, R., and Star, R.A. *Am. J. Physiol.* **267**, F296–F302 (1994).

64. Kriz, W., Sakai, T., and Hosser, H. In: *Nephrology.* Ed. Davison, A.M. Baillière Tindall, London, 1988, pp. 23–32.

65. Deng, A. and Baylis, C. *Am. J. Physiol.* **264**, F212–F215 (1993).

66. Ito, S., Arima, S., Ren, Y.L., Juncos, L.A., and Carretero, O.A. *J. Clin. Invest.* **91**, 2012–2019 (1993).

67. Ito, S. *NIPS* **9**, 115–119 (1994).

68. Hoffend, J., Cavarape, A., Endlich, K., and Steinhausen, M. *Am. J. Physiol.* **265**, F285–F292 (1993).

69. Ganz, M.B., Scott, E., Kasner, E., and Unwin, R.J. *Am. J. Physiol.* **268**, F1081–F1086 (1995).

70. Mundel, P., Gambaryan, S., Bachmann, S., Koesling, D., and Kriz, W. *Histochem. Cell Biol.* **103**, 75–79 (1995).

71. Bosse, H.M. and Bachmann, S. *Hypertension* (in press).

72. Beckman, J.S., Ye, Y.Z., Anderson, P.G., et al. *Biol. Chem. Hoppe–Seyler* **375**, 81–88 (1994).

73. Sigmon, D.H. and Beierwaltes, W.H. *Hypertension* **22**, 237–242 (1993).

74. Biondi, M.L., Dousa, T., Vanhoutte, P., and Romero, J.C. *Am. J. Hypertens.* **3**, 876–878 (1990).

75. Klahr, S. *Lab. Invest.* **72**, 1–3 (1995).

76. Cook, T. and Sullivan, R. *Am. J. Pathol.* **139**, 1047–1052 (1991).

77. Cattel, V., Largen, P., De Heer, E., and Cook, T. *Kidney Int.* **40**, 847–851 (1991).

78. Morrissey, J.J., McCracken, R., Kaneto, H., Vehaskari, M., Montani, D., and Klahr, S. *Kidney Int.* **45**, 998–1005 (1994).

79. Luss, H., Watkins, S.C., Freeswick, P.D., et al. *J. Mol. Cell. Cardiol.* **27**, 2015–2029 (1995).

80. Cattel, V., Lianos, E., Largen, P., and Cook, T. *Exp. Nephrol.* **1**, 36–40 (1993).

81. Nathan, C.F. and Hibbs, J.B., Jr. *Curr. Opin. Immunol.* **3**, 65–70 (1991).

82. Kunz, D., Mühl, H., Walker, G., and Pfeilschifter, J. *Proc. Natl. Acad. Sci. USA* **91**, 5387–5391 (1994).

83. Mühl, H. and Pfeilschifter, J. *J. Clin. Invest.* **95**, 1941–1946 (1995).

84. Saura, M., Lopez, S., Puyol, M.R., Puyol, D.R., and Lamas, S. *Kidney Int.* **47**, 500–509 (1995).

85. Jansen, A., Cook, T., Taylor, G.M., et al. *Kidney Int.* **45**, 1215–1219 (1994).

86. Goto, S., Yamamoto, T., Feng, L., Yaoita, E., Hirose, S., and Fujinaka, H. *Am. J. Pathol.* **147**, 1133–1141 (1995).

87. Buttery, L.D.K., Evans, T.J., Springall, D.R., Carpenter, A., Cohen, J., and Polak, J.M. *Lab. Invest.* **71**, 755–764 (1994).

88. Lau, K., Nakashima, O., Aalund, G.R., et al. *Am. J. Physiol.* **269**, F212–F217 (1995).

89. Bachmann, S. and Kriz, W. *Cell Tissue Res.* **225**, 111–127 (1982).

90. Hwang, S.-M., Wilson, P.D., Laskin, J.D., and Denhardt, D.T. *J. Cell Physiol.* **160**, 61–68 (1994).

91. Ishii, K., Warner, T.D., Sheng, H., and Murad, F. *J. Pharmacol. Exp. Ther.* **259**, 1102–1108 (1991).

92. Levillain, O., Hus-Citharel, A., Morel, F., and Bankir, L. *Am. J. Physiol.* **264**, F1038–F1045 (1993).

93. Radermacher, J., Klanke, B., Schurek, H.J., Stolte, H.F., and Fröhlich, J.C. *Kidney Int.* **41**, 1549–1559 (1992).

94. Majid, D.S.A., Williams, A., and Navar, L.G. *Am. J. Physiol.* **33**, F79–F87 (1993).

95. Alberola, A., Pinilla, J.M., Quesada, T., Romer, J.C., Salom, M.G., and Salazar, F.J. *Hypertension* **19**, 780–784 (1992).

96. Mayeux, P.R., Garner, H.R., Gibson, J.D., and Beanum, V.C. *Biochem. Pharmacol.* **49**, 115–118 (1995).

97. Guzman, N.J., Fang, M.-Z., Tang, S.-S., Ingelfinger, J.R., and Garg, L.C. *J. Clin. Invest.* **95**, 2083–2088 (1995).

98. Amoah-Apraku, B., Chandler, L.J., Harrison, J.K., Tang, S.-S., Ingelfinger, J.R., and Guzman, N.J. *Kidney Int.* **48**, 674–682 (1995).

99. Xie, Q., Kashiwabara, Y., and Nathan, C. *J. Biol. Chem.* **269**, 4705–4708 (1994).

100. Harris, R.C., McKanna, J.A., Akai, Y., Jacobson, H.R., Dubois, R.N., and Breyer, M.D. *J. Clin. Invest.* **94**, 2504–2510 (1994).

101. Hession, C., Decker, J., Sherblom, A., Kumar, S., Yue, C., and Mattaliano, R. *Science* **237**, 1479–1484 (1987).

102. Kone, B.C., Schwöbel, J., Turner, P., Mohaupt, M.G., and Cangro, C.B. *Am. J. Physiol.* **269**, F718–F729 (1995).

103. Wilcox, C.S., Welch, W.J., Murad, F., et al. *Proc. Natl. Acad. Sci. USA* **89**, 11993–11997 (1992).

104. Thorup, C. and Persson, A.E.G. *Am. J. Physiol.* **267**, F606–F611 (1994).

105. Hecker, M., Mülsch, A., and Busse, R. *J. Neurochem.* **62**, 1524–1529 (1994).

106. He, X.R., Greenberg, S.G., Briggs, J.P., and Schnermann, J.B. *Am. J. Physiol.* **268**, 7953–7959 (1995).

107. Ito, S. and Ren, Y.L. *J. Clin. Invest.* **92**, 1093–1098 (1993).

108. Schricker, K., Hamann, M., and Kurtz, A. *Am. J. Physiol.* **269**, F825–F830 (1995).

109. Vallon, V. and Thomson, S. *Am. J. Physiol.* **269**, F892–F899 (1995).

110. Lapointe, J.-Y., Laamarti, A., Hurst, A.M., Fowler, B.C., and Bell, P.D. *Kidney Int.* **47**, 752–757 (1995).

111. Obermüller, N., Kunchaparty, S., Ellison, D.H., and Bachmann, S. *J. Clin. Invest.* **98**, 635–640 (1996).

112. Fischer, E., Schnermann, J., Briggs, J.P., Kriz, W., Ronco, P.M., and Bachmann, S. *Am. J. Physiol.* **268**, F1164–F1176 (1995).

113. Tojo, A., Guzman, N.J., Garg, L.C., Tisher, C.C., and Madsen, K.M. *Am. J. Physiol.* **267**, F509–F515 (1994).

114. Stoos, B.A., Carretero, O.A., and Garvin, J.L. *J. Am. Soc. Nephrol.* **4**, 1855–1860 (1994).

115. Garcia, N.H., Stoos, B.A., Carretero, O.A., and Garvin, J.L. *Hypertension* **27**, 679–683 (1996).

116. Mohaupt, M.G., Schwöbel, J., Elzie, J.L., Kannan, G.S., and Kone, B.C. *Am. J. Physiol.* **268**, F770–F777 (1995).

117. Liu, L. and Barajas, L. *Neurosci. Lett.* **161**, 145–148 (1993).

118. Liu, L., Liu, G.-L., and Barajas, L. *Neurosci. Lett.* **202**, 69–72 (1995).

119. Okamura, T., Yoshida, K., and Toda, N. *Hypertension* **25**, 1090–1095 (1995).

120. Gabbai, F.B., Thomson, S.C., Peterson, O., Wead, L., Malvey, K., and Blantz, R.C. *Am. J. Physiol.* **268**, F1004–F1008 (1995).

121. Burnett, A.L. *Urology* **45**, 1071–1083 (1995).

122. Koushanpour, E. and Kriz, W. In: *Renal Physiology.* Springer-Verlag, New York, 1986, Vol. 9.

PART III

NO and the Regulation of Renal Hemodynamics

8

Vascular Effects of NO

Ingrid Fleming and Rudi Busse

1. The Endothelial Nitric Oxide Synthase: Localization

Nitric oxide synthase (NOS) III is a constitutively expressed 135-kDa protein predominantly associated with the particulate subcellular fraction, suggesting that the native enzyme is a membrane-bound protein [1–4]. Membrane association appears to be achieved by attachment of myristic acid to the amino terminal end of the enzyme and, as such, is consistent with reports that NOS cDNA contains a consensus sequence for cotranslational modification of the enzyme by N-terminal myristoylation [5–9]. Prevention of myristic acid incorporation in site-directed mutagenesis experiments converts the membrane-associated NOS to a cytosolic form [3,10]; however, this intervention has been reported to have little effect on enzyme activity [10]. Myristoylation alone, however, provides barely enough energy to anchor a protein to a lipid bilayer; therefore, it seems likely that other factors are important in determining the fraction of the enzyme associated with the plasma membrane (e.g., additional hydrophobic interactions owing to reversible palmitoylation of a nearby cysteine residue). In cases of such a dual protein acylation, the combined hydrophobic interactions of the two covalently attached lipid moieties anchor the protein firmly to the plasma membrane and its attachment could be regulated by palmitate turnover. Indeed, the membrane-bound NO synthase is palmitoylated, and depalmitoylation of the endothelial NO synthase, along with its translocation to the cytosol, has been reported following exposure to the receptor-dependent agonist bradykinin [11]. Because agonist-induced de-palmitoylation was reportedly maximal only after 15–30 min, this phenomenon would appear to be associated with inactivation rather than activation of NOS activity. The role of actively regulated NOS III palmitoylation in determining its cellular localization remains controversial [12].

Recent detailed analysis of the membrane association of NOS III has demonstrated that this enzyme is localized to specific structures in the plasmalemmal membrane identified as caveolae and that this membrane compartmentalization requires both myristoylation and palmitoylation [13]. The association of NOS III with a subcompartment of the plasma membrane in which several key signal transducing complexes are concentrated (e.g., G-proteins, src-family tyrosine kinases) is likely to have profound repercussions on enzyme activity, as well as its sensitivity to activation by signal transduction cascades other than those resulting in an increase in $[Ca^{2+}]_i$. Indeed, the variety of signal molecules detected within the caveolae render these structure ideal sites for cross-talk between receptor-dependent and receptor-independent signal transduction pathways.

2. The Endothelial Nitric Oxide Synthase: Functional Properties

NOS III is classified as a Ca^{2+}/calmodulin-dependent enzyme which is able to tick over, producing low amounts of NO, at basal levels of intracellular Ca^{2+} ($[Ca^{2+}]_i$). Ca^{2+} concentration-dependently increases activity of the purified NOS III [14], and an increase in $[Ca^{2+}]_i$, such as that observed following agonist stimulation, enhances endothelial NO production (for review, see Ref. 15). Both the agonist-induced NO formation and the subsequent vasodilation are abolished in the absence of extracellular Ca^{2+} [16,17]. A basal enzyme activity, which is sensitive to the NO synthase inhibitor N^G-nitro-L-arginine (L-NNA), is however evident even at Ca^{2+} concentrations as low as 10 nM, indicating that approximately 75% of basally produced NO may be formed via a Ca^{2+}-independent pathway [14]. Little physiological relevance was attributed to this phenomenon, and the identification of a calmodulin-binding domain in the primary structure of the endothelial NO synthase, [6,9,18] together with the finding that calmodulin-binding proteins inhibited enzyme activity, [19] strengthened the hypothesis that the binding of a Ca^{2+}/calmodulin complex is essential to activate the constitutive enzyme. However, recent evidence indicates that mechanical stimulation of native and cultured endothelial cells by fluid shear stress results in the formation of NO via a pathway which comprises Ca^{2+}-dependent as well as Ca^{2+}-independent components [20–24]. In native endothelial cells, the shear stress-induced NO production consists of an initial peak followed by a sustained plateau phase which is maintained as long as shear stress is applied [22,23]. The initial component of this response can be abolished by removal of Ca^{2+} from the extracellular medium and coincides with an increase in $[Ca^{2+}]_i$ observed in cultured endothelial cells exposed to shear stress in a parallel-plate flow chamber. Moreover, the sustained phase of NO production is insensitive to the chelation of extracellular Ca^{2+} as well as to calmodulin antagonists [22,23].

Additional support for the concept of a Ca^{2+}-independent NO production following the application of shear stress is provided by the finding that shear stress results in, at best, a transient increase in $[Ca^{2+}]_i$ attributed to the activation of phospholipase C [25]: an increase in intracellular levels of inositol-1,4,5-trisphosphate (IP$_3$) [26,27] followed by an acute release of intracellularly stored Ca^{2+} [28–34]. Thus, a signaling pathway distinct from the classically accepted Ca^{2+}-dependent pathway regulates the activity of NOS III in response to shear stress. For example, it has been reported that NOS III is highly sensitive to variations in pH$_i$ within a narrow range at basal levels of [Ca^{2+}]$_i$ [35] and that inactivation of the Na$^+$/H$^+$ exchanger attenuates Ca^{2+}-independent NO production in sheared endothelial cells [23]. Other possibilities include the activation of various protein kinases. Indeed, several consensus sequence sites for phosphorylation by protein kinase A, C, and calmodulin kinase II are found in all the cloned NOS isoforms; however, the functional relevance of these sites has not been fully elucidated. Cellular levels of phosphotyrosine appear to have a marked impact on cellular signaling in endothelial cells, and a link between enhanced tyrosine phosphorylation and activation of NOS III has been proposed, although this interaction was originally thought to be attributed to effects on Ca^{2+} signaling. For example, tyrosine kinase inhibitors have been shown to selectively attenuate Ca^{2+} influx into agonist-stimulated endothelial cells [36–38], whereas the tyrosine phosphatase inhibitors, phenylarsine oxide and vanadate, are reported to activate transmembrane Ca^{2+} influx via an IP$_3$-independent mechanism [39]. However, there are fundamental differences in the role played by phosphotyrosine in mediating endothelial responsiveness to agonist-induced stimulation and mechanical stimulation. For example, the tyrosine kinase inhibitor erbstatin A was reported to only slightly attenuate the vasodilator response to acetylcholine in perfused carotid arteries [38], whereas we recently demonstrated that this inhibitor abolished the Ca^{2+}-independent phase of shear stress-induced NO production [23]. The activation of a tyrosine kinase-dependent signal transduction pathway in endothelial cells in response to shear stress was confirmed by Western blot analysis of tyrosine phosphorylated proteins in detergent-soluble and -insoluble cell fractions. In cultured human endothelial cells, the application of shear stress was associated with the transient tyrosine phosphorylation of the mitogen-activated protein kinases [23,40–42] as well as the maintained tyrosine phosphorylation of a series of detergent-insoluble or cytoskeletal proteins [~88, 90, 103, and ~125 kDa) [23]. Enhancing cellular levels of phosphotyrosine by inhibiting protein tyrosine phosphatases not only induced the tyrosine phosphorylation of a similar series of cytoskeletal proteins but also elicited the Ca^{2+}/calmodulin-independent, erbstatin A-sensitive formation of NO [43]. Thus, the endothelial response to the application of tyrosine phosphatase inhibitors exhibits characteristics identical to those previously thought to be exclusive to the shear stress-induced activation of the NO synthase.

Examination of the effects of shear stress and tyrosine phosphatase inhibitors on NOS III revealed that both stimuli result in the time-dependent decrease in the recovery of the enzyme from the Triton X-100-soluble cell fraction and a concomitant increase in its recovery in the Triton X-100-insoluble fraction [43].

A similar effect was observed with three tyrosine phosphorylated proteins [i.e., the focal adhesion kinase (FAK), the cytoskeletal-associated protein paxillin, and the caveolae marker protein caveolin]. The shear stress-induced tyrosine phosphorylation and redistribution of these proteins to the detergent-insoluble fraction was attributed to activation of integrin-coupled signal transduction pathway, as both effects were abolished by interfering with the association of integrins with the extracellular matrix. We, therefore, envisage that in response to shear stress, a multimolecular, detergent-insoluble complex is formed which is comprised of NOS III, tyrosine kinases, and phosphatases, as well as cytoskeletal and caveolar proteins. The formation of this complex results in the maintained Ca^{2+}-independent activation of NOS III. This mechanism may also account for the reported acute increase in NO production at basal $[Ca^{2+}]_i$, following the activation of growth factor receptors containing inherent tyrosine kinase activity, for example, in response to an insulin-like growth factor [44].

3. Transcriptional Regulation of NOS III

Although generally referred to as a constitutive enzyme, NOS III can be induced or upregulated in response to a number of stimuli. Indeed, the promoter of the NOS III gene contains a number of recognition sites for transcription factors, including SP1, AP-1, AP-2, CF-1, and NF-1, as well as a putative acute-phase response/shear stress and sterol-regulatory cis-elements [45–47].

In addition to the above mentioned acute effects on NO production, shear stress plays a role in the chronic regulation of NOS III activity, an effect which may account for the observation that the basal release of NO is significantly greater from native endothelial cells, continuously exposed to shear, than that from cultured endothelial cells maintained under static conditions. Moreover, the expression of NOS in cultured endothelial cells can be upregulated by exposure to shear stress for several hours, suggesting that the continuous activation of the endothelium by shear stress maintains NOS III levels in vivo [4,9,48,49]. Indeed, chronic exercise, which, as a consequence of the increased blood flow, presumably increases shear stress on the endothelium, also results in an increased production of NO in rat skeletal muscle arterioles [50], as well as in the canine coronary circulation, [51,52] and was associated with a significant increase in NOS mRNA in vivo [52]. Although a putative shear stress-responsive element, similar to that previously reported in the promoter region of the gene-encoding platelet-derived growth factor B [63], has been identified in the promoter region of the NOS III gene [45,46], the mechanism by which increases in shear stress modulate NOS III gene expression remains unclear. However, shear stress has been reported to Ca^{2+}-independently activate the mitogen-activated protein kinase (MAP kinase) in cultured endothelial cells [23,40–42,54], and perhaps a more likely mechanism by which shear stress regulates NOS III expression involves activation of the

Ras/MAP kinase pathway, with the subsequent activation of c-Fos and c-Jun which dimerize and bind to the AP-1 transcription factor site of the promoter. Such a proposal fits well with reports that shear stress induces the synthesis and nuclear localization of c-Fos in cultured endothelial cells [55,56] and stimulates AP-1 [57].

Apart from shear stress, NOS III expression can be enhanced by basic fibroblast growth factor [58] and appears to be negatively regulated by protein kinase C [59]. The signal transduction pathway resulting in these effects remains to be clarified but may also be related to activation of MAP kinases.

Cytokines also effect the expression of NOS III. TNF-α, for example, decreases NOS mRNA levels by increasing the rate of mRNA degradation via a process that involves de novo protein synthesis [60]. The combination of IFN-γ plus TNF-α, or IL-1β, on the other hand, has been shown to paradoxically enhance endothelial NOS activity, despite a concurrent decrease in NOS mRNA, an effect which may be explained by increased endogenous BH$_4$ levels [61].

4. Acute Vascular Effects of NO

The shear stress exerted on the endothelium by the circulating blood, which represents the major stimulus for a continuous production of NO in vivo [62,63] is a highly effective and sensitive system for the local control of vascular tone. Selective inhibition of NO synthesis using L-arginine analogs causes an increase in mean arterial blood pressure and a reduction in blood flow, underscoring the significance of shear stress-dependent endothelium-derived NO in global cardiovascular homeostasis. Knocking out the gene encoding the endothelial NO synthase in mice results in significant hypertension, and aortic rings removed ex vivo from these animals display no relaxation to acetylcholine and are unaffected by treatment with NO synthase inhibitors [64]. Surprisingly, the administration of these inhibitors to mice deficient in endothelial NO synthase resulted in a decrease in mean arterial blood pressure which was prevented by L-arginine. Thus, more than one isoform of NO synthase (e.g., the neuronal NO synthase which is present both in vasomotor centres of the central nervous system and in peripheral nerves) are likely to play a role in the global regulation of blood pressure.

Nitric oxide mediates physiological responses in the endothelium itself as well as in other target cells by mechanisms that include the activation of the heme-containing soluble guanylyl cyclase, binding to nonheme iron proteins, as well as NAD$^+$-dependent protein poly-ADP ribosylationlike reactions (for a review, see Ref. 65). Indeed, it has been hypothesized that in hypertension, chronically depressed levels of bioactive NO lead to a decrease in the ADP-ribosylation of G-proteins and their subsequent disinhibition. This reduced ADP-ribosylation could then lead to vasoconstriction, as activation of the G-proteins by agonists would be unopposed [66]. Another example of such a reaction is the NO-induced,

NAD$^+$-dependent modification (and consequent inhibition) of the glycolytic enzyme glyceraldehyde-3-phosphate dehydrogenase (GAPDH) [67]. The cellular mechanism underlying this effect appears to involve the transferral of the nitrosonium ion (NO$^+$) to a thiol group within the active site of the enzyme, which then leads to the covalent modification of the enzyme in the presence of NAD$^+$ [68]. Sulfhydryl (SH)-containing proteins represent an important cellular target for NO, and their interaction results in the generation of biologically active *S*-nitrosoproteins which appear to be implicated in both the beneficial and deleterious actions of NO.

Nitric oxide can also be cytotoxic, an effect not necessarily related to the inhibition of GAPDH or iron-sulfur enzymes, like aconitase. This NO-mediated cytotoxicity is morphologically and biochemically defined as apoptosis rather than necrosis. Indeed apoptosis, as evidenced by the formation of the characteristic DNA fragmentation ladder, nuclear condensation, apoptotic body formation, and p53 accumulation, can be initiated by NO in cytokine-stimulated cells and is sensitive to NOS inhibition. Certain oncogenes have been shown to modulate apoptosis; Bcl-2 expression inhibits apoptosis, whereas p53 and c-Myc protein may induce it. The signaling pathway implicated in NO-induced apoptosis remains to be fully clarified; however, it has been reported that NO-induced apoptosis is associated with the intranuclear accumulation of p53, which leads to apoptosis either directly by acting on DNA or by causing a block in the cell cycle (for a review, see Ref. 65). The latter toxic effects are, however, more likely to be associated with large amounts of NO produced by cytokine-activated cells (macrophages, pancreatic β cells, and possibly endothelial cells) rather than with the relatively modest amounts of NO produced by the endothelium. There are, on the other hand, certain indications that low concentrations of NO may exert antiapoptotic actions. For example, apoptosis was prevented in B lymphocytes exposed to extracellular NO, an effect which correlated with the sustained expression of the proto-oncogene Bcl-2 [69].

In vascular smooth muscle cells NO activates the soluble guanylyl cyclase and enhances formation of guanosine 3′:5′-cyclic monophosphate (cyclic GMP) [70], which has, in turn, well-documented consequences on vascular tone [71–73]. Despite the fact that this signaling cascade is generally accepted as the chief mechanism by which NO regulates smooth muscle tone, NO-mediated vasodilatation may also occur via cyclic GMP-independent processes. Indeed, a cyclic GMP-independent mechanism is responsible for the NO-mediated relaxation in the rat proximal colon in response to nonadrenegic, noncholinergic nerve stimulation [74]. In addition, both exogenous and endogenously produced NO have been shown to activate charybdotoxin (CTX)-sensitive, calcium-dependent potassium channels (K$^+_{Ca}$) and induce hyperpolarization of vascular smooth muscle cells [75,76]. NO is also reported to reversibly inhibit mitochondrial respiration by competing with oxygen at cytochrome C oxidase, an effect which could account for the observed increase in respiration in certain cell systems following NOS inhibition [77]. By increasing the apparent K_m of cytochrome C oxidase for oxygen, NO may act as a physiological affinity regulator of mitochondrial respiration, enabling mitochondria to act as oxygen sensors. Such a principle might be involved in the mechanisms by which local oxygen concentrations influence vessel tone because a decrease in oxygen concentration within vascular smooth muscle cells would be expected to alter the NO/oxygen balance, inhibit the cytochrome C oxidase, decrease cellular levels of ATP, and thereby result in dilatation [77]. An alternative mechanism by which local oxygen concentrations may modulate vascular function in an NO-dependent manner has recently been proposed [78]. In this model, NO formed in the lung interacts with oxyhemoglobin, resulting in the formation of S-nitrosohemoglobin, which is relatively stable at arterial pO_2. Indeed significant levels of S-nitrosohemoglobin can be detected in arterial blood, whereas it is virtually undetectable in venous blood [78]. Thus, S-nitrosohemoglobin appears to be a circulating stabilized NO donor which only relinquishes bound NO upon deoxygenation of hemoglobin within the capillary circulation. The biological role of the NO thus released remains to be clarified. By increasing local concentrations of NO, it may regulate permeability within the capillary bed and may possibly optimize oxygen consumption in adjacent target cells (i.e., skeletal or cardiac muscle).

Nitric oxide is also able to affect the production of other endothelium-derived autacoids. For example, NO increases prostacyclin synthesis both in vitro [79–81] and in vivo [82] by enhancing prostaglandin H synthase [81] and/or cyclooxygenase activity [79] via a cyclic GMP-independent mechanism. In contrast, the release of the endothelium-derived hyperpolarizing factor, a cytochrome P-450-derived metabolite of arachidonic acid [83–87], appears to be suppressed at physiological concentrations of NO [88]. This latter effect may be related to attenuating effects of NO on endothelial $[Ca^{2+}]_i$.

Platelet aggregation and adhesion are also inhibited by NO, effects which are due to guanylyl cyclase stimulation, activation of cyclic GMP-dependent protein kinases, and phosphorylation of specific platelet proteins, including the so-called vasodilator-stimulated phosphoprotein (VASP) [89].

Other cardiovascular effects of NO include enhanced myocardial relaxation, decreased diastolic tone, and reduced peak contraction in different preparations of cardiac muscle [90]. These latter effects are proposed to be mediated via a cyclic GMP-induced reduction in myofilament responsiveness to Ca^{2+} [90]. However, NO-mediated endogenous cyclic GMP production has also been shown to control the regulation of cardiac L-type Ca^{2+} channel activity [91].

Most of the above-mentioned effects of NO are acute; however, NO plays a crucial role in the long-term regulation of the vasculature by altering the expression of genes encoding certain endothelial proteins [92,93], a mechanism which is crucial to the maintenance of the antiatherogenic properties of the endothelium.

5. Chronic Vascular Effects of NO (NO as an Antiatherogenic Principle)

The attribution of antiatherogenic properties to endothelium-derived NO is based mainly on clinical and experimental observations in which an endothelial dysfunction is manifest, or on experimental systems in which NOS activity is inhibited. In the case of atherosclerosis, a wealth of evidence has been accumulated to demonstrate that, although the endothelium in atherosclerotic vessels is seemingly intact, the amount of bioactive NO released from the endothelium is decreased (for a review, see Ref. 94). For example, atherosclerosis is associated with impaired responsiveness to increased blood flow [95] as well as to endothelium-dependent vasodilators such as acetylcholine, bradykinin, and substance P [96–103], effects unrelated to a decreased sensitivity of vascular smooth muscle to the endothelium-derived vasodilators. The concept of an altered functioning of the endothelium as an initiator of the atherosclerotic process has recently gained momentum following reports that endothelial dysfunction can occur prior to any appreciable intimal thickening and is already apparent in patients with a family history of atherosclerosis [102,104–106]. There is strong in vivo evidence that endothelial dysfunction occurs in resistance vessels of the coronary and peripheral circulations, which themselves do not develop overt atherosclerosis [103], therefore suggesting that the endothelial dysfunction in patients with atherosclerosis is a systemic process, not necessarily confined to vessels in which the atheroma develops. The nature of this endothelial dysfunction is unknown; the possibilities include decreased expression of NOS III, imbalance between the production of endothelium-derived constricting and relaxing factors, decreased substrate availability, production of an endogenous NOS III inhibitor, and overproduction of oxygen-derived free radicals. Experimental evidence has been provided to support almost all of these possibilities [94]; however, current opinion favors the concept that impaired endothelium-dependent relaxation in atherosclerosis is due neither to decreased NO synthase activity nor to a deficiency in the availability of L-arginine, but rather to an increased production of the superoxide anion (O_2^-) [107]. Indeed, direct measurements in vessels from cholesterol-fed animals revealed a much higher production of NO in control vessels, although the amount of bioassayable NO was reduced [108], suggesting that the rate of inactivation of NO is increased in hypercholesterolemia. The consequence of this net decrease in NO has sequelae reaching beyond the regulation of vascular tone or platelet regulation because the activity of several redox-sensitive transcription factors may depend on the cellular balance in the production of NO/O_2^-. The crucial role of NO in suppressing neointimal hyperplasia has been elegantly demonstrated in a rat model of carotid artery balloon injury in which NOS III was introduced into the injured vessel wall. In the latter study, balloon injury was associated with an unabated neointimal hyperplasia in control animals, whereas a 70% reduction in neointimal area was apparent in vessels which had been transfected with NOS III cDNA [109].

Additional evidence suggesting a direct antiatherogenic effect of NO have been provided by experiments in which L-arginine analogs or exogenous NO donors were used to manipulate vascular NO levels. Such studies clearly demonstrated that NO production inhibits proatherogenic events such as monocyte adhesion, one of the earliest events in the pathogenesis of atherosclerosis (for a review, see Ref. 110). The recruitment of blood-borne cells to evolving atherosclerotic lesions appears to be specific for monocytes and requires the induction of specific adhesion molecules on the surface of both endothelial cells and the recruited leukocytes. The adhesion process itself is a multistep event. In the initial stages, E- or P-selectin expressed by stimulated endothelial cells binds to carbohydrates borne by the leukocyte surface molecules. Expression of P-selectin on vascular endothelial cells slows white blood cells and causes them to roll along the endothelial surface. Other cell adhesion molecules, including intercellular adhesion molecule-1 (ICAM-1) and vascular cell adhesion molecule-1 (VCAM-1), then latch onto and stop the white cells completely, prior to their migration.

The importance of a continuous production of NO in the maintenance of antiadhesive properties of the endothelium has been demonstrated in mesenteric venules [111]. Pretreatment of these vessels with an NO synthase inhibitor concentration- and time-dependently enhanced the expression of P-selectin and significantly increased leukocyte rolling and adherence. These effects were effectively blocked by neutralizing P-selectin with a monoclonal antibody and were significantly attenuated in the presence of L-arginine and 8-Br-cyclic GMP. Such findings concur with the observations that in platelets, NO can inhibit aggregation and P-selectin expression [112]. Moreover, NO synthase inhibitors enhance, whereas NO donors attenuate, endothelial expression of P-selectin during ischemia and reperfusion [113]. The mechanism by which endothelium-derived NO is able to control P-selectin expression appears to involve an interplay between NO and the vascular production of oxygen-derived free radicals, which is known to be elevated in hypercholesterolemia [114]. In this regard, superoxide dismutase significantly reduced P-selectin-mediated leukocyte rolling and adherence to venular endothelium [115] and to NO synthase inhibitor-treated endothelial cells [111]. These findings imply that the continuous production of NO scavenges O_2^- and protects the endothelium from the deleterious effects of this free radical.

Following adherence, monocytes migrate across the endothelium, in response to chemotactic factors derived from cells inherent to the vascular wall. One of these factors, the monocyte chemoattractant protein-1 (MCP-1) detected in macrophage-rich areas of human and rabbit atherosclerotic lesions [116], is specific for monocytes and accounts for virtually all of the monocyte chemotactic activity secreted by endothelial cells in vitro in response to tumor necrosis factor-α, interleukin-1β [117,118], and low-density lipoprotein [119]. The link between the dysfunctional endothelium and impaired NO production with the increased expression of MCP-1 in atherosclerosis appears to be more than coincidental. Recent experimental evidence suggests that NO acts as an antiatherosclerotic principle partly by continuously suppressing the expression of MCP-1 [92]. Indeed, inhibition of NOS caused a defined increase in MCP-1 mRNA and protein in cultured endothelial cells, which was reversed following the addition of NO donors. Moreover, monocyte migration in response to medium from cells treated with N^G-nitro-L-arginine and tumor necrosis factor-α was increased, whereas migration in response to medium from NO donor-treated cells was significantly less than that observed in response to control medium. These observations are consistent with reports that prolonged exposure of cultured endothelial cells to fluid shear stress, which is associated with an increased expression of NOS III [9,48], resulted in a decrease in MCP-1 gene expression [120]. These observations, taken together, support the concept that a decrease in NO production alleviates this intrinsic inhibitory influence on MCP-1 expression. A similar effect has since been proposed to account for the finding that NO donors also attenuate cytokine-induced expression of VCAM-1 [93] and the macrophage-colony stimulating factor [121].

There are, most likely, a number of diverse molecular pathway(s) by which NO can alter gene expression, but in the case of MCP-1, this pathway has been partially elucidated. In human endothelial cells, inhibition of NO synthesis has been shown to activate NF-κB-like transcriptional regulatory proteins [92], thus suggesting that basally produced endothelium-derived NO inhibits the activation of the NF-κB. More recently, NO has also been reported to induce the expression and prolong the half-life of the NF-κB inhibitory subunit, IκBα [122]. However, NO may also influence the activation of other transcription factors, such as AP-1, which has recently been characterized as an antioxidant-responsive factor [123]. Indeed, in electrophoretic mobility shift assays, the NO donor sodium nitroprusside, but not free NO, has been reported to S-nitrosylate the AP-1 moiety and inhibit its activity [124].

References

1. Pollock, J.S., Förstermann, U., Mitchell, J.A., et al. *Proc. Natl. Sci. USA* **88**, 10484 (1991).

2. Boje, K.M. and Fung, H. *J. Pharmacol. Exp. Ther.* **253**, 20–26 (1990).

3. Busconi, L. and Michel, T. *J. Biol. Chem.* **268**, 8410–8413 (1993).

4. Hecker, M., Mülsch, A., Bassenge, E., Förstermann, U., and Busse, R. *Biochem. J.* **299**, 247–252 (1994).

5. Janssens, S.P., Shimouchi, A., Quertermous, T., Bloch, D.B., and Bloch, K.D. *J. Biol. Chem.* **267**, 14519–14522 (1992).

6. Marsden, P.A., Schappert, K.T., Chen, H.S., Flowers, M., Sundell, C.L., Wilcox, J.N., Lamas, S., and Michel, T. *FEBS Lett.* **307**, 287–293 (1992).

7. Sessa, W.C., Harrison, J.K., Barber, C.M., Zeng, D., Durieux, M.E., D'Angelo, D.D., Lynch, K.R., and Peach, M.J. *J. Biol. Chem.* **267**, 15274–15276 (1992).

8. Lamas, S., Marsden, P.A., Li, G.K., Tempst, P., and Michel, T. *Proc. Natl. Acad. Sci. USA* **89,** 6348–6352 (1992).

9. Nishida, K., Harrison, D.G., Navas, J.P., Fisher, A.A., Dockery, S.P., Uematsu, M., Nerem, R.M., Alexander, R.W., and Murphy, T.J. *J. Clin. Invest.* **90,** 2092–2096 (1992).

10. Sessa, W.C., Barber, C.M., and Lynch, K.R. *Circ. Res.* **72,** 921–924 (1993).

11. Robinson, L.J., Busconi, L., and Michel, T. *J. Biol. Chem.* **270,** 995–998 (1995).

12. Liu, J., García-Cardena, G., and Sessa, W.C. *Biochemistry* **34,** 12333–12340 (1995).

13. Shaul, P.W., Smart, E.J., Robinson, L.J., German, Z., Yuhanna, I.S., Ying, Y.S., Anderson, R.G.W., and Michel, T. *J. Biol. Chem.* **271,** 6518–6522 (1996).

14. Mülsch, A., Bassenge, E., and Busse, R. *Naunyn Schmiedebergs Arch. Pharmacol.* **340,** 767–770 (1989).

15. Busse, R. and Fleming, I. *Ann. Med.* **27,** 331–340 (1995).

16. Singer, A.H. and Peach, M.J. *Hyperten.* **4,** II-19–II-25 (1982).

17. Lückhoff, A., Pohl, U., Mülsch, A., and Busse, R. *Br. J. Pharmacol.* **95,** 189–196 (1988).

18. Bredt, D.S., Huang, P.M., Glatt, C.E., Lowenstein, C., Reed, R.R., and Snyder, S.H. *Nature* **351,** 714–718 (1991).

19. Busse, R. and Mülsch, A. *FEBS Lett.* **265,** 133–136 (1990).

20. Hecker, M., Mülsch, A., Bassenge, E., and Busse, R. *Am. J. Physiol.* **265,** H828–H833 (1993).

21. Macarthur, H., Hecker, M., Busse, R., and Vane, J.R. *Br. J. Pharmacol.* **108,** 100–105 (1993).

22. Kuchan, M.J. and Frangos, J.A. *Am. J. Physiol.* **266,** C628–C636 (1994).

23. Ayajiki, K., Kindermann, M., Hecker, M., Fleming, I., and Busse, R. *Circ. Res.* **78,** 750–758 (1996).

24. O'Neil, W.C. *Am. J. Physiol.* **269,** C863–C869 (1995).

25. Bhagyalakshmi, A., Berthiaume, F., Reich, K.M., and Frangos, J.A. *J. Vasc. Res.* **29,** 443–449 (1992).

26. Nollert, M.U., Eskin, S.G., and McIntyre, L.V. *Biochem. Biophys. Res. Commun.* **170,** 281–287 (1990).

27. Prasad, A.R.S., Logan, S.A., Nerem, R.M., Schwartz, C.J., and Sprague, E.A. *Circ. Res.* **72,** 827–836 (1993).

28. Shen, J., Luscinskas, F.W., Connolly, A., Dewey, C.F., Jr., and Gimbrone, M.A., Jr. *Am. J. Physiol.* **262,** C384–C390 (1992).

29. Schwarz, G., Droogmans, G., and Nilius, B. *Pflügers Arch.* **421,** 394–396 (1992).

30. Sigurdson, W.J., Sachs, F., and Diamond, S.L. *Am. J. Physiol.* **264,** H1745–H1752 (1993).

31. Demer, L.L., Wortham, C.M., Dirksen, E.R., and Sanderson, M.J. *Am. J. Physiol.* **264,** H2094–H2102 (1993).

32. Oike, M., Droogmans, G., and Nilius, B. *Proc. Natl. Acad. Sci. USA* **91**, 2940–2944 (1994).

33. Schwarz, G., Callewaert, G., Droogmans, G., and Nilius, B. *J. Physiol. (Lond.)* **458**, 527–538 (1992).

34. Falcone, J.C., Kuo, L., and Meininger, G.A. *Am. J. Physiol.* **264**, H653–H659 (1993).

35. Fleming, I., Hecker, M., and Busse, R. *Circ. Res.* **74**, 1220–1226 (1994).

36. Fleming, I., Fisslthaler, B., and Busse, R. *Circ. Res.* **76**, 522–529 (1995).

37. Kruse, H.-J., Negrescu, E.V., Weber, P.C., and Siess, W. *Biochem. Biophys. Res. Commun.* **202**, 1651–1656 (1994).

38. Fleming, I., Bara, A., and Busse, R. *J. Vasc. Res.* **33**, 225–234 (1996).

39. Fleming, I., Fisslthaler, B., and Busse, R. *J. Biol. Chem.* **271** (1996).

40. Berk, B.C., Corson, M.A., Peterson, T.E., and Tseng, H. *J. Biomech.* **28**, 1439–1450 (1995).

41. Shyy, J.Y.-J., Lin, M.-C., Han, J., Lu, Y., Petrime, M., and Chien, S. *Proc. Natl. Acad. Sci. USA* **92**, 8069–8073 (1995).

42. Pearce, M.J., McIntyre, T.M., Prescott, S.M., Zimmerman, G.A., and Whatley, R.E. *Biochem. Biophys. Res. Commun.* **218**, 500–504 (1996).

43. Fleming, I., Bauersachs, J., Fisslthaler, B., and Busse, R. *FASEB J.* **10**, A303 (1996) (Abstract).

44. Tsukahara, H., Gordienko, D.V., Tonshoff, B., Gelato, M.C., and Goligorsky, M.S. *Kidney Int.* **45**, 598–604 (1994).

45. Marsden, P.A., Heng, H.H.Q., Scherer, S.W., Stewart, R.J., Hall, A.V., Shi, X.-M., Tsui, L-C., and Schappert, K.T. *J. Biol. Chem.* **268**, 17478–17488 (1993).

46. Nadaud, S., Bonnardeaux, A., Lathrop, M., and Soubrier, F. *Biochem. Biophys. Res. Commun.* **198**, 1027–1033 (1994).

47. Wariishi, S., Miyahara, K., Toda, K., Ogoshi, S., Doi, Y., Ohnishi, S., Mitsui, Y., Yui, Y., Kawai, C., and Shizuta, Y. *Biochem. Biophys. Res. Commun.* **216**, 729–735 (1995).

48. Ranjan, V., Xiao, Z., and Diamond, S.L. *Am. J. Physiol.* **269**, H550–H555 (1995).

49. Uematsu, M., Ohara, Y., Navas, J.P., Nishida, K., Murphy, T.J., Alexander, R.W., Nerem, R.M., and Harrison, D.G. *Am. J. Physiol.* **269**, C1371–C1378 (1995).

50. Sun, D., Huang, A., Koller, A., and Kaley, G. *J. Appl. Physiol.* **76**, 2241–2247 (1994).

51. Wang, J., Wolin, M.S., and Hintze, T.H. *Circ. Res.* **73**, 829–838 (1993).

52. Sessa, W.C., Pritchard, K., Seyedi, N., Wang, J., and Hintze, T.H. *Circ. Res.* **74**, 349–353 (1994).

53. Resnick, N., Collins, T., Atkinson, W., Bonthron, D.T., Dewey, C.F., Jr., and Gimbrone, M.A. Jr., *Proc. Natl. Acad. Sci. USA* **90**, 4591–4595 (1993).

54. Tseng, H., Peterson, T.E., and Berk, B.C. *Circ. Res.* **77**, 869–878 (1995).

55. Ranjan, V. and Diamond, S.L. *Biochem. Biophys. Res. Commun.* **196**, 79–84 (1993).

56. Ranjan, V., Waterbury, R., Xiao, Z., and Diamond, S.L. *Biotechnol. Bioeng.* **49**, 383–390 (1996).

57. Lan, Q., Mercurius, K.O., and Davies, P. *Biochem. Biophys. Res. Commun.* **201**, 950–956 (1994).

58. Kostyk, S.K., Kourembanas, S., Wheeler, E.L., Medeiros, D., McQuillan, L.P., D'Amore, P.A., and Braunhut, S.J. *Am. J. Physiol.* **269**, H1583–H1589 (1995).

59. Ohara, Y., Sayegh, H.S., Yamin, J.J., and Harrison, D.G. *Hypertension* **25**, 415–420 (1995).

60. Yoshizumi, M., Perrella, M.A., Burnett, J.C., Jr., and Lee, M.-E. *Circ. Res.* **73**, 205–209 (1993).

61. Rosenkranz-Weiss, P., Sessa, W.C., Milstien, S., Kaufman, S., Watson, C.A., and Pober, J.S. *J. Clin. Invest.* **93**, 2236–2243 (1994).

62. Lamontagne, D., Pohl, U., and Busse, R. *Circ. Res.* **70**, 123–130 (1992).

63. Bevan, J.A., Kaley, G., and Rubanyi, G.M. *Flow-Dependent Regulation of Vascular Function*, Oxford University Press, New York, 1995.

64. Huang, P.L., Huang, Z., Mashimo, H., Bloch, K.D., Moskowitz, M.A., Bevan, J.A., and Fishman, M.C. *Nature* **377**, 239–242 (1995).

65. Brüne, B., Mohr, S., and Messmer, U.K. *Rev. Physiol. Biochem. Pharmacol.* **127**, 1–30 (1995).

66. Kanagy, N.L., Charpie, J.R., and Webb, R.C. *Med. Hypotheses* **44**, 159–164 (1995).

67. Brüne, B. and Lapetina, E.G. *J. Biol. Chem.* **264**, 8455–8458 (1989).

68. Mohr, S., Stamler, J.S., and Brüne, B. *FEBS Lett.* **348**, 223–227 (1994).

69. Genaro, A.M., Hortelano, S., Alvarez, A., Martinez-A.C., and Boscá, L. *J. Clin. Invest.* **95**, 1884–1890 (1995).

70. Martin, W., White, D.G., and Henderson, A.H. *Br. J. Pharmacol.* **93**, 229–239 (1988).

71. Kobayashi, S., Kanaide, H., and Nakamura, M. *Science* **299**, 533–556 (1985).

72. Hassid, A. *Am. J. Physiol.* **251**, C681–C686 (1986).

73. Kai, H., Kanaide, H., Matsumoto, T., and Nakamura, M. *FEBS Lett.* **221**, 284–288 (1987).

74. Takeuchi, T., Kishi, M., Ishii, T., Nishio, H., and Hata, F. *Br. J. Pharmacol.* **117**, 1204–1208 (1996).

75. Bolotina, V.M., Najibi, S., Palacino, J.J., Pagano, P.J., and Cohen, R.A. *Nature* **368**, 850–853 (1994).

76. Archer, S.L., Huang, J.M.C., Hampl, V., Nelson, D.P., Schultz, P.J., and Weir, E.K. *Proc. Natl. Acad. Sci. USA* **91**, 7583–7587 (1994).

77. Brown, G.C. *FEBS Lett.* **369**, 136–139 (1995).

78. Jia, L., Bonaventura, C., Bonaventura, J., and Stamler, J.S. *Nature* **380**, 221–226 (1996).

79. Salvemini, D., Misko, T.P., Masferrer, J.L., Siebert, K., Currie, M.G., and Needleman, P. *Proc. Natl. Acad. Sci. USA* **90,** 7240–7244 (1993).

80. Salvemini, D., Seibert, K., Masferrer, J.L., Misko, T.P., Currie, M.G., and Needleman, P. *J. Clin. Invest.* **93,** 1940–1947 (1994).

81. Davidge, S.T., Baker, P.N., McLaughlin, M.K., and Roberts, J.M. *Circ. Res.* **77,** 274–283 (1995).

82. Sautebin, L., Ialenti, A., Ianaro, A., and Di Rosa, M. *Br. J. Pharmacol.* **114,** 323–328 (1995).

83. Rubanyi, G.M. and Vanhoutte, P.M. *Circ. Res.* **61** (suppl II), II-61–II-67 (1987).

84. Komori, K. and Vanhouette, P.M. *Blood Vessels* **27,** 238–245 (1990).

85. Bauersachs, J., Hecker, M., and Busse, R. *Br. J. Pharmacol.* **113,** 1548–1553 (1994).

86. Hecker, M., Bara, A.T., Bauersachs, J., and Busse, R. *J. Physiol.* **481,** 407–414 (1994).

87. Fulton, D., McGiff, J.C., and Quilley, J. *Br. J. Pharmacol.* **113,** 954–958 (1994).

88. Bauersachs, J., Popp, R., Hecker, M., Sauer, E., Fleming, I., and Busse, R. *Circulation* (in press).

89. Walter, U. *Rev. Physiol. Biochem. Pharmacol.* **113,** 41–88 (1989).

90. Shah, A.M., Spurgeon, H., Sollott, S.J., Talo, A., and Lakatta, E.G. *Circ. Res.* **74,** 970–978 (1994).

91. Méry, P.-F., Pavoine, C., Belhassen, L., Pecker, F., and Fischmeister, R. *J. Biol. Chem.* **268,** 26286–26295 (1993).

92. Zeiher, A.M., Fisslthaler, B., Schray-Utz, B., and Busse, R. *Circ. Res.* **76,** 980–986 (1995).

93. De Caterina, R., Libby, P., Peng, H.-B., Thannickal, V.J., Rajavashisth, T.B., Gimbrone, M.A., Jr., Shin, W.S. & Liao, J.K. *J. Clin. Invest.* **96,** 60–68 (1995).

94. Fleming, I. and Busse, R. *J. Vasc. Res.* **33,** 181–194 (1996).

95. Nabel, E.G., Selwyn, A.P., and Ganz, P. *J. Am. Col. Cardiol.* **16,** 349–356 (1990).

96. Verbeuren, T.J., Jordaens, F.H., Zonnekeyn, L.L., van Hove, C.E., Coene, M.C., and Herman, A.G. *Circ. Res.* **53,** 63 (1983).

97. Ludmer, P.L., Selwyn, A.P., Shook, T.L., Wayne, R.R., Mudge, G.H., Alexander, R.W., and Ganz, P. *N. Engl. J. Med.* **315,** 1046–1051 (1986).

98. Freiman, P.C., Mitchell, G.G., Heistad, D.D., Armstrong, M.L., and Harrison, D.G. *Circ. Res.* **58,** 783–789 (1986).

99. Bosaller, C., Habib, G.B., Yamamoto, H., Williams, C., Wells, S., and Henry, P.D. *J. Clin. Invest.* **79,** 170–182 (1987).

100. Chester, A.H., O'Neil, G.S., Moncada, S., Tadjkarimi, S., and Yacoub, M.H. *Lancet* **336,** 897–900 (1990).

101. Creager, M.A., Cooke, J.P., Mendelsohn, M.E., Gallagher, S.H., Coleman, S.M., Loscalzo, J., and Dzau, V.J. *J. Clin. Invest.* **86,** 228–234 (1990).

102. Zeiher, A.M., Drexler, H., Wollschläger, H., and Just, H. *Circulation* **84,** 1984–1992 (1991).

103. Zeiher, A.H., Drexler, H., Wollschläger, H., and Just, H. *Circulation* **83**, 391–401 (1991).

104. Egashira, K., Inou, T., Hirooka, Y., Yamada, A., Maruoka, Y., Kai, H., Sugimachi, M., Suzuki, S., and Takeshita, A. *J. Clin. Invest.* **91**, 29–37 (1993).

105. Egashira, K., Inou, T., Hirooka, Y., Yamada, A., Urabe, Y., and Takeshita, A. *N. Engl. J. Med.* **328**, 1659–1664 (1993).

106. Reddy, K.G., Nair, R.N., Sheehan, H.M., and Hodgson, J.M. *J. Am. Coll. Cardiol.* **23**, 833–843 (1994).

107. Mügge, A., Elwell, J.H., Peterson, T.E., Hofmeyer, T.G., Heistad, D.D., and Harrison, D.G. *Circ. Res.* **69**, 1293–1300 (1991).

108. Minor, R.L., Jr., Myer, P.R., Guerra, R., Jr., Bates, J.N., and Harrison, D.G. *J. Clin. Invest.* **86**, 2109–2116 (1990).

109. von der Leyen, H.E., Gibbons, G.H., Morishita, R., Lewis, N.P., Zhang, L., Nakajima, M., Kaneda, Y., Cooke, J.P., and Dzau, V. *J. Proc. Natl. Acad. Sci. USA* **92**, 1137–1141 (1995).

110. Sanders, M. *Pharm. Ther.* **61**, 109–153 (1994).

111. Davenpeck, K.L., Gauthier, T.W., and Lefer, A.M. *Gastroenterology* **107**, 1050–1058 (1994).

112. Rösen, P., Schwippert, P., Kaufman, B., and Tschope, D. *Platelets* **11**, 42–57 (1994).

113. Gauthier, T.W., Davenpeck, K.L., and Lefer, A.M. *Am. J. Physiol.* **267**, G562–G568 (1994).

114. Ohara, Y., Peterson, T.E., and Harrison, D.G. *J. Clin. Invest* **91**, 2546–2551 (1993).

115. Gaboury, J.P., Anderson, D.C., and Kubes, P. *Am. J. Physiol.* **266**, H637–H642 (1994).

116. Ylä-Herttuala, S., Lipton, B.A., Rosenfeld, M.E., Särkioja, T., Yoshimura, T., Leonard, E.J., Witztum, J.L., and Steinberg, D. *Proc. Natl. Acad. Sci. USA* **88**, 5252–5256 (1991).

117. Rollins, B.J., Yoshimura, T., Leonard, E.J., and Pober, J.S. *Am. J. Pathol.* **136**, 1229–1233 (1990).

118. Satriano, J.A., Hora, K., Shan, Z., Stanley, E.R., Mori, T., and Schlondorff, D. *J. Immunol.* **150**, 1971–1978 (1993).

119. Cushing, S.D., Berliner, J.A., Valente, A.J., Territo, M.C., Navab, M., Parhami, F., Gerrity, R., Schwartz, C.J., and Fogelman, A.M. *Proc. Natl. Acad. Sci. USA* **87**, 5134–5138 (1990).

120. Shyy, Y.-J., Hsieh, H.-J., Usami, S., and Chien, S. *Proc. Natl. Acad. Sci. USA* **91**, 4678–4682 (1994).

121. Peng, H.-B., Rajavashisth, T.B., Libby, P., and Liao, J.K. *J. Biol. Chem.* **270**, 17050–17055 (1995).

122. Peng, H.-B., Libby, P., and Liao, J.K. *J. Biol. Chem.* **270**, 14214–14219 (1995).

123. Meyer, M., Schreck, R., and Baeuerle, P.A. *EMBO J.* **12**, 2005–2015 (1993).

124. Tabuchi, A., Sano, K., Oh, E., Tsuchiya, T., and Tsuda, M. *FEBS Lett.* **351**, 123–127 (1994).

9

Role of NO in the Function of the Juxtaglomerular Apparatus

Jürgen Schnermann and Josie P. Briggs

1. Introduction

Juxtaglomerular apparatus (JGA) is the anatomical term for a conglomeration of specialized cells positioned at the vascular pole of the renal glomerulus. The constituents of this cell complex are epithelial, interstitial, and modified smooth muscle cells. The epithelial cells of the JGA are located at the distal end of the thick ascending limbs, a part of the nephron that always returns to the vascular pole of its parent glomerulus. The tubular cells in this contact area, called the macula densa (MD) cells, are cytologically distinct from the surrounding thick ascending limb cells. Underlying the MD cell plaque and filling the space between it and the arterioles is a cushionlike complex of specialized interstitial cells, called the extraglomerular mesangium (EGM). In their fine structure, EGM cells are similar to intraglomerular mesangial cells. EGM cells are coupled by an extensive network of gap junctions with each other. They are also coupled with vascular smooth muscle cells in the afferent arteriole and with the renin-containing granular cells in the media of the arteriolar wall. This anatomical arrangement is the probable route for a communication pathway along which changes in tubular fluid composition in the tubular lumen at the macula densa initiate successive alterations in the functional state of MD cells, EGM cells, and, finally, vascular smooth muscle and granular cells.

2. Function of the Juxtaglomerular Apparatus

The MD cells are found at a location in the nephron where NaCl concentrations are typically hypotonic, and are subject to substantial variation. This variability is almost exclusively a reflection of changes in the rate of fluid delivery from the proximal tubule and, therefore, of the relative rates of glomerular filtration rate (GFR) and proximal tubule reabsorption. The physiological range of salt concentrations at the MD may extend from 15–20 mEq/L in low flow states to 60–70 mEq/L in states where flow rates are high, such as extracellular volume expansion. There is now widespread consensus that variations in NaCl concentration at the MD have at least two distinct physiological effects—a change in afferent arteriolar tone and a change in the rate of renin secretion [1,2]. As shown in Fig. 9-1, increasing salt concentration within the physiological range causes constriction of the afferent arteriole leading to a decline in GFR (left) and an inhibition of renin secretion (right) [3–5]. The effect on vascular tone is commonly called tubuloglomerular feedback (TGF) and has been shown to operate as a local homeostatic feedback loop, stabilizing salt concentration at the MD. The renin-secretory effect appears to be important to maintain activation of the renin–angiotensin system at a level appropriate for the volume state of the organism. Both effects have been observed in isolated tubule/vascular preparations, establishing MD-dependent mechanisms as autonomous responses independent of systemic and nervous inputs [2,4–7]. Thus, vascular smooth muscle cells in the terminal afferent arteriole and renin-secreting granular cells constitute the effector cells for entirely local pathways for the regulation of the rates of glomerular filtration and renin release. The focus of this chapter will be to discuss the possible role of nitric oxide in these two local regulatory responses.

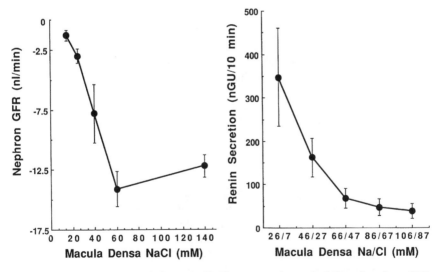

Figure 9-1. *Left:* Relationship between NaCl concentration at the MD and nephron GFR in the in situ rat kidney (data redrawn from Ref. 3). MD NaCl concentrations were altered by retrograde microperfusion. *Right:* Relationship between MD NaCl concentration and renin secretion in the isolated perfused JGA preparation from rabbit kidney (data are from Ref. 5).

3. Nitric Oxide Synthases in the JGA

3.1. Neuronal NOS

Investigation of the role of NO in the dual function of the juxtaglomerular apparatus began with the striking observation in two laboratories that MD cells express the neuronal isoform of NO synthase (ncNOS) at a level much higher than any other cells in the renal cortex [8,9]. Identity of the ncNOS isoform in MD cells was established at the protein level by immunocytochemistry, and at the mRNA level by in situ hybridization and reverse polymerase chain reaction (RT-PCR) [8–12]. The presence of NOS was also confirmed by the positive NADPH-dependent diaphorase reactivity in MD cells, a finding that was reported several decades ago but that has only recently been recognized to reflect NOS activity [13–15]. In the MD cells, the distribution of ncNOS appears to be cytosolic. Occasionally, other epithelial cells in the vicinity of the glomerulus, but not in continuity with the MD, have been noted to show ncNOS positivity [10]. Independent of the exact topography of NOS positive cells, expression of ncNOS and Tamm–Horsfall protein appears to be mutually exclusive [10,11]. Lower levels of ncNOS expression have also been described in the efferent arteriole and occasionally in nerves along preglomerular arterioles [10].

There is some evidence that a constitutive NOS might also be expressed by the mesangium. In cultured mesangial cells, NO release increased in response to a reduction in ambient Cl concentration, an effect that was greatly reduced by a calmodulin inhibitor or by intracellular calcium chelation, evidence for the presence of a constitutive, Ca/calmodulin-dependent NOS [16]. On the other hand, ncNOS immunoreactivity has not been seen in the mesangial cell field in fresh tissue sections, and ncNOS PCR products were very low in glomeruli dissected free of the attached macula densa, indicating that ncNOS expression in native mesangial cells, if present, must be markedly lower than in MD cells [8–12].

3.2. Endothelial NOS

Immunocytochemistry for the endothelial isoform, ecNOS, shows the expected presence in endothelial cells of both afferent and efferent arterioles, as well as in the endothelium of glomerular capillaries and at the luminal surface of larger intrarenal blood vessels [10]. Expression of ecNOS in renal vessels and in glomeruli was confirmed at the mRNA level by RT-PCR [12,17]. This distribution is consistent with the functional evidence showing that the tone of isolated arterioles, particularly of the afferent arteriole, is affected by alterations in basal NO production [18–21]. It is noteworthy that the effect of NO inhibition on the basal tone of efferent arterioles is less prominent even though the immunocytochemistry results indicate that efferent arterioles possess NO synthases [19].

3.3. Inducible NOS

In many tissues, expression of the third NOS isoform, iNOS, requires induction by proinflammatory cytokines. However, recent studies indicate that iNOS is constitutively expressed in a number of sites in the kidney. Expression has been observed in the wall of large renal vessels and of the terminal afferent arteriole [22,23], in glomeruli [24,25], and in several nephron segments, particularly medullary thick ascending limbs [22,24,25]. Sequencing of PCR products established the presence of cDNA for two highly homologous iNOS isoforms which could be differentiated by restriction analysis [24]. One isoform originally cloned from rat vascular smooth muscle cells (vsmNOS) [26] was found at the vascular expression sites, whereas a rat homolog of the other isoform cloned originally from mouse macrophages (macNOS) [27] was found in glomeruli and epithelial cells [24]. Upon stimulation with lipopolysaccharide, iNOS expression increased in all structures in which the enzyme was found under basal conditions [22,24,25] In the context of the current discussion, it is of special interest which cells in the glomerulus express iNOS in a constitutive manner. Because both iNOS isoforms were found to be expressed in cultured mesangial cells [24], it is likely that glomerular expression reflects the presence of iNOS in this cell type. Consistent with the conclusion that iNOS is present in mesangial cells in culture are studies showing a several-fold upregulation of NO production and cGMP generation following lipopolysaccharide treatment [24,28,29].

Figure 9-2 summarizes schematically the distribution of NOS isoforms in the JGA. There is evidence that almost all cell types, including the MD cells, mesangial cells, vascular smooth muscle cells, and endothelial cells, possess some form of NOS and could, therefore, serve as a source for locally active nitric oxide. In

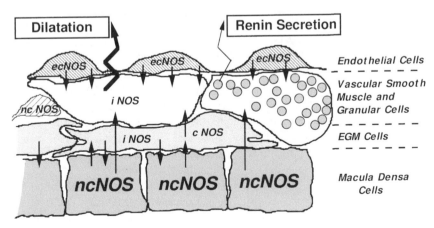

Figure 9-2. Schematic diagram showing the different cell types in the juxtaglomerular region and their expression of NO synthase isoforms.

addition, NO may be released from nerves associated with afferent arterioles. Because each cell type in the JGA appears to express a single dominant isoform (ncNOS in MD cells, iNOS in mesangial and vascular smooth muscle cells, and ecNOS in endothelial cells), differential regulation of NOS synthesis and NO production by the different JG cells seems possible. Nevertheless, release of NO from different sources within the JGA area creates a multiplicity of potential interactions, so that the effect of manipulation of NO formation on the two MD-dependent end points may not always be predictable.

4. Nitric Oxide and Tubuloglomerular Feedback

4.1. NOS Inhibition and Augmentation of TGF

A possible involvement of NO in the tubuloglomerular feedback (TGF) mechanism has been investigated by examining the effect of the pharmacological blockade of NO synthesis on the magnitude of NaCl-dependent vascular responses. In the first of these studies, perfusion of the loop of Henle with N^ω-methyl-L-arginine (L-NMA) or pyocyanin caused a concentration-dependent reduction of stop flow pressure greater than that caused by perfusion with vehicle alone [9]. In a subsequent series of experiments, an assessment of the flow dependency of stop flow pressure in the presence and absence of the NOS blocker N^ω-nitro-L-arginine (L-NNA) showed a dramatic increase in the maximum response magnitude and a significant left shift of $V_{1/2}$, the flow rate associated with the half-maximum response [30]. To obtain these marked effects, an L-NNA concentration of 10^{-3} was required (Fig. 9-3, left). Responses have also been shown to be augmented in an in vitro system for studying TGF which employs perfusion of both afferent arterioles and the tubular segment which includes the MD (Fig. 9-3, right) [31]. In control conditions, changing the tubular perfusate from a low-NaCl (Na 26, Cl 7 mEq/L) to a high-NaCl (Na 144, Cl 122 mEq/L) solution caused a 14% reduction of the afferent arteriolar diameter [31]; when 10^{-5} M of N^ω-nitro-L-arginine methyl ester (L-NAME) was added to the high-NaCl perfusate, the response was significantly augmented, but L-NAME did not affect the afferent diameter when the luminal NaCl concentration was low [31]. The question of the identity of the NO synthase involved in TGF attenuation has been addressed by using agents which reportedly inhibit NO formation in an isoform-specific manner [32,33]. A selective blocker of ncNOS, 7-nitro indazole, had effects comparable to those of the unspecific blockers, whereas aminoguanidine, a blocker of iNOS activity, was without significant effects [34].

The augmentation of TGF responses to elevated NaCl by NOS inhibition has essentially excluded the possibility that NO released by MD cells could be the actual flow-dependent mediator of the vascular response. It has been suggested instead that MD-derived NO tonically downregulates TGF-induced vasoconstriction. Modulation of TGF responses by NO could also be causal in the alteration

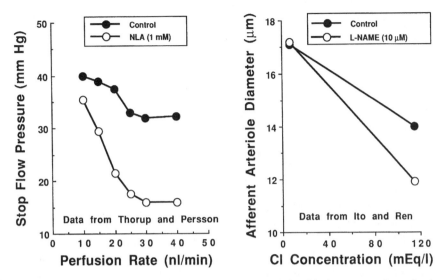

Figure 9-3. *Left:* Effect of NOS inhibition on the relationship between end-proximal perfusion rate and stop flow pressure in the in situ rat kidney (data redrawn from Ref. 30). *Right:* Effect of NOS inhibition on the relationship between MD chloride concentration and afferent arteriolar diameter in the isolated perfused MD/arteriole preparation from rabbit kidney (data redrawn from Ref. 31).

of the TGF response magnitude observed during changes in the volume state of the animal. Salt deprivation typically augments TGF responses, whereas responses are blunted or abolished in volume expansion. Salt-dependent regulation of NO release from MD cells (stimulation in volume expansion and inhibition in volume depletion) could be responsible for this resetting of the TGF response magnitude [9,30]. In fact, preliminary evidence suggests that the TGF-augmenting effect of NO blockade is only demonstrable in high-salt-fed animals, but not in animals on a low-salt diet, a finding in line with the assumption of higher NO formation in volume expansion [35]. Whereas the evidence summarized above provides compelling evidence for a role of locally generated NO in MD-induced changes in vascular tone, a number of questions related to the exact mechanism of action of JGA-derived nitric oxide have remained.

4.2. Is NO Generation in MD Cells Regulated by Tubular NaCl Concentration?

One critical unresolved issue is the question of whether MD cells release NO at a constant rate or whether NO production rate by MD cells is variable and regulated by luminal fluid composition. The functional data summarized in Fig. 9-3 indicate that the effect of NOS blockade on TGF was most pronounced at high flow rates or high NaCl concentrations. In addition, intraperitoneal administration of the ncNOS blocker 7-nitro indazole did not affect resting stop flow pressure, a condition where the MD segment is not perfused [34]. These functional observations would suggest that local NO levels are negligible at low luminal NaCl concentrations and that generation and release of NO from MD cells increases when NaCl concentration at the MD rises. Unfortunately, direct measurements of NO concentrations in the juxtaglomerular interstitium in response to changes in luminal NaCl concentration currently are not available. Although the performance of such determinations would seem to be a formidable task, an NO-sensitive microsensor with appropriate tip dimensions has been described, and it is possible that this electrode may be applicable to measurements of NO levels in the juxtaglomerular interstitium [36,37].

Because the activity of constitutive NO synthases is calcium dependent, measurements of changes in MD cytosolic calcium are another, although more indirect, approach to predict the effect of changes in luminal NaCl concentration on MD NOS activity. Measurements of this type have been performed in two laboratories [38,39]. Because both small increments and small decrements in MD $[Ca]_i$ have been reported with large changes in luminal NaCl concentration, it is probably fair to conclude that calcium in MD cells appears to change very little in response to changing luminal NaCl concentrations. It is also of note that the addition of calmodulin inhibitors to the luminal fluid does not mimic the effect of NOS blockers [40].

Flow-dependent increments in NO production without changes in NOS activity could also occur if the availability of L-arginine was rate limiting and dependent on tubular flow rate. In all perfusion studies discussed above, the luminal perfusate did not initially contain L-arginine. Although L-arginine might diffuse into the tubular lumen, one would predict increases in flow to be associated with lower levels at the MD. Therefore, changes in arginine delivery to the apical membrane of MD cells could not have been responsible for an acute increase in NO production by MD cells under high-flow conditions, even though it might conceivably augment NO production at low flows. Experimental evidence against luminal L-arginine being rate limiting for NO generation is furnished by the observation that the addition of L-arginine in physiological concentrations to the perfusate did not alter the TGF response magnitude, although some TGF blunting was noted at very high concentrations of L-arginine [9,34]. Finally, although L-arginine uptake across the basolateral membrane is possible, it is unclear how such a transport step could be altered by luminal flow rate.

A reduction in extracellular Cl concentration has been reported to cause an increase in NO release from cultured mesangial cells [16]. Sizable variations of Cl concentration which track Cl levels in the luminal fluid occur in the juxtaglomerular interstitium of amphiuma kidneys [14]. If similar changes in extracellular Cl are elicited in the mammalian kidney and if NO synthesis in native extraglomerular mesangial cells is similarly sensitive to Cl concentration, it is possible that NO release from mesangial cells could increase at low luminal Cl concentrations. However, because mesangial NO production would be expected to be enhanced at low luminal Cl, such an interaction should reduce the dilator component of the TGF mechanism, a prediction opposite to the observation that NOS inhibitors enhance its constrictor effect (Fig. 9-3).

4.3. Are the MD Cells the Only Target Cells for Luminally Applied NOS Blockers?

Ideally, pharmacological NOS inhibitors should be targeted to a single cell type of the JGA to identify the cellular source of NO involved in JGA signaling pathways. Because NOS isoform expression in different JGA cells shows a typical cell-specific pattern, studies of the effects of isoform-specific inhibitors could be helpful in localizing the NO-generating cells. The recent findings with 7-nitro indazole, a possible preferential ncNOS inhibitor [34], would support the notion that the NO involved in the TGF response is MD derived. However, this approach is vulnerable to the concern that the agents may show less than complete specificity. In fact, several recent studies have questioned the specificity of 7-nitro indazole and aminoguanidine as selective inhibitors of ncNOS and iNOS, respectively [42–44].

In the perfusion experiments using nonspecific inhibitors, the assumption of selective targeting of MD ncNOS rests solely on the experimental approach of including the agents in the luminal perfusate, thereby providing direct access to the apical side of the MD cells. Although MD involvement is likely in this approach, it is possible that luminally applied NOS inhibitors may be absorbed to some extent and that their subsequent interaction with mesangial or endothelial NO production is responsible for the observed effects. Direct experimental evidence for this possibility comes from a study in which luminal NLA infusion was noted to augment TGF responses in an adjacent nephron [45]. In this situation, one must assume that the inhibitor (NLA) diffused out of the nephron and interacted with an extratubular, probably endothelial NOS at the vascular pole of the neighboring tubule. In addition, the full TGF-enhancing effect developed relatively slowly, consistent with a delayed access of the NOS blocker to the enzyme [45]. Thus, the effect of luminally administered NOS blockers appears to be mediated at least in part by an interaction of the blocking agent with a non-MD NOS. This conclusion is consistent with the results of a study in which the systemic administration of an NOS inhibitor was observed to greatly increase the TGF response magnitude and to lower $V_{1/2}$, even though the luminal perfusate did not contain the inhibitor [46]. The increase in arterial blood pressure associated with systemic NOS blockade, a change that in itself is expected to augment TGF responses [47], could account for only about one-third of the observed TGF enhancements [46]. Although the effect of systemic NOS blockade on MD ncNOS is not known, these data support the notion that blockade of endothelial NOS can enhance TGF responses.

The evidence in favor of a participation of ecNOS in TGF control may also explain why the TGF-enhancing effect appears to be more pronounced at high flow rates. Because in addition to flow velocity a reduction in vessel radius determines fluid shear stress [48] it is conceivable that the vasoconstriction caused by elevated luminal NACl concentrations is responsible for an increase in endothelial NO production mediated by an increase in wall shear stress. Using the degree of vasoconstriction observed in the in vitro studies of Ito and Ren [31] and the reduction in plasma flow observed in vitro [49], one can estimate that afferent arteriolar shear stress probably increases by as much as 25% at high NaCl at the MD. This could be an underestimate if the use of a noncolloidal solution in the in vitro perfusion studies was associated with a smaller reduction in perfusion flow. Although numerous details in the regulation of endothelial NO production are still unclear, there is unequivocal evidence that wall shear stress is an important regulator of both ecNOS mRNA expression, presumably through specific shear stress response elements and other transcription factor-binding sites in the promoter region of the ecNOS gene [50–53], and of NOS enzyme activity, probably at least in part through shear stress-induced changes in cytosolic calcium [54–56]. The assumption of a TGF-stimulated increase in the production of endothelial NOS would be consistent with the blocker effect being greater at higher flows [30,31]. It is also consistent with the observation that the reduction of stop flow pressure caused by peritubular administration of an NOS blocker was marginal in the absence of loop perfusion but increased when the TGF mechanism was activated by high NaCl at the MD [9]. Finally, furosemide was found to eliminate the TGF-augmenting effect of NOS blockade [9]. During the administration of furosemide, afferent arteriolar constriction is blocked, eliminating the cause for increased endothelial NO production. Thus, it is conceivable that the attenuation of the TGF response by nitric oxide at high flows, for the most part, does not reflect increased release of NO from MD cells, but that it is caused by stimulation of NO release from endothelial cells.

In the isolated MD/afferent arteriole double-perfusion approach, it was shown that tubular application of L-NAME did not alter the vasodilatory effect of intravascular acetylcholine in norepinephrine preconstricted arterioles [31]. This finding would seem to indicate that endothelial NOS activity was not inhibited by luminal administration of the nonspecific NOS inhibitor in this preparation. This conclusion would be justified if the vasodilator effect of acetylcholine is fully dependent on NO. However, studies in the isolated hydronephrotic kidney indicate that this may not always be the case [57].

4.4. NO and Renin–Angiotensin System Interactions in TGF

Agents which modify the renin–angiotensin system, angiotensin II itself as well as converting enzyme and angiotensin receptor blockers, have been consistently found to modify both TGF responses and the renal vascular responses to NO. Interference by NO with the local action of angiotensin II may furnish an alternate possible explanation for the apparent flow dependency of the effect of NOS blockade. Several studies have shown that NOS blockade modulates the constrictor effect of angiotensin II in glomerular arterioles. For example, the constrictor effect of L-NNA in the juxtamedullary nephron preparation was attenuated by simultaneous blockade of angiotensin receptors, suggesting that endogenous angiotensin II actively contributed to the vasoconstriction caused by acute NO removal [58]. Furthermore, L-NAME enhanced the effect of angiotensin II in isolated perfused afferent arterioles [20]. Similar interactions between NO and angiotensin II have been observed in the intact kidney [59–61]. On the other hand, angiotensin II infused either systemically or into peritubular capillaries has been found to specifically enhance TGF responses [62,63]. The changes in TGF curve characteristics caused by angiotensin II and by NOS blockers are similar in that both interventions reduce $V_{1/2}$ and increase maximum responses, particularly in the range of elevated flow rates. Thus, it is possible that locally formed NO antagonizes some component of the TGF response that is angiotensin II-modulated. Since this component manifests itself as an enhanced constriction in response to a high NaCL concentration at the MD, NOS blockers may be expected to further augment this constrictor response without affecting the vasodilated state at low NaCl concentrations at the MD, where the angiotensin II component is largely inactivated.

4.5. NO and Resetting of TGF in Volume Expansion

Resetting of the magnitude of the constrictor response to MD NaCl has been identified as one of the characteristics of the TGF system [64]. For example, TGF sensitivity is generally increased in states of extracellular volume depletion and reduced in volume expansion. One mechanism for this adaptation appears to be changes in activity of the renin–angiotensin system, but the participation of other regulatory factors is likely. As alluded to previously, an increase in MD NO formation has been suggested to be in part responsible for the resetting of TGF sensitivity during chronic volume expansion [35,65]. Even though ncNOS is a constitutively active enzyme, the presence of numerous cis-acting regulating elements in its promoter region is a clear indication that its rate of expression is regulated [66]. Consistent with this expectation, marked changes in ncNOS mRNA and protein levels have been reported in both extrarenal and intrarenal tissues. Following a 1-week treatment of rats with a low-NaCl diet, ncNOS mRNA levels in MD cells were found to be about sixfold higher than observed in rats maintained on a high-NaCl diet [12]. Similarly, treatment with a low-NaCl diet increased ncNOS immunostaining of MD cells [23]. Thus, a chronic reduction of NaCl concentration at the MD is likely to be accompanied by increased levels of NO generation by MD cells. This adjustment could be interpreted as an appropriate attempt to limit the constrictor effect of an activation of TGF-sensitizing mechanisms, such as the elevated angiotensin II levels caused by salt depletion. The directional change in NO production predicted from the observed change in ncNOS expression is not consistent with the proposal that MD-derived NO is responsible for the change in TGF sensitivity reported to be caused by changes in salt intake. Increased ncNOS mRNA levels in MD cells during low-NaCl intake are unexpected in view of earlier findings showing that the urinary excretion of nitrite and nitrate, the breakdown products of NO, is higher in rats fed a high-salt diet compared to low-salt-treated animals [67]. Furthermore, NO synthase blockers exert greater effects on renal blood flow and renal vascular resistance in high-salt- than in low-salt-treated animals [65], even though the opposite finding had been reported in an earlier study [68]. These findings are reconcilable if one assumes that overall renal NO production as assessed by urinary NO metabolites is dominated by endothelial-cell-derived NO and that endothelial cells both in the renal and extrarenal circulation respond to volume expansion with increased NO formation. This response could, for example, be a consequence of a higher shear stress due to the increased cardiac output. Thus, cell-specific regulation of constitutive NO synthases may well lead to increased NO production by endothelial cells and to reduced NO generation by MD cells. Recent preliminary data show that the administration of a low-salt diet was associated with an about 40% lower plasma level of L-arginine than seen in high-salt-fed animals [69]. Thus, substrate availability may be responsible for a reduced NO production by renal epithelial cells in states of volume depletion. On the other hand, proximal tubules have a high capacity to generate L-arginine and to transport it into the peritubular capillaries [70,71]. Thus, it is not clear that measurements of systemic L-arginine are a reliable index of intrarenal substrate levels.

4.6. Summary

The addition of various competitive blockers to the fluid perfusing individual loops of Henle is associated with an augmented TGF response to elevations in MD NaCl concentration. This observation is consistent with locally formed NO attenuating the vasoconstrictor limb of the TGF mechanism. Flow-dependence of this effect may be due to increased NO production at high flow rates or to an attenuation of the angiotensin II-dependent component of the TGF constrictor mechanism. In addition, luminally applied NOS blockers are absorbed to some extent and are likely to interact with the endothelial NOS in the terminal afferent arterioles. This interaction may be more pronounced at elevated flow rates, where arteriolar shear stress and NO formation may be augmented as a result of TGF-induced vasoconstriction. These observations indicate that tonic NO release from several cell types at the glomerular vascular pole can attenuate the vasoconstrictor effect of the flow-dependent TGF-mediating agent. Chronic dietary NaCl restriction is associated with an upregulation of MD NOS activity, suggesting that the increased TGF sensitivity induced by a low-salt diet is not simply a result of decreased NO release from MD cells.

5. Nitric Oxide and Renin Secretion

5.1. Effect of NO on Renin Secretion in Complex Systems

The possibility that the release of renin from juxtaglomerular granular cells is regulated by locally produced nitric oxide has been studied using various approaches. Results from these experiments are remarkably discrepant, with even the direction of the change of the renin secretory response to NO being in dispute. Early studies suggested that NO inhibits renin release. In renal cortical slices from dog and rat kidneys, endogenous NO released from carotid arteries or exogenous NO generated from nitroprusside inhibited renin secretion; conversely, inhibition of NOS with L-NMMA stimulated basal renin release [72–74]. In contrast, in isolated perfused kidneys, the administration of NOS blockers caused inhibition of renin secretion, and increased NO production from endogenous or exogenous sources increased renin release [75–77]. Several studies have examined the effect of systemic administration of NOS blockers on renin secretion in intact animals [78–82]. Even though results are difficult to interpret in view of the simultaneous changes in arterial blood pressure and renal resistance, it is noteworthy that, in general, renin release or plasma renin concentration were either unchanged or increased. Because one would expect the increased blood pressure and the suppressed sympathetic nerve activity to enhance any inhibitory effect of NOS blockade, these results seem to contradict the conclusions drawn from the isolated kidney studies that NO is renin stimulatory. Two in vivo studies in which provisions were made to control for the influence of NOS blockade on perfusion pressure or renal sympathetic tone were inconclusive in that both an increase or a decrease in renin secretion were observed [83,84]. We suspect that the apparent complexity of the NO effect is related to the multiple sources of NO in the JGA as well as the possibility of both direct and indirect interactions with granular cells.

5.2. Effect of NO on Renin Secretion in Simpler Systems

To further delineate the effects of NO in the control of renin secretion, simpler experimental systems have been used. There is agreement that short-term exposure of isolated granular cells in primary culture to NO donors causes a dose-dependent inhibition of renin secretion [85–87]. This inhibition was accompanied by an increase in cellular cGMP levels [85]. A functional role of cGMP in NO-induced renin inhibition was suggested by the findings that the effect on renin secretion was prevented by methylene blue and mimicked by exogenous cGMP [86,87]. cGMP dependency is consistent with the earlier observation that atrial natriuretic factor, another compound signaling through cGMP, also reduces renin secretion [88,89]. The use of isolated cells of high purity and the apparent involvement of guanylate cyclase, the most established receptor and second-messenger system of NO, indicate that the direct interaction of NO with granular cells results in inhibition of renin secretion. In agreement with these studies are the results from experiments in the isolated perfused juxtaglomerular apparatus of the rabbit, an in vitro preparation in which renin release is measured under conditions of controlled fluid composition at the MD [90]. In this preparation, agents can be applied either to the external medium superfusing the entire preparation or to the luminal fluid at the macula densa. Addition of nitroprusside or L-arginine to the bath superfusate caused an inhibition of renin secretion when NaCl at the MD was low (i.e., when renin secretion was stimulated). Conversely, bath addition of the NO synthase inhibitor L-NNA caused an increase in renin secretion. The cellular localization of the NOS responding to L-NNA or L-arginine in the superfusate is not known. However, it may be reasonable to speculate, mainly on anatomical grounds, that the most important source of NO with direct access to the granular cells is the endothelium, although NO release from the mesangium is another possibility.

Additional observations in the isolated perfused JGA preparation have yielded results that may explain some of the contradictory results discussed above. Results with direct application of drugs to the macula densa were directionally opposite to those observed with the bath application. Luminal application of L-arginine stimulated renin secretion, and NO synthesis blockade with luminal L-NNA reduced it [90]. This effect was most pronounced at low intratubular NaCl concentrations. In fact, because L-NNA inhibited renin release at low NaCl concentration, NaCl-dependent renin secretion was completely abolished (Fig. 9-4). Due to the design of these studies, one may speculate that NO released by MD cells stimulates renin release, a change directionally opposite to the direct inhibitory effect described above. Changes in renin secretion in this preparation cannot be the secondary consequence of MD-dependent changes in vascular tone because the fragments of arterioles present in this preparation are not perfused. Consistent with these findings are recent results showing that the acute stimulation of renin secretion by furosemide, a stimulus that probably utilizes the MD pathway, is attenuated by the administration of L-NAME or of the ncNOS-specific inhibitor 7-nitroindazole [91,92]. Stimulation of renin secretion by NO may be subsequent to MD transport inhibition, a possibility supported by the transport inhibitory effect of NO seen in other renal epithelial cells [93,94]. Alternatively, NO may act locally to modulate the formation of a secondary paracrine agent capable of altering renin secretion.

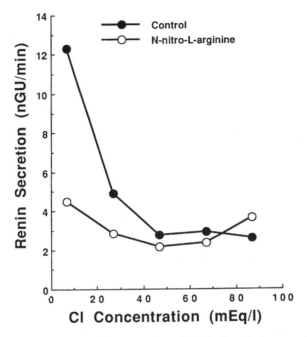

Figure 9-4.   Effect of luminal application of an NOS blocker (*N*-nitro-L-arginine) on the relationship between MD chloride concentration and renin secretion in the isolated perfused MD preparation of rabbit kidney. (Data from Ref. 90.)

5.3. Nitric Oxide–Cyclooxygenase Interactions

Clarification of the signaling mechanisms by which MD-generated NO causes stimulation of renin secretion requires further studies, and an exhaustive discussion of the potential intermediate steps is beyond the scope of this chapter. Nevertheless, it seems pertinent to discuss the recent finding that MD cells show a strikingly high expression of the inducible form of cyclooxygenase (COX_2) [96]. Thus, whereas COX_2 in most tissues is found only at low levels and requires induction by cytokines, MD cells appear to abundantly express this enzyme in a constitutive fashion. The fact that MD cells, a rare cell type of the renal cortex, coexpress specific isoforms of both NO- and prostaglandin-generating enzymes suggests an interaction between these enzymatic products that may be key to understanding MD-dependent regulation. This notion is supported by the observation that COX_2 mRNA and enzyme levels in MD cells, like ncNOS expression [12,23], are markedly upregulated by treatment with a low-NaCl diet [96,97]. In fact, there is abundant functional evidence to show that NO can modulate COX_2 enzyme activity and expression even though the directional change may depend on the cell type under study. For example, NOS inhibition was found to attenuate cytokine- or LPS-induced stimulation of prostanoid production and release by rat pancreatic islets [98], lung tissue [99], vascular smooth muscle cells [100], and a mouse macrophage cell line [101].

In intact animals, L-NMMA and aminoguanidine reduced plasma 6-keto-PGF$_{1\alpha}$ levels and urinary PGE$_2$ and 6-keto-PGF$_{1\alpha}$ excretion during stimulation with LPS [102]. Finally, norepinephrine-stimulated PGE$_2$ release from hypothalamic tissue [103] and bradykinin-induced stimulation of prostanoid production in the isolated perfused kidney [104] was found to be attenuated by NOS blockers, whereas sodium nitroprusside augmented the release of PGE$_2$ and thromboxane from hypothalamic explants [103] or uterine tissue [105], effects that could be prevented by the NO scavenger hemoglobin [103,105]. Although these data suggest a stimulatory influence of NO on the activity or expression of both COX$_2$ and COX$_1$ [101,104], there is also evidence that NO may exert the opposite effect, an inhibition of prostanoid production by COX$_2$ [106,107]. MD-generated NO may interact with COX isoforms in the MD or in mesangial cells and cause a change in the release of prostaglandins and/or thromboxane, agents known to affect vascular tone and renin secretion. This possibility is consistent with our earlier finding that COX inhibitors abolish NaCl-dependent stimulation of renin secretion (Fig. 9-5, left) [108]. Similar to the effect of NOS blockade (Fig. 9-4),

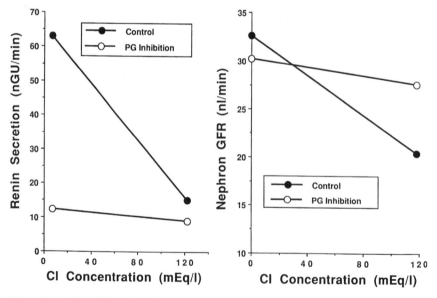

Figure 9-5. Left: Effect of cyclooxygenase inhibitors on the relationship between chloride concentration at the MD and renin secretion in the isolated perfused rabbit JGA preparation (data redrawn from Ref. 108). *Right:* Effect of cyclooxygenase inhibition (indomethacin 0.5 mM) on the relationship between luminal chloride concentration and nephron GFR in the in situ rat kidney. Indomethacin was given by retrograde microperfusion (data redrawn from Ref. 109).

this effect was most pronounced at reduced flow rates. The magnitude of the effect of NOS and COX inhibition on renin secretion suggests that NO and prostaglandins operate in series rather than in parallel. COX inhibitors have also been shown to inhibit TGF responses [109] (Fig. 9-5, right), a finding opposite to that produced by NOS inhibition (Fig. 9-3). Thus, regulation of NaCl-dependent vascular tone by a similar serial interaction between NO and prostaglandins seems less likely.

5.4. Integrating Vascular and Renin Secretory Pathways

Although both MD-mediated mechanisms, the change in vascular tone and the change in renin secretion, are markedly altered by nitric oxide, it is not easy to reconcile the results from these experimental series. Whereas the effect of NOS blockade on TGF responses was most pronounced at high flow rates [9,30,31], renin responses were only affected by NOS blockers at low flow rates [90]. These results are conflicting, as they suggest opposite effects of loop flow rate on NO generation by MD cells. Direct NO measurements have to be awaited to resolve this issue. Nevertheless, there is additional evidence to show that NO contributes to the stimulation of renin secretion during low luminal NaCl concentrations. In contrast to the erratic effect of NOS blockade on renal renin release under control conditions, several studies using widely different approaches indicate that NOS blockers exert a strong inhibitory effect under conditions of reduced renal or afferent arteriolar perfusion pressure [77,80,95]. Some of these results are summarized in Fig. 9-6. Endothelial NO production is likely to be reduced in the low autoregulatory and subautoregulatory blood flow range because of the relative vasodilatation (i.e., reduced cytosolic calcium) and the reduced shear stress. MD NaCl concentration, on the other hand, is probably reduced, resulting in increased NO formation and stimulation of renin secretion. Even though the MD segment was not perfused in the studies shown in the right-hand panel of Fig. 9-5, it is likely that it was present and that the MD cells were exposed to the low luminal NACl concentrations typically generated at very low salt loads.

5.5. Summary

The effect of NO on renin secretion is strikingly complex. Both inhibition and stimulation of renin release has been observed, a divergence that is probably the result of direct and indirect effects of NO on granular cells. Direct interaction of NO with granular cells causes inhibition of renin release, and we assume that the neighboring endothelial cells are the most likely source of this "inhibitory" NO. In contrast, stimulation of renin release may be the result of NO generated and released by MD cells, presumably through an unknown indirect pathway. The stimulatory effect of NO is intricately involved in the increase in renin secretion caused by low luminal NaCl concentration as well as by low renal

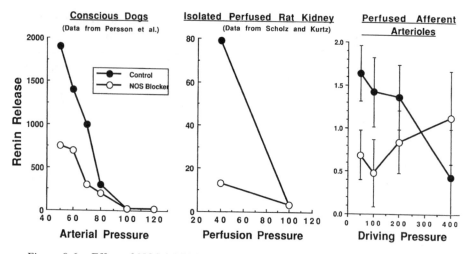

Figure 9-6. Effect of NOS inhibition on the relationship between renin secretion and perfusion pressure. *Left:* Relationship between arterial pressure and renin release (RR) in the conscious dog (data redrawn from Ref. 95; RR calculated from venous–arterial PRA difference and plasma flow with units [ng Ang I/ml h] × [ml/min]). *Middle:* Relationship between perfusion pressure and renin release in isolated perfused rat kidneys (data redrawn from Ref. 77). *Right:* Relationship between driving pressure and renin release from isolated perfused afferent arterioles of the rabbit (unpublished data from He, X.-R., Briggs, J.P., and Schnermann, J.; RR expressed as nano-Goldblatt units (nGU)/min).

perfusion pressure. Due to the opposing actions of NO derived from endothelial and MD cells, net renin release under resting conditions may not be affected markedly by the autocoid. However, when endothelial NO formation is suppressed, as would be expected in states of low plasma flow and reduced renal perfusion pressure, or when MD NO formation is stimulated (as during reduced MD NaCl concentrations), stimulation of renin secretion through the MD-dependent pathway may dominate. On the other hand, in states of high vascular shear stress associated with high MD NaCl (for example, in volume expansion), endothelial NO may dominate contributing to inhibition of renin secretion as well as vasodilatation.

6. Conclusions

The NaCl concentration in the tubular lumen at the MD serves as the signal for a local regulatory pathway controling renin secretion and afferent arteriolar vasomotor tone. The extracellular signaling molecules responsible for this interaction include NO, which is generated by nitric oxide synthases expressed in various cells of the JGA. Locally formed NO attenuates NaCl-dependent vasoconstriction, an effect that is probably subsequent to release of NO from both MD and endothelial cells. NO released from MD cells appears to be responsible for the stimulation of renin secretion by low NaCl, an effect that may be secondary to NO-stimulated formation of prostaglandins. Although substantial progress has been made in implicating NO in the control functions of the JGA, many details have remained unclear. Improved understanding may be expected from technical advances in the measurement of local NO concentrations and in the development of selective NOS and COX inhibitors, as well as from the use of transgenic animals with null mutations in specific NOS or COX isoforms.

References

1. Schnermann, J., Wright, F.S., Davis, J.M., v. Stackelberg, W., and Grill, G. *Pflügers Arch.* **318**, 147–175 (1970).

2. Skøtt, O., and Briggs, J.P. *Science* **237**, 1618–1620 (1987).

3. Schnermann, J., Ploth, D.W., and Hermle, M. *Pflügers Arch.* **362**, 229–240 (1976).

4. Lorenz, J.N., Weihprecht, H., Schnermann, J., Skøtt, O., and Briggs, J.P. *Am. J. Physiol.* **259**, F186–F193 (1990).

5. He, X.-R., Greenberg, S.G., Briggs, J.P., and Schnermann, J. *Hyperten.* **26**, 137–142 (1995).

6. Casellas, D., and Moore, L.C. *Am. J. Physiol.* **258**, F660–F669 (1990).

7. Ito, S. and Carretero, O.A. *Kidney Int.* **38**, 1206–1210 (1990).

8. Mundel, P., Bachmann, S., Bader, M., Fischer, A., Kummer, W., Mayer, B., and Kriz, W. *Kidney Int.* **42**, 1017–1019 (1992).

9. Wilcox, C.S., Welch, W.J., Murad, F., Gross, S.S., Taylor, G., Levi, L., and Schmidt, H.H.W. *Proc. Natl. Acad. Sci. USA* **89**, 11993–11997 (1992).

10. Bachmann, S., Bosse, H.M., and Mundel, P. *Am. J. Physiol.* **268**, F885–F898 (1995).

11. Fischer, E., Schnermann, J., Briggs, J.P., Kriz, W., Ronco, P.M., and Bachmann, S. *Am. J. Physiol.* **268**, F1164–F1176 (1995).

12. Singh, L., Grams, M., Wang, W.-H., Yang, T., Killen, P., Smart, A., Schnermann, J., and Briggs, J.P. *Am. J. Physiol.* **270**, F1027–F1037 (1996).

13. Wachstein, M. and Meisel, E. *Am. J. Pathol.* **35**, 1189–1205 (1959).

14. Brown, J.J., Davies, D.L., Lever, A.F., Parker, R.A., and Robertson, J.I.S. *Clin. Sci.* **30**, 223–235 (1966).

15. Bredt, D.S., Hwang, P.M., Glat, C.E., Lowenstein, C., Reed, R.R., and Snyder, S.H. *Nature* **351**, 714–718 (1991).

16. Tsukuhara, H., Krivenko, Y., Moore, L.C., and Goligorsky, M. *Am. J. Physiol.* **267**, F190–F195 (1994).

17. Ujiie, K., Yuen, J., Hogarth, L., Danziger, R., and Star, R.A. *Am. J. Physiol.* **267**, F296–F302 (1994).

18. Imig, J.D. and Roman, R.J. *Hypertension* **19**, 770–774 (1992).

19. Edwards, R.M. and Trizna, W. *J. Am. Soc. Nephrol.* **4**, 1127–1132 (1993).

20. Ito, S., Arima, S., Ren, Y.L., Juncos, L.A., and Carretero, O.A. *J. Clin. Invest.* **91**, 2012–2019 (1993).

21. Tamaki, T., Hasui, K., Kimura, S., and Abe, Y. *Jpn. J. Pharmacol.* **62**, 231–237 (1993).

22. Tojo, A., Gross, S.S., Zhang, L., Tisher, C.C., Schmidt, H.H.W., Wilcox, C.S., and Madsen, K.M. *J. Am. Soc. Nephrol.* **4**, 1438–1447 (1994).

23. Tojo, A., Madsen, K.M., and Wilcox, C.S. *Jpn. Heart J.* **36**, 389–398 (1995).

24. Mohaupt, M.G., Elzie, J.L., Ahan, K.Y., Clapp, W.L., Wilcox, C.S., and Kone, B.C. *Kidney Int.* **46**, 653–665 (1994).

25. Morrissey, J.J., McCracken, R., Kaneto, H., Vehaskari, M., Montani, D., and Klahr, S. *Kidney Int.* 45, 998–1005 (1994).

26. Nunokawa, Y., Ishida, N., and Tanaka, S. *Biochem. Biophys. Res. Commun.* 191, 89–94 (1993).

27. Xie, Q.-W., Cho, H.J., Calaycay, J., Mumford, R.A., Swiderek, K.M., Lee, T.D., Ding, A., Troso, T., and Nathan, C. *Science* **256**, 225–228 (1992).

28. Shultz, P.J., Tayeh, M.A., Marletta, M.A., and Raij, L. *Am. J. Physiol.* **261**, F600–F606 (1991).

29. Pfeilschifter, J., Rob, P., Mulsch, A., Fandrey, J., Vosbeck, K., and Busse, R. *Eur. J. Biochem.* **203**, 251–255 (1992).

30. Thorup, C. and Persson, A.E.G. *Am. J. Physiol.* **267**, F606–F611 (1994).

31. Ito, S. & Ren, Y.L. J. Clin. Invest. **92**, 1093–1098 (1993).

32. Moore, P.K., Babbedge, R.C., Wallace, P., Gaffen, Z.A., and Hart, S.L. *Br. J. Pharmacol.* **108**, 296–297 (1993).

33. Misko, T.P., Moore, W.M., Kasten, T.P., Nickols, G.A., Corbett, J.A., Tilton, R.G., McDaniel, M.L., Williamson, J.R., and Currie, M.G. *Eur. J. Pharmacol.* **233**, 119–125 (1993).

34. Thorup, C. and Persson, A.E.G. *Kidney Int.* **49**, 430–436 (1996).

35. Welch, W.J. and Wilcox, C.S. *J. Am. Soc. Nephrol.* **4**, 572 (1993).

36. Malinski, T. and Taha, Z. *Nature* **358**, 676–678 (1992).

37. Malinski, T., Taha, Z., Grunfeld, S., Patton, S., Kapturczak, M., and Tomboulian, P. *Biochem. Biophys. Res. Commun.* **193**, 1076–1082 (1993).

38. Bell, P.D., Franco-Guevara, M., Abrahamson, D.R., Lapointe, J.-Y., and Cardinal, J. In: *The Juxtaglomerular Apparatus.* Eds. Persson, A.E.G. and Boberg, U. Elsevier, Amsterdam, 1988, pp. 63–77.

39. Salomonsson, M., Gonzalez, E., Sjölin, L., and Persson, A.E.G. *Acta Physiol. Scand.* **138**, 425–426 (1990).

40. Bell, P.D. *Am. J. Physiol.* **250**, F715–F719 (1986).

41. Persson, B.E., Sakai, T., Ekblom, M., and Marsh, D.J. *Am. J. Physiol.* **254**, F445–F449 (1988).

42. Wolff, D.J., Lubeskie, A., and Umansky, S. *Arch. Biochem. Biophys.* **314,** 360–366 (1994).

43. Wolff, D.J. and Gribin, B.J. *Arch. Biochem. Biophys.* **311,** 300–306 (1995).

44. Laszlo, F., Evans, S.M., and Whittle, B.J. *Eur. J. Pharmacol.* **272,** 169–175 (1995).

45. Braam, B. and Koomans, H.A. *Kidney Int.* **47,** 1252–1257 (1995).

46. Thorup, C., Sundler, F., Ekblad, E., and Persson, A.E.G. *Acta Physiol. Scand.* **148,** 359–360 (1993).

47. Schnermann, J. and Briggs, J.P. *Am. J. Physiol.* **256,** F421–F429 (1989).

48. Hecker, M., Mülsch, A., Bassenge, E., and Busse, R. *Am. J. Physiol.* **265,** H828–H833 (1993).

49. Briggs, J.P. and Wright, F.S. *Am. J. Physiol.* **236,** F40–F47 (1979).

50. Resnick, N., Collins, T., Atkinson, W., Bonthron, D.T., Dewey, C.F., and Gimbrone, M.A. *Proc. Natl. Acad. Sci. USA,* **90,** 4591–4595 (1993).

51. Robinson, L.J., Weremowicz, S., Morton, C.C., and Michel, T. *Genomics,* **19,** 350–357 (1994).

52. Venema, R.C., Nishida, K., Alexander, R.W., Harrison, D.G., and Murphy, T. *J. Biochim. Biophys. Acta* **1218,** 413–420 (1994).

53. Lan, Q., Mercurius, K.O., and Davies, P.F. *Biochem. Biophys. Res. Commun.* **201,** 950–956 (1994).

54. Mo, M., Eskin, S.G., and Schilling, W.P. *Am. J. Physiol.* **260,** H1698–H1707 (1991).

55. Moncada, S., Palmer, R.M.J., and Higgs, E.A. *Pharmacol. Rev.* **43,** 109–142 (1991).

56. Buga, G.M., Gold, M.E., Fukuto, J.M., and Ignarro, L.J. *Hypertension* **17,** 187–193 (1991).

57. Hayashi, K., Loutzenhiser, R., Epstein, M., Suzuki, H., and Saruta, T. *Circ. Res.* **75,** 821–828 (1994).

58. Ohishi, K., Carmines, P.K., Inscho, E.W., and Navar, L.G. *Am. J. Physiol.* **263,** F900–F906 (1992).

59. Sigmon, D.H., Carretero, O.A., and Beierwaltes, W.H. *Hypertension* **20,** 643–650 (1992).

60. Takanaka, T., Mitchell, K.D., and Navar, L.G. *Am. J. Soc. Nephrol.* **4,** 1046–1053 (1993).

61. Alberola, A.M., Salazar, F.J., Nakamura, T., and Granger, J.P. *Am. J. Physiol.* **267,** R1472–R1478 (1994).

62. Mitchell, K.D. and Navar, L.G. *Am. J. Physiol.* **255,** F383–F390 (1988).

63. Schnermann, J. and Briggs, J.P. *Miner. El. Metab.* **15,** 103–107 (1989).

64. Schnermann, J. and Briggs, J.P. In: *The Kidney: Physiology and Pathophysiology,* 2nd ed. Eds. Seldin, D.W. and Giebisch, G. Raven Press, New York, 1992, pp. 1249–1289.

65. Deng, X., Welch, W.J., and Wilcox, C.S. *Kidney Int.* **46,** 639–646 (1994).

66. Hall, A.V., Antoniou, H., Wang, Y., Cheung, A.H., Arbus, A.M., Olson, S.L., Lu, W.C., Kau, C.L., and Marsden, P.A. *J. Biol. Chem.* **269,** 33082–33090 (1994).

67. Shultz, P.J. and Tolins, J.P. *J. Clin. Invest.* **91**, 642–650 (1993).

68. Sigmon, D.H., Carretero, O.A., and Beierwaltes, W.H. *Am. J. Soc. Nephrol.* **3**, 1288–1294 (1992).

69. Deng, X., Welch, W.J., and Wilcox, C.S. *J. Am. Soc. Nephrol.* **6**, 658 (1995).

70. Levillain, O., Hus-Citharel, A., Morel, F., and Bankir, L. *Am. J. Physiol.* **264**, F1038–F1045 (1993).

71. Hus-Citharel, A., Levillain, O., and Morel, F. *Pflügers Arch.* **429**, 485–493 (1995).

72. Vidal, M.J., Romero, J.C., and Vanhoutte, P.M. *Eur. J. Pharmacol.* **149**, 401–402 (1988).

73. Beierwaltes, W.H. and Carretero, P.M. *Hypertension,* **19** (Suppl II), 68–73 (1992).

74. Beierwaltes, W.H. *Hypertension* **23**, I-40–I-44 (1994).

75. Münter, K. and Hackenthal, E. *J. Hypertens.* **9** (Suppl. 6), S236–S237 (1991).

76. Gardes, J., Poux, J.M., Gonzales, M.F., Alhenc-Gelas, F., and Menard, J. *Life Sci.* **50**, 987–993 (1992).

77. Scholz, H. and Kurtz, A. *J. Clin. Invest.* **91**, 1088–1094 (1993).

78. Elsner, D., Muntze, A., Kromer, E.P., and Riegger, G.A. *Am. J. Hyperten.* **5**, 288–291 (1992).

79. Salazar, F.J., Pinilla, J.M., Lopez, F., Romero, J.C. and Quesada, T. *Hypertension* **20**, 113–117 (1992).

80. Naess, P.A., Christensen, G., Kirkeboen, K.A., and Kiil, F. *Acta Physiol. Scand.* **148**, 137–142 (1993).

81. El Karib, A.O., Sheng, J., Betz, A.L., and Malvin, R.L. *Clin. Exp. Hyperten.* **15**, 819–832 (1993).

82. Manning, R.D. and Hu, L. *Hypertension* **23**, 619–625 (1994).

83. Sigmon, D.H., Carretero, O.A., and Beierwaltes, W.H. *Am. J. Physiol.* **263**, F256–F261 (1992).

84. Johnson, R.A. and Freeman, R.H. *Am. J. Physiol.* **266**, R1723–R1729 (1994).

85. Schricker, K. and Kurtz, A. *Am. J. Physiol.* **265**, F180–F186 (1993).

86. Greenberg, S.G., He, X.-R., Schnermann, J.B., and Briggs, J.P. *Am. J. Physiol.* **268**, F948–F952 (1995).

87. Galle, J., Schini, V.B., Stunz, P., Wanner, C., and Schollmeyer, P. *Nephrol. Dial. Transplant.* **10**, 191–197 (1995).

88. Kurtz, A., Della Bruna, R., Pfeilschifter, J., Taugner, R., and Bauer, C. *Proc. Natl. Acad. Sci. USA* **83**, 4769–4773 (1986).

89. Henrich, W.L., McAllister, E.A., Smith, P.B., and Campbell, W.B. *Am. J. Physiol.* **255**, F474–F478 (1988).

90. He, X.-R., Greenberg, S.G., Briggs, J.P., and Schnermann, J.B. *Am. J. Physiol.* **268**, F953–F959 (1995).

91. Reid, I.A. and Chou, L. *Clin. Sci.* **88,** 657–663 (1995).

92. Beierwaltes, W.H. *Am. J. Physiol.* **269,** F134–F139 (1995).

93. Stoos, B.A., Carretero, O.A., Farhy, R.D., Scicli, G., and Garvin, J. *J. Clin. Invest.* **89,** 761–765 (1992).

94. Stoos, B.A., Garcia, N.H., and Garvin, J. *J. Am. Soc. Nephrol.* **6,** 1–6 (1995).

95. Persson, P.B., Baumann, J.E., Ehmke, H., Hackenthal, E., Kirchheim, H.R., and Nafz, B. *Am. J. Physiol.* **264,** F943–F947 (1993).

96. Harris, R.C., McKanna, J.A., Akai, Y., Jacobson, H.R., Dubois, R.N., and Breyer, M.D. *J. Clin. Invest.* **94,** 2504–2510 (1994).

97. Singh, I.J., Yang, T., Smart, A., Schnermann, J., and Briggs, J.P. *J. Am. Soc. Nephrol.* **6,** 761 (1995).

98. Corbett, J.A., Kwon, G., Turk, J., and McDaniel, M.L. *Biochemistry* **32,** 13767–13770 (1993).

99. Sautebin, L. and Di Rosa, M. *Eur. J. Pharmacol.* **262,** 193–196 (1994).

100. Inoue, T., Fukuo, K., Morimoto, S., Koh, E., and Ogihara, T. *Biochem. Biophys. Res. Commun.* **194,** 420–424 (1993).

101. Salvemini, D., Misko, T.P., Masferrer, J.L., Seibert, K., Currie, M.G., and Needleman, P. *Proc. Natl. Acad. Sci. USA* **90,** 7240–7244 (1993).

102. Salvemini, D., Settle, S.L., Masferrer, J.L., Seibert, K., Currie, M.G., and Needleman, P. *Bri. J. Pharmacol.* **114,** 1171–1178 (1995).

103. Rettori, V., Gimeno, M., Lyson, K., and McCann, S.M. *Proc. Natl. Acad. Sci. USA* **89,** 11543–11546 (1993).

104. Salvemini, D., Seibert, K., Masferrer, J.L., Misko, T.P., Currie, M.G., and Needleman, P.J. *Clin. Invest.* **93,** 1940–1947 (1994).

105. Franchi, A.M., Chaud, M., Rettori, V., Suburo, A., McCann, S.M., and Gimeno, M. *Proc. Natl. Acad. Sci. USA* **91,** 539–543 (1994).

106. Swierkosz, T.A., Mitchell, J.A., Warner, T.D., Botting, R.M., and Vane, J.R. *Br. J. Pharmacol.* **114,** 1335–1342 (1995).

107. Stadler, J., Harbrecht, B.G., Di Silvio, M., Curran, R.D., Jordan, M.L., Simmons, R.L., and Billiar, T.R. *J. Leuk. Biol.* **53,** 165–172 (1993).

108. Greenberg, S.G., Lorenz, J.N., He, X.-R., Schnermann, J.B., and Briggs, J.P. *Am. J. Physiol.* **265,** F578–F583 (1993).

109. Schnermann, J., Schubert, G., Hermle, M., Herbst, R., Stowe, N.T., Yarimizu, S., and Weber, P.C. *Pflügers Arch.* **379,** 269–286 (1979).

10

NOS in Mesangial Cells: Physiological and Pathophysiological Roles

Josef Pfeilschifter and Heiko Mühl

1. Introduction

The mesangium is a highly specialized pericapillary tissue that is involved in most pathological processes in the renal glomerulus. Three prominent proinflammatory features of intrinsic mesangial cells evolve as a result of the cross-talk with invading professional immune cells: (i) increased mediator production, (ii) increased matrix production by mesangial cells, and (iii) increased mesangial cell proliferation [1]. Resting mesangial cells do not synthesize any inflammatory mediator constitutively but require a triggering by factors secreted by professional inflammatory cells invading the glomerulus, such as macrophages or leukocytes or by factors present in serum. However, once activated, mesangial cells have the potential to become autonomous in terms of mediator production and they start to secrete a myriad of biologically active substances [2]. The orchestration of the glomerular wounding response must be exact to initiate a sophisticated orderly process of repair. The aberrant production of these mediators, however, may sustain connective tissue accumulation and result in irreversible alteration in glomerular structure and function, to end finally in what pathologists describe as glomerulosclerosis. This review focuses on a highly versatile member of this orchestra of inflammatory mediators, nitric oxide (NO), that plays a crucial role in the pathogenesis of inflammatory and autoimmune diseases.

In this chapter we will focus on NO synthesis by glomerular mesangial cells and its possible roles in glomerular physiology and pathophysiology.

2. Mesangial Cells as a Target for NO

Mesangial cells are a major determinant in the regulation of glomerular filtration rate. Morphologically, mesangial cells resemble vascular smooth muscle cells, able to contract upon stimulation by vasoactive hormones like angiotensin II or vasopressin. These hormones trigger a rapid phospholipase C-mediated hydrolysis of membrane phosphoinositides with a subsequent mobilization of intracellular Ca^{2+} and cell contraction. Vasodilators that stimulate cGMP formation in mesangial cells, such as atrial natriuretic peptides or sodium nitroprusside, as well as cell-permeable cGMP, analogs, attenuate angiotensin II-induced cell contraction. This is due to a cGMP-mediated inhibition of angiotensin II-stimulated phosphoinositide turnover and Ca^{2+} mobilization [3].

In coincubation experiments, it has been demonstrated that NO release from aortic endothelial cells [4] and, more importantly, from glomerular endothelial cells [5] increases the cGMP in mesangial cells. In addition, NO produced by endothelial cells inhibited angiotensin II-induced mesangial cell contraction [4]. Therefore, NO may be an important signaling molecule in the cross-communication between glomerular endothelial and mesangial cells, contributing to normal glomerular physiology.

This view is supported by the report of Garg and Hassid [6], demonstrating that NO and cell-permeable cGMP analogs dose-dependently inhibited serum-stimulated DNA synthesis and mesangial cell proliferation. This growth inhibitory effect of NO may help to preserve the structure of the glomerulus under conditions of increased growth factor production, as it is typically found in certain forms of glomerulonephritis.

Mattana and Singhal [7] observed that supplementing media with additional L-arginine (4 mM) significantly antagonizes the mitogenic effect of endothelin-1 and angiotensin II on mesangial cells. This effect was attenuated by the NOS inhibitor N^{G}-monomethyl-L-arginine (L-NMMA) as well as by the guanylate cyclase inhibitor Methylene Blue, suggesting that the effect of L-arginine is mediated via cGMP generation. The NO donor sodium nitroprusside was also found to suppress thymidine incorporation in mesangial cells exposed to angiotensin II or endothelin-1. Furthermore, the inhibitor of protein kinase G, Rp-8-Br-cGMPS, attenuated the antiproliferative action of L-arginine [7]. These findings may, in part, account for the reported ability of dietary L-arginine supplementation to retard the development of glomerulosclerosis.

In contrast, NO did not inhibit DNA synthesis in logarithmically growing cells [8] and Mohaupt and colleagues [9] observed that endogenously as well as exogenously supplied NO does not suppress platelet-derived growth-factor-stimulated mesangial cell proliferation. In this context, it is important to note that the platelet-derived growth factor has been shown to inhibit inducible NO synthase (iNOS) expression in mesangial cell [10] and, thus, may not allow to generate amounts of NO that are sufficient to block mesangial cell proliferation.

3. Regulation of NO Production by Mesangial Cells

In addition to their contractile, smooth-muscle-like properties, glomerular mesangial cells also have features common to macrophages; that is, they are able to phagocytose and they produce oxygen radicals, cytokines, growth factors, and eicosanoids [2]. It was, therefore, of interest to investigate whether cytokines are able to induce NOS activity in mesangial cells, as it has been observed for activated macrophages. Indeed, mesangial cells are not only a target for NO but are themselves an important source for NO. In 1990, Pfeilschifter and Schwarzenbach [11] produced the first evidence for NOS activity in rat mesangial cells exposed to inflammatory cytokines like interleukin 1β (IL-1β) or tumor necrosis factor α (TNFα). Subsequently, NO formation by rat mesangial cells has been confirmed [12], and NOS activity also reported in bovine [13] and human mesangial cells [14].

Interleukin-1β, TNFα or interferon γ, and bacterial endotoxin were found to stimulate soluble guanylate cyclase in rat and bovine mesangial cells via the NO producing L-arginine pathway. IL-1β and TNFα-stimulated formation of cGMP was dose dependent, with a significant stimulation occurring already at 10 pM [11]. The induction of cGMP formation by cytokines required a lag period of 4–8 h and was inhibited by actinomycin D and cycloheximide, suggesting dependence on new RNA and protein synthesis. Coincubation with L-NMMA attenuated cytokine-induced cGMP formation, an effect that was reversed by L-arginine, thus demonstrating that NO is responsible for IL-1β- and TNFα-stimulated cGMP production in mesangial cells [11–13]. To identify the type of NOS induced in mesangial cells, we characterized its cofactor requirement. Mesangial cell NOS was completely dependent on the presence of L-arginine and NADPH, was weakly enhanced by tetrahydrobiopterin (BH$_4$), and, most importantly, was not significantly affected by Ca^{2+} and calmodulin [15]. These features clearly identify the mesangial cell NOS as being of the macrophage type (iNOS). Western blot analysis with a specific polyclonal antibody confirms that IL-1β treatment of mesangial cells dramatically upregulates the appearance of a single band of 130 kDa, consistent with iNOS protein [16]. Furthermore, Northern blot analysis shows the dose- and time-dependent increase of iNOS mRNA steady-state levels in mesangial cells after stimulation with cytokines as shown in Fig. 10-1 [16,17]. The iNOS mRNA was present as a single band of approximately 4.5 kb. In unstimulated mesangial cells, there was no detectable iNOS mRNA (Fig. 10-1).

Induction of iNOS has also been shown in human mesangial cells [14]. However, human mesangial cells require multiple cytokines, unlike rat mesangial cells which require only single stimulants to produce NO. IL-1β and interferon γ must be present together to elicit NO production, whereas TNFα augments the response [14]. Induction of iNOS expression by a diverse number of agents is triggered in a species- and tissue-specific manner [18] as summarized in Table 10-1 for mesangial cells.

The iNOS expression in renal mesangial cells is controlled by at least two separate signaling pathways—one involving adenosine 3′,5′-cyclic monophosphate (cAMP) and the other triggered by cytokines such as IL-1β or TNFα. Both

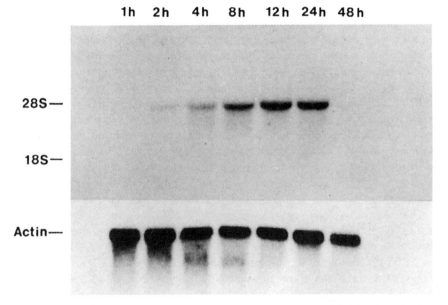

Figure 10-1. Time course of induction of iNOS mRNA in mesangial cells after stimulation with a combination of IL-1β and TNFα. Mesangial cells were treated with IL-1β (1 n*M*) plus TNFα (1 n*M*) for the indicated time periods. Total cellular RNA (20 µg) was successively hybridized to ^{32}P-labeled iNOS and β-actin cDNA probes. Reprinted by permission of Blackwell Science, Inc. from Ref. 59.

pathways act synergistically and thus potently upregulate the expression of iNOS in mesangial cells [19,20]. Nuclear run-on transcription experiments suggested that IL-1β and cAMP synergistically interact to increase iNOS gene expression at the transcriptional level. Furthermore, cAMP exposure markedly prolonged the half-life of iNOS mRNA. The decay of iNOS mRNA after stimulation with

Table 10-1. Agents That Have Been Shown to Induce iNOS Expression in Mesangial Cells

Agent(s)	Species	Ref.
Interleukin 1	Rat	11, 15–17, 23, 24
Tumor necrosis factor α	Rat	11, 15, 25
Interferon γ	Rat	12
Lipopolysaccharide	Rat	12, 22, 25
Cyclic AMP	Rat	19, 20, 23
Tumor necrosis factor α	Bovine	13
Interleukin 1	Human	14, 21
Tumor necrosis factor α	Human	14, 21
Interferon γ	Human	14

IL-1β alone corresponded to a half-life of about 1 h. After costimulation with IL-1β and forskolin, a direct activator of adenylate cyclase, the mRNA half-life was increased markedly to about 3.3 h [20]. Thus, iNOS mRNA seems to be a very labile mRNA species comparable to those mRNAs coding for many transiently expressed genes, including some cytokines, oncogenes, and transcription factors. The presence of A+U-rich sequences in the 3'-untranslated region of certain mRNAs may function as a destabilizing element which targets these mRNAs for rapid cytoplasmic degradation. We do not yet know if 3'-untranslated or other specific mRNA sequences are important for iNOS mRNA stabilization by cAMP.

The control of transcription in response to inflammatory cytokines has been intensively studied, and it has become clear that a number of ubiquitous transcription factors are involved in many different cellular systems. Very prominent among these factors is nuclear factor κB (NFκB), a multisubunit transcription factor that is activated in response to various stimuli and plays a pivotal role in the development of the cellular immune and inflammatory responses [26]. Recently, we obtained evidence that pyrrolidine dithiocarbamate (PDTC), an inhibitor of NFκB activation, potently suppresses IL-1β-induced but not cAMP-stimulated iNOS expression in mesangial cells [27]. Obviously, IL-1β uses NFκB to trigger iNOS transcription, whereas cAMP activates transcription factors different from NFκB. This finding has also been confirmed by others [25]. Whether the cooperation between these two factors provides the basis of the synergistic activation of iNOS transcription is not yet known.

Recently, Xie and colleagues [28] reported on the cloning of the 5'-flanking region of the iNOS gene from a mouse genomic library. The mRNA transcription start site is preceded by a TATA box and at least 24 consensus sequences for the binding of transcription factors involved in the induction of other genes by cytokines or lipopolysaccharide. Lipopolysaccharide inducibility of iNOS depends on an unique NFκB sequence (nucleotides −85 to −76) that is recognized by a cycloheximide-sensitive complex containing both p50/c-Rel and p50/RelA heterodimers of NFκB [29]. We have cloned and characterized a 1.8-kb fragment containing a part of the promoter region of the rat iNOS gene. The sequence revealed a multitude of possible cis-acting elements homologous to consensus sequences for the binding of different transcription factors, including the interferon γ response element, the NFκB response element, the TNF response element, binding sites for nuclear factor interleukin 6 (NF-IL6), the cAMP response element, and one γ-activated site. The involvement of NFκB, NF-IL6, and cAMP response element binding proteins in the induction of iNOS gene expression by IL-1β and cAMP has been investigated using electrophoretic mobility shift assays with specific DNA probes. DNA–protein complexes were further characterized by supershift experiments with specific antibodies. Moreover, the functional relevance of distinct binding sites was examined by transient transfection studies using deletions of the 5'-flanking region linked to a reporter gene. Our preliminary data suggests that IL-1β and cAMP use different sets of transcription factors to signal stimulation of iNOS gene expression in rat mesangial cells [30].

The signaling pathways of IL-1 linking receptor occupancy to activation of transcription factors and cellular responses are not yet known. The IL-1 receptor has no intrinsic tyrosine kinase activity and it has been demonstrated that IL-1 does not activate protein kinase C or protein kinase A in mesangial cells [31]. Nevertheless, one of the earliest cellular responses observed after IL-1 exposure is the phosphorylation of a variety of proteins in mesangial cells and other cell types. Recently, we have demonstrated that IL-1 triggers the production of ceramide from sphingomyelin and subsequent activation of the mitogen-activated protein kinase cascade [32]. This signaling module may link IL-1 receptor activation to the activation of NFκB and subsequent iNOS gene expression and may explain why tyrosine kinase inhibitors like genistein or herbimycin A block IL-1β-induced nitrite production in mesangial cells [33].

Most intriguing observation came from studies demonstrating that NO strongly augments IL-1β-stimulated iNOS gene expression in mesangial cells [34]. Amplification of iNOS expression is an action of NO that is not mediated by cGMP and is selective for inflammatory cytokines like IL-1β and TNFα without affecting cAMP induction of iNOS. To determine whether such a positive feedback loop triggered by NO is physiologically relevant, we used different compounds known to modulate NO formation in cytokine-stimulated mesangial cells. IL-1β induction of iNOS mRNA was substantially reduced by inclusion in the culture media of L-NMMA, a guanidino-N-substituted L-arginine analog that acts as a competitive inhibitor of iNOS. Conversely, changes of L-arginine concentration markedly modulate iNOS mRNA levels. In an alternative approach, we inhibited iNOS production by blocking the synthesis of tetrahydrobiopterin (BH$_4$), an essential cofactor of iNOS. Inhibition of the tetrahydrobiopterin synthetic enzyme GTP-cyclohydrolase I by 2,4-diamino-6-hydroxy-pyrimidine (DAPH) has been shown to prevent NO synthesis in mesangial cells [35]. DHAP markedly reduced IL-1β-stimulated iNOS mRNA levels in a concentration-dependent manner. Addition of sepiapterin, which is intracellularly converted into tetrahydrobiopterin, completely reversed the inhibitory action of DAHP and even potentiated iNOS mRNA levels [34]. In this context, it is worth noting that sepiapterin dose-dependently augments IL-1β-stimulated NO synthesis, indicating that the availability of tetrahydrobiopterin limits the production of NO in stimulated mesangial cells [35]. Nuclear run-on experiments suggest that NO acts to augment IL-1β-induced iNOS gene expression at the transcriptional level without changing the half-life of iNOS mRNA. Our data suggest that NO acts as an autocrine mediator that upregulates IL-1β-induced iNOS gene expression in mesangial cells and thus leads to an optimal generation of NO by the cells. This positive feedback mechanism of NO serving to maximally amplify its own production is not restricted to renal mesangial cells but is also observed in vascular smooth muscle cells, derived from rat aorta, and thus may be a general mechanism operative in cells and tissues that are able to express iNOS.

These results raise several questions as to (i) the mechanism of NO-induced iNOS expression, (ii) the physiological role of this positive feedback loop, and (iii) the possible termination of this positively regulated process. The precise mechanism by which NO amplifies iNOS expression remains to be clarified but may be related to the ability of NO to activate the transcription factor NFκB. Lander and colleagues [36] found that NO-generating compounds such as sodium nitroprusside and *S*-nitroso-*N*-acetyl-D,L-penicillamine (SNAP) induce NFκB binding activity in peripheral blood mononuclear cells, an effect that is not mimicked by a cell-permeable cyclic GMP analog.

The physiological role of the described potent amplification mechanism of NO generation may be to rapidly provide injured cells with a powerful host defense mechanism that also may form the basis for the dramatic production of NO in acute and chronic inflammatory diseases.

Powerful negatively acting regulatory pathways are required to terminate amplification loops such as the one described above. Recently, Moncada and co-workers [37] reported that NO generators markedly inhibit macrophage iNOS activity in vitro in an irreversible fashion. However, high concentrations of NO donors were required to get significant inhibition of the enzyme. Incubation of cytosolic fraction of IL-1β-stimulated-mesangial cells with SNAP gave a dose-dependent inhibition of iNOS activity as measured by the citrulline assay. Significant reduction of iNOS activity required concentrations of SNAP as high as 500 μ*M*. These data show that NO donors not only potentiate the effects of IL-1β on the expression of iNOS mRNA, synthesis of iNOS, and production of NO, but they also inhibit IL-1β-induced activity of the enzyme. Therefore, NO may provide an additional level of modulation with an effect on the amount of the enzyme combined with the opposite effect on its activity. Can these seemingly disparate results be reconciled? We suggest that NO acts as a positive feedback regulator to rapidly amplify its own synthesis, but as soon as a critical threshold of NO production is reached, NO now functions as a negative feedback modulator, to finally terminate its production in an irreversible manner. In an alternative scenario, NO-triggered amplification of iNOS expression is suppressed simply because of substrate depletion. The remarkably high turnover of L-arginine may reduce extracellular L-arginine concentration to an extent that it becomes rate limiting for iNOS activity and thus provides a stop signal for further iNOS expression. In this context, it is worth noting that mesangial cells express high arginase activity, an enzyme that converts L-arginine to ornithine and urea and thus may compete with iNOS for the common substrate L-arginine [38,39]. High arginase activity in wounds is known to cause a marked depletion of L-arginine in the extracellular space [40].

Moreover, we have observed that NO can trigger apoptotic cell death in glomerular mesangial and endothelial cells [41] (Fig. 10-2). This suicide pathway

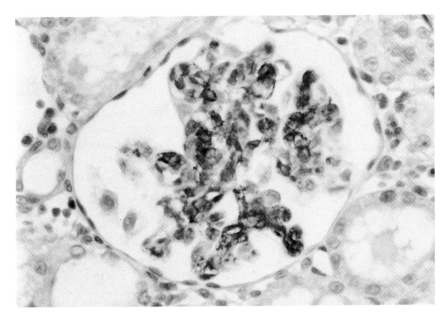

Figure 15-3   Acute unilateral glomerulonephritis induced by cationized IgG—μ day 4. Immunohistochemistry shows strong glomerular staining for iNOS. Immunoperoxidase/hematoxylin. (From Ref. 40 with permission.) See page 314.

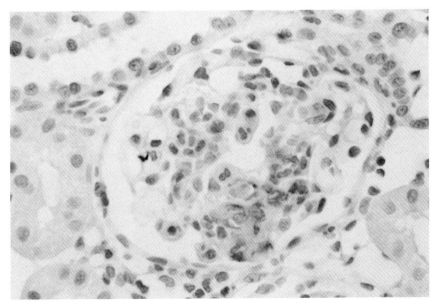

Figure 15-4   Nephrotoxic nephritis in the WKY rat μ—day 4. Positive immunohistochemistry for iNOS; immunoperoxidase/hematoxylin. See page 315.

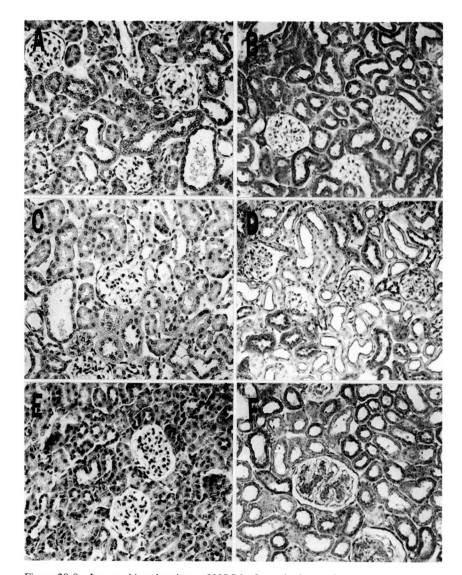

Figure 20-8   Immunohistochemistry of NOS isoforms in the renal cortex under room air (A–D) and hypoxia (E–F). (A) NOS III in cortical tubules of normal mouse. (B) Note the increased NOS III immunostaining in cortical tubules of $\alpha^H \beta^S [\beta^{MDD}]$ mouse. (C) Absence of NOS II in cortex of normal mouse. Nonspecific staining resulting from endogenous peroxi-dase was also observed in control sections incubated with normal rabbit serum. (D) Strong NOS II immunostaining in distal/collecting tubules of $\alpha^H \beta^S [\beta^{MDD}]$ mouse. (E) Moderate NOS II immunostaining in cortical tubules of normal mouse under hypoxia. (F) Increased NOS II immunostaining in cortical tubules and glomerular mesangium of $\alpha^H \beta^S [\beta^{MDD}]$ mouse. Scale bar = 100 µm. See page 410.

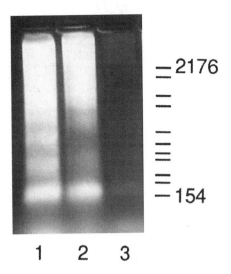

Figure 10-2. Electrophoretic analysis of cytoplasmic DNA in mesangial cells. Glomerular mesangial cells were exposed for 20 h to spermine-NO 500 μ*M* (lane 1) or 250 μ*M* (lane 2) or vehicle (control, lane 3), and cytoplasmic DNA was separated on an 1.5% agarose gel.

may provide a delayed shut-off mechanism for the production of NO by mesangial cells. The phenomenon of NO-modulated iNOS expression and activity may have broad implications for understanding the regulation of iNOS activity and may help to develop new pharmacological approaches applicable to conditions of pathological NO overproduction.

A number of cytokines, growth factors, and drugs have been reported to suppress induction of iNOS in mesangial cells. These compounds are compiled in Table 10-2.

Suppression of iNOS expression was first observed in macrophages exposed to transforming growth factor β (TGFβ) [51] and was confirmed and extended in mesangial cells [43]. The potency TGFβ as a deactivator of interferon γ-induced NO production in mouse peritoneal macrophages is due to three distinct mechanisms: (i) TGFβ reduces iNOS mRNA stability without affecting the transcription of the iNOS gene; (ii) TGFβ reduces translation efficiency of iNOS mRNA; (iii) TGFβ accelerates the degradation of iNOS protein [52]. Whether these mechanisms also operate in mesangial cells remains to be elucidated. Subsequently, suppression of iNOS expression was shown with platelet-derived growth factor-treated mesangial cells [10]. Recent data from our laboratory have shown that whereas a platelet-derived growth factor reduces IL-1β-induced nitrite production, iNOS protein expression, and iNOS mRNA levels, the basic fibroblast growth factor potently amplifies all of these cytokine-induced responses in mesangial cells. Nuclear run-on experiments suggested that a platelet-derived growth

Table 10-2.   Agents That Have Been Shown to Inhibit iNOS Expression in
Mesangial Cells

Agent(s)	Stimulators	Ref.
Dexamethasone	IL-1β, TNFα, cAMP,	11, 16, 19, 22, 24, 25
	LPS, INFγ	14, 42, 48
Prednisolone	IL-1β	42
Hydrocortisone	IL-1β	42
Corticosterone	IL-1β	42
Actinomycin D	IL-1β, TNFα, cAMP	11, 19, 20, 22, 24, 25
Cycloheximide	IL-1β, TNFα, cAMP	11, 19, 20, 24, 25
Transforming growth factor β	IL-1β, TNFα	13, 15, 24, 43, 44
Platelet-derived growth factor	IL-1β, TNFα, LPS	10, 45
Cyclosporin A, G, H	IL-1β	17, 46
PDTC	IL-1β, TNFα, LPS	25, 27
Phorbol esters	IL-1β	47
N-acetylserotonin	IL-1β	35
DAHP	IL-1β	34, 35
Genistein	IL-1β	33
Herbimycin A	IL-1β	33
Staurosporine	LPS, INFγ	49
H-7	LPS, INFγ	49
Tetranactin	IL-1β, cAMP	50
Postaglandin E₂	IL-1β, TNFα	23
Endothelin-1	IL-1β, TNFα	44
Angiotensin II	IL-1β	a
Miconazole	IL-1β	a

[a]From Mühl and Pfeilschifter, unpublished observations.

factor inhibits whereas basic fibroblast growth factor potentiates IL-1β-stimulated
iNOS gene expression at the transcriptional level [53]. These observations suggest
that two growth factors which are crucially involved in the pathogenesis of
mesangioproliferative glomerulonephritis [54] differentially affect iNOS expression in mesangial cells.

A possible explanation for the disparate effects of the two growth factors that
both bind to and activate receptor tyrosine kinases is the different degree of
activation of protein kinase C. A platelet-derived growth factor potently stimulates
phosphoinositide turnover and subsequent activation of protein kinase C in mesangial cells [55]. In contrast, the basic fibroblast growth factor is a poor activator
of this signaling pathway (Pfeilschifter, unpublished observations). Such a difference is important, as we have shown recently that the ε-isoenzyme of protein
kinase C tonically suppresses iNOS expression in mesangial cells [47]. Indeed,
coincubation of mesangial cells with platelet-derived growth factor and calphostin
C, a potent and specific inhibitor of protein kinase C, not only reverses the
inhibitory effect of platelet-derived growth factor but even causes a potentiation
of IL-1β-induced iNOS expression in a manner comparable to the effect of

the basic fibroblast growth factor (Kunz, Eberhardt, Walker, and Pfeilschifter, unpublished observations). Possible targets for protein kinase C-ε that eventually could result in inhibition of iNOS gene transcription are (i) the IL-1 receptor, (ii) the signaling cascade triggering the activation of NFκB or other transcription factors involved in iNOS gene expression, and (iii) the induction of a putative inhibitory factor that suppresses iNOS gene transcription. This may also explain why endothelin-1 [44] and angiotensin II (Mühl and Pfeilschifter, unpublished observations), two potent activators of protein kinase isoenzymes in mesangial cells [56,57], also potently antagonize iNOS induction by IL-1β. This hypothesis is currently evaluated in our laboratory.

4. Physiological and Pathophysiological Roles of NO Produced by Mesangial Cells

As mesangial cells do not express a constitutive type of NO synthase, they are not able to respond to Ca^{2+}-mobilizing agonists by an increased NO production [58–60]. This conforms well to the lack of detection of neuronal NOS and endothelial NOS in mesangial cells by histochemical and immunohistochemical methods and by in situ hybridization [61]. There have been a few reports on rapid NO production by mesangial cells in response to endothelin-3 [62], acetylcholine [63], bradykinin [63], or to changes in medium chloride content [64]. Whether this is due to contaminations of mesangial cells cultures with endothelial or epithelial cells that do express constitutive types of NOS or to contamination of cell culture constituents, especially fetal calf serum with endotoxin or other factors, remains to be investigated. In this context, it is noteworthy that Beck et al. [44] noted that endothelin does not induce NO formation in quiescent mesangial cells, but maintaining the cells in medium containing 10% fetal calf serum enables the ET_B receptor-selective agonist sarafotoxin S6c to trigger NO production. Recently, Mohaupt et al. [65] have observed that two iNOS isoforms are constitutively and heterogeneously expressed in the normal rat kidney and cultured glomerular mesangial cells. Importantly, endotoxin treatment caused a much greater increase in the vascular smooth-muscle-type iNOS as compared to the macrophage-type iNOS [65]. However, it remains to be proven whether the amounts of iNOS mRNA detected in this study, after 40 cycles of polymerase chain reaction (PCR) amplification, are of any physiological importance. Our present view is that mesangial cells do not express physiologically relevant amounts of NO under basal conditions and only function as targets for NO produced by adjacent glomerular endothelial or epithelial cells.

To better understand the pathophysiological relevance of mesangial cell NO formation, it was of interest to investigate whether culture medium conditioned by cytokine-treated cells is able to increase cGMP in untreated companion cells. A rapid transfer of conditioned medium to untreated mesangial cells evoked an immediate and pronounced formation of cGMP, thus demonstrating that cytokine-treated mesangial cells released a soluble factor that increases cGMP in adjacent cells [13,15]. cGMP is a potent mediator of mesangial cell relaxation and is responsible for the relaxing effects of atrial natriuretic peptides and sodium nitroprusside [3]. Furthermore, NO formed by glomerular endothelial cells has been shown to increase mesangial cell cGMP concentration [5]. Therefore, we were interested whether chronic treatment with IL-1β or TNFα modulates the contractile responsiveness of mesangial cells. A 24-h exposure of mesangial cells to IL-1β and TNFα drastically attenuated the contractile effect of angiotensin II. Indomethacin had no significant effect on angiotensin II-stimulated contraction of control or cytokine-treated cells, thus excluding prostaglandins as mediators of IL-1β- and TNFα-induced inhibition of cell contraction. In contrast, N^G-nitro-L-arginine significantly reversed the inhibitory action of IL-1β, thus indicating that the IL-1β effect is predominantly mediated by NO formation [15].

Such a mechanism could cause an excessive vasodilatation and thus lead to a state of glomerular hyperfiltration. This condition seems to be an important pathogenetic mechanism for the development of glomerular sclerosis. Moreover, the excessive formation of NO and cGMP in mesangial cells not only may alter glomerular filtration but may also cause tissue injury and thus contribute to the pathogenesis of certain forms of glomerulonephritis (Fig. 10-3) [59].

Cattell and colleagues have reported that glomeruli isolated from rats with four different models of immune complex-mediated glomerular injury produce nitrite ex vivo [66–69]. These authors also demonstrated by immunohistochemistry and in situ hybridization that iNOS is expressed in kidneys of rats with acute unilateral immune complex glomerulonephritis induced by cationized IgG. There was no expression of iNOS in normal rat kidneys or in glomeruli obtained from the non-nephritic kidneys of experimental rats [70]. Expression of iNOS mRNA in accelerated nephrotoxic nephritis was studied by quantitative reverse transcriptase polymerase chain reaction. iNOS expression was present at low levels in normal glomeruli and was markedly enhanced within 6 h and peaked at 24 h after the induction of glomerulonephritis. Elevated levels of iNOS mRNA persisted to day 7 [71]. These were the first studies demonstrating an in vivo induction of iNOS in immune complex glomerulonephritis and indicating that the onset of iNOS gene expression is closely related to the initial formation of immune complexes. The results, however, do not allow the discrimination between mesangial cell or macrophage expression of iNOS in nephritic glomeruli and thus it remains debatable whether infiltrating macrophages or resident mesangial cells are the major source of nitrite production in nephritic glomeruli [58,72].

There is an extensive cross-talk between macrophages and mesangial cells in the inflamed glomeruli in that macrophages secrete a variety of cytokines that trigger mesangial cell activation. Once activated, mesangial cells have the capacity to become autonomous in terms of cytokine production and NO synthesis [2]. Conversely, mesangial cells have been reported to selectively control macro-

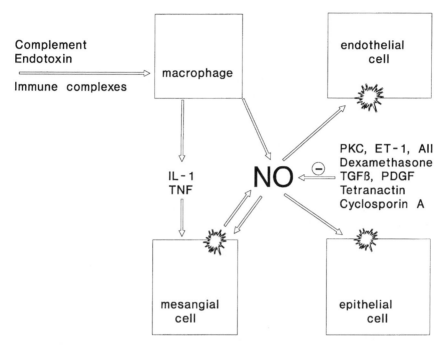

Figure 10-3. Schematic presentation of the inflammatory process in glomerulonephritis with focus on the role of NO generated by mesangial cells and macrophages. PKC, protein kinase C; ET-1, endothelin-1; AII, angiotensin II; TGFβ, transforming growth factor β; PDGF, platelet-derived growth factor. Reprinted by permission of Blackwell Science, Inc. in modified form from Ref. 59.

phage-induced glomerular injury by secreting chemotactic factors like the transforming growth factor β, and thus participating in the recruitment and proliferation of infiltrating macrophages and suppressing macrophage NO generation [73].

Recently, Weinberg et al. [74] reported that oral administration of L-NMMA prevents the development of glomerulonephritis and reduces the intensity of inflammatory arthritis in MRL-lpr/lpr mice that develop a spontaneous autoimmune disease. It is tempting to suggest that the regional or systemic use of compounds that block the production of NO may be of value in the therapy of autoimmune diseases.

Narita et al. [75] have shown that in antithymocyte serum-induced glomerulonephritis, the complement-dependent mesangial cell lysis, can be prevented by pretreatment of rats with L-NMMA. Thus, glomerular injury in this model also seems to be NO mediated. Increased iNOS mRNA expression associated with immunological tissue injury was previously described in several other animal models of immune-mediated diseases [76–78].

However, NO appears to be a double-edged sword in that it is not only a proinflammatory mediator but it can also act to protect cells and tissues from damage. NO may exert both of these effects, depending on the redox state of the microenvironment. By increasing cellular levels of cGMP and by dilatation of blood vessels, it can prevent severe hypoperfusion of the pancreas, small and large intestine, and the kidney during acute endotoxemia in rats [79]. Within solid organism, iNOS expression in endotoxemia is cytoprotective, inhibiting microthrombosis by blocking platelet adhesion and oxygen-radical-mediated injury [80,81]. Garg and Hassid [6] reported that NO and cell-permeable cGMP analogs dose-dependently inhibit serum-stimulated DNA synthesis and mesangial cell proliferation. This growth-inhibitory effect of NO may help to preserve the structure of the glomerulus under conditions of increased growth factor production, as it is typically found in certain forms of glomerulonephritis.

Trachtman et al. [82] have shown that NO release by mesangial cells modulates the synthesis of extracellular matrix proteins such as collagen, fibronectin, and laminin in glomerular disease. Enhanced NO formation in early phases of diabetic nephropathy may be an adaptive response that limits matrix deposition and fibrosis and attenuates progressive glomerulosclerosis.

5. Therapeutic Strategies for the Inhibition of Inducible NO Synthase

A number of strategies have emerged with regard to a pharmacological control of pathological NO overproduction and the reader is referred to two recent comprehensive reviews on this topic [59,60]. Herein, we shall only highlight the mechanisms of action of immunosuppressive drugs that have provided exciting new insights into the treatment of inflammatory diseases.

5.1. Glucocorticoids

Recent observations suggest that at least some of the antiinflammatory effects of glucocorticoids are mediated by the inhibition of iNOS induction. We were the first to demonstrate that dexamethasone completely inhibits IL-1β- and TNFα-induced NO production in mesangial cells [11]. Half-maximal inhibition of IL-1β-stimulated cGMP formation was observed at 4 nM of dexamethasone. The rank order of potency for different steroids to inhibit NO production corresponded well with their anti-inflammatory potencies. Furthermore, steroids did not directly inhibit iNOS but interfered with the induction process of the enzyme by IL-1β [42]. In subsequent studies, we observed that dexamethasone differentially inhibits IL-1β- and cAMP-induced iNOS mRNA expression in mesangial cells. Nanomolar concentrations of dexamethasone suppressed IL-1β-induced as well as cAMP-induced iNOS protein expression and production of nitrite. In contrast, dexamethasone prevented induction of iNOS mRNA in response to cAMP without affecting IL-1β-triggered increase in iNOS mRNA levels [16]. Negative and positive regulation of cAMP-induced genes by glucocorticoids have been reported in several instances and may involve DNA binding of the hormone receptor to a glucocorticoid response element overlapping a cAMP response element. Alternatively, an interaction between the glucocorticoid receptor and the cAMP response element binding protein may modulate gene expression. Inhibition of IL-1β-induced iNOS protein expression by dexamethasone without inhibition of mRNA levels suggests that dexamethasone blocks iNOS expression at a posttranscriptional level, possibly by interfering with iNOS mRNA translation or iNOS protein degradation. Surprisingly, nuclear run-on transcription experiments demonstrate that dexamethasone markedly attenuates IL-1β-induced iNOS gene transcription. However, this is counteracted by a prolongation of the half-life of iNOS mRNA. Moreover, dexamethasone drastically decreases the amount of iNOS protein by reducing iNOS mRNA translation and increasing iNOS degradation [48]. The examples described above highlight that several levels of control exist that must be considered in further investigations. Glucocorticoid hormones inhibit inflammatory processes and limit the proliferative response of cells in chronic destructive diseases. Inhibition of cytokine and cAMP induction of iNOS in mesangial cells may be one aspect of the beneficial glucocorticoid action seen in certain renal diseases.

5.2. Cyclosporin

Recently, we have shown that cyclosporin derivatives inhibit IL-1β induction of iNOS in mesangial cells [17]. Addition of cyclosporin A, cyclosporin G, or cyclosporin H dose-dependently inhibited interleukin-1β-induced nitrite generation. Half-maximal inhibition was observed at concentrations of 0.9 μM, 2.0 μM, and 3.8 μM of cyclosporin A, cyclosporin G, and cyclosporin H, respectively. Time-course studies indicated that cyclosporin A could be added up to 6 h after the IL-1β stimulus and still caused maximal inhibition of nitrite production. Furthermore, IL-1β increased iNOS mRNA levels in mesangial cells and this effect was potently suppressed by all three cyclosporin derivatives. As cyclosporin H has no immunosuppressive activity, these data indicate that the inhibitory effect of the cyclosporin derivatives on iNOS expression is not related to the immunosuppressive action of the drugs. This suggestion is further substantiated by the observation that the potent immunosuppressants rapamycin and FK506 did not alter IL-1β-induced iNOS mRNA levels or nitrite generation in mesangial cells.

Western and Northern blot analyses reveal that the inhibition of IL-1β-induced nitrite formation by cyclosporin A is due to decreased iNOS protein and iNOS mRNA levels. Using nuclear run-on experiment, we showed that the transcription rate of the IL-1β-induced iNOS gene is reduced. Furthermore, by electrophoretic mobility shift analysis, we observed reduced DNA binding of the nuclear factor κB, an essential component of the cytokine-dependent upregulation of iNOS gene transcription [46].

6. Perspectives

A lot more work will be required before the regulation of NO formation and the functions of NO are fully understood. Especially, the iNOS deserves all research efforts of investigators interested in acute and chronic inflammatory processes in the kidney. One of the key challenges for nephrological research is the development of effective treatment for patients with chronic renal diseases that, once established, tend to progress to end-stage renal failure.

Only time and further experimentation will establish whether inhibition of iNOS transcription and/or activity will yield effective therapeutic agents for the treatment and prevention of renal diseases.

Acknowledgments

This work was supported by Swiss National Science Foundation Grant 31-43090.95, by a grant from the Commission of the European Union (Biomed 2, PL950979), by a grant from the cancer league of Basel, by a grant from the Wilhelm Sander-Stiftung, and by a grant of the Roche Research Foundation.

Note Added in Proof

The following recently published papers are pertinent to the subject of this chapter.

Eberhardt, W., Kunz, D., Hummel, R., and Pfeilschifter, J. *Biochem. Biophys. Res. Commun.* **223**, 752–756 (1996).

Beck, K.-F. and Sterzel, R.B. FEBS Lett. **394**, 263–267 (1996).

Huwiler, A., Brunner, J., Hummel, R., Vervoordeldonk, M., Stabel, S., van den Bosch, H., and Pfeilschifter, J. *Proc. Natl. Acad. Sci.* USA **93**, 6959–6963 (1996).

References

1. Kashgarian, M. and Sterzel, R.B. *Kidney Int.* **41,** 524–529 (1992).

2. Pfeilschifter, J. *News Physiol. Sci.* **9,** 271–276 (1994).

3. Pfeilschifter, J. *Eur. J. Clin. Invest.* **19,** 347–361 (1989).

4. Shultz, P.J., Schorer, A.E., and Raij, L. *Am. J. Physiol.* **258,** F162–F167 (1990).

5. Marsden, P.A., Brock, T.A., and Ballermann, B.J. *Am. J. Physiol.* **258,** F1295–F1303 (1990).

6. Garg, U. and Hassid, A. *Am. J. Physiol.* **257,** F60–F66 (1989).

7. Mattana, J. and Singhal, P.C. *Cell Physiol. Biochem.* **5,** 176–192 (1995).

8. Firnhaber, C. and Murphy, M.E. *Am. J. Physiol.* **265,** R518–R523 (1993).

9. Mohaupt, M., Schoecklmann, H.O., Schulze-Lohoff, E., and Sterzel, R.B. *J. Hyperten.* **12,** 401–408 (1994).

10. Pfeilschifter, J. *Eur. J. Pharmacol.* **208**, 339–340 (1991).

11. Pfeilschifter, J. and Schwarzenbach, H. *FEBS Lett.* **273**, 185–187 (1990).

12. Shultz, P.J., Tayeh, M.A., Marletta, M.A., and Raji, L. *Am. J. Physiol.* **261**, F600–F606 (1991).

13. Marsden, P.A. and Ballermann, B.J. *J. Exp. Med.* **172**, 1842–1852 (1990).

14. Nicolson, A.G., Haites, N.E., McKay, N.G., Wilson, H.M., MacLeod, A.M., and Benjamin, N. *Biochem. Biophys. Res. Commun.* **193**, 1269–1274 (1993).

15. Pfeilschifter, J., Rob, P., Mülsch, A., Fandrey, J., Vosbeck, K., and Busse, R. *Eur. J. Biochem.* **203**, 251–255 (1992).

16. Kunz, D., Walker, G., and Pfeilschifter, J. *Biochem. J.* **304**, 337–340 (1994).

17. Mühl, H., Kunz, D., Rob, P., and Pfeilschifter, J. *Eur. J. Pharmacol.* **49**, 95–100 (1993).

18. Kröncke, K.-D., Fehsel, K., and Kolb-Bachofen, V. *Biol. Chem. Hoppe–Seyler* **376**, 327–343 (1995).

19. Mühl, H., Kunz, D., and Pfeilschifter, J. *Br. J. Pharmacol.* **112**, 1–8 (1994).

20. Kunz, D., Mühl, H., Walker, G., and Pfeilschifter, J. *Proc. Natl. Acad. Sci. USA* **91**, 5387–5391 (1994).

21. Brown, Z., Robson, R.L., and Westwick, J. In: *The Chemokines.* Ed. Lindley, I.J.D. Plenum Press, New York, 1993.

22. Shultz, P.J., Archer, S.L., and Rosenberg, M.E. *Kidney Int.* **46**, 683–689 (1994).

23. Tetsuka, T., Daphna-Iken, D., Srivastava, S.K., Baier, L.D., DuMaine, J., and Morrison, A.R. *Proc. Natl. Acad. Sci. USA* **91**, 12168–12172 (1994).

24. Ikeda, M., Ikeda, U., Ohkawa, F., Shimida, K., and Kano, S. *Cytokine* **6**, 602–607 (1994).

25. Saura, M., Lopez, S., Puyol, M.R., Puyol, D.R., and Lamas, S. *Kidney Int.* **47**, 500–509 (1995).

26. Baeuerle, P.A. and Henkel, T. *Annu. Rev. Immunol.* **12**, 141–179 (1994).

27. Eberhardt, W., Kunz, D., and Pfeilschifter, J. *Biochem. Biophys. Res. Commun.* **200**, 163–170 (1994).

28. Xie, Q.-W., Whisnant, R., and Nathan, C. *J. Exp. Med.* **177**, 1779–1784 (1993).

29. Xie, Q.-W., Kashiwabara, Y., and Nathan, C. *J. Biol. Chem.* **269**, 4705–4708 (1994).

30. Eberhardt, W., Kunz, D., and Pfeilschifter, J. Proceedings of the 5th International Symposium on Biological Reactive Intermediates, Munich, 1995.

31. Pfeilschifter, J., Leighton, J., Pignat, W., Märki, F., and Vosbeck, K. *Biochem. J.* **273**, 199–204 (1991).

32. Huwiler, A. and Pfeilschifter, J. *FEBS Lett.* **350**, 135–138 (1994).

33. Tetsuka, T. and Morrison, A.R. *Am. J. Physiol.* **269**, C55–C59 (1995).

34. Mühl, H. and Pfeilschifter, J. *J. Clin. Invest.* **95**, 1941–1946 (1995).

35. Mühl, H. and Pfeilschifter, J. *Kidney Int.* **46**, 1302–1306 (1994).

36. Lander, H.M., Sehajpal, P., Levine, D.M., and Novogrodsky, A. *J. Immunol.* **105**, 1509–1515 (1993).

37. Assreny, J., Cunha, F.Q., Liew, F.Y., and Moncada, S. *Br. J. Pharmacol.* **108,** 833–837 (1993).

38. Jansen, A., Lewis, S., Cattell, V., and Cook, H.T. *Kidney Int.* **42,** 1107–1112 (1992).

39. Cook, H.T., Jansen, A., Lewis, S., Largen, P., O'Donnell, M., Reaveley, D., and Cattell, V. *Am. J. Physiol.* **267,** F646–F653 (1994).

40. Albina, J.E., Mills, C.D., Henry, W.L., Jr., and Cladwell, M.D. *J. Immunol.* **144,** 3877–3880 (1990).

41. Mühl, H., Sandau, K., Brüne, B., Briner, V.A., and Pfeilschifter, J. *Eur. Z. Pharmacol.* **317,** 137–149 (1996).

42. Pfeilschifter, J. *Eur. J. Pharmacol.* **195,** 179–180 (1991).

43. Pfeilschifter, J. and Vosbeck, K. *Biochem. Biophys. Res. Commun.* **175,** 372–379 (1991).

44. Beck, K.F., Mohaupt, M.G., and Sterzel, R.B. *Kidney Int.* **48,** 1893–1899 (1995).

45. Hishikawa, K., Nakaki, T., Hirahashi, J., Marumo, T., and Saruta, T. *Eur. J. Pharmacol.* **291,** 435–438 (1995).

46. Kunz, D., Walker, G., Eberhardt, W., and Pfeilschifter, J. *Biochem. Biophys. Res. Commun.* **216,** 438–446 (1995).

47. Mühl, H. and Pfeilschifter, J. *Biochem. J.* **303,** 607–612 (1994).

48. Kunz, D., Walker, G., Eberhardt, W., and Pfeilschifter, J. *Proc. Natl. Acad. Sci. USA* **93,** 255–259 (1996).

49. Sharma, K., Danoff, T.M., DePiero, A., and Ziyadeh, F.N. *Biochem. Biophys. Res. Commun.* **207,** 80–88 (1995).

50. Kunz, D., Walker, G., Wiesenberg, I., and Pfeilschifter, J. *Br. J. Pharmacol.* **118,** 1621–1626 (1996).

51. Ding, A., Nathan, C.F., Graycar, J., Derynck, R., Stuehr, D.J., and Srimal, S. *J. Immunol.* **145,** 940–944 (1990).

52. Vodovotz, Y., Bogdan, C., Paik, J., Xie, Q.-W., and Nathan, C. *J. Exp. Med.* **178,** 605–613 (1993).

53. Kunz, D., Walker, G., Eberhardt, W., and Pfeilschifter, J. *Experientia* **50,** A47 (1994).

54. Floege, J., Eng, E., Young, B.A., Alpers, C.E., Barrett, T.B., Bowen-Pope, D.F., and Johnson, R.J. *J. Clin. Invest.* **92,** 2952–2962 (1993).

55. Pfeilschifter, J. and Hosang, M. *Cell Signaling* **3,** 413–424 (1991).

56. Huwiler, A., Fabbro, D., and Pfeilschifter, J. *Biochem. J.* **279,** 441–445 (1991).

57. Wang, Y., Simonson, M.S., Poussegur, J., and Dunn, M.J. *Biochem. J.* **287,** 589–594 (1992).

58. Pfeilschifter, J., Kunz, D., and Mühl, H. *Nephron* **64,** 518–525 (1993).

59. Pfeilschifter, J. *Kidney Int.* **48,** S50–S60 (1995).

60. Pfeilschifter, J., Eberhardt, W., Hummel, R., Kunz, D., Mühl, H., Nitsch, D., Plüss, C., and Walker, G. *Cell Biol. Int.* **20,** 51–58 (1996).

61. Bachmann, S., Bosse, H.M., and Mundel, P. *Am. J. Physiol.* **268**, F885–F898 (1995).

62. Owada, A., Tomita, K., Terada, Y., Sakamoto, H., Nonoguchi, H., and Marumo, F. *J. Clin. Invest.* **93**, 556–563 (1994).

63. Ganz, M.B., Kasner, S.E., and Unwin, R.J. *Am. J. Physiol.* **268**, F1081–F1086 (1995).

64. Tsukahara, H., Krivenko, Y., Moore, L.S., and Goligorsky, M.S. *Am. J. Physiol.* **267**, F190–F194 (1994).

65. Mohaupt, M.G., Elzie, J.L., Ahn, K.Y., Clapp, W.L., Wilcox, C.S., and Kone, B.C. *Kidney Int.* **46**, 653–665 (1994).

66. Cattell, V., Cook, T., and Moncada, S. *Kidney Int.* **38**, 1056–1060 (1990).

67. Cattell, V., Largen, P., De Heer, E., and Cook, T. *Kidney Int.* **40**, 847–851 (1991).

68. Cook, H.T. and Sullivan, R. *Am. J. Pathol.* **139**, 1047–1052 (1991).

69. Cattell, V., Lianos, E., Largen, P., and Cook, T. *Exp. Nephrol.* **1**, 36–40 (1993).

70. Jansen, A., Cook, T., Taylor, M., Largen, P., Riveros-Moreno, V., Moncada, S., and Cattell, V. *Kidney Int.* **45**, 1215–1219 (1994).

71. Cook, H.T., Ebrahim, H., Jansen, A.S., Foster, G.R., Largen, P. and Cattell, V. *Clin. Exp. Immunol.* **97**, 315–320 (1994).

72. Cattell, V. and Cook, H.T. *Exp. Nephrol.* **1**, 265–280 (1993).

73. Largen, P.J., Tam, F.W.K., Rees, A.J., and Cattell, V. *Exp. Nephrol.* **3**, 34–39 (1995).

74. Weinberg, J.B., Granger, D.L., Pisetsky, D.S., Seldin, M.F., Misukonis, M.A., Mason, S.N., Pippen, A.M., Ruiz, P., Wood, E.R., and Gilkeson, G.S. *J. Exp. Med.* **179**, 651–660 (1994).

75. Narita, I., Border, W.A., Ketteler, M., and Nobel, N.A. *Lab. Invest.* **72**, 17–24 (1995).

76. Mulligan, M.S., Hevel, J.M., Marletta, M.A., and Ward, P.A. *Proc. Natl. Acad. Sci. USA* **88**, 6338–6342 (1991).

77. Corbett, J.A., Mikhael, A., Shimizu, J., Frederick, K., Misko, T.P., McDaniel, M.L., Kanagawa, O., and Unanue, E.R. *Proc. Natl. Acad. Sci. USA* **90**, 8992–8995 (1993).

78. Worrall, N.K., Lazenby, W.D., Misko, T.P., Lin, T.S., Rodi, C.P., Manning, P.T., Tilton, R.G., Williamson, J.R., and Ferguson, T.B., Jr. *J. Exp. Med.* **181**, 63–70 (1995).

79. Mulder, M.F., Van Lambalgen, A.A., Huisman, E., Visser, J.J., Van den Bos, G.C., and Thijs, L.G. *Am. J. Physiol.* **266**, H1558–H1564 (1994).

80. Morris, S.M., Jr. and Billiar, T.R. *Am. J. Physiol.* **266**, E829–E839 (1994).

81. Ferrario, R., Takahashi, K., Fogo, A., and Badr, K.F. *J. Am. Soc. Nephrol.* **4**, 1847–1854 (1994).

82. Trachtman, H., Futterweit, S., and Singhal, P. *Biochem. Biophys. Res. Commun.* **207**, 120–125 (1995).

11

Tubuloglomerular Feedback and Macula Densa-Derived NO

William J. Welch and Christopher S. Wilcox

Tubuloglomerular feedback (TGF) is an intrarenal regulatory mechanism that stabilizes the glomerular filtration rate (GFR) and distal delivery of NaCl and fluid during periods in which homeostasis is challenged, such as changes in systemic blood pressure. The system acts at the single-nephron level primarily by increasing tone at the afferent arteriole, thus reducing glomerular capillary pressure (P_{GC}), in response to increased delivery of NaCl and fluid to the macula densa segment of the nephron. The macula densa is the site of signal transduction, where changes in solute delivery are sensed and initiation or release of vasoactive mediators occurs. TGF may also maintain homeostasis through regulation of sodium excretion, because increased delivery of sodium chloride to the macula densa activates TGF and decreases single-nephron GFR, thus adjusting the rate of sodium excretion.

The primary mediator of the TGF-associated vasoconstriction has not been discovered, but several studies have identified vasoactive agents that can mediate or modulate the response. Angiotensin II [1,2], adenosine [3,4], thromboxane [5–7], and cytochrome P-450 metabolic products of arachidonic acid [8] have all been implicated in either mediation or modulation of the vasoconstriction associated with TGF. In these studies, the blockade of each vasoconstrictor led to reduction of the TGF response, whereas local enhancement or delivery of these agents or their agonists increased TGF. This subject was recently surveyed in a comprehensive review by Navar et al. [9]. These reports have characterized TGF as primarily a vasoconstrictor response. However, evidence from our laboratory and others suggests the important vasodilator, nitric oxide (NO), acts on TGF as a modulating influence, to offset the primary vasoconstrictor pathway.

1. Nitric Oxide and Tubuloglomerular Feedback

L-Arginine (L-arg)-derived nitric oxide (NO) can be synthesized by a group of three isoenzymes of nitric oxide synthases (NOSs) present in several tissues. All three have been located in the kidney. The constitutive NOS I, originally identified in rat brain and subsequently referred to as brain or neuronal NOS (nNOS), was localized in the kidney. This was described simultaneously in reports from two laboratories [10,11]. In both studies, the major immunocytochemical staining was specifically localized to the macula densa cells of the normal rat, although there were additional sites located. NOS II, an inducible isoform (iNOS), and NOS III, a second constitutive form primarily located in endothelial cells (eNOS), are also found in or near the juxtaglomerular apparatus (JGA), the site of TGF sensing and action. Based on the specific localization of NOS I and the response to microperfusion of NOS inhibitors into the macula densa area, we proposed that nitric oxide generated in the macula densa cells as a result of NOS I activity regulated TGF [10].

To test this hypothesis, we performed a series of studies that blocked endogenous production of NO by local perfusion of mono-methyl-L-arginine (L-NMA), a competitive inhibitor of NOS. The TGF response was measured by in vivo micropuncture. The loop of Henle (LH) of a single nephron was perfused with artificial tubular fluid (ATF) containing either N-mono-methyl-L-arginine (L-NMA) or N-monomethyl-D-arginine (D-NMA). Glomerular capillary pressure was estimated from the proximal stop flow pressure (P_{SF}) of the same nephron during LH perfusions at 40 nl/min. P_{SF} (at LH perfusion of 40 nl/min) was decreased in a dose-dependent manner by the addition of L-NMA, when compared to perfusion of the inert enantiomer, D-NMA (Fig. 11-1). The specificity of L-NMA was demonstrated in this study by three experiments. First, only the L enantiomer was effective. Second, L-NMA did not inhibit the vasodilation associated with the endothelium-independent NO donor sodium nitroprusside (SNP). Third, LH perfusion of a NO scavenger, pyocyanin, increased the maximal TGF, identical to the effect of L-NMA. Pyocyanin reduced the vasodilation caused by SNP, confirming that it was acting as an NO scavenger. To study the mechanism of action of L-NMA on TGF, we coperfused the loop diuretic furosemide which blocks macula densa reabsorption and thus prevents activation of TGF. During coperfusion with furosemide, L-NMA had no effect on TGF. This suggests that NO generation is closely linked to the reabsorptive process that initiates TGF.

We also tested the effects of alternate perfusion sites of L-NMA on TGF. During retrograde perfusion from the early distal tubule, L-NMA again reduced PSF, similar to orthograde perfusion. Peritubular capillary perfusions (PTC) of L-NMA also lowered P_{SF}. These data demonstrate that the inhibited NOS pool is accessible by both luminal and peritubular delivery of L-NMA.

Subsequent studies in rats have confirmed these early observations (Table 11-1) and have further shown that the TGF response is reduced at both maximal (a measure of responsiveness) and physiological (a measure of sensitivity) LH perfusion rates by local blockade of NOS. Thorup et al. [12], using the in vivo microperfused nephron preparation, showed that intravenous infusion of nitro-L-arginine (NLA) increased the maximal TGF sensitivity. In a subsequent study,

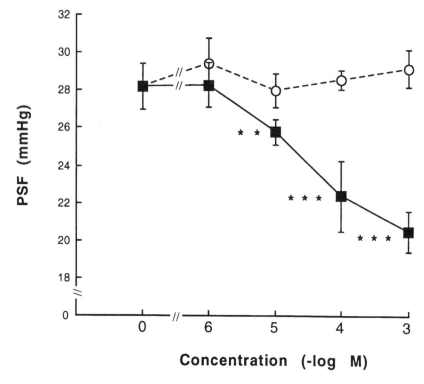

Figure 11-1. Proximal stop flow pressure (P_{SF}, mm Hg) in nephrons with the loops of Henle perfused at 40 nl/min with artificial tubular fluid alone (0) or with increasing concentrations of either N-monomethyl-L-arginine (closed squares) or N-monomethyl-D-arginine (open circles). Compared to D-NMA: *$p<.05$, **$p<.01$.

Thorup and Persson [13] used the more direct loop of Henle microperfusion technique and showed that NLA was equally effective in increasing TGF. These studies were made after 10 min of microperfusion with NLA at 5 nl/min., whereas in our original studies with L-NMA, we found that prolonged perfusions were not required, because the response was almost instantaneous. Braam and Koomans [14], using a similar prolonged perfusion of 10–15 min, confirmed that NLA increased TGF maximal response. In a companion study, these authors [15] perfused the surrounding peritubular capillary with NLA and, after 10 min, showed the maximal response of TGF was increased. These experiments suggest that although both L-NMA and NLA enhance TGF, there are important differences between the responses to microperfusion of these two NOS inhibitors. NLA must be absorbed into the interstitium and diffuse to a site of action to be effective, whereas the very rapid response to L-NMA suggests that it likely is absorbed directly into the macula densa.

Vallon and Thomson [16], using the in vivo open-loop technique to assess

Table 11-1. Summary of studies that tested the effects of nitric oxide inhibition on TGF

Author	NO Inhibitor	Route	Parameter	TGF Sensitivity	TGF Responsiveness
Wilcox et al. (1992)	L-NMA	LH orthograde	P_{SF}		+
	L-NMA	LH retrograde	P_{SF}		+
	L-NMA	PTC	P_{SF}		+
Thorup et al. (1993)	NLA	IV	P_{SF}	+	+
Thorup and Persson (1994)	NLA	LH orthograde	P_{SF}	+	+
	NLA	LH orthograde	EPFR		+
Braam and Koomans (1995)	NLA	LH orthograde	P_{SF}		+
Braam and Koomans (1995)	NLA	LH orthograde	P_{SF}		+
Vallon and Thomson (1995)	L-NMA	LH orthograde	SNGFR	+	+
Thorup and Persson (1996)	NLA	LH orthograde	P_{SF}	+	+
	AG	LH orthograde	P_{SF}	0	0
	MG	LH orthograde	P_{SF}	+	+
	7-NI	LH orthograde	P_{SF}	0	+
Wilcox and Welch (1996)	L-NMA	LH orthograde	P_{SF}	+	+

TGF, showed that LH microperfusion of N-monomethyl-L-arginine (L-NMA) reduced single-nephron GFR (SNGFR) and increased TGF. Measurements for SNGFR were made during a 2–3-min period in which L-NMA was perfused into the loop of Henle at one of four graded rates (10–40 nl/min). This aspect of the study confirms our earlier observations that perfusion with L-NMA through the LH yields a rapid response in increasing TGF compared to the more delayed response seen with NLA.

The study by Vallon and Thomson [16] also used the closed-loop method which, unlike the above studies, does not block flow from the glomerulus. With this technique, homeostatic control of late proximal ambient flow is measured in the same nephron during changes in the loop of Henle flow by free-flow perturbations (adding or removing volumes of 4 or 8 nl/min). These authors contend this method reflects the homeostatic capabilities of the integrated TGF and glomerular–tubular balance (GTB) systems at the single nephron. The advantage of this method is that the "homeostatic system" can be evaluated at about its natural operating point, although the disadvantage is that the response does not separate the TGF and GTB components. In these studies, L-NMA perfusions made control of the ambient flow more responsive when negative perturbations (reducing late proximal flow by 4 and 8 nl/min) were made. Under control conditions, this compensation was about 0.5. However, when L-NMA was perfused, this compensation increased to 0.9. The authors point out that this result is counterintuitive, as NO should potentiate vasodilation, yet the system responded with greater vasodilation during NOS blockade. Compensation with positive perturbations during L-NMA were not different than control, suggesting vasoconstriction associated with NO inhibition was not sensed in this model. This interpretation is in sharp contrast to studies using both open-loop techniques and isolated perfused segments that have shown net local vasoconstriction with NO blockade. These results suggest that other complex influences of NO may be involved when the total homeostatic response is considered in vivo.

Recently, Thorup and Persson [17] investigated a series of NOS antagonists to determine, by pharmacological methods, the NOS type involved in TGF. LH perfusion with aminoguanidine (a relatively selective inhibitor of inducible NOS) had no effect on TGF. NLA and methylguanidine (MG), both nonspecific inhibitors of all NOS isoforms, increased TGF responsiveness and sensitivity, similar to their earlier results. The specific neuronal NOS inhibitor, 7-nitroindazole (7-NI) was perfused into the LH and increased maximal TGF. Intraperitoneal infusion of 7-NI led to increases in both TGF responsiveness and sensitivity. These results confirm that neuronal NOS located in the macula densa is responsible for the effect of NO blockade on TGF in this model. These observations with open-loop microperfusion methods show that NO is a modulator of TGF in normal physiological conditions, as well as modifying the maximal effect of TGF. The constitutive enzyme nNOS is located at the macula densa and is an integral component of this intrarenal regulatory system.

Work performed in both in vitro and in vivo models of perfused nephrons support the hypothesis that NO inhibition increases the tone of the afferent arteriole, the target tissue for TGF. Ito and colleagues added nitro-L-arginine methyl ester (L-NAME) to the arteriolar lumen of an isolated perfused nephron and reported that the diameter of the afferent arteriole, but not the efferent arteriole, was reduced, consistent with TGF-related vasoconstriction [18]. This effect was apparent, when the macula densa was perfused with a physiologic salt solution but was abolished by perfusion with a low-NaCl perfusate, consistent with the hypothesis that increased NaCl delivered to the macula densa initiates TGF [19]. Deng and Baylis [20] used an in vivo microperfusion model and showed that intravenous L-NAME increased both afferent arteriolar resistance and, to a lesser extent, efferent arteriolar resistance. However, with direct LH perfusion of L-NAME, they showed only afferent arteriolar resistance was increased, suggesting that local NO blockade targets TGF-associated NOS.

2. Salt-Intake Influences on NO and TGF

The physiological importance of macula densa (MD) NO was not well defined by the above studies. The ability of MD-derived NO to compensate for excess vasoconstriction associated with TGF may be critical during perturbations to normal renal function, such as variations in salt balance. We therefore performed a series of experiments to determine the effect of changes in salt intake on TGF and the role of MD-derived NO. We have shown that maximal TGF is enhanced in low-salt fed rats (Fig. 11-2). We suggest that this difference in TGF is related to MD NO generation associated with salt intake. We tested the hypothesis that high-salt intake generates more MD NO, which maintains a higher single-nephron GFR by suppression of TGF. Some earlier observations support this hypothesis that salt may influence NO production. Studies in which NO levels were estimated by urinary excretion of nitrate and nitrite showed greater NO generation in rats maintained on a high-salt intake compared to those on a low-salt diet [21], which we have confirmed using direct chemiluminescence (unpublished results). This study [21] also showed higher excretion of cyclic guanosine monophosphate (cGMP) in salt-loaded rats. We have recently shown that plasma levels of the NO precursor L-arginine are higher in high salt rats [22]. These results led to a series of experiments in which we contrasted the effect of blockade of NO with L-NMA in rats on a high-salt (HS) and low-salt (LS) diet.

First, we tested the effects of systemic infusion of L-NMA in anesthetized rats, maintained on HS and LS diets for 7–10 days [23]. TGF was assessed by free-flow micropuncture evaluating the SNGFR measured at the proximal tubule and at the distal tubule. The difference between the values of SNGFR at these two sites represents the influence of the macula densa, because tubular fluid

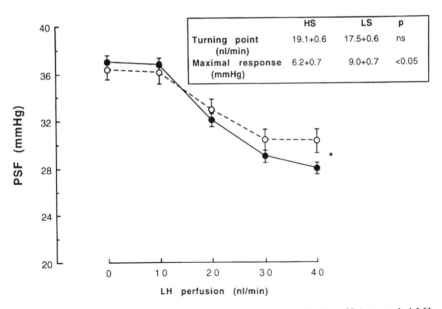

Figure 11-2. Tubuloglomerular feedback response curves: P_{SF} (mm Hg) at graded LH perfusions of ATF in high salt (HS, closed circles, $n=11$) and low salt (LS, open circles, $n=10$) rats. Compared to HS: *$p<.05$. The turning points and maximal responses for each response curve are shown in the inset.

collected for the clearance of inulin in the distal tubule is regulated by the MD, whereas flow to the MD is interrupted during proximal collections. Intravenous L-NMA raised the mean arterial pressure (MAP) in both LS and HS rats, but the rise was not different between groups. The proximal SNGFR was similar in LS and HS rats. The distal SNGFR was lower than the proximal SNGFR in both groups, demonstrating the influence of the macula densa in limiting SNGFR at each level of salt intake. The effect of TGF on filtration as assessed by differences in the proximal–distal SNGFR was greater in LS than HS rats. After L-NMA infusion, TGF was increased in HS rats but was not significantly changed in LS rats (Fig. 11-3).

We subsequently tested more directly the actions of MD-derived NO by comparing loop of Henle (LH) perfusions of L-NMA in HS and LS rats [24]. In these studies with in vivo microperfusion of the single nephron, TGF was evaluated with measurement of SNGFR or proximal stop flow pressure (P_{SF}) changes in response to grade perfusions of the LH. NOS blockade in LS rats with L-NMA had no effect on the TGF response curve, assessed from changes in P_{SF} (Fig. 11-4a). However,

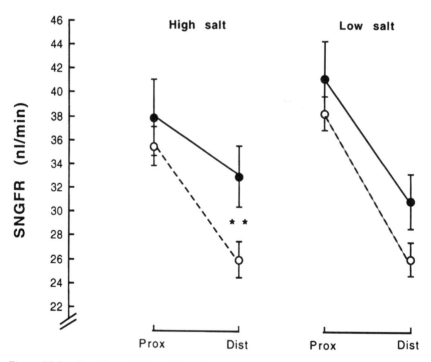

Figure 11-3. Free-flow single-nephron glomerular filtration rate (SNGFR, nl/min) measured in the proximal and distal tubule in HS (*n*=8) and LS (*n*=9) rats, during intravenous (IV) vehicle infusion (closed circles) and during IV L-NMA (open circles, 30 mg/kg) infusion. Compared to vehicle: **p<.01.

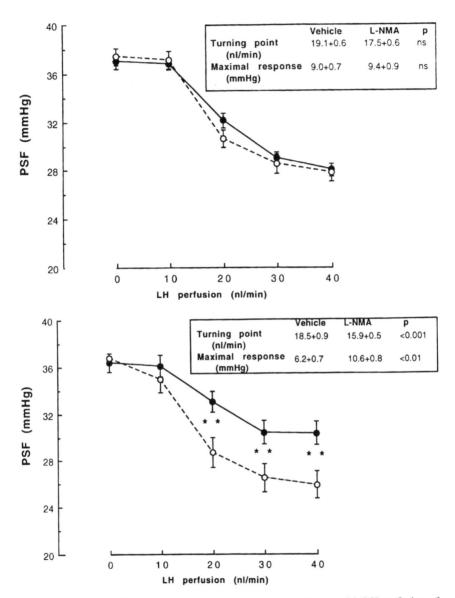

Figure 11-4. TGF (P_{SF}, mm Hg) response curves, in (a) LS rats, with LH perfusion of ATF + vehicle (closed circles, $n=11$) and ATF + L-NMA (10^{-3} M, open circles, $n=11$) and (b) HS rats, with LH perfusion of ATF + vehicle (closed circles, $n=10$) and ATF + L-NMA (10^{-3} M, open circles, $n=11$). Compared to vehicle: **$p<.01$. The turning points (nl/min) and maximal responses (mm Hg) for each response curve are shown in the inset.

in HS rats, L-NMA reduced the P_{SF} at 20, 30, and 40 nl/min (Fig. 11-4b), thus increasing TGF at rates of perfusion in the physiological range as well as at maximal rates. The maximal TGF was not changed in LS rats but was increased by more than 60% in HS rats. The turning point of the response, which reflects sensitivity, was decreased ($p<.05$) by L-NMA in HS rats but was unchanged in LS rats. In a separate group of animals, TGF was evaluated from changes in SNGFR with LH perfusions of L-NMA, in ATF delivered at 0 and 40 nl/min. Again, there was no effect of microperfusions with L-NMA in LS rats, but L-NMA increased TGF in HS rats. To test whether the failure of LS rats to respond to L-NMA was due to decreased responsiveness to NO, nephrons were perfused with the NO donor compound 3-morpholinylsydnoneimine (SIN-1), at 40 nl/min in ATF. SIN-1 increased P_{SF}, consistent with vasodilation of the afferent arteriole, but the maximal responses were greater in LS than HS rats (Fig. 11-5). This suggests that reduced responsiveness to NO does not account for the results with L-NMA in LS rats. Indeed, the greater response to SIN-1 in LS rat nephrons is consistent with upregulation of the response due to reduced NO action during salt restriction.

The in vivo microperfusion techniques used in these studies block flow from the early proximal tubule, and TGF was stimulated by LH perfusion of artificial tubular fluid. This raises an important question relevant to NO: Are there any

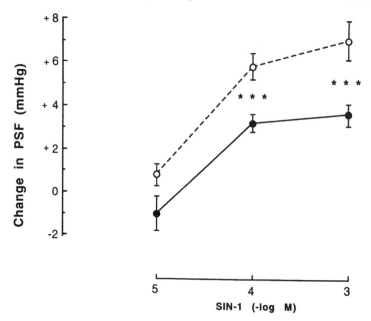

Figure 11-5. PSF (mm Hg) response to LH perfusion of ATF and increasing doses of 3-morpholinysydnoneimine (SIN-1) in HS (closed circles, $n=8$) and LS (open circles, $n=9$) rats. Compared to HS: ***$p<.001$.

native constituents in tubular fluid that could be missing from ATF that would alter the response of the MD NO system? Haberle and Davis [25] perfused the LH with harvested native tubular fluid (NTF). They found NTF from HS rats reduced TGF to a greater extent than ATF. These authors suggested NTF from HS rats contained some unknown factor that suppressed TGF. Therefore, studies limited to ATF perfusions would fail to account for this factor. Schnermann et al. [26] compared ATF and NTF perfusates in several models. They found that NTF was more effective in blunting TGF than ATF in rats with HS intake, confirming the results of Haberle and Davis. Therefore, to validate the method of using artificial tubular fluid perfusate to measure TGF, we tested the effects of LH perfusion of L-NMA with NTF harvested from the proximal tubules of the same animal [23]. Due to time limitations that prevent large collections of NTF, the LH was perfused only at 20 and 40 nl/min. In these experiments, L-NMA had no effect on P_{SF} at 20 and 40 nl/min in LS rats, exactly as was shown previously for L-NMA perfused in ATF. In contrast, L-NMA reduced P_{SF} at both 20 and 40 nl/min in HS rats (Fig. 11-6). The maximal TGF response was increased in HS by $44 \pm 5\%$ ($p<.001$) by L-NMA and was not altered in LS rats. These results with addition of L-NMA to NTF were similar to our studies using L-NMA added to ATF. Therefore, we subsequently used ATF in defining the role of NO in LH perfusions of L-NMA.

In our attempt to understand the pathophysiological role of NO on TGF, we studied the response to NOS blockade in the salt-sensitive Dahl/Rapp rat. Chen and Sanders [27] have suggested that hypertension in the Dahl/Rapp rat is related to a reduced generation of NO from L-arginine, as increases in blood pressure can be prevented or delayed by exogenous L-arginine therapy. Urinary excretion of nitrite and nitrate, a parameter of endogenous NO generation, was elevated in HS rats [28]. To determine the physiological importance of MD NO, we compared LH perfusions of L-NMA in HS and LS Dahl/Rapp rats with our results in Sprague-Dawley (SD) rats [23]. The MAP was higher in the salt-sensitive Dahl/Rapp (ssDR) rats after 7–10 days of high salt intake than SD rats (ssDR-HS: 152 ± 6 versus SD-HS: 123 ± 3 mm Hg, $p<.005$). The MAP did not differ between strains after 7–10 days of low salt intake (ssDR-LS: 111 ± 7 versus SD-HS: 125 ± 9 mm Hg, ns). As shown before, SD-LS rats had a greater maximal TGF response than SD-HS rats (LS: 8.2 ± 0.3 versus HS: 4.6 ± 0.4 mm Hg, $p<.01$). However, ssDR-LS rats had only a slightly larger maximal TGF response than ssDR-HS (LS: 6.4 ± 0.3 versus HS: 5.2 ± 0.3 mm Hg, $p<.05$). There was also a strain difference in basal TGF in LS (SD: 8.2 ± 0.3 versus ssDR: 6.4 ± 0.3 mm Hg, $p<.05$) but not HS (SD: 4.6 ± 0.4 versus ssDR: 5.2 ± 0.3 mm Hg, ns) conditioned rats. LH perfusion of L-NMA increased TGF in SD-HS (4.6 ± 0.4 to 6.7 ± 0.5 mm Hg, $p<0.1$) but had no effect on SD-LS rats (8.2 ± 0.3 to 8.3 ± 0.5 mm Hg, ns). L-NMA had no effect on TGF in ssDR on either LS or HS intakes. (Fig. 11-8). Thus, compared to SD rats, ssDR rats had a blunted TGF during LS and no response to LH L-NMA during HS. These results

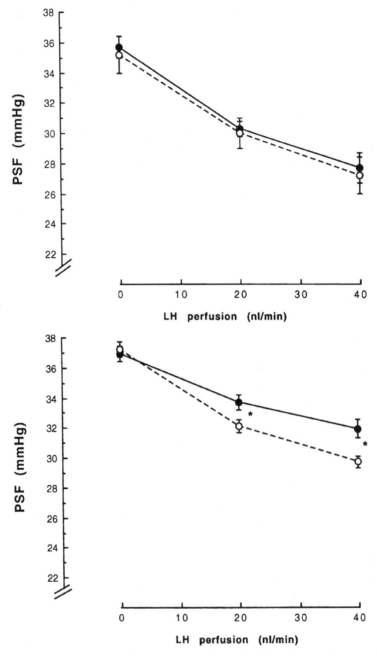

Figure 11-6. TGF (P_{SF}, mm Hg) response curves with LH perfusion of native tubular fluid (NTF) at 0, 20, and 40 nl/min: (a) In LS rats, perfusion of NTF + vehicle (closed circles, $n=12$) and NTF + L-NMA (open circles, $n=15$). (b) In HS rats, perfusion of NTF + vehicle (closed circles, $n=11$) and NTF + L-NMA (open circles, $n=13$). Compared to vehicle: *$p<.05$.

suggest that TGF is defective in ssDR rats due to a deficiency in local NO generation and is incapable of adjusting properly to increased salt loads. Such a defect could contribute to the development of salt-sensitive hypertension seen in this strain.

3. Salt Intake, NO, and Angiotensin II

Because changes in salt intake have profound effects on the renin–angiotensin system, the possible role of angiotensin in mediating the effect of salt intake on the response to NOS blockade was considered. The role of angiotensin II (Ang II) was evaluated in LS and HS rats. TGF response to LH perfusion of L-NMA was measured in rats in which either angiotensin-type 1 (AT_1) receptors were blocked with losartan or in which exogenous angiotensin II was administered. One set of LS and HS rats was infused with the AT_1 receptor antagonist losartan in osmotic minipumps for 3 days prior to and throughout experimentation. A second set of LS and HS rats was infused with angiotensin II (200 ng/kg per min) for 3 days. Three days of losartan infusion led to a lower MAP in LS (LS: 128 ± 5 versus LS + Los: 106 ± 4 mm Hg, $p<.05$) rats but was unchanged in HS rats. The basal maximal TGF response in LS rats was lower in losartan pretreated compared to LS alone. LH perfusion of L-NMA now *increased* TGF in LS + Los (5.3 ± 0.5 to 6.9 ± 0.6 mm Hg, $p<.01$) (Fig. 11-7). There was no effect of losartan on basal TGF in HS rats, and LH perfusion of L-NMA had no effect on TGF in HS + Los.

The basal maximal TGF in LS rats was not altered by Ang II but was increased in HS rats. LH perfusion of L-NMA had no effect on maximal TGF responses in LS + Ang II rats, similar to LS alone. L-NMA increased TGF in HS + Ang II (7.9 ± 0.6 mm Hg to 9.4 ± 0.5 mm Hg, $p<.01$). However, this increase was significantly less than HS alone (HS $52 \pm 5\%$ versus Hs + Ang II $22 \pm 4\%$, $p<.05$). There was a reduced response to L-NMA in LS+Los compared to HS (both of which have similarly suppressed Ang II activity) and a greater effect of L-NMA on HS+Ang II than LS (both with elevated Ang II activity). Both results suggest an Ang II independent effect of NO blockade. The maximal TGF data from these two experiments are summarized in Fig. 11-7. From these results, we can draw three conclusions. First, Ang II and NO regulate TGF responsiveness, which is related to changes in dietary salt intake. Second, these two systems can function relatively independently. Thus, during low salt intake, the renin–angiotensin system is stimulated and Ang II enhances TGF response. During salt loading, the macula densa/afferent arteriolar L-arginine-NO pathway is stimulated with NO-blunting TGF. Third, the effects of salt intake on the response of TGF to local microperfusion of L-NMA is a specific response that is independent of the basal TGF response. Thus, the failure of the maximal TGF response to be further enhanced by L-NMA in LS rat nephrons is not due to a system that has

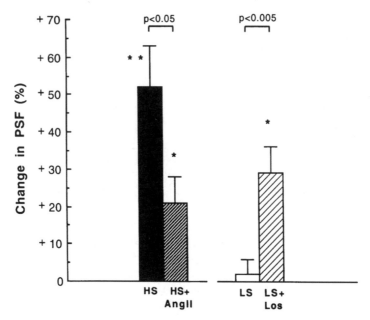

Figure 11-7.   Increase in TGF (% changes in P_{SF}) in response to LH perfusion of L-NMA in four groups of rats: HS (solid bar, $n=6$); HS + angiotensin II (dark hatched bar, 200 ng/kg per min, by osmotic minipump for 3 days, $n=6$); LS (open bar, $n=7$); LS + losartan (light hatched bar, 10–15 μg/day in drinking water for 3 days, $n=6$). Compared to response to vehicle: *$p<.05$; **$p<.01$.

become unresponsive to vasoconstrictors. Ang II infusion into HS rats enhances TGF to the level of LS rat nephrons, yet does not lead to a response to L-NMA microperfusion. In a study in which the relationship between Ang II and NO was evaluated with local perfusions, Braam and Koomans [29] suggested that the systems may be independent. With PTC perfusions of Ang II, NLA was similarly effective in increasing TGF as compared to control. The complex relationship between NO and angiotensin II is reviewed in Chapter 9.

4. Potential Mechanisms

These reports from several laboratories support a conclusion that NO has a distinct role in setting tubuloglomerular feedback sensitivity and responsiveness. We further suggest that NO appears to be physiologically important in adjusting TGF to differing salt intakes. The blockade of macula densa NO by intravenous infusion or by loop of Henle perfusion of monomethyl-L-arginine (L-NMA) increased maximal TGF in normal or HS rats but was ineffective in LS rats. This suggests that HS rats have a greater generation of NO to maintain a reduced

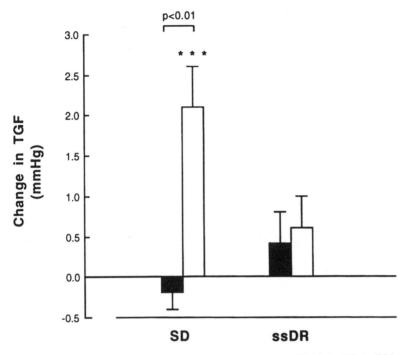

Figure 11-8. Change in TGF (P_{SF}, mm Hg) in response to L-NMA in HS (solid bars) and LS (open bars) rats from two strains: Sprague-Dawley (SD; HS: $n=12$, LS; $n=10$) and salt-sensitive Dahl/Rapp (ssDR; HS; $n=6$, LS; $n=7$). Compared to vehicle: ***$p<.001$.

TGF under conditions of increased NaCl intake. A reduced TGF permits ongoing NaCl excretion despite increased delivery to the macula densa and contributes to the homeostatic response of the kidney to increased NaCl intake. This role implies that macula densa-afferent arteriolar L-arginine NO pathway functions homeostatically to adjust the kidney's ability to retain or excrete sodium chloride appropriately at varying salt intakes. A defective macula densa NO system would lead to an unresponsive TGF that could contribute to excessive salt retention, fluid volume expansion, and volume-dependent hypertension. Preliminary data in animal models suggests that such a deficiency may be related to salt-sensitive hypertension [27,30].

There are several possible mechanisms that may relate salt intake to macula densa nitric oxide. First, nitric oxide synthase may be regulated in such a way that NOS is suppressed in low salt and increased during high salt. Salt intake has profound effects on the renin–angiotensin–aldosterone system and alters renin synthesis and release [5], as well as other related enzymes (converting enzyme, renal epithelial Na-K ATPase). However, several attempts at enzyme quantitation have not supported this hypothesis. Mundel et al. [11] and Tojo et al. [31] have shown quantitatively greater staining for nNOS in the MD with low salt intake. More recently, Singh et al. [32] showed that mRNA for cNOS was greater or not different in low-salt rats compared to high-salt rats. Several studies have shown greater NO_2 and NO_3 levels in urine of HS rats ([20]; Welch and Wilcox, unpublished). There is some concern, however, that urine NO_2 and NO_3 may not reflect changes in renal (and specifically macula densa) production of NO but represent the total body output of arginine-derived NO.

A second mechanism by which salt could regulate NO is the effect of endogenous inhibitors of NOS. Dimethylarginine (DMA) is an example of an endogenous inhibitor to NOS [33]. Studies in which DMA or another inhibitor has been correlated with changes in salt intake are currently lacking but could represent a major salt-dependent regulator of nitric oxide generation.

The third possible mechanism whereby salt intake could regulate NO is by changes in substrate availability. Salt-intake changes lead to considerable salt-handling changes along the nephron. Arginine, a basic amino acid, is normally transported by both sodium-dependent and sodium-independent pathways. A recent report suggests that arginine is transported into the loop of Henle and that concentrations near the macula densa approach 35 µmol/L in normal rats that have plasma levels of 116 µmol/L [34]. This value is greater than arginine at the end of the proximal tubule (5–10 µmol/L). We have also shown that plasma arginine is greater in HS than LS (HS—165 ± 18 versus LS—108 ± 21 µmol/ L). These studies suggest that arginine availability to the macula densa may be dependent on salt intake and could lead to physiologically important differences in local NO.

5. Conclusions

Figure 11-9 summarizes our working hypothesis for the role NO plays in regulation of TGF in changes in salt intake. TGF regulates the SNGFR by controlling the tone of the afferent arteriole (AA) inverse to the amount of signal input sensed from NaCl delivery and reabsorption. During high salt intake, increased NO generation in the MD offsets positive input to the AA, reduces vasoconstriction, and acts to maintain a higher SNGFR. During low salt intake, which stimulates renin release, angiotensin II (Ang II) acts to further increase vasoconstriction and results in a lower SNGFR. The action of Ang II is enhanced by increased thromboxane(TxA_2) and the endoperoxide PGH_2, which are activated in high Ang II states [7]. We suggest that the combination of these systems acts to alter the efficiency of TGF and contributes to the appropriate homeostatic response to salt-intake extremes.

The discovery of this important relaxing factor in the kidney has led many investigators to speculate on the role it plays in control of afferent arteriolar tone mediated by TGF. Based on the studies reviewed here and our own work, we

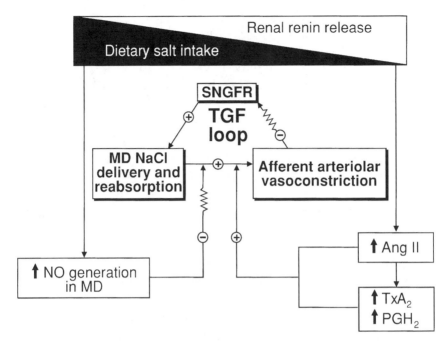

Figure 11-9. Hypothesis for the relationship between nitric oxide and salt intake in the regulation of TGF.

conclude that nitric oxide modulates TGF by offsetting the net vasoconstriction and may be important in adjusting TGF to changes in salt intake.

References

1. Mitchell, K.D. and Navar, L.G. *Am. J. Physiol.* **255,** F383–F390 (1987).

2. Ploth, D.W., Rudolph, J., LaGrange, R., and Navar, L.G. *J. Clin. Invest.* **64,** 1325–1335 (1979).

3. Schnermann, J. *Am. J. Physiol.* **255,** F33–F42 (1988).

4. Schnermann, J., Weihprecht, H., and Briggs, J.P. *Am. J. Physiol.* **258,** F553–F561 (1990).

6. Welch, W.J. and Wilcox, C.S. *J. Clin. Invest.* **81,** 1843–1849 (1988).

6. Welch, W.J., Wilcox, C.S., and Dunbar, K.R. *Am. J. Physiol.* **257,** F554–F560 (1989).

7. Welch, W.J. and Wilcox, C.S. *Am. J. Physiol.* **258,** F457–F466 (1990).

8. Zou, A-P, Imig, J.D., Ortiz de Montellano, P.R., Sui, Z., Flack, J.R., and Roman, R.J. *Am. J. Physiol.* **266,** F934–F941 (1994).

9. Navar, L.G., Inscho, E.W., Majid, D.S.A., Imig, J.D., Harrison-Bernard, L.M., and Mitchell, K.D. *Physiol. Rev.* **76**(2), 425–536 (1996).

10. Wilcox, C.S., Welch, W.J., Murad, F., Gross, S.S., Taylor, G., Levi, R., and Schmidt, H.H. *Proc. Natl. Acad. Sci. USA* **89**, 11993–11997 (1992).

11. Mundel, P., Bachmann, S., Bader, M., Fischer, A., Kummer, W., Mayer, B., and Kirz, W. *Kidney Int.* **42**, 1017–1019 (1992).

12. Thorup, C., Sundler, F., Ekblad, E., and Persson, A.E.G. *Acta Physiol. Scand.* **148**, 359–360 (1993).

13. Thorup, C. and Persson, E.G. *Am. J. Physiol.* **267**, F606–F611 (1994).

14. Braam, B. and Koomans, H.A. *Kidney Int.* **47**, 1253–1257 (1995).

15. Braam, B. and Koomans, H.A. *Hypertension* **3**, 531 (1994) (Abstract).

16. Vallon, V. and Thomson, S. *Am. J. Physiol.* **269**, F892–F899 (1995).

17. Thorup, C. and Persson, E.G. *Kidney Int.* **49**, 430–436 (1996).

18. Ito, S., Arima, S., Ren, Y.L., Juncos, L.A., and Carretero, O.A. *J. Clin. Invest.* **91**, 2012–2019 (1993).

19. Ito, S. and Ren, Y. *J. Clin. Invest.* **92**, 1093–1098 (1993).

20. Deng, A. and Baylis, C. *Am. J. Physiol.* **264**, F212–F215 (1993).

21. Shultz, P.J. and Tolins, J.P. *J. Clin. Invest.* **91**, 642–650 (1993).

22. Welch, W.J. and Wilcox, C.S. *J. Am. Soc. Nephrol.* **5**(3), 601 (1994) (Abstract).

23. Wilcox, C.S. and Welch, W.J. *Kidney Int.* **49**(Suppl 55), 59–513 (1996).

24. Welch, W.J. and Wilcox, C.S. *Clin. Exp. Pharm. Phys.* (in press).

25. Haberle, D.A. and Davis, J.M. *Am. J. Physiol.* **246**, F495–F500 (1984).

26. Schnermann, J., Schubert, G., and Briggs, J. *Am. J. Physiol.* **250**, F16–F21 (1986).

27. Chen, P.Y. and Sanders, P.W. *J. Clin. Invest.* **88**, 1559–1567 (1991).

28. Chen, P.Y. and Sanders, P.W. *J. Clin. Invest.* **22**, 812–818 (1993).

29. Braam, B. and Koomans, H.A. *Kidney Int.* **48**, 1406–1411 (1995).

30. Salazar, F.J., Alberola, A., Pinilla, J.M., Romero, J.C., and Quesada, T. *Hypertension* **19**, 333–338 (1993).

31. Tojo, A., Madsen, K.M., and Wilcox, C.S. *Jpn. Heart J.* **36**(3), 389–398 (1995).

32. Singh, J.J, Graham, M., Schnermann, J., and Briggs, J.P. *FASEB J.* **9**(4), a843 (1995) (Abstract).

33. Vallance, P., Leone, A., Calver, A., Collier, J., and Moncada, S. *J. Cardiovasc. Pharmacol.* **20**(Suppl. 12), S60–S62 (1992).

34. Silbernagl, S., Volker, K., and Dantzler, W.H. *Pfluegers Arch.* **429**, 210–215 (1994).

12

NO And The Renin System

Armin Kurtz and Karin Schricker

1. Renal Juxtaglomerular Epitheloid Cells Are the Main Site of Renin Synthesis and Secretion

The renin–angiotensin–aldosterone cascade plays an important role in the blood pressure, electrolyte, and fluid homeostasis of the organism. The activity of the renin–angiotensin system in the circulation is mainly dependent on the activity of the protease renin, which is considered the key regulator of the system. Renin found in the circulation comes predominantly from the kidneys, where renin is produced primarily by the so-called juxtaglomerular epitheloid (JGE) cells. These cells are located in the media layer of the afferent arterioles adjacent to the vascular poles of the glomeruli [1,2]. JGE cells develop from vascular smooth cells by a reversible metaplastic transformation [1,2]. This differentiation is associated by a marked change of cell morphology in a way that numerous granular (renin storage) vesicles of various size and shape appear while the number of myofilaments disappear [1]. The morphologic appearance of the cells becomes more epitheloid rather than smooth muscle cell-like. Which intracellular events trigger and control the shift of smooth muscle cells into JGE cells and back is not yet known. The JGE cells are directly neighbored to four cell types: smooth muscle cells of the afferent arterioles, endothelial cells covering the interior of the afferent arterioles, mesangial cells of the glomeruli, and the macula densa cells. It is conceivable, therefore, that the functions of JGE cells, namely renin synthesis and renin secretion, are essentially modulated by these neighboring cells. In fact, it is known that the macula densa cells exert influence on JGE cells by a yet undefined "macula densa signal" which acts inhibitorily on renin secretion and renin synthesis [3]. This rather elusive signal is probably generated dependent on the salt transport activity of the macula densa cells, in a way that the increase of salt transport increases the generation of the renin inhibitory macula densa signal and vice versa [4].

Apart from this tubular signal, renin expression and renin secretion in JGE cells are controlled by an apparently rather heterogeneous group of factors, such as the amount of salt intake, intrarenal perfusion pressure, the renal nerve activity, and by hormones, in particular by angiotensin II (ANGII), which inhibits renin expression and renin secretion in the sense of a short feedback loop [5]. There is good evidence that the effects of ANGII and of renal nerves are mediated by ANGII-AT$_1$ [6] and by β-adreno receptors [7] in JGE cells, respectively. Along with transcellular and intracellular pathways, the perfusion pressure and the rate of salt intake influence on renin secretion and renin gene expression in JGE cells is less well understood.

2. The Endothelium Modulates the Function of Renal Juxtaglomerular Epitheloid Cells

In view of the accumulating evidence that the endothelium is not only a passive cover of the lumen of blood vessels, but is also actively involved in the blood control by modulating the function of vascular smooth muscle cells, a possible interaction between endothelial cells and JGE cells attracted interest. JGE cells are modified vascular smooth cells and they are directly neighbored to endothelial cells, thus being a possible target for endothelial signals. First evidence for a substantial modulation of renin expression and renin secretion by endothelial cells was derived from co-culture experiments of JGE cells with endothelial cells [8]. This pilot study provided preliminary evidence that endothelial prostaglandins and the endothelium-derived relaxing factor (EDRF) stimulate renin secretion, whereas endothelins act inhibitorily. Meanwhile, a number of studies exist to support these findings, namely that prostaglandins and EDRF are stimulators and endothelins are inhibitors of the renin system.

This chapter aims to focus on the present knowledge about the role of EDRF, which has been identified as nitric oxide and which was named as endothelium-derived NO (NO), in the regulation of renin expression and renin secretion in the renal JGE cells. To specifically consider the role of NO in the control of the renin system in more detail, we may assume that the renal juxtaglomerular regions are sites of high production capacity for NO. As outlined elsewhere in this book, all currently known types of NO synthases (NOS) are expressed in this region (Chapter 7). The endothelial type is expressed by the endothelium [9], the neuronal type is highly expressed in the macula densa cells [10,11], mesangial cells are a known expression site for the inducible form of NO synthases [12–14], and preliminary information exists that JGE cells themselves may contain a special form of inducible type of NOS [15]. Quantitative production of nitric oxide in the juxtaglomerular region has already been indirectly assessed [9].

3. NO Is a Stimulator of the Renin System in Vivo

The in vivo approach to study the influence of NO on renin gene expression and renin secretion comprises the use of NO donors on the one hand and NO synthase inhibitors on the other hand. There is information that NO donors increase plasma renin activity, suggesting a stimulation of renal renin secretion [16,17]. The use of NO donors in vivo, however, is limited by the hemodynamic effects of these compounds which cause vasodilation and, consequently, a decrease of blood pressure. Secondary to the fall of blood pressure, NO donors will also induce an activation of sympathetic outflow. There is also increasing evidence that renal NO may be of importance for salt excretion [18] and it cannot be ruled out, therefore, that systemic administration of NO donors could lead to salt loss. These side effects of NO donors are of major importance for studies on renin secretion and renin gene expression, because the renal perfusion pressure, the sympathetic nerve activity, and the salt content themselves are physiologically important determinants for renin secretion and renin gene expression in JGE cells. It is difficult to distinguish, therefore, whether the renin stimulatory effects of NO donors are due to a direct effect of NO on JGE cells or due to the systemic side effects.

A number of in vivo studies on the role of NO on the renin system have been performed using analogs of L-arginine which cannot be metabolized to nitric oxide and which, consequently, decrease NO formation. These NOS inhibitors have been found to lower plasma renin activity [18–21] in dogs and rats. An increase of plasma renin activity upon treatment has not yet been described. When comparing these studies, it appears that treatment with NOS inhibitors lowers plasma renin activity preferentially in animals which display a relatively high basal plasma renin activity, whereas animals with an already low basal renin secretion show a smaller inhibitory response. Nonetheless, all of the studies are compatible with the conclusion that NO plays a stimulatory role for renin secretion in vivo. This conclusion is supported by rat studies, suggesting that NOS inhibitors also moderately decrease the basal level of renin mRNA in the kidney [21,22]. Application of NOS inhibitors is frequently associated with increases of systemic blood pressure [23], and, consequently, reflex inhibition of sympathetic outflow, again creating problems in the interpretation of the results because an increase of blood pressure and a suppression of sympathetic nerve activity per se would be expected to inhibit the renin system. Moreover, a possible sodium retention by NOS inhibitors could also account for the inhibitory effect of these compounds on the renin system. Because NOS inhibitors have been found to act acutely on renin secretion in vivo [18,20,24,25], it appears less likely that their effects on the renin system were mediated by perturbations of the sodium balance. Moreover, there is accumulating information that an inhibition of renin secretion and renin gene expression by NOS inhibitors in vivo can also occur without changes of intrarenal blood pressure and, consequently, without reflex changes of the sympathetic nervous system [18,24,25], suggesting that a pressure-independent mechanism also has to be involved in the effects of NOS inhibitors on the renin system.

The renin inhibitory effects of NO synthase inhibitors in vivo becomes the more prominent if the renin system is prestimulated. Although, as already stated, the effects of NOS inhibitors on basal renin secretion and renin gene expression are rather moderate, NOS inhibitors substantially attenuate or even blunt the stimulation of renin secretion and of renin gene expression induced by a variety of maneuvers (Fig. 12-1). The best examined stimulator of the renin system in this context is unilateral renal hypoperfusion induced by unilateral renal artery stenosis. There are studies performed in dogs [24,26] and rats [21], showing that NOS inhibitors substantially attenuate the rise of renin secretion in response to unilateral renal artery stenosis. Notably, the inhibitory effect of NOS inhibitors was evident at subnormal renal perfusion pressures but disappeared at high perfusion pressures [24,26]. There also exist, however, studies reporting that NOS inhibitors do not change the response of renin secretion to renal artery stenosis [18,25,27]. Although there is not yet a clear explanation for these divergent findings, it should be mentioned in this context that those studies reporting no effect on hypoperfusion-induced renin secretion show some particularities. Thus, one study reported that basal, but not hypoperfusion-stimulated, renin secretion was sensitive to NOS inhibitors [18]. Another study reported that neither basal, nor hypoperfusion-stimulated (but furosemide-stimulated) renin secretion was sensitive to NOS inhibitors [25]. In this study, a b-NOS specific inhibitor was used and a possible explanation could be that b-NOS (macula densa-derived NO) is not relevant for hypoperfusion-induced renin secretion. The third study, in which NOS inhibitors did not interfere with hypoperfusion-induced renin secretion, reported normal values of plasma renin activity in animals which became severely hypertensive upon chronic treatment with NOS inhibitors, suggesting that the pressure control of renin secretion was disturbed in these animals [27].

A series of experiments done in conscious rats provided evidence that NOS inhibitors not only decreased renin secretion but also renin gene expression during unilateral renal artery stenosis [21].

Nitric oxide synthase inhibition was found to markedly attenuate the rise of plasma renin activity and of renin mRNA levels due to disinhibition induced by ANGII antagonists such as converting enzyme antagonists or ANGII-AT$_1$ receptor blockers [28]. The macula densa control of the renin system appears to be sensitive to NOS inhibition. To inhibit the macula densa function in vivo, animals were chronically infused with the loop-diuretic furosemide, which blocks salt transport at the macula densa site. To avoid a negative salt and water balance, the conscious animals were supplemented with salt and water. These furosemide-treated animals displayed elevated plasma renin activity and increased levels of renin mRNA.

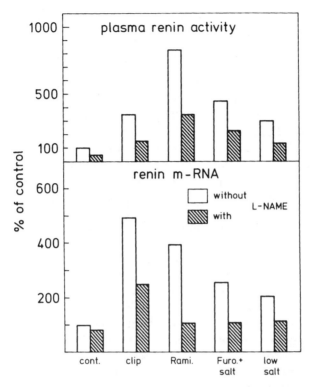

Figure 12-1. Influence of L-NAME treatment on plasma renin activity (upper panel) and renal renin mRNA levels (lower panel) in conscious rats subjected to different maneuvers to stimulate the renin system. For this purpose, animals received an unilateral 0.2-mm artery clip for 2 days (clip), were fed with the angiotensin I-converting-enzyme inhibitor rampril by gavage (5 mg/kg per day) for 2 days (ramipril), were subcutaneously infused with furosemide (dose 12 mg/kg) having free access to food, salt, and water (furo+salt), or were kept on a low-salt diet (0.02% w/w) for 10 days (low salt). L-NAME (40 mg/kg, twice a day) was administered intraperitoneally during the last 2 days of the experiments. Data represent the means of five to seven animals per experimental groups. The effect of L-NAME on plasma renin activity and on renin mRNA were significant ($p<.05$ versus vehicle) in all experimental groups. Data are given as percentage relative to respective controls consisting of animals receiving no treatment. (Data adapted from Refs. 21 and 28–30.)

NOS inhibitors blunted the increases of plasma renin activity and renin mRNA levels in these animals [29]. Finally, the renin system can also be stimulated by low salt intake. NOS inhibitors again potently attenuate the increase of plasma renin activity and renin mRNA levels in rats on a low-salt diet [30]. In summary, the overall effect of NO on renin secretion and renin gene expression in the kidney in vivo appears to be a stimulatory one.

4. NO Is a Stimulator of Renin Secretion in the Isolated Kidney

Already mentioned the effects of NO donors or NOS inhibitors on the renin system in vivo are conflicted by the side effects on blood pressure, sympathetic nerve activity, or sodium excretion. To minimize possible indirect systemic effects of NO on the renin system, experiments with isolated perfused kidneys have been performed to gain more direct information about the role of nitric oxide in renin secretion. Unfortunately, studies on renin gene expression with isolated perfused kidneys are hampered by the fact that changes of renin mRNA occur rather slowly, with a delay of several hours, which is longer than the functional lifetime of an isolated perfused kidney. Therefore, only information about the role of NO in renin secretion can be obtained from experiments with the isolated perfused kidney. Acetylcholine is known to evoke the release of EDRF from the endothelium [31]. The addition of acetylcholine to the perfusate of isolated perfused kidneys stimulates renin secretion and this stimulation can be blunted by adding inhibitors of NO formation to the perfusate [32].

The addition of the NO donor sodium nitroprusside to the perfusate of an isolated rat kidney also causes a prompt stimulation of renin secretion [32]. At a normal perfusion pressure, the magnitude of stimulation of renin secretion by NO releasers is relatively small when compared with β-adrenoreceptor agonists, which are classic stimulators of renin secretion. Notably, the stimulatory effect of the NO releasers is dependent on the perfusion pressure in such a way that the effect on secretion is markedly stronger at a low perfusion pressure and becomes blunted at high perfusion pressures (Fig. 12-2) [32].

Inhibitors of NO formation such as L-NAME, L-NMMA, or L-NAG decrease the basal release of renin from isolated rat kidneys [32–34]. They also decrease renin secretion from kidneys taken from animals kept on low- or high-salt diet [35]. Moreover, we have found that inhibitors of NO formation markedly attenuate the stimulation of renin secretion evoked by a fall of the renal perfusion pressure, which is attributed to the so-called "baroreceptor" control of renin secretion (Fig. 12-2) [32]. Notably, NOS inhibitors do not inhibit renin secretion at supranormal perfusion pressures in the isolated perfused kidney, which is in a good agreement with in vivo observations [24,25].

Taking all these findings together suggests that on the level of the isolated kidney NO appears to act as a stimulator of renin secretion.

5. The Direct Effect of NO on JGE Cells Is Not Yet Clear

Although the regulation of renin secretion from isolated perfused kidneys is more controllable than from the kidneys in vivo, renin secretion from isolated kidneys still results from a complex interference of factors acting either directly or indirectly on JGE cells. Such an indirect mediation in the isolated kidney could occur

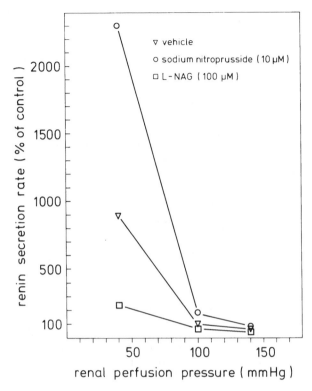

Figure 12-2. Effect of the NO donor sodium nitroprusside (SNP, 10 μ*M*) and of the NOS inhibitor L-NAG (100 μ*M*) on the pressure-dependent renin secretion from isolated perfused rat kidneys. Values are given relative to the renin secretion at a perfusion pressure of 100 mm Hg in the absence of drugs. Data represent means of six kidneys each. Renin secretion is significantly stimulated by SNP at 100 and 40 mm Hg and is significantly inhibited by L-NAME at 40 mm Hg. (Data adapted from Ref. 32 and from unpublished work of our laboratory.)

via the macula densa, changes of the arteriolar resistance, endothelial cell function, and so forth. Therefore, it appeared reasonable to consider the influence of NO on renin secretion and renin gene expression directly at the level of renal juxtaglomerular cells.

First, experiments in this context were done in kidney slices and the results of these experiments showed that NO inhibits renin secretion [36] and NOS inhibitors stimulate [37] the basal release of renin from this preparation.

Two studies [8,38] with cultured JGE cells showed that inhibitors of NO formation did not influence renin secretion (Fig. 12-3). If JGE cells, however, were co-cultured with endothelial cells, NOS inhibitors or arginine deprivation not only inhibited NO formation in the co-cultures but also led to a decrease of

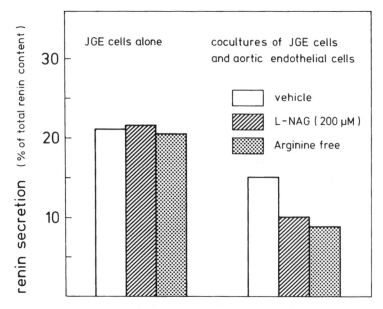

Figure 12-3. Influence of the NOS inhibitor L-NAG (200 μM) and of arginine-free medium (for 24 h) on 20-h renin secretion from cultured renal JGE cells (left) and from cocultures of renal juxtaglomerular cells with aortic endothelial cells (right). Renin secretion is given as the fractional release of renin (i.e., the percentage of renin that had been released by the cells within 20 h). Data are means of six juxtaglomerular cell culture preparations. The basal renin secretion from cocultures was significantly ($p<.05$) lower than from JGE cells alone. Moreover, renin secretion from cocultures was significantly ($p<.05$) by L-NAG and by arginine-free medium. (Data adapted from Ref. 38.)

basal renin secretion (Fig. 12-3). [8,38], suggesting that NO is a stimulator of renin secretion in this system. The direct effect of NO donors on renin secretion from JGE cells is somewhat complex. When added to the culture medium of isolated JGE cells, NO donors caused a transient inhibition of renin secretion lasting for about 1 h, followed by a continuous stimulation of renin secretion thereafter [39,40] (Fig. 12-4). Prolonged administration of NO donors to cultured JGE cells also induced a moderate increase of the steady-state levels of renin mRNA [41].

A third model that was used to examine the direct effect of NO on renin secretion from JGE cells was the use of microdissected juxtaglomerular apparatus containing the macula densa. If L-arginine was administered via microperfusion to the macula densa of microdissected juxtaglomerular apparatus, an almost immediate stimulation of renin secretion from JGE cells was observed [42]. The addition of NOS inhibitors to the macula densa perfusate attenuated renin secretion and, in particular, reduced the sensitivity of renin secretion toward changes of the salt concentration at the macula densa [42]. Notably, macula densa-induced

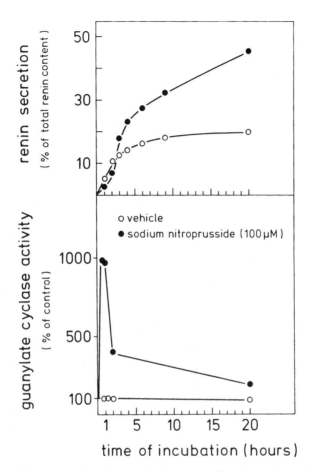

Figure 12-4. Time course of renin secretion (upper panel) and guanylate cyclase activity (lower panel) in cultured renal JGE cells in the presence of the NO donor sodium nitroprusside (100 μM). Renin secretion is given as the fractional release of stored renin. Guanylate cyclase activity was determined by measuring cGMP levels after a 15-min incubation time with the phosphodiesterase inhibitor IBMX (500 μM). Data represent the means of five JGE cell preparations. Renin secretion was significantly stimulated by SNP after 3 h and, later, guanylate cyclase activity was significantly increased by SNP at all time points examined. (Data adapted from Ref. 39.)

stimulation of renin secretion in this preparation was markedly inhibited if NO donors were added directly to JGE cells.

6. Which Second Messenger Mediates the Effects of NO in JGE Cells?

The best established cellular effect of NO is the stimulation of soluble guanylate cyclase activity leading to the enhanced formation of cGMP [43]. When considering the temporal relationship between guanylate cyclase activity and renin secretion in isolated JGE cells (Fig. 12-4), it becomes obvious that maximal activation of cGMP formation is associated with inhibition of renin secretion, whereas renin secretion appears to increase when cGMP formation declines. This would suggest an inhibitory effect of cGMP on renin secretion. In fact, it has been found that membrane-permeable cGMP [39] analogs or guanylate cyclase activators such as atrial natriuretic peptide [44] inhibit renin secretion from isolated JGE cells. The mediation of the delayed stimulatory effect of NO donors on renin secretion from JGE cells remains yet unclear. There is, however, accumulating evidence that NO may induce intracellular events, such as calcium release, which are independent of cGMP formation [45] and which stimulate secretory events in other cells, as, for example, insulin secretion from pancreatic β cells [46].

Considering all available data on the effects of NO donors and NO inhibitors on renin secretion from isolated JGE cells does not allow a clear inference about the physiological direct effect of NO on renal JGE cells. What are possible explanations for the clear stimulatory effects of NO on renin secretion and renin gene expression in the kidney, on the one hand, and the unclear effects of NO in JGE cells on the other? There arise some arguments that could help explain this complex situation. For instance, NO may exert both direct and indirect effects on renal JGE cells, which, together, account for the overall stimulatory effect of NO observed in the whole kidney. Indirect effects could result from an action of NO within or on neighboring cells, such as macula densa cells or endothelial cells, which both harbor substantial NO synthase activity. Second, it could be that the transient inhibition by NO of renin secretion from isolated JGE cells is a cell culture artifact. The latter possibility is confirmed by the observation that native JGE cells contain two types of cGMP-dependent protein kinases: GKI and GKII [47,48] and that GKII rapidly disappears from JGE cells in vitro. Thus, cGMP, which is considered the classic second mesenger of NO, could exert effects in JGE cells that are mediated via the two types of G-kinases. Nonphysiologic disappearance of one type of G-kinase could therefore lead to an imbalance of the intracellular effects of cGMP and, consequently, change the physiological response of JGE cells to NO. Obviously, more work is needed for the understanding of the direct effect of NO on JGE cells.

7. The Physiological Role of NO in the Control of the Renin System

Let us return to the in vivo situation or situation of the isolated perfused kidney where NO appears as a stimulator of the renin system. Given such an overall stimulatory effect of NO on renin secretion and renin gene expression in the whole kidney, one raises the question about the role of NO in the physiological control of renin secretion and renin gene expression. In particular, does NO act as a regulator of the renin system or does it serve as a permissive stimulator of the renin system? The intrarenal and, in particular, the local production of NO in the juxtaglomerular region has not yet been assayed because of technical limitations. It is, therefore, still unknown whether changes of local NO formation could underly changes of renin secretion and renin gene expression. However, there is recent evidence from three laboratories, including our own, that the expression of the b-NOS gene and the expression of the b-NOS protein in the macula densa cells changes under certain conditions of altered renin secretion and renin gene expression. Thus, it has been found that b-NOS expression in the macula densa increases during a low-salt diet and decreases during a high-salt diet [22,49] (Fig. 12-5). If an altered local production of NO can be deduced

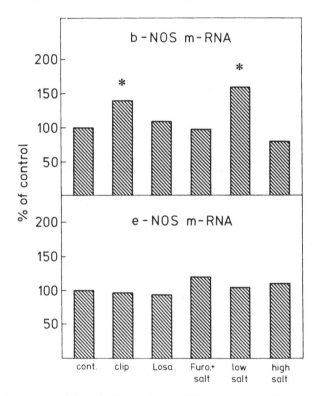

Figure 12-5. mRNA levels of b-NOS and of eNOS in the kidneys of in animals subjected to different maneuvers to change renin secretion and renin gene expression. These maneuvers comprised a unilateral 0.2-mm renal artery clip for 2 days, treatment with the angiotensin I-coverting enzyme inhibitor ramipril (5 mg/kg per day) for 2 days, subcutaneous infusion of furosemide (12 mg/day) for 6 days and concomitant substitution of salt and water, and low-salt (0.02%) and high-salt (4%) diet for 10 days. Data represent means of six animals in each group. b-NOS mRNA was significantly ($p<.05$) changed in animals with clips and in animals on a low-salt diet. (Data are unpublished material from our laboratory.)

from the altered expression of b-NOS, it is conceivable that it may contribute to the enhancement of the renin system during a low-salt diet and to the suppression of the renin system during a high-salt intake. An enhancement of b-NOS expression has also been observed during renal hypoperfusion, which is also a well-known stimulus for renin secretion and renin gene expression [22] (Fig. 12-5). The mechanism of changes in macula densa b-NOS expression under the aforementioned conditions remain to be clarified. One critical issue is that b-NOS expression may be secondary to changes of the renin system, rather than being the reason for the change of the renin system. This argument, however, appears less valid in view of the findings that stimulation of renin secretion and renin gene expression by ANGII antagonists or by furosemide in salt-balanced rats is not associated with an altered b-NOS expression (Fig. 12-5). If local NO production would change in proportion to the changes of b-NOS expression, one could imagine that changes of NO production could account for the changes in renin secretion and renin gene expression during states of altered salt intake. This inference would be compatible with the finding that the stimulation of renin secretion and renin gene expression by low-salt intake can be blunted by NOS inhibitors [30].

More difficult to explain is the strong stimulation of the renin system during renal hypoperfusion by a rather moderate increase of b-NOS expression. In fact, NOS inhibitors only attenuate but do not blunt the stimulation of the renin system in this situation [21]. Obviously, other factors aside from NO have to contribute to the activation of the renin system during renal hypoperfusion.

As mentioned earlier, the stimulation of the renin system by ANGII antagonists or by furosemide in salt supplemented rats is not associated with changes of b-NOS expression. Nonetheless, the increases of renin secretion and of renin gene expression induced by these drugs can be substantially attenuated by NOS inhibitors. If an altered local NO production is not involved in the stimulations of the renin system under these conditions, it appears reasonable to assume that it is the normal release of NO that is essential for the stimulation of renin secretion, suggesting a tonical stimulatory role of NO for the renin system. In this vein, locally produced NO would play an important permissive stimulatory role for the renin system that is effectively counteracted by inhibitors of the renin system such as the perfusion pressure, by ANGII, or by the macula densa signal. Once the inhibitors become weakened during a fall of the renal perfusion pressure, blockade of ANGII receptors or during inhibition of the macula densa mechanism, the stimulatory effect of NO then becomes apparent. Supportive of this concept is the observation that the renin stimulatory effect of NO in the isolated perfused kidney is dependent on the perfusion pressure, in a way that its effect is substantially greater at lower than at higher perfusion pressures [32].

Thus far, we have only considered possible changes of juxtaglomerular NO production that arise from changes of NOS gene or NOS protein expression. It is well known that both b-NOS and e-NOS activity are allosterically regulated by calcium. We do not yet know whether b-NOS or e-NOS activity in the juxtaglomerular region are changed under conditions associated with an altered renin system.

The macula densa has long been known as a link between tubular fluid, tubular function, and the renin system. The pathway along which the macula densa cells signal the neighboring JGE cells is still poorly understood. The demonstration of high-level NOS expression in macula densa cells makes it tempting to speculate that NO could be involved in the macula densa signaling. As outlined in more detail elsewhere in this book (Chapters 9 and 11), evidence has already been obtained for a positive role of NO in the macula densa signaling.

8. Relevance of NO for the Development and Recruitment of Renin-Producing Cells

Given a stimulatory role of NO for renin secretion and renin gene expression and given the fact that the macula densa is an important site of NO production, the question arises of whether NO formation in the macula densa is of importance for the development of renin-producing JGE cells. This question has been addressed by comparing the ontogeny of renin-producing cells and b-NOS-expressing macula densa cells [50]. It was found that b-NOS expression temporally precedes the onset of renin expression in the juxtaglomerular apparatuses [50]. This finding would be compatible with the idea that NO formation in the macula densa cells could be of relevance for the development of renin-producing JGE cells. As already stated, renin-producing JGE cells develop from vascular smooth muscle cells by a reversible metaplastic transformation process. This process is active throughout life, and states of a stimulated renin system lead to a recruitment of renin-producing cells in the afferent arterioles, whereas states of suppressed renin activity are associated with a reduction in the number of renin-producing cells [1]. A strong retrograde recruitment of renin-producing cells in afferent arterioles is found, for example, upon treatment of animals with ANGII antagonists. This recruitment of renin-producing cells can be blunted by NOS inhibitors [28], suggesting that the basal release of NO is of importance for the capability of vascular smooth muscle cells to transform into renin-secreting cells.

This possible influence of NO on the development of renin-producing cells from vascular smooth muscle cells could be of major importance, because renin-producing cells are not restricted to the kidney but are also found in the walls of blood vessels in which a substantial formation of NO may take place.

9. The Origin of Juxtaglomerular NO

As already stated, JGE cells are surrounded by cells capable of producing NO. Thus, the macula densa cells express the b-NOS type [10,11]; the endothelial cells the eNOS type [9], the mesangial cell the iNOS type [12–14], and even JGE cells themselves have once been mentioned to contain a special form of iNOS [15]. This raises the question about the physiological relevance of different NO synthases for renin expression and secretion. As mesangial cells are known to express the iNOS predominantly during inflammatory stimulation [12–14] and NOS-inhibitors do not influence renin secretion in isolated JGE cells [8,38], one may infer that the physiological role of the iNOS for the renin system is a minor one. Consequently, the interest focuses on the b-NOS in the macula densa and the eNOS in the endothelial cells. In the last few years, inhibitors of NOS activity have been developed that are more specific for b-NOS than for eNOS. It was found that inhibitors of this type suppressed intrarenal renin secretion stimulated via macula densa mechanism [25]. This is the first direct evidence that NO production in the macula densa, is of functional importance for renin secretion from JGE cells. Meanwhile, it is well established that the biological range of NO is restricted. It is difficult, therefore, to imagine that macula densa-derived NO could account for the retrograde recruitment of renin-producing cells, which may occur at distances of 100 μm or even more from the macula densa. It is reasonable to assume a major role of the eNOS for the renin system, in particular for the recruitment of renin-producing cells. Whether mesangial-derived NO produced by iNOS during inflammation is of functional relevance for the renin system remains to be clarified.

Acknowledgments

The authors wish to thank Karl-Heinz Götz and Marlies Hamann for doing the artwork and Hannelore Trommer for secretarial help. The authors' work is financially supported by the Deutsche Forschungsgemeinschaft.

References

1. Taugner, R., Bührle, C.P., Hackenthal, E., Mannek, E., and Nobiling, R. Morphology of the juxtaglomerular apparatus. *Contr. Nephrol.* **43**, 76–101 (1984).

2. Barajas, L. Anatomy of the juxtaglomerular apparatus. *Am. J. Physiol.* **236**, F240–F246 (1979).

3. Briggs, J.P. and Schnermann, J. Macula densa control of renin secretion and glomerular vascular tone: evidence for common cellular mechanisms. *Renal Physiol.* **9**, 193–203 (1986).

4. Skott, O. and Briggs, J.P. Direct demonstration of macula densa mediated renin release. *Science* **237**, 1618–1620 (1987).

5. Hackenthal, E., Paul, M., Ganten, D., Taugner, R. Morphology, physiology, and molecular biology of renin secretion. *Physiol. Rev.* **70**, 1067–1116 (1990).

6. Tufro-McReddie, A., Chevalier, R.L., Everett, A.D., and Gomez, R.A. Decreased perfusion pressure modulates renin and ANGII type 1 receptor gene expression in the rat kidney. *Am. J. Physiol.* **264**, R696–R702 (1993).

7. Atlas, D., Melamend, E., and Lahav, M. Beta-adrenergic receptors in rats kidney. *Lab. Invest.* **36**, 464–468 (1977).

8. Kurtz, A., Kaissling, B., Busse, R., and Baier, W. Endothelial cells modulate renin secretion from isolated juxtaglomerular cells. *J. Clin. Invest.* **88**, 1147–1154 (1991).

9. Bachmann, S., Bosse, H.M., and Mundel, P. Topography of nitric oxide synthesis by localizing constitutive NO synthases in mammalian kidney. *Am. J. Physiol.* **268**, F885–F898 (1995).

10. Wilcox, C.S., Welch, W.J., Murad, F., Gross, S.S., Taylor, G., Levi, R., Schmidt, H.H.H.W. Nitric oxide synthase in macula densa regulates glomerular capillary pressure. *Proc. Natl. Acad. Sci. USA* **89**, 11993–11997 (1992).

11. Mundel, P., Bachmann, S., Bader, M., Fischer, A., Kummer, W., Mayer, B., and Kriz, W. Expression of nitric oxide synthase in kidney macula densa cells. *Kidney Int.* **42**, 1017–1019 (1992).

12. Shultz, P.J., Archer, S.L., and Rosenberg, M.E. Inducible nitric oxide synthase mRNA and activity in glomerular mesangial cells. *Kidney Int.* **46**, 683–689 (1994).

13. Ahn, K.Y., Mohaupt, M.G., Madsden, K.M., and Kone, B.C. In situ hybridization of mRNA encoding inducible nitric oxide synthase in rat kidney. *Am. J. Physiol.* **267**, F748–F757 (1994).

14. Morrissey, J.J., McCracken, R., Kaneto, H., Vehaskari, M., Montani, D., and Klahr, S. Location of an inducible nitric oxide synthase mRNA in the normal kidney. *Kidney Int.* **45**, 998–1005 (1994).

15. Tojo, A., Gross, S.S., Zhang, L., Tisher, C.C., Schmidt, H.H.H.W., Wilcox, C.S., and Madsden, K.M. Immunocytochemical localization of distinct isoforms of nitric oxide synthase in the juxtaglomerular apparatus of normal rat kidney. *J. Am. Soc. Nephrol.* **4**, S1438–S1447 (1994).

16. Hof, R.P., Evenou, J.P., and Hof-Miyashita, A. Similar increases in circulating renin after equihypotensive doses of nitroprusside, diihydralazine or isradipine in conscious rabbits. *Eur. J. Pharmacol.* **136**, 251–254 (1987).

17. Johnson, R.A., and Freeman, R.H. 1994. Renin release in rats during blockade of nitric oxide synthesis. *Am. J. Physiol.* **266**, R1723–R1729 (1994).

18. Shultz, P.J. and Tolins, J.P. Adaptation to increased dietary salt intake in the rat. *J. Clin. Invest.* **91**, 642–650 (1993).

19. Dewan, S., Majid, A., and Navar, G.L. Suppression of blood flow autoregulation plateau during nitric oxide blockade in canine kidney. *Am. J. Physiol.* **262**, F40–F46 (1992).

20. Sigmon, D.H., Carretero, O.A., and Beierwaltes, W.H. Endothelium-derived relaxing factor regulates renin release in vivo. *Am. J. Physiol.* **263**, F256–F261 (1992).

21. Schricker, K., Della Bruna, R., Hamann, M., and Kurtz, A. Endothelium derived relaxing factor is involved in the pressure control of renin gene expression in the kidney. *Pflügers Arch.* **428**, 261–268 (1994).

22. Bosse, H.M., Böhm, R., Resch, S., and Bachmann, S. Parallel regulation of constitutive nitric oxide synthase and renin at the juxtaglomerular apparatus of rat kidney under various stimuli. *Am. J. Physiol.* **269**, F793–F805.

23. Johnson, R.A. and Freeman, R.H. Sustained hypertension in the rat induced by chronic blockade of nitric oxide production. *Am. J. Hyperten.* **5**, 919–922 (1992).

24. Naess, P.A., Christensen, G., Kirkeboen, K.A., and Kiil, F. Effect on renin release of inhibiting renal nitric oxide synthsis in anaesthetized dogs. *Acta Physiol. Scand.* **148**, 137–142 (1993).

25. Beierwaltes, W.H. Selective neuronal nitric oxide synthase inhibition blocks furosemide-stimulated renin secretion in vivo. *Am. J. Physiol.* **269**, F134–F139 (1995).

26. Persson, P.B., Baumann, J.E., Ehmke, H., Hackenthal, E., Kirchheim, H.R., and Nafz, B. Endothelium-derived NO stimulates pressure-dependent renin release in conscious dogs. *Am. J. Physiol.* **264**, F943–F947 (1993).

27. Delacretaz, E., Zanchi, A., Nussberger, J., Hayoz, D., Aubert, J.F., Brunner, H.R., and Waeber, B. Chronic nitric oxide synthase inhibition and carotid artery distensability in renal hypertensive rats. **26**, 332–336 (1995).

28. Schricker, K., Hegyi, I., Hamann, M., Kaissling, B., and Kurtz, A. Tonic stimulation of renin gene expression by nitric oxide is counteracted by tonic inhibition through angiotensin II. *Proc. Natl. Acad. Sci. USA* **92**, 8006–8010 (1995).

29. Schricker, K., Hamann, M., and Kurtz, A. Nitric oxide and prostaglandins are involved in the macula densa control of the renin system. *Am. J. Physiol.* **269**, F825–F830.

30. Schricker, K. and Kurtz, A. Blockade of nitric oxide formation inhibits the stimulation of the renin system by low salt intake. *Pflügers Arch.* **432**, 187–189.

31. Furchgott, R.E. Role of endothelium in responses of vascular smooth muscle cells. *Circ. Res.* **53**, 557–573 (1983).

32. Scholz, H. and Kurtz, A. Endothelium derived relaxing factor is involved in the pressure control of renin secretion from the kidneys. *J. Clin. Invest.* **91**, 1088–1094 (1993).

33. Münter, K. and Hackenthal, E. The participation of the endothelium in the control of renin release. *J. Hyperten.* **9**, S236–S237 (1991).

34. Gardes, J., Poux, J.M., Gonzales, M.F., Alhenc-Gelas, F., and Menard, J. Decreased renin release and constant kallikrein secretion after injection of L-Name in isolated perfused rat kidney. *Life Sci.* **50**, 987–993 (1992).

35. Gardes, J., Gonzales, M.F., Alhenc-Gelas, F., and Menard, J. Influence of sodium diet on L-Name effects on renin release and renal vasoconstriction. *Am. J. Physiol.* **267**, F798–F804 (1994).

36. Vidal, M.J., Romero, J.C., and Vanhoutte, P.M. Endothelium-derived relaxing factor inhibits renin release. *Eur. J. Pharmacol.* **149**, 401–402 (1988).

37. Beierwaltes, W.H. and Carretero, O.A. Nonprostanoid endothelium-derived factors inhibit renin release. *Hypertension* **19**, II-68–II-73 (1992).

38. Schricker, K., Ritthaler, T., Krämer, B.K., and Kurtz, A. Effect of endothelium-derived relaxing factor on renin secretion from isolated mouse renal juxtaglomerular cells. *Acta Physiol. Scand.* **140**, 347–354 (1993).

39. Schricker, K. and Kurtz, A. Liberators of NO exert a dual effect on renin secretion from isolated mouse renal juxtaglomerular cells. *Am. J. Physiol.* **265**, F180–F186 (1993).

40. Greenberg, S.G., He, X.-R., Schnermann, J.B., and Briggs, J.P. Effect of nitric oxide on renin secretion. I. Studies in isolated juxtaglomerular granular cells. *Am. J. Physiol.* **268**, F948–F952 (1995).

41. Della Bruna, R., Pinet, F., Corvol, P., and Kurtz, A. Opposite regulation of renin gene expression by cyclic AMP and calcium in isolated mouse juxtaglomerular cells. *Kidney Int.* **47**, 1266–1273 (1995).

42. He, X.-R., Greenberg, S.G., Briggs, J.P., and Schnermann, J.B. Effect of nitric oxide on renin secretion. II. Studies in the perfused juxtaglomerular apparatus. *Am. J. Physiol.* **268**, F953–F959 (1995).

43. Gnarro, L.J. Endothelium-derived nitrix oxide: actions and properties. *FASEB J.* **3**, 31–36 (1989).

44. Kurtz, A., Della Bruna, R., Pfeilschifter, J., Taugner, R., and Bauer, C. Atrial natriuretic peptide inhibits renin release from isolated renal juxtaglomerular cells by a cGMP mediated process. *Proc. Natl. Acad. Sci. USA* **83**, 4769–4773 (1986).

45. Laffranchi, R., Gogvadze, V., Richter, C., and Spinas, G.A. Nitrix oxide stimulates insulin secretion by inducing calcium release from mitochondria. *Biochem. Biophys. Res. Commun.* **217**, 584–591 (1995).

46. Schmidt, H.H.H.W., Warner, T.D., Ishii, K., Sheng, H., and Murad, F. Insulin secretion from pancreatic B cells caused by L-arginine-derived nitriogen oxide. *Science* **255**, 721–723 (1992).

47. Schmidt, H.H.H.W., Lohmann, S.M., and Walter, U. The nitric oxide and cGMP signal transduction system: regulation and mechanismen of action. *Biochem. Biophys. Acta* **1178**, 153–175 (1993).

48. Gambarian, S., Lohmann, S., and Walter, S. Personal communication.

49. Singh, I.J., Graham, M., Wang, W.H., Young, T., Killen, P., Smart, A., Schnermann, J., Briggs, J.P. 1996. Coordinate regulation of renal expression of nitric oxide synthase, renin, and angiotensinogen oxide mRNA by dietary salt. *Am. J. Physiol.* **270**, F1027–F1037.

50. Fischer, E., Schnermann, J., Briggs, J.P., Kriz, W., Ronco, P.M., and Bachmann, S. Ontogeny of NO synthase and renin in juxtaglomerular apparatus of rat kidneys. *Am. J. Physiol.* **268**, F1164–F1176 (1995).

13

NO and the Medullary Circulation

Yoram Agmon and Mayer Brezis

1. Introduction

Important data have accumulated over the last decade regarding the role of nitric oxide (NO) in the regulation of renal function [1]. Studies from our laboratory and others have examined the specific role of nitric oxide in the regulation of blood flow and oxygenation of the renal medulla, noted for its vulnerability to hypoxic damage. Our observations and a literature review are presented, along with experimental data regarding the role of nitric oxide in the pathogenesis of radiocontrast nephropathy, a prototype of acute renal failure resulting from medullary hypoxia.

2. The Renal Medulla

2.1. Anatomy of the Medullary Vasculature (Fig. 13-1)

The unique anatomy of the medullary vasculature allows efficient urine concentration and water conservation, an obligatory task in terrestrial mammals. The medullary vessels arise from the juxtamedullary cortex, where efferent arterioles originating from the juxtamedullary glomeruli descend into the medulla [2,3]. These efferent arterioles are typically larger than those located in the mid and superficial cortex, which do not supply the medulla but rather form the cortical peritubular capillary network. The juxtamedullary efferent arterioles give rise to the medullary vasa recta, consisting of long hairpinlike capillary loops extending deep into the medulla. In the outer medulla, an extensive capillary network connects the ascending and descending vessels, whereas the inner medullary vasa recta are unbranched [2].

Venous blood from the inner medulla drains directly from the vasa recta into

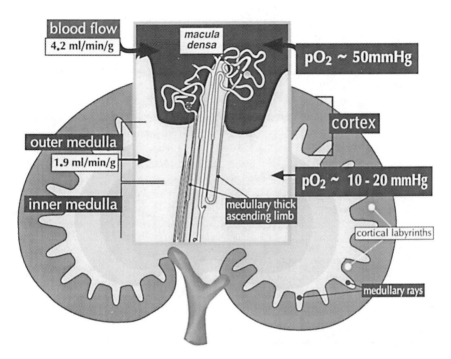

Figure 13-1. Anatomical and physiological features of the kidney comparing cortex and medulla. The cortex, which receives an ample blood supply to optimize glomerular filtration, is generally well oxygenated, except for the medullary rays areas devoid of glomeruli, which are supplied by venous blood ascending from the medulla. The medulla, which receives a meager blood supply to optimize concentration of the urine, is poorly oxygenated. Medullary hypoxia results both from countercurrent exchange of oxygen within the vasa recta and from oxygen consumption by the medullary thick ascending limbs. Renal medullary hypoxia is an obligatory price the mammalian kidney pays for successful urinary concentration.

the arcuate veins. By contrast, the outer medullary venous vessels ascend into the medullary rays, which extend into the cortex and are placed between the cortical labyrinths [2,4]. The medullary rays are devoid of glomeruli and are often viewed as expansions of the outer medulla into the cortex.

Overall, the medulla is supplied by approximately one-tenth of the total renal blood flow [2]. Medullary blood flow is typically low in order to preserve the osmotic gradient generated by the tubular countercurrent mechanism.

2.2. Medullary Oxygenation

Active solute reabsorption by the medullary thick ascending limb (mTAL) of Henle's loop generates the medullary osmotic gradient necessary for urine concentration. The anatomic design of the medullary vasa recta allows a countercurrent exchange to build and preserve osmotic gradients, but at the price of impaired medullary oxygenation. Oxygen diffuses from the arterial to venous vasa recta, thus limiting the oxygen supply to the medulla. Oxygen deficiency is mostly pronounced in the outer medulla, where the mTALs are located. This nephron segment performs active, oxygen-consuming solute reabsorption in an oxygen-deprived milieu and is therefore prone to hypoxic damage [5,6].

We have measured the oxygen tension in the renal parenchyme, applying pO_2 microelectrodes inserted into the kidney [7–11]. Partial pressure of oxygen in the cortex is about 50 mm Hg, whereas outer medullary pO_2 is in the range 10–20 mm Hg. This degree of tissue hypoxia is compatible with intact tubular function. However, additional worsening of medullary oxygenation by disruption of the delicate balance of oxygen supply (by the medullary vasculature) and demand (solute reabsorption by the mTAL) may result in mTAL dysfunction and eventually frank necrosis of mTALs [5].

Numerous protective mechanisms normally serve to defend the renal medulla from hypoxic damage [4,5,12,13]. Medullary oxygenation may improve by increasing medullary oxygen supply (i.e., increasing medullary blood flow) or by decreasing medullary oxygen demand (i.e., decreasing mTAL transport activity).

Medullary Vasodilators

The renal medulla is rich in endogenous vasodilators, including prostanoids (primarily prostaglandin E_2) [3,14], adenosine [15], nitric oxide (to be discussed), and dopamine [3]. These counteract a variety of vasoconstrictors, such as catecholamines, vasopressin, angiotensin II, and endothelin [4,12]. Enhanced synthesis of these vasodilators when medullary oxygen demand is augmented protects the medulla from hypoxic damage.

Modification of Tubular Transport

Medullary oxygenation may also improve by decreasing solute reabsorption, and thus oxygen utilization, by the mTAL. Direct inhibition of mTAL sodium reabsorption results from the local action of endogenous prostaglandin E_2 [16] and adenosine [17], as well as dopamine, the platelet-activating factor, and cytochrome P-450-dependent arachidonic acid metabolites [4]. Transport activity by the mTAL may decrease indirectly by lowering the glomerular filtration rate. For example, renal hypoperfusion results in cortical oligemia and a reduction of glomerular filtration, although medullary blood flow is usually well maintained (so-called corticomedullary "redistribution"). The reduction of glomerular filtration decreases mTAL reabsorptive activity and improves medullary oxygenation. The outer medulla is also normally protected by the tubuloglomerular feedback [4]. Hypoxic impairment of mTAL solute reabsorption increases solute delivery to the macula densa and results in activation of this feedback mechanism which decreases glomerular filtration and thus additional solute delivery to the mTAL.

Multiple mechanisms usually act in concert to protect the outer medulla from hypoxic damage. For example, endogenous adenosine decreases cortical blood flow [15] and glomerular filtration, increases medullary blood flow [15], and decreases mTAL solute reabsorption [17], all acting to improve medullary oxygenation [9,19]. The diversity of protective mechanisms allows compensation when specific mechanisms are limited by disease or drugs. Therefore, experimental medullary injury in vivo and clinical renal failure are usually prevented unless a combination of insults occurs, impairing multiple protective mechanisms [20].

Experimental Modification of Medullary Oxygenation

A variety of experimental manipulations may disrupt the delicate balance of medullary oxygen supply and demand, and predispose the outer medulla to hypoxic damage. Medullary blood flow and oxygenation may decline in response to synthesis inhibition of vasodilator prostaglandins [8,14] or nitric oxide (see below). Enhanced solute reabsorption by the mTAL during osmotic diuresis {mannitol [10], contrast media (to be discussed)}, polyene antibiotics administration (which increase tubular membrane permeability [21]), or compensatory renal hypertrophy [22] decreases medullary oxygenation and may result in mTAL necrosis. By contrast, inhibition of mTAL reabsorption by furosemide improves medullary oxygenation [10].

In summary, the renal medulla is remarkably prone to hypoxic injury due to its delicate balance between oxygen supply and demand. Nitric oxide plays a major role in the protection of the outer medulla from hypoxic damage, as will be further discussed.

3. Medullary Nitric Oxide Production

Various structures in the renal medulla are capable of synthesizing NO, including blood vessels, tubules, and interstitial cells. Indirect evidence supporting medullary NO synthesis is presented in the studies of Biondi et al. [23–25] in which canine kidney slices were tested for their content of cyclic guanidine 3',5'-monophosphate (cGMP), a second messenger in the NO pathway. Overall, cGMP content in the medulla is higher than in the cortex [24,25]. Within the medulla, basal cGMP production is maximal at its inner portion. Stimulation of NO synthesis by acetylcholine (ACh) or bradykinin increases cGMP production in the medulla, an observation most pronounced in the mid and outer portions of the inner medulla. Inhibition of NO synthesis by N^G-monomethyl-L-arginine (L-NMMA) significantly reduces both basal and ACh- or bradykinin-stimulated cGMP production.

Recent studies have localized the exact medullary sites of NO synthesis by molecular techniques. These studies present conflicting data due to the existence of constitutive and inducible NO synthase (NOS) in various isoforms.

3.1. Medullary Vascular NO Synthesis

As expected, the vasa rectae are capable of NO synthesis. Terada et al. [26] have shown by reverse transcription-polymerase chain reaction that microdissected vasa rectae possess a constitutive nitric oxide synthase (NOS), specifically isoform I of NOS, as well as a soluble guanylate cyclase, the site for NO binding.

3.2. Nonvascular Medullary Sources of NO Synthesis

Various medullary nephron segments are also capable of NO synthesis, as shown by Terada et al. in microdissected nephrons [26]. The inner medullary collecting duct possesses large amounts of constitutive NOS (isoform I), the largest amount of all nephron segments. The outer medullary collecting duct and the inner medullary thin limbs also possess small quantities of constitutive NOS, but the medullary thick ascending limb of Henle's loop does not contain any of this NOS isoform. By contrast, studies by Ujiie et al. [27] revealed the presence of isoform III of constitutive NOS, the main isoform in endothelial cells, in both the mTAL and the outer medullary collecting duct.

In addition to constitutive NOS activity, various renal medullary structures possess the capacity to express an inducible NOS when stimulated by cytokines. These include the mTAL [28,29], the inner medullary collecting duct [30], and medullary interstitial cells [31]. In the mTAL, the regulation of the inducible NOS appears to be mediated by the nuclear transcription factor NFκB [29]. The principal sources and regulation of medullary NO synthesis have not been clarified at the time of this writing. In addition, inflammatory cells invading the medulla may serve as a source of NO in various renal pathologies. In summary, it is clear from the above data that various medullary structures are capable of synthesizing NO.

4. Medullary Effect of Nitric Oxide

4.1. Effect of Nitric Oxide on the Medullary Circulation

The role of nitric oxide (NO) in the regulation of medullary hemodynamics has been recently studied using laser–Doppler flowmetry (LDF) and videomicroscopy of the medullary vasculature. The role of basal NO synthesis was evaluated by specific inhibitors of NO synthase (NOS), and the potential for NO-mediated vasodilation assessed pharmacologically by the administration of acetylcholine (ACh) or bradykinin, and physiologically by increasing renal perfusion pressure.

We recently published observations on medullary hemodynamics (specifically, in the *outer medulla*), using LDF, with a fine-needle probe inserted into the outer medulla of anesthetized rats [10,14,15,32]. The role of nitric oxide in the control of the basal tone of the outer medullary vessels was evaluated by measuring the changes in outer medullary blood flow following NO synthase inhibition by the intravenous injection of N^ω-nitro-L-arginine (L-NAME) (10 mg/kg bolus injection) [32]. Outer medullary blood flow declined to 61±7% of baseline within 30 min of injection ($n=6$; $p<.005$) (Fig. 13-2), indicating significant basal NO-mediated medullary vasodilation. Although systematic dose-response studies were not performed, our experience suggests that NO synthesis is not completely inhibited using the above dose, suggesting underestimation of the role of NO in basal

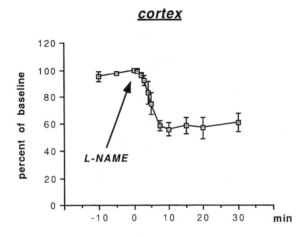

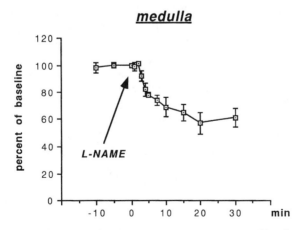

Figure 13-2. Nitric-oxide-dependent cortical and medullary vasodilatation. Cortical and outer medullary blood flow following in anesthetized rats, before and after inhibition of nitric oxide synthesis by L-NAME (10 mg/kg, IV $n=6$).

medullary vasodilation. The intravenous injection of L-NAME also lowered superficial cortical blood flow to a similar extent. Because the medullary vessels arise from the juxtamedullary cortex, L-NAME may cause juxtamedullary vasoconstriction, with secondary lowering of the medullary blood flow. The exact role of the medullary versus cortical effects of NO has yet to be more clearly defined.

Mattson et al. studied the effects of direct medullary interstitial infusions of L-NAME and ACh via chronically implanted medullary interstitial catheters in rats [33]. Outer medullary blood flow was measured by LDF via implanted intrarenal optical probes. Interstitial L-NAME infusion (120 µg/0.5 ml/h) lowered outer medullary blood flow to 63% of baseline (LDF signal change from 0.95±0.09 V to 0.60±0.07 V; n=15; p<.05) within 2 h of infusion. During the chronic (5 days) infusion of L-NAME, medullary blood flow averaged ~70% of baseline. The medullary interstitial infusion did not affect cortical blood flow throughout the prolonged experimental period, emphasizing the direct role of NO in medullary vasodilation. Interstitial ACh infusion (1-h infusions, 200 µg/h) increased outer medullary blood flow by 19±4% (n=5, p<.05). This effect of ACh was abolished by concomitant L-NAME infusion.

Earlier studies by Mattson et al. evaluated the effects of similar interstitial infusions on papillary blood flow, applying LDF with hemodynamic measurements from an exposed papilla [34]. They assumed that changes in papillary blood flow were representative of changes in the inner parts of the renal medulla, an assumption which may not be always correct [35]. The medullary interstitial infusion of N^G-nitro-L-arginine (L-NA) (120 µg/0.5 ml/h) lowered papillary blood flow by 24±4% from baseline (n=7, p<.05) following 75 min of infusion. ACh infusion (200 µg/h) raised papillary blood flow by 13±2% (n=7, p<.05). Superficial cortical blood flow did not change following L-NAME or ACh infusion into the medullary interstitium. In a separate set of studies, L-NA infusion in similar doses into the renal artery lowered papillary blood flow to 66±3% of baseline (n=5, p<.05 with a similar reduction in cortical blood flow. Mattson et al. also evaluated the medullary effects of bradykinin, a NO-mediated vasodilator [36]. Bradykinin or captopril (which inhibits bradykinin degradation) were infused directly into the medullary interstitium through chronically implanted catheters, and papillary blood flow was measured by LDF of exposed papillas in rats. A 30-min infusion of bradykinin (0.1 µg/min) or captopril (17 µg/min) raised papillary blood flow to 117±3% (n=6, p<.05) or 121±5% (n=6, p<.05) of control values, respectively. Interstitial infusion of L-NAME (2 µg/min) both lowered papillary blood flow to 78±3% of baseline (n=7, p<.05) and completely abolished the papillary blood flow response to bradykinin or captopril.

Comparable results of papillary blood flow studies in rats were obtained by other investigators [37,38]. Atucha et al. [37] revealed a ~15% decline in papillary blood flow measured by LDF, within 1 h following intravenous infusion of L-NAME (10 µg/kg/min). Of interest is their observation that L-NAME blunted the normal rise in papillary blood flow following volume expansion. Lockhart et al. assessed the changes in papillary blood flow by fluorescence videomicros-copy of the papillary vasa recta [38]. Within 15 min of intravenous L-NAME infusion (a 15-mg/kg bolus followed by 500 μg/kg/min), *descending* vasa recta flow declined by ~47% (8.43±1.81 to 4.46±1.19 nl/min; $n=17$, $p<.01$) and *ascending* vasa recta flow by ~44% (4.28±0.80 to 2.38±0.55 nl/min; $n=27$, $p<.01$).

The role of NO in the response of the papillary vasculature to changes in renal perfusion pressure was evaluated by Fenoy et al. [39]. Papillary blood flow was measured by LDF of exposed papillas in rats during aortic clamping. In control animals, increasing renal perfusion pressure (from 60 to 140 mm Hg) caused a near-linear rise in papillary blood flow, indicating the poor autoregulation of medullary blood flow. Systemic pretreatment with L-NAME blunted the rise of papillary blood flow at high perfusion pressures, indicating a role for NO in pressure-induced medullary vasodilation. Interestingly, L-arginine infusion did not change the relationship between renal perfusion pressure and papillary blood flow, suggesting that the availability of L-arginine is not a limiting step in NO synthesis in this experimental setting. However, L-arginine completely abolished the hemodynamic effects of subsequent L-NAME administration.

Taken together, the above studies highlight a major role for NO in the control of the medullary vascular tone, both in the basal state and in response to diverse physiologic stimuli.

4.2. Nonvascular Medullary Effects of NO

Preliminary experiments suggest that NO may exhibit direct tubular effects [40,41]. NO inhibits tubular transport in cortical collecting duct cells [40] but enhances proximal tubular sodium reabsorption [41]. No direct medullary tubular effects have been described at the time of this writing. However, the demonstration of soluble guanylate cyclase in medullary tubular structures such as the mTAL and the medullary collecting duct [26] suggests that NO may have a direct tubular effect.

5. Role of Nitric Oxide in the Pathogenesis of Acute Renal Failure

5.1. Radiocontrast Nephropathy

Studies from our laboratory evaluated the role of NO in the pathogenesis of radiocontrast-induced acute renal dysfunction, characterized by hypoxic medul-lary injury. Whereas direct tubular toxicity from the radiocontrast material has been suggested in the pathogenesis of radiocontrast nephropathy (RCN), clinical characteristics [42,43] and experimental data [44–50] suggest that systemic and/or intrarenal hemodynamic changes predispose to ischemic renal damage from contrast media (CM).

Defective endothelial-dependent, NO-mediated vasodilation has been recently observed in diabetes mellitus [51], hypertension [52], atherosclerosis [53], and heart failure [54]. Because these conditions are often present in patients developing RCN [42], we speculated that endothelial dysfunction may play a role in the vulnerability of the kidney to contrast-induced renal failure. We therefore developed an in vivo model of RCN and evaluated the hemodynamic basis of contrast-induced renal failure, with special interest in a potential role for NO.

In Vivo Model of RCN

One of the first in vivo models of contrast nephropathy, developed in our laboratory by Heyman [55], combined uninephrectomy, salt depletion, prostaglandin synthesis inhibition, and contrast media administration to produce acute renal failure in rats. This model resembles the clinical syndrome in humans, which is characterized by multiple clinical risk factors [20,56,57], such as preexisting renal failure, volume depletion, or concomitant administration of other nephrotoxins. Histologically, tubular necrosis was observed, confined to the medullary thick ascending limbs of Henle's loop, emphasizing the susceptibility of the renal outer medulla to hypoxic damage. Additional NO synthesis inhibition by N^G-monomethyl-L-arginine (L-NMMA) resulted in aggravation of renal function and histologic mTAL damage [7]. Without L-NMMA, creatinine clearance (CC) decreased from 0.24 ± 0.03 ml/min per 100 g body weight at baseline to 0.08 ± 0.1 ($n=22$; $p<.001$) the day after CM administration. Adding L-NMMA (15 mg/kg), CC decreased from 0.17 ± 0.01 to 0.03 ± 0.01 ml/min per 100 g ($n=12$; $p<.002$), significantly more than without L-NMMA ($p<.005$ for the comparison between final creatinine clearances). Pathologically, mTAL necrosis was significantly more severe after the addition of L-NMMA (approximately 25% of mTALs without L-NMMA versus more than 50% with L-NMMA; $p<.01$).

We recently developed a simpler model of RCN in rats [32], in which chronic preparations (i.e., uninephrectomy and chronic salt depletion) were not necessary. The model consisted of NO synthesis inhibition prior to CM injection with or without additional inhibition of the synthesis of prostaglandins, which are potent medullary vasodilators [14]. Multiple protocols were examined, consisting of injections of CM (intraarterial injection of iothalamate 80%, 6 ml/kg), NO synthesis inhibition (intravenous L-NAME, 10 mg/kg), and prostaglandin synthesis inhibition (intravenous indomethacin, 10 mg/kg), alone and in various combinations.

As previously described [55], the sole administration of CM does not impair renal function or cause tubular necrosis. This resembles the clinical experience, in which CM usually does not cause renal failure in the absence of additional risk factors. Similarly, L-NAME did not impair renal function when given alone. However, the injection of CM following L-NAME caused significant renal failure, with CC decreasing from 1.22 ± 0.08 at baseline to 0.48 ± 0.10 ml/min after 24 h ($n=8$; $p<.001$) and necrosis of $18\pm6\%$ of mTALs, figures resembling the results in Heyman's original model [55]. Additional indomethacin injection along with

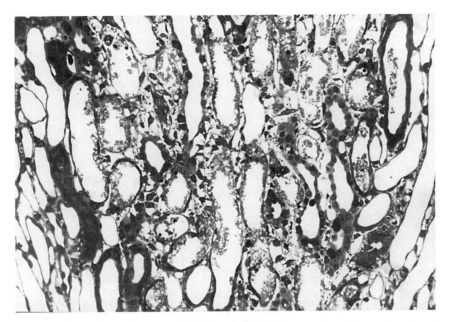

Figure 13-3. Histology of a model of acute renal failure produced by radiocontrast injection following inhibition of nitric oxide and prostaglandin synthesis. Outer medulla of a rat kidney 24 h after an injection of the radiocontrast iothalamate following L-NAME and indomethacin. Extensive tubular necrosis of mTALs is observed in the center, whereas tubules immediately adjacent to vasa recta (on each side of the figure) are better preserved.

L-NAME, prior to CM administration, resulted in severe renal failure, with CC decreasing from 1.05±0.10 to 0.27±0.05 ml/min ($n=12$; $p<.001$) and diffuse necrosis of 49±9% of mTALs (Fig. 13-3). In all protocols, there was no necrosis of cortical nephron segments, similar to Heyman's previous observations [55]. These data suggest that inhibition of nitric oxide synthesis is both *sufficient* and *necessary* to allow radiocontrast-mediated medullary necrosis and renal failure. Additional prostaglandin synthesis inhibition aggravates these pathologies.

Changes in Intrarenal Hemodynamics Following CM Administration

Previous hemodynamic studies have shown that contrast media transiently reduce *total* renal blood flow. However, because the medulla receives only approximately 10% of the total renal blood flow, these measurements may overlook significant intrarenal hemodynamic changes, specifically in the *outer medulla*. We measured directly the changes in the outer medullary circulation, the target of histologic injury in our model of RCN, in an attempt to explain the hemodynamic basis of our in vivo model. We applied LDF in anesthetized rats to study intrarenal hemodynamics, with laser–Doppler probes placed in the kidney upon the superficial cortex and within the outer medulla (a needle probe is inserted into the medulla through the cortex). Hemodynamic responses to CM were observed in intact rats and in rats pretreated with L-NAME or indomethacin prior to CM administration [32].

In intact rats, iothalamate injection transiently reduces *cortical* blood flow (Fig. 13-4), resembling the changes in total renal blood flow observed in previous studies [44–50]. Pretreatment with either L-NAME or indomethacin does not qualitatively affect the cortical blood flow response to iothalamate. In all cases, a transient reduction in cortical blood flow is observed.

By contrast, the medullary blood flow response to iothalamate depends on preexisting renal physiology (Fig. 13-4). In intact rats, medullary blood flow increased to 196±25% of baseline (*n*=6; *p*<.001). This hyperemic response per-

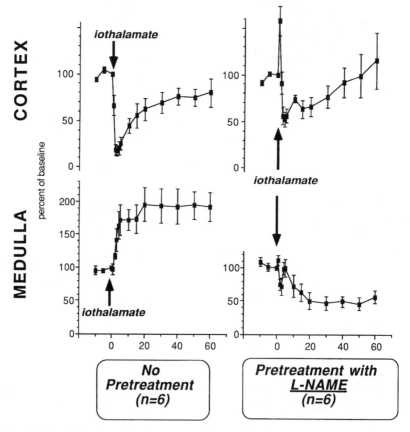

Figure 13-4. Hemodynamic response to radiocontrast injection before or after inhibition of nitric oxide synthesis. Cortical and outer medullary microcirculatory flow in response to iothalamate before or after inhibition of nitric oxide synthesis by L-NAME.

sisted for over 1 h following iothalamate injection. As mentioned earlier, pretreatment with L-NAME reduced medullary blood flow to 61±7% of baseline (*n*=6; *p*<.005). In addition, the previously observed hyperemic response to iothalamate was replaced by prolonged vasoconstriction, with a further reduction of outer medullary flow to 47±9% of baseline (*n*=6; *p*<.005) (and this vasoconstrictive response persisted even longer than our formal 1-h experimental observations). Pretreatment with indomethacin reduced medullary blood flow to half of baseline [14]. In addition, the previously observed hyperemic response to iothalamate was similarly replaced by prolonged vasoconstriction, with a reduction of outer medullary flow to 45±6% of baseline (*n*=6; *p*<.001). The combined administration of L-NAME, indomethacin, and iothalamate lowered medullary blood flow to 12±4% of baseline (*n*=3; *p*<.005).

It appears, therefore, that medullary hyperemia is the response to iothalamate under normal conditions. When nitric oxide or prostaglandin synthesis is inhibited, iothalamate reduces medullary blood flow. These data suggest that iothalamate tends to lower medullary blood flow, a tendency normally counteracted by medullary vasodilators, such as nitric oxide and prostaglandins. Restriction of this vasodilator potential, by inhibition of nitric oxide or prostaglandin synthesis, and possibly in diseases characterized by endothelial dysfunction, unmasks this vasoconstrictive tendency. A possible mediator of the observed vasoconstriction is endothelin, which is secreted in response to contrast media, as previously demonstrated in both in vivo and in vitro studies [48,49]. This reversal of the vasoactive response is reminiscent of the effect of endothelial dysfunction in other vascular beds. For example, atherosclerotic coronary arteries paradoxically constrict in response to acetylcholine, in contrast to the vasodilation observed in normal arteries [53]. Radiocontrast nephropathy may be an important example of the pathophysiological implications of endothelial dysfunction within the kidney.

Changes in Intrarenal Oxygenation Following CM Administration

We previously evaluated the CM-induced changes in medullary oxygenation by pO_2 microelectrodes inserted into the outer medulla in anesthetized rats [8]. Iothalamate administration reduced medullary pO_2 from 26±3 to 9±2 mm Hg within 10 min of CM administration (*n*=15; *p*<.005). Our hemodynamic data showing CM-induced medullary vasodilation suggests that enhanced pO_2 consumption by the outer medulla following CM administration, rather than decreased pO_2 delivery, aggravates medullary hypoxia. This is supported by our observation that furosemide (10 mg/kg) injection following contrast administration completely reverses the CM-induced worsening of medullary oxygenation [8]. Our experience that CM alone does not cause renal failure despite severe medullary hypoxia suggests that this degree of hypoxia is not sufficient to cause significant renal damage. The combination of CM administration along with inhibition of NO and prostaglandin synthesis further decreases outer medullary pO_2 toward zero, predisposing to tubular necrosis.

Proposed Pathogenesis of RCN

Combining the above experimental data, the following mechanisms of RCN are proposed:

- Contrast media induce osmotic diuresis, leading to enhanced solute reabsorption, and therefore oxygen consumption, by the medullary thick ascending limb.

- In addition, contrast media tend to induce renal vasoconstriction, a phenomenon possibly mediated by endothelin. This response is transient in the cortex, but prolonged in the medulla.

- The combination of the above medullary effects aggravates outer medullary hypoxia.

- The normal medulla reacts to hypoxia by protective vasodilation, mediated by a combination of vasodilators, including nitric oxide, prostaglandins, adenosine [14,15,32], and possibly other vasoactive substances. This normal hyperemic response partially improves medullary oxygenation and prevents tubular damage.

- The attenuation of these vasodilator mechanisms, as a result of systemic, renal, or endothelial disorders, or by drugs (for instance, nonsteroidal anti-inflammatory drugs) leads to contrast media-induced vasoconstriction and medullary ischemia, aggravation of medullary hypoxia and frank tubular necrosis (Fig. 13-5).

Miscellaneous Models of RCN

Touati et al. have studied the role of nitric oxide in the pathogenesis of RCN, including pathologic studies of the renal medulla [58]. Their in vivo rat model of RCN consisted of uninephrectomy, renal ischemia (by aortic clamping), and injection of CM. This protocol resulted in mild renal failure, with CC reduction of approximately 25% from baseline at 24 h. The addition of L-NAME to the protocol markedly decreased CC to less than 50% of baseline. Pathologically, necrosis of the outer medullary structures was observed. Schwartz et al. have also emphasized the protective role of nitric oxide in the pathogenesis of RCN [59]. However, their studies concentrated on measurements of total renal blood flow and CC, without specifically addressing the pathophysiology of the outer medulla.

6. Role of NO in Other Models of Acute Renal Failure

6.1. Cyclosporine A Nephrotoxicity

Previous experiments from our laboratory have shown that chronic cyclosporine A (CyA) administration to rats consistently resulted in renal failure [60]. Histolog-

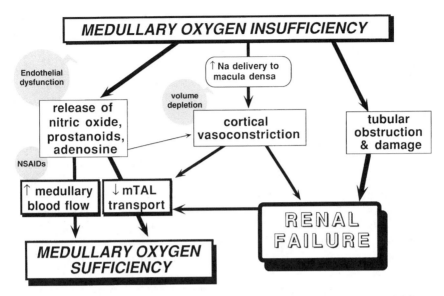

Figure 13-5. Putative mechanisms leading from medullary cell hypoxia to renal failure. On the left are shown physiological homeostatic signals, such as nitric oxide, that improve medullary oxygenation. In the center and on the right are shown some pathophysiological consequences of more advanced medullary hypoxia, such as cortical vasoconstriction and tubular damage. The potential adverse effects of endothelial dysfunction (with defective release of nitric oxide), of some nephrotoxins, and of volume depletion are shown. Nonsteroidal anti-inflammatory agents (NSAIDs) disable the beneficial prostanoid-mediated medullary vasodilatatory response to local hypoxia, and volume depletion enhances the tubuloglomerular feedback reflex decrease in glomerular filtration.

ically, degenerative changes were observed in the mTAL and in S2–S3 nephron segments localized in the medullary rays.

Recent studies have examined the role of nitric oxide in the pathogenesis of CyA nephrotoxicity [61–63]. However, all of these studies have emphasized the changes in glomerular hemodynamics induced by CyA, without focusing on the outer medulla or medullary rays, histologic sites of CyA toxicity [60]. Some of these studies [61,62], but not all [63], suggest that renal cortical NO production is diminished in CyA-treated rats. It is conceivable that CyA induces similar effects in the medulla and in the medullary rays, thereby decreasing medullary blood flow and O_2 supply to these regions, resulting in hypoxic damage. Further studies in the medulla are necessary to clarify this hypothesis.

6.2. Renal Ischemia

The protective role of nitric oxide has been studied in a model of renal ischemic renal failure employing complete renal ischemia via renal artery occlusion and reperfusion [64–68]. These studies uniformly suggest that NO has a renal protective role in this model. NOS inhibition exacerbated the deleterious effects of ischemia/reperfusion [64], whereas L-arginine administration had the opposite effect [64,66–68]. Insulinlike growth factor I had a protective effect in this model, an effect possibly mediated by NO [65]. Interestingly, Sabbatini et al. present data suggesting that renal impairment in old rats submitted to ischemia/reperfusion is greater than in young rats, possibly due to reduced basal NO production in aged rats [67].

None of the above studies addressed the role of NO in the medullary pathology following renal ischemia. This model of renal ischemia has limited clinical relevance, as acute renal failure is rarely the result of complete cessation of renal blood flow—except in the setting of renal transplantation and aortic clamping during vascular surgery [12]. Interestingly, total cessation of renal blood flow may theoretically have a protective role upon the renal medulla, because glomerular filtration is completely abolished and the mTAL performs no reabsorptive work. This has been previously observed in nonfiltering rat kidneys perfused by a hyperoncotic medium [18].

Yu et al. has demonstrated an adverse effect of NO upon isolated proximal tubules exposed to hypoxia and reoxygenation [69]. Hypoxia stimulated NO release from the tubules. NOS inhibition reduced and L-arginine increased hypoxic tubular damage.

7. Summary

The experimental data presented above emphasize the significant role of nitric oxide in maintaining adequate medullary oxygenation via medullary vasodilation. Impaired endothelial-dependent vasodilation predisposes the kidney to acute renal failure secondary to hypoxic medullary damage.

References

1. Romero, J.C., Lahera, V., Salom, M.G., and Biondi, M.L. *J. Am. Soc. Nephrol.* **2**, 1371–1387 (1992).

2. Dworkin, L.D. and Brenner, B.M. In: *The Kidney,* 4th ed. Eds. Brenner, B.M. and Rector, F.C. W.B. Saunders, Philadelphia, 1991, pp. 164–204.

3. Chou, S.Y., Porush, J.G., and Faubert, P.F. *Kidney Int.* **37**, 1–13 (1990).

4. Brezis, M. and Rosen, S. *N. Engl. J. Med.* **332**, 647–655 (1995).

5. Brezis, M., Rosen, S., Silva, P., and Epstein, F.H. *Kidney Int.* **26**, 375–383 (1984).

6. Brezis, M., Rosen, S., Silva, P., and Epstein, F.H. *J. Clin. Invest.* **73**, 182–190 (1984).

7. Brezis, M., Heyman, S.N., Dinour, D., Epstein, F.H., and Rosen, S. *J. Clin. Invest.* **88**, 390–395 (1991).

8. Heyman, S.N., Brezis, M., Epstein, F.H., Spokes, K., Silva, P., and Rosen, S. *Kidney Int.* **40,** 632–642 (1991).

9. Dinour, D. and Brezis, M. *Am. J. Physiol.* **261,** F787–F791 (1991).

10. Brezis, M., Agmon, Y., and Epstein, F.H. *Am. J. Physiol.* **267,** F1059–F1062 (1994).

11. Brezis, M., Heyman, S.N., and Epstein, F.H. *Am. J. Physiol.* **267,** F1063–F1066 (1994).

12. Brezis, M., Rosen, S., and Epstein, F.H. In: *The Kidney,* 4th ed. Eds. Brenner, B.M. and Rector, F.C. W.B. Saunders, Philadelphia, 1991, pp. 993–1061.

13. Brezis, M., Rosen, S., and Epstein, F.H. *Am. J. Kidney Dis.* **13,** 253–258 (1989).

14. Agmon, Y. and Brezis, M. *Exp. Nephrol.* **1,** 357–363 (1993).

15. Agmon, Y., Dinour, D., and Brezis, M. *Am. J. Physiol.* **265,** F802–F806 (1993).

16. Lear, S., Silva, P., Kelley, V.E., and Epstein, F.H. *Am. J. Physiol.* **258,** F1372–F1378 (1990).

17. Beach, R.E. and Good, D.W. *Am. J. Physiol.* **263,** F482–F487 (1992).

18. Brezis, M., Rosen, S., Silva, P., and Epstein, F.H. *Kidney Int.* **25,** 65–72 (1984).

19. Dinour, D., Agmon, Y., and Brezis, M. *Exp. Nephrol.* **1,** 152–157 (1993).

20. Agmon, Y. and Brezis, M. *Contrib. Nephrol.* **102,** 23–36 (1993).

21. Brezis, M., Rosen, S., Silva, P., Spokes K., and Epstein, F.H. *Science* **224,** 66–68 (1984).

22. Epstein, F.H., Silva, P., Spokes, K., Brezis, M., and Rosen, S. *Kidney Int.* **36,** 768–772 (1989).

23. Biondi, M.L., Dousa, T., Vanhoutte, P., and Romero, J.C. *Am. J. Hyperten.* **3,** 876–878 (1990).

24. Biondi, M.L. and Romero, J.C. *J. Vasc. Med. Biol.* **2,** 294–298 (1990).

25. Biondi, M.L., Bolterman, R.J., and Romero, J.C. *Renal Physiol. Biochem.* **15,** 16–22 (1992).

26. Terada, Y., Tomita, K., Nonoguchi, H., and Marumo, F. *J. Clin. Invest.* **90,** 659–665 (1992).

27. Ujiie, K., Yuen, J., Hogarth, L., Danziger, R., and Star, R.A. *Am. J. Physiol.* **267,** F296–F302 (1994).

28. Morrissey, J.J., McCracken, R., Kaneto, H., Vehaskari, M., Montani, D., and Klahr, S. *Kidney Int.* **45,** 998–1005 (1994).

29. Kone, B.C., Schwöbel, J., Turner, P., Mohaupt, M.G., and Cangro, C.B. *Am. J. Physiol.* **269,** F718–F729 (1995).

30. Markewitz, B.A., Michael, J.R., and Kahan, D.E. *J. Clin. Invest.* **91,** 2138–2143 (1993).

31. Lau, K.S., Aalund, G.R., Hogarth, L., Ujiie, K., Yuen, J., and Star, R.A. *J. Am. Soc. Nephrol.* **4,** 557A (1993).

32. Agmon, Y., Peleg, H., Greenfeld, Z., Rosen, S., and Brezis, M. *J. Clin. Invest.* **94,** 1069–1075 (1994).

33. Mattson, D.L., Lu, S., Nakanishi, K., Papanek, P.E., and Cowley, A.W. *Am. J. Physiol.* **266,** H1918–H1926 (1994).

34. Mattson, D.L., Roman, R.J., and Cowley, A.W. *Hypertension* **19,** 766–769 (1992).

35. Hellberg, O., Kallskog, O., and Wolgast, M. *Kidney Int.* **40,** 625–631 (1991).

36. Mattson, D.L. and Cowley, A.W. *Hypertension* **21,** 961–965 (1993).

37. Atucha, N.M., Ramirez, A., Quesada, T., and Garcia-Estan, J. *Clin. Sci.* **86,** 405–409 (1994).

38. Lockhart, J.C., Larson, T.S., and Knox, F.G. *Circ. Res.* **75,** 829–835 (1994).

39. Fenoy, F.J., Ferrer, P., Carbonell, L., and Garcia-Salom, M. *Hypertension* **25,** 408–414 (1995).

40. Stoos, B.A., Carretero, O.A., Farhy, R.D., Scicli, G., and Garvia, J.L. *J. Clin. Invest.* **89,** 761–765 (1992).

41. DeNicola, L., Blantz, R.C., and Gabbai, F.B. *J. Clin. Invest.* **89,** 1248–1256 (1992).

42. Berns, A.S. *Kidney Int.* **36,** 730–740 (1989).

43. Weisberg, L.S., Kurnik, P.B., and Kurnik, B.R.C. *Kidney Int.* **41,** 1408–1415 (1992).

44. Larson, T.S., Hudson, K., Mertz, J.I., Romero, J.C., and Knox, F.G. *J. Lab. Clin. Med.* **101,** 385–391 (1983).

45. Bakris, G.L. and Burnett, J.C. *Kidney Int.* **27,** 465–468 (1985).

46. Margulies, K.B., McKinley, L.J., Cavero, P.G., and Burnett, J.C. *Kidney Int.* **38,** 1101–1108 (1990).

47. Arend, L.J., Bakris, G.L., Burnett, J.C., Megerian, C., and Spielman, W.S. *J. Lab. Clin. Med.* **110,** 406–411 (1987).

48. Margulies, K.B., Hildebrand, F.L., Heublein, D.M., and Burnett, J.C. *J. Am. Soc. Nephrol.* **2,** 1041–1045 (1991).

49. Heyman, S.N., Clark, B.A., Kaiser, N., Spokes, K., Rosen, S., Brezis, M., and Epstein, F.H. *J. Am. Soc. Nephrol.* **3,** 58–65 (1992).

50. Cantley, L.G., Spokes, K., Clark, B., McMahon, E.G., Carter, J., and Epstein, F.H. *Kidney Int.* **44,** 1217–1223 (1993).

51. Calver, A., Collier, J., and Vallance, P. *J. Clin. Invest.* **90,** 2548–2554 (1992).

52. Panza, J.A., Quyyumi, A.A., Brush, J.E., and Epstein, S.E. *N. Engl. J. Med.* **323,** 22–27 (1990).

53. Ludmer, P.L., Selwyn, A.P., Shook, T.L., Wayne, R.R., Mudge, G.H., Alexander, R.W., and Ganz, P. *N. Engl. J. Med.* **315,** 1046–1051 (1986).

54. Kubo, S.H., Rector, T.S., Bank, A.J., Williams, R.E., and Heifetz, S.M. *Circulation* **84,** 1589–1596 (1991).

55. Heyman, S.N., Brezis, M., Reubinoff, C.A., Greenfeld, Z., Lechene, C., Epstein, F.H., and Rosen, S. *J. Clin. Invest.* **82,** 401–412 (1988).

56. Taliercio, C.P., Vlietstra, R.E., Fisher, L.D., and Burnett, J.C. *Ann. Intern. Med.* **104,** 501–504 (1986).

57. Rich, M.W. and Crecelius, C.A. *Arch. Intern. Med.* **150,** 1237–1242 (1990).

58. Touati, C., Idee, J.M., Deray, G., Santus, R., Balut, C., Beaufils, H., Jouanneau, C., Bourbouze, R., Doucet, D., and Bonnemain, B. *Invest. Radiol.* **28,** 814–820 (1993).

59. Schwartz, D., Blum, M., Peer, G., Wollman, Y., Maree, A., Serban, I., Grosskopf, I., Cabili, S., Levo, Y., and Iaina, A. *Am. J. Physiol.* **267,** F374–F379 (1994).

60. Rosen, S., Greenfeld, Z., and Brezis, M. *Transplantation* **49,** 445–452 (1990).

61. Takenaka, T., Hashimoto, Y., and Epstein, M. *J. Am. Soc. Nephrol.* **3,** 42–50 (1992).

62. DeNicola, L., Thomson, S.C., Wead, L.M., Brown, M.R., and Gabbai, F.B. *J. Clin. Invest.* **92,** 1859–1865 (1993).

63. Bobadilla, N.A., Tapia, E., Franco, M., Lopez, P., Mendoza, S., Garcia-Torres, R., Alvarado, J.A., and Herrera-Acosta, J. *Kidney Int.* **46,** 773–779 (1994).

64. Chintala, M.S., Chiu, P.J., Vemulapalli, S., Watkins, R.W., and Sybertz, E.J. *Naunyn Schmiedebergs Arch. Pharmacol.* **348,** 305–310 (1993).

65. Noguchi, S., Kashihara, Y., Ikegami, Y., Morimoto, K., Miyamoto, M., and Nakao, K. *J. Pharmacol. Exp. Ther.* **267,** 919–926 (1993).

66. Schramm, L., Heidbreder, E., Schmitt, A., Kartenbender, K., Zimmermann, J., Ling, H., and Heidland, A. *Renal Failure* **16,** 555–569 (1994).

67. Sabbatini, M., Sansone, G., Ucello, F., De-Nicola, L., Giliberti, A., Sepe, V., Margri, P., Conte, G., and Andreucci, V.E. *Kidney Int.* **45,** 1355–1361 (1994).

68. Dagher, F., Pollina, R.M., Rogers, D.M., Gennaro, M., and Ascer, E. *J. Vasc. Surg.* **21,** 453–458 (1995).

69. Yu, L., Gengaro, P.E., Niederberger, M., Burke, T.J., and Schrier, R.W. *Proc. Natl. Acad. Sci. USA* **91,** 1691–1695 (1994).

PART IV

Role of NO system in Renal Pathophysiology

14

NO in Septic Shock

Colin G.M. Millar and Christoph Thiemermann

1. Introduction

Despite significant improvements in critical care, septic shock remains the major cause of death in noncoronary intensive care units with an estimated mortality (gram-negative and gram-positive sepsis) ranging between 25% and 75%. Traditionally recognized as a consequence of gram-negative bacteremia, septic shock is also caused by gram-positive organisms, fungi, and probably viruses and parasites. The pattern of prevalence of nosocomial infections has changed over the past 40 years. Gram-negative infections increased notably throughout the 1950s and 1960s, and by the early 1970s, they were responsible for most cases of bacteremia in adults. In the 1980s, there was a resurgence in the frequency of reported gram-positive bacteremia and sepsis. Today, between 30% and 50% of all cases of sepsis (approximately 500,000 per year in the United States) are caused by gram-positive organisms [1]. It has been postulated that the marked increase in the incidence of sepsis and septic shock recognized over the last 15–20 years is secondary to (i) improved life-support technology which keeps intensive care patients with high risk of infection alive for prolonged periods, (ii) the prevalence of immunocompromised patients due to an increased incidence of acquired immunodeficiency syndrome or due to chemotherapy and immunotherapy, and (iii) the increased use of invasive medical procedures.

Septic shock can be defined as sepsis (systemic response to infection) with hypotension resulting in impaired tissue perfusion despite adequate fluid resuscitation. The multiple-organ dysfunction syndrome (MODS) is defined as impaired organ function in acutely ill patients such that homeostasis cannot be maintained without intervention. MODS can be primary as a direct result of a specific insult to an organ (e.g., acute lung injury or adult respiratory distress syndrome following gastric aspiration) or secondary as a consequence of the host response to the insult.

This, often exaggerated, host response was recently termed systemic inflammatory response syndrome (SIRS), which may occur in response to varied insults including infection, multiple trauma, hemorrhage, ischemia, and immune-mediated organ injury. SIRS is manifested by two or more of the following criteria of inflammation: pyrexia, tachycardia, tachypnea, or hypocapnea, and leukocytosis or leukopenia—providing that these signs have occurred acutely in the absence of any other obvious cause. In order to meet the criteria for septic shock, SIRS has to be accompanied by hypotension (systolic blood pressure <90 mm Hg or a reduction of 40 mm Hg from baseline) and one or more of the following indices of organ hypoperfusion: arterial hypoxemia, increase in plasma lactate, acute renal failure, and/or acute alterations in mental status (see Table 14-1) [2].

Presenting clinical signs of sepsis include fever, tachycardia, hypotension, peripheral vasodilatation, and oliguria. Current therapeutic approaches include antimicrobial chemotherapy, volume resuscitation, inotropic and vasopressor support, oxygen therapy and mechanical ventilation, and hemodialysis or hemofiltration. These have, however, failed to make a substantial impact on the high

Table 14-1.   Definitions for Sepsis and Multiple-Organ Failure

Infection: Microbial phenomenon characterized by an inflammatory response to the presence of microorganisms or the invasion of normally sterile host tissue by those organisms.

Bacteremia: The presence of viable organisms in the blood.

Systemic inflammatory response syndrome (SIRS): The systemic inflammatory response to a variety of severe clinical insults. The response is manifested by two or more of the following conditions: (1) temperature >38°C or <36°C; (2) heart rate >90 beats/min; (3) respiratory rate >20 breaths/min or $PaCO_2$ <32 mm Hg; and (4) white blood cell count >12000/mm^3, <4000/mm^3, or >10% immature (band) forms.

Sepsis: The systemic response to infection, manifested by two or more of the following conditions as a result of infection: (1) temperature >38°C or <36°C; (2) heart rate >90 beats/min; (3) respiratory rate >20 breaths/min or $PaCO_2$ <32 mm Hg; and (4) white blood cell count >12000/mm^3, <4000/mm^3, or >10% immature (band) forms.

Severe sepsis: Sepsis associated with organ dysfunction, hypoperfusion, or hypotension. Hypoperfusion and perfusion abnormalities may include, but are not limited to, lactic acidosis, oliguria, or an acute alteration in mental status.

Septic shock: Sepsis induced with hypotension despite adequate fluid resuscitation along with the presence of perfusion abnormalities that may include, but are not limited to, lactic acidosis, oliguria, or an acute alteration in mental status. Patients who are receiving inotropic or vasopressor agents may not be hypotensive at the time perfusion abnormalities are measured.

Sepsis-induced hypotension: A systolic blood pressure <90 mm Hg or a reduction of ≥40 mm Hg from baseline in the absence of other causes for hypotension.

Multiple-organ dysfunction syndrome (MODS): Presence of altered organ function in an acutely ill patient such that homeostasis cannot be maintained without intervention.

Source: ACCP/SCCM consensus conference committee (Ref. 2).

mortality associated with sepsis, and there remains a great need to explore novel approaches toward understanding the events leading to tissue ischemia, MODS, and death in circulatory shock. Greater insight into the pathophysiology of septic shock will hopefully result in the rational design of novel therapies and, ultimately, improved survival of critically ill patients.

In 1990, several groups independently discovered that an enhanced formation of endogenous nitric oxide (NO) contributes to the hypotension [3,4], vascular hyporeactivity to vasoconstrictor agents [5,6], or the reduced hepatic protein synthesis [7] observed in animal models of endotoxic shock. We know today that circulatory shock of various etiologies is associated with an enhanced formation of NO due to the early activation of the constitutive NO synthase located in the vascular endothelium (eNOS or NOS III) and the delayed expression of the calcium-independent or "cytokine-inducible" isoform of NO synthase (iNOS or NOS II). The enhanced formation of NO, particularly by iNOS, may contribute to circulatory failure, myocardial dysfunction, and organ injury, ultimately leading to the development of MODS and death. Thus, it has been argued that the inhibition of NO synthesis may be of therapeutic benefit in patients with septic shock. However, an enhanced formation of NO in septic shock may also exert beneficial effects including inhibition of platelet and leukocyte adhesion to the endothelium, improved microcirculatory blood flow and tissue oxygen extraction, and augmentation of host defense. These conflicting roles have fueled the controversy regarding potential therapeutic approaches with some colleagues advocating strategies to reduce NO formation, whereas others propose that NO synthesis should be augmented. Although experimental approaches have thus far failed to reach a consensus, a Phase II clinical trial evaluating the use of a nonisoenzyme-selective NOS inhibitor in patients with septic shock has already been commenced (for reviews, see Refs. 8–14).

This chapter will discuss (i) the role of NO in the pathophysiology of gram-negative and gram-positive septic shock, (ii) the advantages and disadvantages of therapeutic approaches which modulate the biosynthesis of NO, (iii) the pathophysiological events leading to acute renal dysfunction in sepsis, and, most importantly, (iv) review the role of NO in the acute renal failure associated with septic shock.

2. Role of Nitric Oxide in the Pathophysiology of Endotoxic Shock

2.1. Hypotension

The circulatory failure associated with shock of various etiologies is characterized by severe hypotension (peripheral vasodilatation), hyporeactivity of the vasculature to vasoconstrictor agents, myocardial dysfunction, maldistribution of organ blood flow, and reduced tissue oxygen extraction. There is now good evidence that an enhanced formation of NO contributes to several of these pathophysiological features of septic shock. For instance, an enhanced formation of NO due to activation of eNOS (acute phase of shock) and particularly following the induction of iNOS in the vascular wall (late phase of shock) importantly contributes to the hypotension in animals (rat, dog, pig, sheep) and man with septic shock [10,14]. Similarly, an enhanced formation of NO by iNOS also contributes to (i) the delayed hypotension in animals and patients exposed to immunotherapy with interleukin (IL)-2 [15,16], (ii) the delayed vascular decompensation (excessive peripheral vasodilatation) in hemorrhagic shock [17,18], and (iii) possibly to the hyperdynamic circulatory failure associated with liver cirrhosis [19]. Although the circulatory failure associated with anaphylactic shock, traumatic shock, or burns also results in an increase in the plasma levels of nitrite, it is unclear whether this is merely a surrogate marker of these disorders or a cause of the associated hemodynamic alterations [18].

2.2. Vascular Hyporeactivity

The peripheral vascular failure in animals and man with septic shock also results in a progressive attenuation of the pressor effects afforded by norepinephrine and other vasoconstrictor agents (epinephrine, vasopressin, angiotensin II, serotonin, histamine, calcium, potassium). This phenomenon, which has also been termed "vasoplegia," also contributes to the therapy–refractory hypotension in septic shock. Clearly, the hyporeactivity of blood vessels obtained from animals exposed to endotoxic or hemorrhagic shock (for several hours) to catecholamines is largely—but not exclusively—due to an enhanced formation of NO secondary to the induction of iNOS [5,6,17,20]. In endotoxemia, an NO-mediated vascular hyporeactivity occurs in conductance, resistance, as well as venous vessels (see Ref. 14).

2.3. Cardiac Dysfunction

The question as to whether an enhanced formation of NO contributes to the myocardial dysfunction associated with shock is still controversial [21–23]. Large amounts of exogenous NO (authentic NO or NO donors) reduce cardiac contractility by a cGMP-dependent mechanism. It is, however, unclear whether the amounts of endogenous NO generated under physiological or even pathophysiological conditions are sufficient to exert this effect. Isolated (rat) cardiac myocytes or papillary muscles exposed to tumor necrosis factor α (TNFα) or Interleukin-1 (IL-1) express iNOS; NOS inhibitors attenuate the impairment of contractility in isolated cardiac myocytes obtained from animals with endotoxic shock [24,25]. Although the iNOS activity in the heart of animals exposed to endotoxemia (for 3–6 h) is relatively small when compared with the activity found in other organs [26], longer periods of septic shock may well result in a more pronounced elevation in expression of iNOS protein and activity. It has been difficult to demonstrate that NOS inhibition in animals with septic shock results in an improvement in cardiac contractility or cardiac output, as most studies have employed nonselective NOS inhibitors or even eNOS-selective inhibitors (L-NAME; N^G-nitro-L-arginine methyl ester) which cause a reduction in cardiac output by reducing myocardial blood flow (eNOS inhibition). Thus, further studies are warranted aimed at evaluating the effects of iNOS-selective inhibitors on cardiac function in septic shock.

2.4. Impairment in Tissue Oxygen Extraction

Circulatory shock often results in a marked defect in tissue oxygen extraction, resulting in tissue hypoxia and an increase in venous oxygen concentration. As the local generation of large amounts of NO (e.g., by activated macrophages) serves to kill bacteria or tumor cells as part of the host defense [27], it is not surprising that the generation of NO by iNOS in other cells is cytotoxic (suicide mechanism). Indeed, large amounts of NO cause an autoinhibition of mitochondrial respiration by inhibiting several key enzymes in the mitochondrial respiratory chain (aconitase, NADH-ubiquinone reductase, succinate-ubiquinone oxidoreductase) or in the Krebs' cycle (e.g., *cis*-aconitase), resulting in a shift in glucose metabolism from aerobic to anaerobic pathways [9]. NO (like other radicals and oxidants) also causes DNA strand breakage, which triggers a futile, energy-consuming repair cycle by activating the nuclear enzyme poly(ADP)ribosyltransferase (PARS). Activation of PARS results in the rapid depletion of the intracellular concentration of NAD^+ (its substrate) slowing the rate of glycolysis, electron transfer, and ATP formation, which ultimately results in cell death ("PARS suicide hypothesis") [28,29]. NO and peroxynitrite anions also cause DNA strand breaks and, hence, activation of PARS [30–32]. Most notably, inhibitors of PARS activity (e.g., 3-aminobenzamide, nicotinamide) attenuate the inhibition of cellular respiration caused by peroxynitrite [32]. Thus, the generation of large amounts of NO by iNOS may contribute to the defect in oxygen extraction and, ultimately, cell hypoxia and death by causing (i) maldistribution of regional blood flow (reduced oxygen supply), (ii) formation of a diffusion barrier for oxygen within the vascular wall (reduced oxygen transport) (see Ref. 18), (iii) inhibition of the generation of ATP (reduced oxygen utilization), and (iii) excessive and futile consumption of ATP. In concert with the severe hypotension, these effects of the local overproduction of NO may importantly contribute to the organ injury and dysfunction associated with septic shock.

2.5. Endothelial Dysfunction

Prolonged periods of septic shock also result in the development of an endothelial dysfunction, which is characterized by the impairment of "endothelium-dependent vasodilatation" and, therefore, presumably eNOS activity. The mechanism(s) of this endothelial dysfunction may include the downregulation of the expression of the eNOS gene by pro-inflammatory cytokines such as TNFα, endothelial cell damage due to cytotoxic effects of NO, peroxynitrite or oxygen-derived radicals, and (to a lesser extent) the inactivation of NO by oxygen radicals (for review, see Ref. 10).

2.6. Host Defense

The recent generation of mice deficient in iNOS (iNOS mutant or "iNOS knock-out" mice) has helped to shed a further light into the physiological and/or patho-physiological importance of the generation of NO by iNOS [33,34]. For instance, MacMicking and colleagues [34] demonstrate that iNOS-deficient mice failed to restrain the replication of Listeria monocytogenes in vivo or lymphoma cells in vitro. Moreover, the hypotension and early mortality caused by endotoxic shock was reduced in iNOS deficient mice, whereas the degree of liver injury was unaltered [34]. In a separate study, Wei and colleagues show that iNOS-deficient mice were resistant to the mortality caused by endotoxin and exhibited reduced nonspecific inflammatory responses to carageenin. Moreover, iNOS-deficient mice (but not wild-type or heterozygous mice) were highly susceptible to infection with the protozoa parasite Leishmania major [33]. Taken together, these studies support the view that an enhanced formation of NO by iNOS (in macrophages) defends the host against infectious agents and tumor cells, whereas an excessive induction of iNOS in other tissues (e.g., vasculature) may cause shock and tissue destruction.

3. Principles of the Regulation of the Expression of iNOS

As the regulation of the expression of iNOS has recently been comprehensively reviewed elsewhere [8,9,11], the following paragraphs provide only a brief sum-mary of some of the key features of the regulation of iNOS expression. The induction of iNOS caused by endotoxin in vitro and in vivo is secondary to the release of the pro-inflammatory cytokines TNFα, IL-1, and interferon-γ (IFNγ), which either alone or in concert with each other ("cytokine network") or with platelet-activating factor (PAF) activate cells to express iNOS protein and activity. The signal-transduction events leading to the induction of iNOS are less well investigated. There is, however, evidence that the activation of tyrosine kinase (e.g., by cytokines) and of the nuclear transcription factor NFκB play an important role (see Fig. 14-1). The expression of iNOS (in vivo) is tonically inhibited by endogenous glucocorticosteroids, and an increase in the plasma levels of steroids contributes to the development of cardiovascular tolerance to repetitive injections of endotoxin [35]. Interestingly, the inhibition by dexamethasone of the expression of iNOS (but not of cyclooxygenase-2) is mediated by lipocortin-1 [36]. Poly-amines, such as spermine, are metabolized by polyamine oxidase to aldehydes, which, in turn, also inhibit the expression of iNOS caused by endotoxin. Thus, it is conceivable that endogenous polyamines or their active metabolites exert a

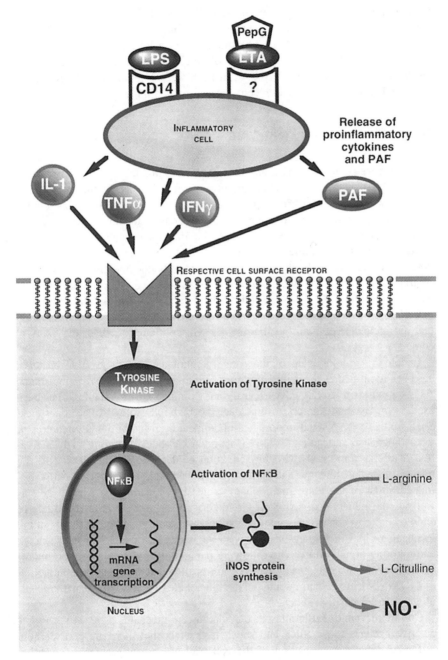

Figure 14-1. Outline of events leading to iNOS induction in sepsis and SIRS.

tonic inhibition of the expression of the iNOS gene [37]. Moreover, regulatory cytokines including IL-4, IL-10, or IL-13 inhibit the expression of iNOS (in vitro), whereas IL-6 or IL-8 are without effect (see Ref. 11). As limitations in space do not allow an extensive review of all of the factors and cellular mechanisms which are involved in the expression and/or regulation of iNOS, the interested reader is referred to recent reviews of this topic [8,9,11].

4. Role of NO in Gram-Positive Septic Shock

The incidence of nosocomial gram-negative bacteremia increased notably in the 1950s and 1960s, whereas it was relatively rare in the 1940s. By the early 1970s, gram-negative bacteria were responsible for most of the cases of bacteremia in adults. The resurgence in gram-positive infections including gram-positive bacteremia and gram-positive sepsis marked a dramatic change in the prevalence patterns for nosocomial infections which was first noted in the early 1980s. Most notably, in the late 1980s there were several reports documenting severe group A streptococcal infections which exhibited features of toxic shock syndrome with an increased incidence of shock and death. Today, between one-third and one-half of all cases of sepsis are caused by gram-positive organisms, and it is likely that the incidence of gram-positive sepsis will continue to rise in the years to come [1].

Endotoxin, a component of the outer membrane of gram-negative bacteria, has been identified as the prime initiator of gram-negative bacterial septic shock. As endotoxin is relatively easy to purify and commercially available; it has been used by many investigators to elucidate the pathophysiology of septic shock. Such studies were of vital importance in delineating some of the key features of the pathogenesis and pathophysiology of endotoxic shock including (i) the activation of blood-borne cells, (ii) the activation of the cytokine network, (iii) the release of endogenous mediators, and (iv) the activation of humoral pathways. Indeed, most of the above-mentioned studies investigating the role of NO as well as the induction of iNOS in circulatory shock employ animal models of endotoxemia. In contrast to endotoxic shock, we know relatively little about the mechanisms by which gram-positive bacteria, which lack endotoxin, cause shock and multiple-organ failure. The following paragraphs review our limited understanding of the mechanisms by which gram-positive bacteria cause the induction of iNOS in order to gain a better understanding of the role of NO in the pathophysiology of gram-positive shock.

One of the major difficulties in elucidating the mechanism by which gram-positive bacteria cause SIRS and shock is the fact that there are considerable differences in the composition of the cell wall among different gram-positive organisms. Specific components of the cell wall may be structurally different from species to species (or even within the same species) and the composition of these components may change during maturation of the organism. In principle, the cell wall of gram-positive bacteria comprises (from inside to outside) a cell membrane (a single phospholipid membrane), cell wall, and capsule. The cell wall, which provides gram-positive bacteria with the ability to retain the Gram stain, contains primarily peptidoglycan (40–60%) and teichoic acids (20–30%). These teichoic acids are divided into teichoic acid (covalently linked to peptidoglycan) and structurally different lipoteichoic acids (LTA) which are anchored in the phospholipid membrane and extend through the peptidoglycan multilayer to the cell surface (see Fig. 14-2). There is now good evidence that LTA from *Staphylococcus aureus* induces iNOS activity and protein in vascular smooth muscle cells and macrophages in vitro [38,39]. The signal transduction pathway leading to the expression of iNOS protein and activity by LTA in cultured macrophages (J774.2) involves the activation of tyrosine kinase, as this expression is inhibited by the tyrosine kinase inhibitors genistein, erbstatin, and tyrphostin AG126 [40]. The induction of iNOS by LTA is also reduced by certain antioxidants (PDTC, rotenone, butylated hydroxyanisole) or inhibitors of IκB-protease (TPCK), suggesting the involvement of the activation of the nuclear transcription factor NFκB [40]. In contrast to LTA, however, teichoic acid from *S. aureus* does not induce iNOS activity in cultured macrophages (unpublished observations). In addition, LTA (from *S. aureus*) also causes (i) a vascular hyporeactivity to vasoconstrictor agents in vitro and in vivo, (ii) hypotension, and (iii) induction of iNOS protein and activity in lung (in the rat) [41]. Prevention of the expression of iNOS protein and activity with dexamethasone, on the other hand, attenuates the circulatory failure caused by LTA [41]. Although these findings indicate that an enhanced formation of NO accounts for the circulatory failure caused by LTA in the rat, LTA does not cause MODS (e.g., renal failure, liver failure, lung dysfunction, or metabolic acidosis) in this model. Thus, unlike endotoxin, LTA alone is not sufficient to cause MODS and death in rodents, suggesting that other components of the cell wall of gram-positive bacteria are involved in the pathogenesis of gram-positive septic shock.

Peptidoglycan (PepG) consists of long sugar chains of alternating two *N*-acetylglucosamine (NAG) and *N*-acetylmuramic acid (NAM) residues which are linked by tetra- or penta-peptides (consisting of alternating L- and D-amino acids) (see Fig. 14-3). The peptide subunit is highly variable between different strains of gram-positive bacteria. PepG from *S. aureus* causes some of the features of septic shock, including leukopenia, thrombocytopenia, hemolysis, complement activation and release of TNFα. However, PepG alone is not able to cause a severe circulatory failure, MODS, or death in the rat (see Ref. 42). Although PepG (from *S. aureus*) does not induce iNOS activity in cultured macrophages, PepG synergizes with LTA (from *S. aureus*) in causing the expression of iNOS protein and activity in vitro and in the rat in vivo [43]. Moreover, PepG and LTA synergize to cause hypotension, tachycardia, and vascular hyporeactivity to norepinephrine and, hence, circulatory failure in the anesthetized rat. In addi-

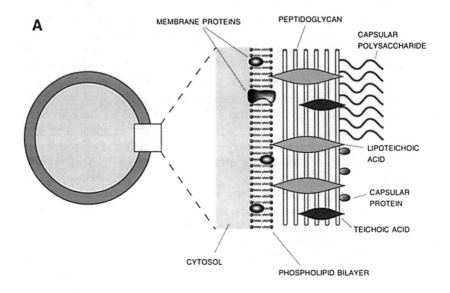

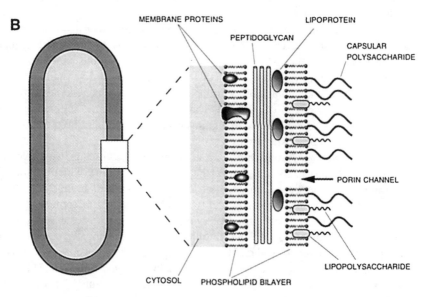

Figure 14-2. Schematic representation of the components of the cell wall of (A) gram-positive and (B) gram-negative bacteria.

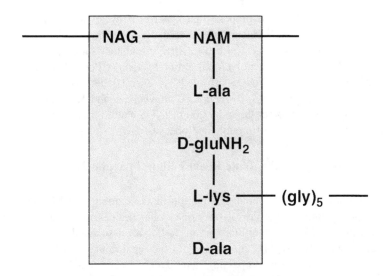

Figure 14-3. Structure of peptidoglycan (PepG). The monomer consists of long sugar chains of alternating *N*-acetyl-glucosamine (NAG) and *N*-acetyl-muramic acid (NAM) residues, and a peptide subunit (stem peptide) linked to the next monomer by a pentaglycine bridge.

tion, PepG and LTA also synergize to cause renal failure, hepatocellular dysfunction and injury, pancreatic dysfunction, as well as metabolic acidosis in this species [43] (see Table 14-2). Prevention of iNOS induction with dexamethasone or inhibition of iNOS activity with aminoguanidine attenuates the circulatory failure as well as the liver dysfunction caused by PepG plus LTA, suggesting that these abnormalities are mediated by an enhanced formation of NO by iNOS (unpublished observations). The induction of iNOS protein and activity caused by these wall fragments appears to be mediated by pro-inflammatory cytokines, because PepG and LTA also synergize to cause the release of TNFα and inter-

Table 14-2. Lipoteichoic Acid (LTA) and Peptidoglycan (PepG), Two Cell-Wall Components of the Gram-Positive Bacterium S. aureus, Synergized to Cause Multiple-Organ Dysfunction Syndrome in vivo

	LPS (from *E. coli*)	LTA (from *S. aureus*)	PepG (from *S. aureus*)	LTA/PepG
Hypotension	+++	++	+	+++
Expression of iNOS	+++	+	−	+++
Renal failure	+++	+	−	+++
Liver dysfunction	+++	−	−	+++
Metabolic acidosis	+++	−	−	+++
Release of TNFα	+++	+	+	+++

feron-γ in the rat [43]. We propose that synergy between PepG and LTA to release cytokines and to induce iNOS activity is a key mechanism by which pathogenic gram-positive bacteria cause SIRS, circulatory failure, MODS, and, ultimately, death. Indeed, a polyclonal antibody against PepG from *S. aureus* attenuates the induction of iNOS afforded by several killed pathogenic, gram-positive bacteria, including *S. pyogenes, S. pneumonia,* and *S. aureus,* whereas this antibody does not affect the induction of iNOS caused by the gram-negative bacterium *Escherichia coli* (unpublished observations).

5. Pathophysiology of the Acute Renal Failure in Septic Shock

The development of acute renal dysfunction in association with MODS carries an ominous prognosis. Failure of only one organ besides the kidney carries a mortality around 75%; three or more organs failing are associated with a mortality rate approaching 100% [44]. The etiology of the renal injury in SIRS and septic shock is multifactorial, encompassing hemodynamic causes as well as humoral and cellular factors [45].

5.1. Renal Hemodynamic Alterations in Sepsis

In nonseptic hypotension, such as that secondary to hypovolemia, the initial renal response to a reduced cardiac output is vasodilatation, thereby maintaining renal perfusion by the process of autoregulation [46]. When the cardiac output falls further, however, such that the mean arterial pressure falls below the autoregulatory threshold, renal vasoconstriction develops [47]. This vasoconstriction is secondary to (i) activation of the renin–angiotensin–aldosterone axis, (ii) elevations in the plasma concentration of antidiuretic hormone (ADH), and (iii) increased sympathetic activity. The heterogeneity of intrarenal perfusion requires complex control and this increase in vasoconstrictor tone is balanced by vasodilator influences including NO, the kallikrein–kinin system, and vasodilator prostaglandins [48].

Limited information is available on the pathophysiology of renal dysfunction in sepsis in man. There is evidence that septic shock in man results in a reduction in renal blood flow despite adequate fluid resuscitation [49]. The hemodynamic alterations associated with animal models of sepsis vary greatly, exhibiting a hypodynamic to hyperdynamic circulation, depending on the model employed (endotoxin versus bacteremia, conscious versus anesthetized animals, etc.). Hence, it is not entirely surprising that the alterations in total renal blood flow observed in these models range from severe reductions to significant increases in blood flow [50–54]. Animal models of hyperdynamic sepsis with an elevated cardiac output probably best simulate the majority of cases seen clinically [55]. Although alterations in the total renal blood flow are undoubtedly important, the intrarenal distribution of perfusion carries major implications both for renal function and protection against injury. Under normal circumstances, the kidneys receive one of the highest rates of blood flow per unit tissue mass with the overall oxygen delivery greatly exceeding oxygen demand [56]. Despite this, the kidneys are particularly susceptible to injury from underperfusion. Demonstration of the great heterogeneity of intrarenal blood flow and tissue oxygen tension has provided an insight into the susceptibility of certain regions to ischemic injury and, in particular, highlighted the vulnerability of the relatively hypoxic yet metabolically active medulla [57–61]. Blood flow and oxygen delivery are directed mainly to the cortex, so as to maximize solute clearance by glomerular filtration, whereas in the medulla, perfusion and oxygenation are minimized, the low rate of blood flow helping to maintain the steep osmolality gradient required for urinary concentration via the countercurrent mechanism. This distribution is controlled by complex interactions between vasoactive mediators acting on the renal vasculature, the influence of renal nerves on vascular tone, and factors controlling the metabolic activity of the tubule epithelium, and is altered in situations with abnormal systemic hemodynamics [48,50] (see also Chapter 13).

Although changes in total renal blood flow may be minor, septic shock may still lead to a maldistribution of regional blood flow resulting in focal ischemia and renal injury. Sepsis results in a corticomedullary redistribution of blood flow, with a reduction in outer corticomedullary perfusion and an increase in juxtamedullary and medullary flow [53,62]. Although this may cause a reduction in glomerular hydrostatic pressure and thereby reduce the glomerular filtration rate (GFR), it may also be a mechanism whereby the kidney strives to protect the outer medullary tubule epithelium from regional ischemia. Here, an important distinction between nonseptic and septic renal failure can be observed. In nonseptic shock, there is an increase in filtration fraction due to an increase in efferent glomerular vascular tone, for example, mediated by the potent vasoconstrictor angiotensin II [47,63,64], whereas in septic shock, filtration fraction may be unchanged or reduced [63,65,66]. It is tempting to speculate that NO is an important mediator of this response, by postglomerular vasodilatation, although other mediators of intrarenal blood flow such as adenosine and vasodilator prostaglandins must also be considered.

The maldistribution of perfusion and oxygenation leads to an imbalance between oxygen delivery and demand [57,61]. Renal oxygen extraction and cellular energetics may be impaired, further exacerbating the inadequacy of perfusion [67,68]. Variable alterations in cellular metabolic demand secondary to the effect of endotoxin have been shown in vitro, with decreased oxygen consumption observed in renal cortical cells, whereas renal medullary cells had a lower baseline consumption, yet were stimulated to increased oxygen consumption following endotoxin [69].

5.2. Direct Effects of Endotoxin on the Kidney

Lipopolysaccharide (LPS), and particularly the lipid-A moiety of LPS, is directly toxic to cell membranes [70]. Thus, it has been proposed that LPS would be directly injurious to the kidney in sepsis. A study in the isolated perfused kidney, however, showed no changes in GFR, sodium reabsorption, or renal tissue potassium content in kidneys perfused with LPS compared to kidneys that received timed-control perfusion. In comparison, infusion of comparable doses of LPS in vivo led to a prompt and marked reduction in GFR, with a fall in sodium reabsorption and tissue potassium content [71]. This, therefore, implies that the renal dysfunction in sepsis is predominantly due to hemodynamic, humoral, and cellular factors rather than direct toxicity from LPS. However, due to the limited viability of the kidney ex vivo, the time course of this study was short, incorporating only 80 min of LPS infusion in graded doses. This would be insufficient time to allow for gene transcription and protein synthesis and, hence, later effects resulting directly from LPS exposure would not be assessed.

Other studies have investigated the effect of LPS on isolated renal segments or cells in culture. When challenged with LPS, isolated medullary thick ascending limb (mTAL) tubule segments express the gene for TNFα, and synthesize and release this cytokine [72]. The TNFα produced in this in vitro system is cytotoxic and exerts effects on ion transport in the mTAL. LPS also stimulates the synthesis of TNFα in isolated glomeruli [73] and mesangial cells [74]. However, the importance of the relative contribution of renal TNFα synthesis compared with TNFα production from macrophages is not yet known. Further studies have addressed the effect of LPS, or its components, on tubule epithelial viability. In a porcine proximal tubule cell line (LLC-PK$_1$), lipid-A produces a time- and concentration-dependent cytotoxicity associated with increased intracellular calcium concentrations [75].

5.3. Humoral Factors in the Renal Dysfunction in Sepsis

Cytokines

Cytokines, including TNFα, interleukins, and interferon-γ, play a pivotal role in the development of SIRS and septic shock following endotoxin [76]. Clinical studies have shown that persistence of elevated levels of TNFα and IL-6, rather than peak levels, correlate negatively with outcome [77].

Endogenous TNFα contributes to the pathology associated with experimental nephrotoxic nephritis, and anti-TNF antibodies provide protection against renal injury [78]. Similarly, anti-TNF antibodies prevent the augmented injury seen when LPS administration is superimposed on nephrotoxic nephritis [79]. Administration of TNFα in vivo leads to the deposition of fibrin in glomerular capillary loops, leukocyte sequestration, and damage to glomerular endothelial cells [80]. Pretreatment of rats with TNFα prior to the induction of antiglomerular basement membrane nephritis exacerbates glomerular injury [81]. The renal effects of TNFα are widespread and include (i) direct mesangial cell cytotoxicity, (ii) recruitment and adherence of neutrophils to the endothelium, (iii) activation of renal plasminogen activator inhibitor leading to fibrin deposition, (iv) stimulation of endothelial and mesangial cells to release platelet-activating factor (PAF) and endothelin (ET)-1, (v) production of reactive oxygen species by mesangial cells, (vi) induction of iNOS activity in mesangial cells leading to NO release, and (vii) the release of prostaglandin E_2 (PGE_2) from mesangial cells presumably due to induction of cyclooxygenase-2 [82].

The interleukins IL-1, IL-2, IL-6, IL-8 and IL-10 also play an important role in septic shock, and their release can be stimulated by endotoxin [83]. In clinical practice, IL-2, when administered as a cancer chemotherapeutic agent, causes a clinical syndrome analogous to SIRS [84]. The hypotension associated with IL-1 and IL-2 administration is mediated by NO [15,16,85,86]. Acute renal dysfunction is a common limiting side effect during interleukin chemotherapy, and both hemodynamic factors and intrinsic renal injury have been invoked in its pathophysiology [87–90].

Platelet-activating factor is released both systemically and locally within the kidney; for example, in glomerular mesangial cells [91]. PAF may lead to a reduction in GFR by causing hypotension [92], or secondary to mesangial cell contraction. Both of these mechanisms can be inhibited by the use of PAF antagonists [92,93]. PAF stimulates the production of a series of potentially injurious and protective mediators, including thromboxane A_2 (TXA_2), prostaglandin E_2, and leukotrienes [94]. PAF also triggers the release of TNFα, potentiates the release of TNFα stimulated by endotoxin [95], and causes the expression of iNOS [96]. PAF also causes microvascular injury, which can result in hypovolemia following extravasation, and thrombin release from platelets leading to activation of the coagulation cascade [92].

Vasoactive Mediators

Numerous vasoactive mediators and endothelial factors are involved in the renal response to systemic sepsis. These include the products of arachidonic acid metabolism, the renin–angiotensin axis, endothelin-1, adenosine, bradykinin, vasopressin, atrial natriuretic factor, and NO (see Ref. 97). Endotoxin stimulates the release of the vasodilatory prostaglandins PGE_2 and PGI_2. PAF stimulates PGE_2 release from mesangial cells, where it acts to oppose the mesangial contraction induced by PAF [92]. The vasodilatory action of PGE_2 mediates a reduction in renovascular resistance [98]. Similarly, PGI_2 also mediates a reduction in renovascular resistance [98,99] and, in addition, may contribute to systemic hypotension [50]. In contrast to the proposed renoprotective roles of PGE_2 and PGI_2, TXA_2 may play a role in the pathophysiology of acute renal failure in endotoxemia [100]. TXA_2 is a potent vasoconstrictor, stimulates platelet aggregation, has prothrombotic effects, and mediates PAF-induced vasoconstriction [94]. Increased levels of TXA_2 occur in the renal cortex following exposure to endotoxin [101], and TXA_2 synthetase inhibitors attenuate the renal dysfunction associated with sepsis [102]. In addition, the TXA_2 receptor appears to transduce the signal from a newly recognized class of arachidonic acid metabolites that are formed by the nonenzymatic oxidation of arachidonic acid by oxygen-derived free radicals [103]. These eicosanoids resemble prostaglandin $F_{2\alpha}$, and this has coined the term "F_2-isoprostanes" used to describe this class of metabolites. One such compound, 8-epi-$PGF_{2\alpha}$, is a potent, selective renal vasoconstrictor [104], and F_2-isoprostanes may therefore also contribute to the renal dysfunction in sepsis.

The potent vasoconstrictor peptide endothelin-1 (ET-1) is generated from its precursor bigET-1 by endothelin-converting enzyme(s) in vascular endothelial cells (large vessels) [105], in glomerular endothelial cells, and in renal epithelial cell lines [105–108]. Recently, human proximal tubule cells have been shown to synthesize ET-1 [109,110]. Three isoforms of ET have been identified, ET-1, ET-2 and ET-3, and their effects are mediated via two receptor subtypes, ET_A and ET_B [111,112]. The vasoconstrictor effects of ET-1 are primarily due to activation of ET_A receptors located on vascular smooth muscle cells, whereas activation of ET_B receptors on endothelial cells leads to the formation of NO and vasodilator prostaglandins which mediate the vasodilator effects of ET-1 in the renal medulla of the rat [113]. In addition to vascular effects, ET receptors have also been recognized on nonvascular cells, notably glomerular mesangial and renal epithelial cells [114]. The renal vasculature is particularly sensitive to the vasoconstrictor effects of ET-1, with a 10-fold greater response than that observed in bronchial, femoral, and coronary arteries [115]. In the kidney, ET-1 causes a reduction in GFR and an increase in renovascular resistance. The vasoconstriction is equipotent in the afferent and efferent arterioles, such that the glomerular hydrostatic pressure, the driving force for glomerular filtration, is unaltered [116]. Thus, it has been proposed that the reduction in GFR caused by ET-1 is due to mesangial cell contraction [116]. The plasma levels of ET-1 are elevated in animals and man with endotoxemia and septic shock, and (in humans) the plasma levels of ET-1 correlate positively with the severity of endotoxemia and mortality [117,118]. In a rodent model of endotoxemia, anti-ET antiserum was infused selectively into the left kidney via the renal artery, preventing the fall in GFR in the ipsilateral kidney but not affecting the fall in GFR in the contralateral, control kidney [116]. The release of endogenous ET-1 by endotoxin also serves to maintain peripheral vascular resistance and, hence, perfusion pressure to the renal vascular bed [119]. The release of ET-1 in endotoxemia may serve to attenuate hypotension (beneficial effect), whereas excessive rises in the plasma levels of ET-1 may cause excessive vasoconstriction in some vascular beds [119].

5.4. Cellular Factors in the Renal Dysfunction in Sepsis

Endotoxemia results in the activation of circulating leukocytes, which are major contributors to host defense. Activated neutrophils may also lead to tissue injury due to the release of toxic mediators, including oxygen-derived free radicals and proteinases. However, studies addressing the role of neutrophils in postischemic renal failure have yielded conflicting results [120–122], although this may be due to differences in the stage of neutrophil activation [123]. In the isolated perfused rat kidney, LPS and neutrophils act in concert to cause acute renal failure, which is (at least in part) mediated by neutrophil-derived oxidants and elastase [124]. In particular, neutrophils primed with LPS cause severe functional injury in the setting of mild renal ischemia [123].

Activation of the coagulation cascade may lead to fibrin deposition, disseminated intravascular coagulation, and thrombocytopenia. This can lead to accumulation of microthrombi and entrapment of leukocytes in glomeruli which may impair blood flow and cause injury through the release of inflammatory mediators. Endogenous NO acts to inhibit the adhesion of platelets to the vascular endothelium, and NOS inhibition with L-NAME greatly exacerbates glomerular thrombosis in experimental endotoxemia [125].

Clearly, the pathogenesis of the acute renal failure in sepsis and endotoxemia is multifactorial. In addition, the administration of nephrotoxic drugs such as aminoglycoside antibiotics to patients with sepsis may often worsen—or even cause—acute renal failure. Particularly, the risk of aminoglycoside nephrotoxicity is increased in endotoxemia and under conditions of renal hypoperfusion [126].

6. Role of Nitric Oxide in Sepsis-Induced Acute Renal Failure

6.1. Expression of Nitric Oxide Synthases in the Kidney in Sepsis

As the renal distribution of all three isoforms of NOS has already been reviewed in detail (see Chapter 7), the following section will concentrate on the differences in isoform expression between normal kidneys and those exposed to endotoxemia.

Neuronal NOS

The neuronal isoform of NOS (nNOS or NOS I) is located predominantly in the macula densa [127–129] where it is thought to act to regulate glomerular capillary pressure [127]. In addition, nNOS has been demonstrated in the efferent arteriolar endothelial cells [130], and in the medullary collecting duct, glomerulus, and thin limb by RT-PCR [131]. There is no evidence that the expression of nNOS is altered by exposure to LPS [129].

Inducible NOS

Unlike many other tissues, the antibodies against inducible NOS (iNOS or NOSII) recognize this (or a similar protein) in the normal kidney, implying that it is either constitutively expressed or subject to continuous low-grade induction. In the normal rat kidney, iNOS mRNA is found mainly in the outer medulla, particularly in the medullary thick ascending limb, with minor amounts in the medullary collecting duct and vasa recta. It is also found to a lesser extent in glomeruli as well as tubules in the cortex and inner medulla [132]. Exposure to LPS (0.1 mg/kg) for 24 h results in a 3-fold increase in iNOS expression in the outer medulla and a 20-fold increase in the glomeruli and inner medulla [132]. Using in situ hybridization for iNOS mRNA, intense labeling can be found in the proximal straight tubule, cortical and medullary thick ascending limbs, the distal convoluted tubule, and the cortical and inner medullary collecting duct, with weak staining in other segments of the nephron and medullary interstitial cells. Following LPS, labeling is seen in mesangial cells and the papillary surface epithelium with an increase in the intensity of the medullary interstitial cell staining [133]. Although the aforementioned studies have not demonstrated iNOS in the renal vasculature, immunoreactivity against iNOS has been found specifically in the afferent arteriole, occasionally in the efferent arteriole, but not in the arcade or interlobular arteries, with an increase in the afferent arteriolar staining following LPS [129]. Cytokines stimulate the expression of iNOS protein and activity to produce NO in mesangial cells [134], rat proximal tubule and inner medullary collecting duct epithelial cells [135], and in cell lines of inner medullary collecting duct and medullary thick ascending limb epithelial cells [136,137].

Endothelial NOS

In the normal kidney, endothelial NOS (eNOS or NOS III) is found throughout the renal arterial circulation and in the glomerular capillaries [130,138]. Furthermore, there is some evidence for the presence of eNOS in tubule epithelium of various segments of the nephron, including the proximal tubule, thick ascending limb, and collecting duct [138]. In LLC-PK$_1$ cells, a porcine proximal tubule cell line, calcium-dependent NO synthesis can be demonstrated [139] from an enzyme identified as eNOS [140]. Isolated rat proximal tubules demonstrate an increase in eNOS activity 6 h following IV injection of endotoxin [141]. Interestingly, in isolated rat proximal tubules subjected to hypoxia and reoxygenation, a toxic role for NO produced from eNOS has been proposed [142].

6.2. Regional Hemodynamic Effects of Nitric Oxide Synthase Inhibition

Inhibition of NO synthesis with nonselective NOS inhibitors improves systemic arterial blood pressure and restores vascular responsiveness to vasopressor agents in animals (see Refs. 10 and 14) and man [143,144] with septic shock. Despite improvement in systemic hemodynamics, inhibition of NOS activity, particularly with either nonselective inhibitors of all isoforms of NOS (e.g., N^G-methyl-L-arginine; L-NMMA) or even selective inhibitors of eNOS activity (e.g., N^G-nitro-L-arginine methyl ester; L-NAME), may have unfavorable effects on mortality, regional perfusion of certain vascular beds, or organs or on blood flow distribution within a certain organ [53,145–149]. These adverse effects arise primarily from inhibition of eNOS activity which may result in excessive vasoconstriction and may predispose to adhesion of platelets and polymorphonuclear leukocytes (PMNs) to the endothelium (see Ref. 14). Thus, the systemic administration of L-NMMA or L-NAME in animals or man with endotoxic or septic shock results in (i) a fall in cardiac output (albeit in some instances from elevated to normal levels) which has been attributed to an increase in systemic vascular resistance and/or excessive vasoconstriction within the coronary vascular bed and, hence, myocardial ischemia and (ii) an increase in pulmonary vascular resistance [14,145–149]. In addition, there is evidence (in animals but not in patients) that nonselective NOS inhibitors reduce renal blood flow in septic shock [150]. This is not surprising, as the basal release of NO by eNOS is of particular importance in regulating renal cortical blood flow. Indeed, systemic administration of very low doses of L-NMMA (30–100 μg/kg/min) cause a pronounced reduction in renal cortical blood flow in rats without causing a significant increase in blood pressure [151]. In a porcine model of endotoxemia, L-NAME improves mesenteric blood flow and gastric mucosal pH_i (a marker of splanchnic perfusion) but reduces cardiac index, carotid, and renal blood flow [54]. In the anesthetized rat, bacteremia leads to a reduction in cortical flow of 19% and an increase in medullary flow of 36% (measured by laser–Doppler flowmetry), whereas effective renal plasma flow (PAH clearance) was slightly reduced (−9%). In this model, NOS inhibition with L-NAME augmented the reduction in cortical flow (−43%) caused by bacteremia, reversed the increase in medullary flux (−42%), and further impaired effective renal plasma flow (−28%) [53]. Using in vivo microscopy of the hydronephrotic rat kidney, Spain and colleagues demonstrate that the constriction of interlobular and afferent arteries caused by bacteremia is worsened by NOS inhibition, emphasizing the role of NO in determining preglomerular tone [150]. Thus, there is a large body of evidence demonstrating that endogenous NO plays an important role in the control of regional blood flow and intrarenal hemodynamics.

In addition to its vasoactive properties, NO is an important inhibitor of platelet aggregation and thrombosis [12,152]. The intricate network of capillary loops within the glomerulus is a potential target for vascular mediated injury, and renal cortical necrosis is known to occur in association with disseminated intravascular coagulation (DIC), although, fortunately, this is now a rare complication. Although endotoxemia in the rat results in only a minor degree (2%) of glomerular thombosis, this is substantially augmented by L-NAME (55%) [125]. This excessive thrombosis of the glomerular capillaries caused by this NOS inhibitor in rats with endotoxic shock was reversed by L-arginine and glyceryl trinitrate (GTN, which is metabolized within vascular smooth muscle cells to release NO) but not by the (NO-independent) vasodilators hydralazine and atrial natriuretic peptide [153]. This protective role of NO may be particularly relevant in septic complications of pregnancy. Classically, renal cortical necrosis has been recognized as a complication of septic abortion, or abruptio placentae, and although

Table 14-3. Examples of Inhibitors of Nitric Oxide Synthase with Variable
Isoform Selectivity

Inhibitor	Class	Selectivity	Ref.
L-NMMA	L-arginine analog	eNOS > iNOS	160, 182
L-NAME	L-arginine analog	eNOS > iNOS	163, 182
Aminoguanidine	Guanidine	iNOS > eNOS	155, 163
1-Hydroxy-2-aminoguanidine	Guanidine	iNOS > eNOS	156
S-Methyl-isothiourea	Isothiourea	iNOS > eNOS	157–159
S-Aminoethyl-isothiourea	Isothiourea	iNOS > eNOS	157, 159, 160
2-Imino piperidine	Amidine	iNOS > eNOS	162
2-Amino piperidine	Amidine	iNOS > eNOS	162
Butyramidine	Amidine	iNOS > eNOS	162

the mechanism for the apparent increased susceptibility to glomerular thrombosis has not been properly elucidated, there is some evidence from experimental studies that a limited reserve capacity for NO production in pregnancy may be implicated [154].

Unfortunately, all of the above studies have employed inhibitors of NOS activity which are relatively nonselective for all NOS isoforms (L-NMMA) or weak inhibitors of iNOS activity (L-NAME). Indeed, in the rat, L-NAME causes a significant rise in blood pressure and, hence, inhibition of eNOS activity without attenuating the rise in nitrite (from iNOS) caused by endotoxin. Thus, L-NAME is a more potent inhibitor of eNOS than iNOS activity in this species (see Ref. 14). There are now several chemical classes of novel, competitive NOS inhibitors, including guanidines, isothiourea derivatives, thiazines, and certain amidines. Among these agents there are potent and relatively selective iNOS inhibitors, including aminoguanidine [155], 1-hydroxy-2-amino-guanidine [156], S-methyl-isothiourea [157–159], aminoethyl-isothiourea [157,159,160], 2-amino-5,6-dihydro-6-methyl-4H-1,3-thiazine [161], and 2-iminopiperidine (amidine) [162], to name but a few (see Table 14-3). Unfortunately, there are, to our knowledge, no studies reporting on the effects of these NOS inhibitors on renal hemodynamics in animals with or even without shock.

6.3. Effects of Nitric Oxide Synthase Inhibitors on Sepsis-Induced Renal Dysfunction

We have recently compared the effects of nonselective and iNOS-selective NOS inhibitors on hemodynamics and multiple-organ dysfunction syndrome in the anesthetized rat. In this model, endotoxaemia for 6 h resulted in circulatory failure comprised of hypotension, tachycardia, and hyporeactivity of the vasculature to the pressor responses elicited by norepinephrine. This was associated with (i) liver dysfunction (rise in the serum levels of bilirubin and γ-GT) and hepatocellular injury (rise in the serum levels of GPT and GOT), (ii) acute renal failure (substantial rise in the serum levels of urea and creatinine), and (iii) induction of iNOS activity (measured in lung and liver homogenates) resulting in an increase in plasma nitrite. Therapeutic administration (e.g., continuous infusion starting at 2 h after injection of endotoxin) of the iNOS-selective NOS inhibitors aminoguanidine [163], 1-hydroxy-2-amino-guanidine [156], and aminoethyl-isothiourea [160] attenuates the liver dysfunction and hepatocellular injury which developed between 4 and 6 h after the injection of endotoxin. Interestingly, infusion of low doses of L-NMMA (up to 3 mg/kg/h IV) also reduced the liver dysfunction, but not the hepatocellular injury [160]. In contrast, low doses of L-NAME (0.3 mg/kg/h) increased blood pressure and, hence, reduced the degree of hypotension without affecting the degree of liver dysfunction or injury caused by endotoxin in this model [163]. The effects of these NOS inhibitors on the renal dysfunction caused by prolonged periods of endotoxemia were more difficult to assess, as 2 h of endotoxemia already resulted in pronounced increases in the serum levels of urea (from 4 to 13 mmol/L; maximum at 6 h: 15 mmol/L) and creatinine (from 30 to 50 μmol/L; maximum at 6 h: 80 μmol/L). Infusion of aminoguanidine [163], 1-amino-2-hydroxy-guanidine [156], aminoethyl-isothiourea [160], or L-NMMA [160] commencing at 2 h after endotoxin neither improved nor worsened the renal dysfunction associated with endotoxemia in this model. As a substantial degree of the renal dysfunction in this model occurs within the first 2 h of endotoxemia, further studies evaluating the effects of NOS inhibitors on early alterations in renal function and, particularly, intrarenal blood flow distribution (see above) are warranted.

The finding that various NOS inhibitors exert beneficial effects on hemodynamics and MODS in a rodent model of hypodynamic shock caused by endotoxin raises the questions as to whether similar effects can also be documented in models of hyperdynamic sepsis in larger animals. In contrast to rodents, sheep are very sensitive to small doses of endotoxin [164] in a manner similar to humans [165]. Indeed, infusion of either endotoxin or bacteria into sheep leads (within 24 h) to a hyperdynamic circulation with a fall in peripheral vascular resistance, an increase in cardiac output, and increases in organ blood flow associated with a reduction in oxygen extraction [55]. In this model, prolonged periods of endotoxemia or bacteremia (*P. aeruginosa*) are also associated with increases in total renal blood flow and the development in precapillary arteriovenous shunts resulting in (i) regional maldistribution of renal blood flow, (ii) fall in glomerular filtration pressure, and, ultimately, (iii) GFR. Interestingly, administration (at 24 h after the onset of endotoxemia) of L-NAME or L-NMMA increased urine output and reversed the impairment in creatinine clearance caused by infusion of bacteria, without causing a significant fall (below baseline) of renal blood flow [166; see Ref. 55 for a review]. Thus, in contrast to other studies demonstrating either a reduction or no effect of NOS inhibitors on renal function in animal models of hypodynamic shock, inhibition of NOS activity improves hemodynamics and renal function in an ovine model of hyperdynamic shock. In addition to these beneficial effects on renal blood flow and function, NOS inhibition also resulted in an increase in oxygen extraction, a fall in organ blood flow from elevated to normal levels (in brain, heart, jejunum, ileum), and an increase in peripheral vascular resistance, but no significant increase in lactate, indicating a normalization of hemodynamic parameters in the absence of excessive vasoconstriction [167,168].

7. Involvement of Nitric Oxide in Humans with Septic Shock

Although our understanding of the role of NO in animal models of circulatory shock has improved substantially over the past years, our knowledge regarding the biosynthesis and importance of NO in the pathophysiology of shock (of various etiologies) in patients is still very limited. Indeed, a Medline search covering the time period from 1987 to November 1995 revealed that only 8–14% of all of the publications which included the key word "nitric oxide" also included the key word "human" [169]. What, then, is the evidence that septic shock in man is associated with an enhanced formation of NO? Elevated plasma and urine levels of nitrite/nitrate have been reported in adults and children with severe septic shock [170–173] as well as in patients with burn injuries who subsequently developed sepsis [169]. Moreover, elevated plasma levels of nitrite/nitrate occur in patients receiving IL-2 chemotherapy [174]. In contrast, there is also evidence that the plasma levels of nitrite/nitrate are lower in patients after trauma and surgery [170,174] and in patients with HIV infection [175]. Interestingly, the increase in iNOS activity in leukocytes obtained from patients with sepsis appears to correlate with the number of failing organs but not with blood pressure [176]. Taken together, these studies support the conclusion that septic shock in man is associated with an enhanced formation of NO. It should, however, be stressed that the increase in the plasma levels of nitrite/nitrate elicited by endotoxin, cytokines, or bacteria in rodents (10-fold) is substantially higher than the observed increases in the plasma levels of these metabolites of NO in other animal species (pig, sheep, etc.) or humans (see Ref. 169). Moreover, our understanding of (i) the biosynthesis of NO, (ii) the regulation of and the mechanism involved in the expression of iNOS, and (iii) the role of NO in MODS in shock are largely based on animal experiments of endotoxic shock in rodents. In contrast, we know relatively little about the role of NO in patients with septic and other forms of circulatory shock.

There is evidence that endotoxin and cytokines (when given in combination) causes the expression of iNOS as well as the formation of NO in various human cells (primary or cell lines), including hepatocytes [177], mesangial cells [178], retinal pigmented epithelial cells [179], and lung epithelial cells [180]. Interestingly, IL-1 causes the hyporeactivity of human hand veins (in situ) to the constrictor effects elicited by exogenous or endogenous norepinephrine, and this vascular hyporeactivity is largely attenuated by L-NMMA, suggesting that it is mediated by NO (P. Vallance, personal communication). Early reports of beneficial hemodynamic effects of L-NMMA in humans with septic shock [143,144] stimulated a Phase I, multicenter, open-label, dose-escalation (1, 2.5, 5, 10, or 20 mg/kg/h for up to 8 h) study using L-NMMA (546C88) in 32 patients with septic shock. In this study, L-NMMA sustained blood pressure and enabled a reduction in vasopressor (norepinephrine) support. The cardiac index fell (possibly due to an increase in peripheral vascular resistance) and left ventricular function was well maintained. Moreover, L-NMMA increased oxygen extraction, whereas the pulmonary shunt was not worsened [181]. Although the development of thrombocytopenia has been documented in septic patients treated with L-NMMA, it is unclear whether this effect is due to the NOS inhibitor or caused by the underlying illness [144]. A multicenter clinical trial evaluating the effects of L-NMMA on morbidity and mortality in patients with septic shock is ongoing, and the results of this trial are awaited with interest.

References

1. Bone, R.C. *Arch. Intern. Med.* **154**, 26–34 (1994).

2. Bone, R.C., Balk, R.A., Cerra, R.B., Dellinger, R.P., Fein, A.M., Knaus, W.A., Schein, R.M.H., and Sibbald, W.J. *Chest* **101**, 1644–1655 (1992).

3. Thiemermann, C. and Vane, J.R. Inhibition of nitric oxide synthesis reduces the hypotension induced by bacterial lipopolysaccharide in the rat. *Eur. J. Pharmacol.* **182**, 591–595 (1990).

4. Kilbourn, R.G., Jubran, A., Gross, S.S., Griffith, O.W., Levi, R., Adams, J., and Lodato, R.F. Reversal of endotoxin mediated shock by N^G-monomethyl-L-arginine, an inhibitor of nitric oxide synthase. *Biochem. Biophys. Res. Commun.* **172**, 1132–1138 (1993).

5. Julou-Schaeffer, G., Gray, G.A., Fleming, I., Schott, C., Parratt, J.R., and Stoclet, J.C. Loss of the vascular responsiveness induced by endotoxin involves L-arginine pathway. *Am. J. Physiol.* **259**, H1038–H1043 (1990).

6. Rees, D.D., Cellek, S., Palmer, R.M.J., and Moncada, S. Dexamethasone prevents the induction by endotoxin of a nitric oxide synthase and the associated effects on vascular tone: an insight into endotoxin shock. *Biochem. Biophys. Res. Commun.* **173**, 541–546 (1990).

7. Curran, R.D., Billiar, T.R., Stuehr, D.J., Ochoa, J.B., Harbrecht, B.G., Flint, S.G., and Simmons, R.L. Multiple cytokines are required to induce hepatocyte nitric oxide production and inhibit total protein biosynthesis. *Ann. Surg.* **212**, 462–489 (1990).

8. Dinerman, J.L., Lowenstein, C.J., and Snyder, S.H. Molecular mechanism of nitric oxide regulation: Potential relevance to cardiovascular disease. *Circ. Res.* **73**, 217–222 (1993).

9. Morris, S.M. and Billiar, T.R. New insights into the regulation of inducible nitric oxide synthesis. *Am. J. Physiol.* **266**, E829–E839 (1994).

10. Thiemermann, C. The role of the arginine–nitric oxide pathway in circulatory shock. *Adv. Pharmacol.* **28**, 45–79 (1994).

11. Szabo, C. and Thiemermann, C. Regulation of the expression of the inducible form of nitric oxide synthase. *Adv. Pharmacol.* **34**, 113–154 (1995).

12. Moncada, S. and Higgs, A. The L-arginine–nitric oxide pathway. *N. Engl. J. Med.* **329**, 2002–2012 (1993).

13. Kilbourn, R.G. and Griffith, O.W. Overproduction of nitric oxide in cytokine-mediated and septic shock. *J. Natl. Cancer Inst.* **84**, 827–831 (1992).

14. Thiemermann, C. Inhibition of induction or activity of nitric oxide synthase: Novel approaches for the therapy of circulatory shock. In: *Shock, Sepsis and Organ Failure—Nitric Oxide.* Eds. Schlag, G. and Redl, H. Springer-Verlag, Berlin, 1995, pp. 30–58.

15. Higgs, J., Jr., Westenfelder, C., Taintor, R., Vavrin, Z., Kablitz, C., Baranowski, R.L., Ward, J.H., Menlove, R.L., McMurry, M.P., Kushner, J.P., and Samlowski, W.E. Evidence for cytokine-inducible nitric oxide synthesis from L-arginine in patients receiving interleukin-2 therapy. *J. Clin. Invest.* **89**, 867–877 (1992).

16. Miles, D., Thomsen, L., Balkwill, F., Tharasu, P., and Moncada, S. Association between biosynthesis of nitric oxide and changes in immunological and vascular parameters in patients treated with interleukin-2. *Eur. J. Clin. Invest.* **24**, 287–290 (1994).

17. Thiemermann, C., Szabo, C., Mitchell, J.A., and Vane, J.R. Vascular hyporeactivity to vasoconstrictor agents and hemodynamic decompensation in hemorrhagic shock is mediated by nitric oxide. *Proc. Natl. Acad. Sci. USA* **90**, 267–271 (1993).

18. Szabo, C. and Thiemermann, C. Invited opinion: Role of nitric oxide in hemorrhagic, traumatic and anaphylactic shock and thermal injury. *Shock* **2**, 145–155 (1994).

19. Vallance, P. and Moncada, S. Hyperdynamic circulation in cirrhosis: A role for nitric oxide? *Lancet* **337**, 776–777 (1991).

20. Wu, C.C., Szabo, C., Chen, S.J., Thiemermann, C., and Vane, J.R. Activation of soluble guanyl cyclase by a factor other than nitric oxide or carbon monoxide contributes to the vascular hyporeactivity to vasoconstrictor agents in the aorta of rats treated with endotoxin. *Biochem. Biophys. Res. Commun.* **201**, 436–442 (1994).

21. Kumar, A. and Parrillo, J.E. Nitric oxide in the heart in sepsis. In: *Role of Nitric Oxide in Sepsis and ARDS. Update in Intensive Care and Emergency Medicine.* Eds. Fink, M.P. and Payen, D. Springer-Verlag, Berlin, 1995, pp. 73–99.

22. Shah, A.M. Influence of nitric oxide on cardiac systolic and diastolic function. In: *Role of Nitric Oxide in Sepsis and ARDS. Update in Intensive Care and Emergency Medicine.* Eds. Fink, M.P. and Fayen, D. Springer-Verlag, Berlin, 1995, pp. 100–113.

23. Lefer, A.M. Cellular actions of nitric oxide on the circulatory system. In: *Role of Nitric Oxide in Sepsis and ARDS. Update in Intensive Care and Emergency Medicine.* Eds. Fink, M.P. and Payen, D. Springer-Verlag, Berlin, 1995, pp. 114–124.

24. Brady, A.J.P., Poole-Wilson, P.A., Harding, S.E., and Warren, J.B. Nitric oxide production within cardiac myocytes reduces their contractility in endotoxemia. *Am. J. Physiol.* **263**, H1963–H1966 (1992).

25. Finkel, M.S., Oddis, C.V., Jacob, T.D., Watkins, S.C., Hattler, B.G., and Simmons, R.L. Negative inotropic effects of cytokines on the heart mediated by nitric oxide. *Science* **257**, 387–389 (1992).

26. Szabo, C., Mitchell, J.A., Thiemermann, C., and Vane, J.R. Nitric oxide mediated hyporeactivity to noradrenaline precedes nitric oxide synthase induction in endotoxin shock. *Br. J. Pharmacol.* **108**, 786–792 (1993).

27. Nathan, C. Nitric oxide as a secretory product of mammalian cells. *FASEB J.* **6**, 3051–3064 (1992).

28. Schraufstatter, I.U., Hinshaw, D.B., Hyslop, P.A., Spragg, R.G., and Cochrane, C.G. Oxidant injury of cells. DNA strand-breaks activate polyadenosine diphosphate–ribose polymerase and lead to depletion of nicotinamide adenine dinucleotide. *J. Clin. Invest.* **77**, 1312–1320 (1986).

29. Thies, R.L. and Autor, A.P. Reactive oxygen injury to cultured pulmonary endothelial cells: Mediation by poly (ADP-ribose) polymerase activation causing NAD depletion and altered energy balance. *Arch. Biochem. Biophys.* **286**, 353–363 (1991).

30. Zhang, J., Dawson, V.L., Dawson, T.M., and Snyder, S.H. Nitric oxide activation of poly (ADP-ribose) synthetase in neurotoxicity. *Science* **263**, 687–689 (1994).

31. Heller, B., Wang, Z.-Q., Wagner, E.F., Radons, J., Bürkle, A., Fehsel, K., Burkart, V., and Kolb, H. Inactivation of the poly (ADP-ribose) polymerase gene affects oxygen radical and nitric oxide toxicity in islet cells. *J. Biol. Chem.* **270**, 11176–11180 (1995).

32. Zingarelli, B., O'Connor, M., Wong, H., Salzman, A., and Szabo, C. Peroxynitrite-mediated DNA strand breakage activates poly-adenosine diphosphate ribosyl synthetase and causes cellular energy depletion in macrophages stimulated with bacterial lipopolysaccharide. *J. Immunol.* **156**, 350–358 (1996).

33. Wei, X.Q., Charles, I.G., Smith, A., Ure, J., Feng, G.J., Huang, F.P., Xu, D., Muller, W., Moncada, S., and Liew, F.Y. Altered immune responses in mice lacking inducible nitric oxide synthase. *Nature (London)* **375**, 408–411 (1995).

34. MacMicking, J.D., Nathan, C., Hom, G., Chartrain, N., Fletcher, D.S., Traumbauer, M., Stevens, K., Xie, O.W., Sokol, K., Hutchinson, N., Chen, H., and Mudgett, J.S. Altered response to bacterial infection and endotoxic shock in mice lacking inducible nitric oxide synthase. *Cell* **82**, 641–650 (1995).

35. Szabo, C., Thiemermann, C., Wu, C.C., Perretti, M., and Vane, J.R. Attenuation of the induction of nitric oxide synthase by endogenous glucocorticoids accounts for endotoxin tolerance in vivo. *Proc. Natl. Acad. Sci. USA* **91**, 271–275 (1994).

36. Wu, C.C., Croxtall, J.D., Perretti, M., Bryant, C.E., Thiemermann, C., Flower, R.J., and Vane, J.R. Lipocortin-1 mediates the inhibition by dexamethasone of the induction by endotoxin of nitric oxide synthase in the rat. *Proc. Natl. Acad. Sci. USA* **92**, 3473–3477 (1994).

37. Szabo, C., Southan, G., Thiemermann, C., and Vane, J.R. The mechanism of induction of the inhibitory effect of polyamines on the induction by endotoxin of nitric oxide synthase. Role of aldehyde metabolites. *Br. J. Pharmacol.* **113**, 757–766 (1994).

38. Auguet, M., Longchampt, M.O., Delafotte, S., Gouline-Schulz, J., Chabrier, P.E., and Braquet, P. Induction of nitric oxide synthase by lipoteichoic acid from *Staphylococcus aureus* in vascular smooth muscle cells. *FEBS Lett.* **297**, 183–185 (1992).

39. Cunha, F.Q., Moss, D.W., Leal, L.M., Moncada, S., and Liew, F.Y. Induction of a macrophage parasiticidal pathway by *Staphylococcus aureus* and exotoxins through the nitric oxide synthesis pathway. *Immunology* **78**, 563–567 (1993).

40. Kengatharan, M., De Kimpe, S., and Thiemermann, C. Analysis of the signal transduction in the induction of nitric oxide synthase by lipoteichoic acid in macrophages. *B. J. Pharmacol.* **117**, 1163–1170 (1996).

41. De Kimpe, S.J., Hunter, M., Bryant, C.E., Thiemermann, C., and Vane, J.R. Delayed circulatory failure due to the induction of nitric oxide synthase by lipoteichoic acid from *Staphylococcus aureus* in anesthetized rats. *Br. J. Pharmacol.* **114**, 1317–1323 (1995).

42. Thiemermann, C., Kengatharan, M., and De Kimpe, S.J. Role of nitric oxide in the pathogenesis of gram-positive shock. In: *1996 Yearbook of Intensive Care and Emergency Medicine.* Ed. Vincent, J.L. Springer-Verlag, Berlin, 1996, pp. 345–357.

43. De Kimpe, S.J., Kengatharan, M., Thiemermann, C., and Vane, J.R. The cell wall components peptidoglycan and lipoteichoic acid from *Staphylococcus aureus* act in synergy to cause shock and multiple organ failure. *Proc. Natl. Acad. Sci. USA* **92**, 10359–10363 (1996).

44. Cameron, J.S. Acute renal failure in the ITU: The nephrologist's view. In: *Acute Renal Failure in the Intensive Therapy Unit.* Eds. Bihari, D. and Neild, G.H. Springer-Verlag, Berlin, 1990, pp. 3–12.

45. Groenwald, A.B.J. Pathogenesis of acute renal failure during sepsis. *Nephrol. Dial. Transplant* **9**, 47–51 (1994).

46. Vatner, S.F. Effects of hemorrhage on regional blood flow distribution in dogs and primates. *J. Clin. Invest.* **54**, 225–235 (1974).

47. Lucas, C.E. The renal response to acute injury and sepsis. *Surg. Clin. North Am.* **56**, 953–975 (1976).

48. Arendshorst, W.J. and Navar, L.G. Renal circulation and glomerular hemodynamics. In: *Diseases of the Kidney,* 5th ed. Eds. Schrier, R.W. and Gottschalk, C.W. Little, Brown & Co., Boston, 1993, pp. 65–117.

49. Brenner, M., Schaer, G.L., Mallory, D.L., Suffredini, A.F., and Parrillo, J.E. Detection of renal blood flow abnormalities in septic and critically ill patients using a newly designed indwelling thermodilution renal vein catheter. *Chest* **98**, 170–179 (1990).

50. Henrich, W.L., Hamasaki, Y., Said, S.I., Campbell, W.B., and Cronin, R.E. Dissociation of systemic and renal effects in endotoxemia. *J. Clin. Invest.* **69**, 691–699 (1982).

51. Badr, K., Kelley, V.E., Rennke, H.G., and Brenner, B.M. Roles for thromboxane A_2 and leukotrienes in endotoxin-induced acute renal failure. *Kidney Int.* **30**, 474–480 (1986).

52. Meyer, J., Hinder, F., Stothert, J., Jr., Traber, L.D., Herndon, D.N., Flynn, J.T., and Traber, D.L. Increased organ blood flow in chronic endotoxemia is reversed by nitric oxide synthase inhibition. *J. Appl. Physiol.* **76,** 2785–2793 (1994).

53. Garrison, R.N., Wilson, M.A., Matheson, P.J., and Spain, D.A. Nitric oxide mediates redistribution of intrarenal blood flow during bacteremia. *J. Trauma* **39,** 90–96 (1995).

54. Offner, P.J., Robertson, F.M., and Pruitt, B.A., Jr. Effects of nitric oxide inhibitors on regional blood flow in a porcine model of endotoxic shock. *J. Trauma* **39,** 338–343 (1995).

55. Booke, M.B., Meyer, J., Lingnau, W., Hinder, F., Traber, L.D., and Traber, D.L. Use of nitric oxide synthase inhibitors in animal models of sepsis. *New Horizons* **3,** 123–138 (1995).

56. Cohen, J.J. and Kamm, D.E. Renal metabolism: relation to renal function. In: *The Kidney,* 2nd ed. Eds. Brenner, B.M. and Rector, F.C. W.B. Saunders Co., Philadelphia, 1981, p. 147.

57. Brezis, M., Rosen, S., Silva, P., and Epstein, F.H. Renal ischemia: A new perspective. *Kidney Int.* **26,** 375–383 (1984).

58. Brezis, M., Heyman, S.N., Dinour, D., Epstein, F.H., and Rosen, S. Role of nitric oxide in renal medullary oxygenation. *J. Clin. Invest.* **88,** 390–395 (1991).

59. Heyman, S.N., Brezis, M., Epstein, F.H., Spokes, K., Silva, P., and Rosen, S. Early renal medullary hypoxic injury from radiocontrast and indomethacin. *Kidney Int.* **40,** 632–642 (1991).

60. Agmon, Y. and Brezis, M. Effects of non-steroidal anti-inflammatory drugs on intrarenal blood flow: Selective medullary hypoperfusion. *Exp. Nephrol.* **1,** 357–363 (1993).

61. Brezis, M. and Rosen, S. Hypoxia of the renal medulla—its implication for disease. *N. Engl. J. Med.* **332,** 647–655 (1995).

62. Cronenwett, J.L. and Lindenauer, S.M. Distribution of intrarenal blood flow during bacterial sepsis. *J. Surg. Res.* **28,** 132–141 (1978).

63. Lambalgen, A.A., van, Kraats, A.A., van, Bosse, G.C., van den, Stel, H.V., Straub, J., Donker, A.J., and Thijs, L.G. Renal function and metabolism during endotoxemia in rats—Role of hypoperfusion. *Circ. Shock* **35,** 164–173 (1991).

64. Edouard, A.R., Degremont, A.C., Duranteau, J., Pussard, E., Berdeaux, A., and Samii, K. Heterogeneous regional vascular responses to simulated transient hypovolemia in man. *Intens. Care Med.* **20,** 414–420 (1994).

65. Churchill, P.C., Bidani, A.K., and Schwartz, M.M. Renal effects of endotoxin in the male rat. *Am. J. Physiol.* **253,** F244–F250 (1987).

66. Lugon, J.R., Boim, M.A., Ramos, O.L., Ajzen, H., and Schor, N. Renal function and glomerular hemodynamics in male endotoxemic rats. *Kidney Int.* **36,** 570–575 (1989).

67. Gullichsen, E., Nellimarka, O., Halkola, L., and Niinikoski, J. Renal oxygenation in endotoxin shock in dogs. *Crit. Care Med.* **17,** 547–550 (1989).

68. Shimahara, Y., Kono, Y., Tanaka, J., Ozawa, K., Sato, T., Jones, R.T., Cowley, R.A., and Trump, B.F. Pathophysiology of acute renal failure following living *Escherichia coli* injections in rats: High energy metabolism and renal functions. *Circ. Shock* **21**, 197–205 (1987).

69. James, P.E., Jackson, S.K., Grinberg, O.Y., and Swartz, H.M. The effects of endotoxin on oxygen consumption of various cell types in vitro: An EPR oximetry study. *Free Rad. Biol. Med.* **18**, 641–647 (1995).

70. Linares, H.A. Sepsis related renal morphological alterations and the functional correlates. In: *Pathophysiology of Shock, Sepsis and Organ Failure*. Eds. Schlag, G. and Redl, H. Springer-Verlag, Berlin, 1993, pp. 961–972.

71. Cohen, J.J., Black, A.J., and Wertheim, S.J. Direct effects of endotoxin on the function of the isolated perfused rat kidney. *Kidney Int.* **37**, 1219–1226 (1990).

72. Macica, C.M., Escalante, B.A., Conners, M.S., and Ferreri, N.R. TNF production by the medullary thick ascending limb of Henle's loop. *Kidney Int.* **46**, 113–121 (1994).

73. Fouqueray, B., Philippe, C., Herbelin, A., Perez, J., Ardaillou, R., and Baud, L. Cytokine formation within rat glomeruli during experimental endotoxemia. *J. Am. Soc. Nephrol.* **3**, 1783–1791 (1993).

74. Baud, L., Oudinet, J.P., Bens, M., Noe, L., Peraldi, M.N., Rondeau, E., Etienne, J., and Ardaillou, R. Production of tumor necrosis factor by rat mesangial cells in response to bacterial lipopolysaccharide. *Kidney Int.* **35**, 1111–1118 (1989).

75. Mayeux, P.R. and Shah, S.V. Intracellular calcium mediates the cytotoxicity of Lipid-A in LLC-PK$_1$ cells. *J. Pharm. Exp. Ther.* **266**, 47–51 (1993).

76. Pinsky, M.R. Clinical studies on cytokines in sepsis: role of serum cytokines in the development of multiple-systems organ failure. *Nephrol. Dial. Transplant.* **9**, 94–98 (1994).

77. Pinsky, M.R., Vincent, J.L., Deviere, J., Alegre, M., Content, J., and Dupont, E. Serum cytokine levels in human septic shock: Relation to multiple-systems organ failure and mortality. *Chest* **103**, 565–575 (1993).

78. Hruby, Z.W., Shirota, K., Hothy, S., and Lowry, R.P. Antiserum against tumor necrosis factor-alpha and a protease inhibitor reduce immune glomerular injury. *Kidney Int.* **40**, 43–51 (1991).

79. Karkar, A.M., Koshino, Y., Cashman, S.J., Dash, A.C., Bonnefoy, J., Meager, A., and Rees, A.J. Passive immunisation against tumour necrosis factor-alpha and IL-1 beta protects from LPS enhancing glomerular injury in nephrotoxic nephritis in rats. *Clin. Exp. Immunol.* **90**, 312–318 (1992).

80. Bertani, T., Abbate, M., Zoja, C., Corna, D., Perico, N., Ghezzi, P., and Remuzzi, G. Tumor necrosis factor induces glomerular damage in the rabbit. *Am. J. Pathol.* **134**, 419–430 (1989).

81. Tomosugi, N.I., Cashman, S.J., Hay, H., Pusey, C.D., Evans, D.J., Shaw, A., and Rees, A.J. Modulation of antibody-mediated glomerular injury in vivo by bacterial lipopolysaccharide, tumor necrosis factor, and IL-1. *J. Immunol.* **142**, 3083–3090 (1989).

82. Kohan, D.E. Role of endothelin and tumour necrosis factor in the renal response to sepsis. *Nephro. Dial. Transplant.* **9** (suppl. 4), 73–77 (1994).

83. Pinsky, M.R. Sepsis and inflammation: The process of dying from a critical illness. In: *1996 Yearbook of Intensive Care and Emergency Medicine.* Ed. Vincent, J.L. Springer-Verlag, Berlin, 1996, pp. 3–10.

84. Ognibene, F.P., Rosenberg, S.A., Lotze, M., Skibber, J., Parker, M.M., Shelhamer, J.H., and Parrillo, J.E. Interleukin-2 administration causes reversible hemodynamic changes and left ventricular dysfunction similar to those seen in septic shock. *Chest* **94,** 750–754 (1988).

85. Fonseca, G.A. and Kilbourn, R.G. Cardiovascular alterations associated with inter-leukin-2 therapy. In: *Role of Nitric Oxide in Sepsis and ARDS. Update in Intensive Care and Emergency Medicine.* Eds. Fink, M.P. and Payen, D. Springer-Verlag, Berlin, 1995, pp. 232–252.

86. Kilbourn, R.G., Gross, S.S., Lodato, R.F., Adams, J., Levi, R., Miller, L.L., Lach-mann, L.B., and Griffith, O.W. Inhibition of interleukin-1-alpha-induced nitric oxide synthase in vascular smooth muscle and full reversal of interleukin-1-alpha-induced hypotension by *N*-omega-amino-L-arginine. *J. Natl. Cancer Inst.* **84,** 1008–1016 (1992).

87. Shalmi, C.L., Dutcher, J.P., Feinfeld, D.A., Chun, K.J., Saleemi, K.R., Freeman, L.M., Lynn, R.I., and Wiernik, P.H. Acute renal dysfunction during interleukin-2 treatment: Suggestion of an intrinsic renal lesion. *J. Clin. Oncol.* **8,** 1839–1846 (1991).

88. Mercatello, A., Hadj-Aissa, A., Negrier, S., Allaouchiche, B., Coronel, B., Tognet, E., Bret, M., Favrot, M., Pozet, N., Moskovtchenko, J.F., and Philip, T. Acute renal failure with preserved renal plasma flow induced by cancer immunotherapy. *Kidney Int.* **40,** 309–314 (1991).

89. Feinfeld, D.A., D'Agati, V., Dutcher, J.P., Werfel, S.B., Lynn, R.I., and Wiernik, P.H. Interstitial nephritis in a patient receiving adoptive immunotherapy with recom-binant interleukin-2 and lymphokine-activated killer cells. *Am. J. Nephrol.* **11,** 489–492 (1991).

90. Memoli, B., De Nicola, L., Libetta, C., Scialo, A., Pacchiano, G., Romano, P., Palmieri, G., Morabito, A., Lauria, R., Conte, G., and Andreucci, V.E. Interleukin-2-induced renal dysfunction in cancer patients is reversed by low-dose dopamine infusion. *Am. J. Kidney Dis.* **26,** 27–33 (1995).

91. Schlondorff, D., Goldwasser, P., Neuwirth, R., Satriano, J.A., and Clay, K.L. Production of platelet activating factor in glomeruli and cultured glomerular mesan-gial cells. *Am. J. Physiol.* **250,** F1123–F1127 (1986).

92. Braquet, P., Paubert-Braquet, M., Bourgain, R., Bussolino, F., and Hosford, D. PAF/cytokine auto-generated feedback networks in microvascular immune injury: Consequences in shock, ischemia and graft rejection. *J. Lipid Med.* **1,** 75–112 (1989).

93. Wang, J. and Dunn, M.J. Platelet activating factor mediates endotoxin-induced acute renal insufficiency. *Am. J. Physiol.* **253,** F1283–F1289 (1987).

94. Badr, K.F., DeBoer, D.K., Takahashi, K., Harris, R.C., Fogo, A., and Jacobson, H.R. Glomerular responses to platelet activating factor in the rat: Role of thromboxane A_2. *Am. J. Physiol.* **256,** F35–F43 (1989).

95. Maier, R.V., Hahnel, G.B., and Fletcher, J.R. Platelet activating factor augments tumor necrosis factor and procoagulant activity. *J. Surg. Res.* **52,** 258–264 (1992).

96. Szabo, C., Mitchell, J.A., Gross, S.S. Thiemermann, C., and Vane, J.R. Platelet activating factor contributes to the induction of nitric oxide synthase by bacterial lipopolysaccharide. *Circ. Res.* **73,** 991–999 (1993).

97. Tonneson, A.S. The kidney in sepsis. In: *Pathophysiology of Shock, Sepsis and Organ Failure.* Eds. Schlag, G. and Redl, H. Springer-Verlag, Berlin, 1993, pp. 973–995.

98. Jackson, E.K., Heidemann, H.T., Branch, R.A., and Gerkens, J.F. Low dose intrarenal infusions of PGE_2, PGI_2 and 6-keto PGE_1 vasodilate and in vivo rat kidney. *Circ. Res.* **51,** 67–72 (1982).

99. Bolger, P.M., Eisner, G.M., Ramwell, P.W., Slotkoff, L.M. and Corey, E.J. Renal actions of prostacyclin. *Nature (London)* **271,** 467–469 (1978).

100. Collins, D. and Klotman, P.E. Renin–angiotensin system and arachidonic acid metabolites in acute renal failure. In: *Acute Renal Failure,* 3rd ed. Eds. Lazarus, J.M. and Brenner, B.M. Churchill Livingston, New York, 1993, pp. 69–106.

101. Badr, K.F., Kelley, V.E., Rennke, H.G., and Brenner, B.M. Roles for thromboxane A_2 and leukotrienes in endotoxin-induced acute renal failure. *Kidney Int.* **30,** 474–480 (1986).

102. Cumming, A.D., McDonald, J.W., Lindsay, R.M., Solez, K., and Linton, A.L. The protective effect of thromboxane synthetase inhibition on renal function in systemic sepsis. *Am. J. Kidney Dis.* **13,** 114–119 (1989).

103. Morrow, J.D., Hill, K.E., Burk, R.F., Nammour, T.M., Badr, K.F., and Roberts, L.J., II. A series of prostaglandin F_2-like compounds are produced in vivo in humans by a non-cyclooxygenase, free radical-catalyzed mechanism. *Proc. Natl. Acad. Sci. USA* **87,** 9383–9387 (1990).

104. Takahashi, K., Nammour, T.M., Fukunaga, M., Ebert, J., Morrow, J.D., Roberts, L.J., II, Hoover, R.L., and Badr, K.F. Glomerular actions of a free radical-generated novel prostaglandin, 8-epi-prostaglandin $F_2\alpha$ in the rat: Evidence for interaction with thromboxane A_2 receptors. *J. Clin. Invest.* **90,** 136–141 (1992).

105. Yanigasawa, M., Kurihara, H., Kimura, S., Tomobe, Y., Kobayashi, M., Mitsui, Y., Yazaki, K., Goto, Y., and Masaki. A novel potent vasoconstrictor peptide produced by vascular endothelial cells. *Nature (London)* **332,** 411–415 (1988).

106. Marsden, P.A., Dorfman, D.M., Collins, T., Brenner, B.M., Orkin, S., and Ballerman, B.J. Regulated expression of endothelin-1 in glomerular capillary epithelial cells. *Am. J. Physiol.* **261,** F117–F125 (1991).

107. Wilkes, B.M., Susin, M., Mento, P.F., Macica, C.M., Girardi, E.P., Boss, E., and Nord, E.P. Localization of endothelin-like immunoreactivity in rat kidneys. *Am. J. Physiol.* **260,** F913–F920 (1991).

108. Kohan, D.E. Endothelin synthesis by rabbit renal tubule cells. *Am. J. Physiol.* **261,** F21–F226 (1991).

109. Ong, A.C., Jowett, T.P., Scoble, J.E., O'Shea, J.A., Varghese, Z., and Moorhead, J.F. Effect of cyclosporon A on endothelin synthesis by cultured human renal cortical epithelial cells. *Nephrol. Dial. Transplant.* **8** 748–753 (1993).

110. Ong, A.C., Jowett, T.P., Firth, J.D., Burton, S., Karet, F.E., and Fine, L.G. An endothelin-1 mediated autocrine growth loop involved in human renal tubule regeneration. *Kidney Int.* **48,** 390–401 (1995).

111. Arai, H., Hori, S., Aramori, I., Ohkubo, H., and Nakamishi, S. Cloning and expression of a cDNA encoding an endothelin receptor. *Nature (London)* **348,** 730–732 (1990).

112. Sakamoto, A., Yanigasawa, M., Sakurai, T., Takuwa, Y., Yanigasawa, H., and Masaki, T. Cloning and functional expression of human cDNA for the ET_B endothelin receptor. *Biochem. Biophys. Res. Commun.* **178,** 656–663 (1991).

113. Gurbanov, K., Rubinstein, I., Hoffman, A., Better, O.S., and Winaver, J. Effects of endothelin on the distribution of intrarenal blood flow in the rat kidney. (Abstract). *J. Am. Soc. Nephrol.* **6**(3), 679 (1995).

114. Sugiara, M., Snajdar, R.M.., Schwartzberg, M., Badr, K.F., and Inagami, T. Identification of two types of specific endothelin receptors in rat mesangial cells. *Biochem. Biophys. Res. Commun.* **162,** 1396–1401 (1989).

115. Pernow, J., Bouther, J.-F., Franco-Cereceda, A., Lacroix, J.S., Matran, R., and Lundberg, J.M. Potent selective vasoconstrictor effects of endothelin in the pig kidney in vivo. *Acta Physiol. Scand.* **134,** 573–574 (1988).

116. Kon, V. and Badr, K.F. Biological actions and pathophysiologic significance of endothelin in the kidney. *Kidney Int.* **40,** 1–12 (1991).

117. Pittet, J.F., Morel, D.R., Hemsen, A., Gunning, K., Lacroix, J.S., Suter, P.M., and Lundberg, J.M. Elevated plasma endothelin-1 concentrations are associated with the severity of illness in patients with sepsis. *Ann. Surg.* **213,** 261–264 (1991).

118. Takukawa, T., Endo, S., Nakae, H., Kikichi, M., Suzuki, T., Inada, K., and Yoshida, M. Plasma levels of TNFα, endothelin-1 and thrombomodulin in patients with sepsis. *Res. Commun. Chem. Pathol. Pharmacol.* **84,** 261–269 (1994).

119. Ruetten, H., Thiemermann, C., and Vane, J.R. Blockade of endothelin receptors with SB209670 aggravates the circulatory failure and the organ injury in endotoxic shock in the anesthetized rat. *Br. J. Pharmacol.* (in press).

120. Hellberg, P.O.A. and Källskog, T.O.K. Neutrophil-mediated post-ischemic tubular leakage in the rat kidney. *Kidney Int.* **36,** 555–561 (1989).

121. Paller, M.S. Effect of neutrophil depletion on ischemic renal injury in the rat. *J. Lab. Clin. Med.* **113,** 379–386 (1989).

122. Thorton, M.A., Winn, R., Alpers, C.E., and Zager, R.A. An evaluation of the neutrophil as a mediator of in vivo renal ischemic-reperfusion injury. *Am. J. Pathol.* **135,** 509–515 (1989).

123. Linas, S.L., Whittenburg, D., Parsons, P.E., and Repine, J.E. Mild renal ischemia activates primed neutrophils to cause renal failure. *Kidney Int.* **42,** 610–616 (1992).

124. Linas, S.L., Whittenburg, D., and Repine, J.E. Role of neutrophil-derived oxidants and elastase in lipopolysaccharide-mediated renal injury. *Kidney Int.* **39**, 618–623 (1991).

125. Shultz, P.J., and Raij, L. Endogenously synthesized nitric oxide prevents endotoxin-induced glomerular thrombosis. *J. Clin. Invest.* **90**, 1718–1725 (1992).

126. Zager, R.A. Endotoxemia, renal hypoperfusion, and fever: Interactive risk factors for aminoglycoside and sepsis-associated acute renal failure. *Am. J. Kidney Dis.* **20**, 223–230 (1992).

127. Wilcox, C.S., Welch, W.J., Furad, F., Gross, S.S., Taylor, G., Levi, R., and Schmidt, H.H.H.W. Nitric oxide synthase in the macula densa regulates glomerular capillary pressure. *Proc. Natl. Acad. Sci. USA* **89**, 11993–11997 (1992).

128. Mundel, P., Bachmann, S., Bader, M., Fischer, A., Kummer, W., Mayer, B., and Kriz, W. Expression of nitric oxide synthase in kidney macula densa cells. *Kidney Int.* **42**, 1017–1019 (1992).

129. Tojo, A., Gross, S.S., Zhang, L., Tischer, C.C., Schmidt, H.H.H.W., Wilcox, C.S., and Madsen, K.M. Immunocytochemical localisation of distinct isoforms of nitric oxide synthase in the juxtaglomerular apparatus of normal rat kidney. *J. Am. Soc. Nephrol.* **4**, 1438–1447 (1994).

130. Bachmann, S., Bosse, H.M., and Mundel, P. Topography of nitric oxide synthesis by localizing constitutive nitric oxide synthases in mammalian kidney. *Am. J. Physiol.* **268**, F885–F898 (1995).

131. Terada, Y., Tomita, K., Nonoguchi, H., and Marumo, F. Polymerase chain reaction localization of constitutive nitric oxide synthase and soluble guanylate cyclase messenger RNAs in microdissected rat nephron segments. *J. Clin. Invest.* **90**, 659–665 (1992).

132. Morrissey, J.J., McCracken, R., Kaneto, H., Vehaskari, M., Montani, D., and Klahr, S. Location of an inducible nitric oxide synthase mRNA in the normal kidney. *Kidney Int.* **45**, 998–1005 (1994).

133. Ahn, K.Y., Mohaupt, M.G., Madsen, K.M., and Kone, B.C. In situ hybridization localization of mRNA encoding inducible nitric oxide synthase in rat kidney. *Am. J. Physiol.* **267**, F748–F757 (1994).

134. Pfeilschifter, J., Rob, P., Mülsch, A., Fandrey, J., Vosbeck, K., and Busse, R. Interleukin 1β and tumour necrosis factor α induce a macrophage-type of nitric oxide synthase in rat renal mesangial cells. *Eur. J. Biochem.* **203**, 251–255 (1992).

135. Markewitz, B.A., Michael, J.R., and Kohan, D.E. Cytokine-induced expression of a nitric oxide synthase in rat renal tubule cells. *J. Clin. Invest.* **91**, 2138–2143 (1993).

136. Mohaupt, M.G., Schwöbel, J., Elzie, J.L., Kanna, G.S., and Kone, B.C. Cytokines activate inducible nitric oxide synthase gene transcription in inner medullary collecting duct cells. *Am. J. Physiol.* **268**, F770–F777 (1995).

137. Kone, B.C., Schwöbel, J., Turner, P., Mohaupt, M.G., and Cangro, C.B. Role of NF-κB in the regulation of inducible nitric oxide synthase in an MTAL cell line. *Am. J. Physiol.* **269**, F718–F729 (1995).

138. Ujiie, K., Yuen, J., Hogarth, L., Danziger, R., and Star, R.A. Localization and regulation of endothelial NO synthase mRNA expression in rat kidney. *Am. J. Physiol.* **267,** F296–F302 (1994).

139. Ishii, K., Warner, T.D., Sheng, H., and Murad, F. Endothelin increases cyclic GMP levels in LLC-PK₁ porcine kidney epithelial cells via formation of an endothelium-derived relaxing factor-like substance. *J. Pharm. Exp. Ther.* **259,** 1102–1108 (1991).

140. Tracey, W.R., Pollock, J.S., Murad, F., Nakane, M., and Förstermann, U. Identification of an endothelial-like type III NO synthase in LLC-PK₁ kidney epithelial cells. *Am. J. Physiol.* **266,** C22–C28 (1994).

141. Mayeux, P.R., Garner, H.R., Gibson, J.D., and Beanum, V.C. Effect of lipopolysaccharide on nitric oxide synthase activity in rat proximal tubules. *Biochem. Pharmacol.* **49,** 115–118 (1995).

142. Yu, L., Gengaro, P.E., Niederberger, M., Burke, T.J., and Schrier, R.W. Nitric oxide: A mediator in rat tubular hypoxia/reoxygenation injury. *Proc. Natl. Acad. Sci. USA* **91,** 1691–1695 (1994).

143. Petros, A., Bennett, D., and Vallance, P. Effect of nitric oxide synthase inhibitors on hypotension in patients with septic shock. *Lancet* **338,** 1557–1558 (1991).

144. Petros, A., Lamb, G., Leone, A., Moncada, S., Bennett, D., and Vallance, P. Effects of a nitric oxide synthase inhibitor in humans with septic shock. *Cardiovasc. Res.* **28,** 34–39 (1994).

145. Hutcheson, I.R., Whittle, B.J.R., and Boughton-Smith, N.K. Role of nitric oxide in maintaining vascular integrity in endotoxin-induced acute intestinal damage in the rat. *Br. J. Pharmacol.* **101,** 815–820 (1990).

146. Wright, C.E., Rees, D.D., and Moncada, S. Protective and pathological roles of nitric oxide in endotoxin shock. *Cardiovasc. Res.* **26,** 48–57 (1992).

147. Robertson, F.M., Offner, P.J., Ciceri, D.P., Becker, W.K., and Pruitt, B.A., Jr. Detrimental hemodynamic effects of nitric oxide synthase inhibition in septic shock. *Arch. Surg.* **129,** 149–156 (1994).

148. Henderson, J.L., Statman, R., Cunningham, J.N., Cheng, W., Damiani, P., Siconolfi, A., and Horovitz, J.H. The effects of nitric oxide inhibition on regional hemodynamics during hyperdynamic endotoxemia. *Arch. Surg.* **129,** 1271–1274 (1994).

149. Cobb, Q.P., Natanson, C., Hoffmann, W.D., Lodato, R.F., Banks, S., Koer, C.A., Salomon, M.A., Elin, R.Y., Hosseini, J.M., and Danner, R.L. *N*-Amino-arginine, an inhibitor of nitric oxide synthase, raises vascular resistance but increases mortality rates in awake canines challenged with endotoxin. *J. Exp. Med.* **176,** 1175–1182 (1992).

150. Spain, L.A., Wilson, M.A., and Garrison, R.N. Nitric oxide synthase inhibition exacerbates sepsis-induced renal hypoperfusion. *Surgery* **116,** 322–330 (1994).

151. Walder, C.E., Thiemermann, C., and Vane, J.R. The involvement of endothelium-derived relaxing factor on the regulation of renal cortical flow in the rat. *Br. J. Pharmacol.* **102,** 967–973 (1991).

152. Moncada, S., Palmer, R.M.J., and Higgs, E.A. Nitric oxide: Physiology, pathophysiology and pharmacology. *Pharm. Rev.* **43,** 109–142 (1991).

153. Westberg, G., Shultz, P.J., and Raij, L. Exogenous nitric oxide prevents endotoxin-induced glomerular thrombosis in rats. *Kidney Int.* **46,** 711–716 (1994).

154. Raij, L. Glomerular thrombosis in pregnancy: Role of L-arginine and nitric oxide pathway. *Kidney Int.* **45,** 775–781 (1994).

155. Corbett, J.A., Tilton, R.G., Chang, K., Hasan, K.S., Ido, Y., Wang, J.L., Sweetland, M.A., Lancaster, J.R., Williamson, J.R., and McDaniel, M.L. Aminoguanidine, a novel inhibitor of nitric oxide formation, prevents diabetic vascular dysfunction. *Diabetes* **41,** 552–558 (1992).

156. Ruetten, H., Southan, G.J., Abate, A., and Thiemermann, C. Attenuation of endotoxin-induced organ dysfunction by 1-amino-2-hydroxy-guanine, a potent inhibitor of inducible nitric oxide synthase. *Br. J. Pharmacol.* (in press).

157. Garvey, P.E., Oplinger, J.A., Tanoury, G.J., Sherman, P.A., Fowler, M., Marshall, S., Marmon, M.F., Path, J.E., and Furfine, E.S. Potent selective inhibition of human nitric oxide synthases. Inhibition by non-amino acid isothioureas. *J. Biol. Chem.* **269,** 26669–26676 (1994).

158. Szabo, C., Southan, G.J., and Thiemermann, C. Beneficial effects and improved survival in rodent models of septic shock with *S*-methyl-isothiourea sulphate, a potent and selective inhibitor of inducible nitric oxide synthase. *Proc. Natl. Acad. Sci. USA* **91,** 12472–12476 (1994).

159. Southan, G.J., Szabo, C., and Thiemermann, C. Isothioureas: potent inhibitors of nitric oxide synthases with variable isoform selectivity. *Br. J. Pharmacol.* **114,** 510–516 (1995).

160. Thiemermann, C., Ruetten, H., Wu, C.C., and Vane, J.R. The multiple organ dysfunction syndrome caused by endotoxin in the rat: Attenuation of liver dysfunction by inhibitors of nitric oxide synthase. *Br. J. Pharmacol.* **166,** 2845–2851 (1995).

161. Nakane, M., Klinghofer, V., Kuk, J.E., Donnelly, J.L., Budzik, G.P., Pollock, J.S., Basha, F., and Carter, G.W. Novel potent and selective inhibitors of inducible nitric oxide synthase. *Mol. Pharmacol.* **47,** 831–834 (1995).

162. Southan, G.J., Szabo, C., O'Connor, M.P., Salzmann, A., and Thiemermann, C. Amidines are potent inhibitors of nitric oxide synthase: Preferential inhibition of the inducible isoform. *Eur. J. Pharmacol.* **291,** 311–318 (1995).

163. Wu, C.C., Ruetten, H., and Thiemermann, C. Comparison of the effects of amino-guanidine and *N*-nitro-L-arginine methyl ester on the multiple organ dysfunction caused by endotoxaemia. *Eur. J. Pharmacol.* (in press).

164. Traber, D.L., Redl, H., Schlag, G., Herndon, D.L., Kimura, R., Prien, T., and Traber, L.D. Cardiopulmonary responses to continuous administration of endotoxin. *Am. J. Physiol.* **25,** H833–H839 (1988).

165. Suffredini, A.F., Fromm, R.E., Parker, M.M., Brenner, M., Kovacs, J.A., Wesley, R.A., and Parrillo, J.E. The cardiovascular responses of normal humans to the administration of endotoxin. *N. Engl. J. Med.* **32,** 280–287 (1989).

166. Lingnau, W., McGuire, R., Traber, L.D., and Traber, D.L. Renal function after nitric oxide inhibition in ovine bacterial sepsis (Abstract). *Crit. Care Med.* **22,** A122 (1994).

167. Meyer, J., Hinder, F., Stothert, J., Jr., Traber, L.D., Herndon, D.N., Flynn, J.T., and Traber, D.L. Increased organ blood flow in chronic endotoxemia is reversed by nitric oxide synthase inhibition. *J. Appl. Physiol.* **76,** 2785–2793 (1994).

168. Meyer, J., Lentz, C.W., Stothert, J.C., Traber, L.D., Herndon, D.N., and Traber, D.L. Effects of nitric oxide synthesis inhibition in hyperdynamic endotoxemia. *Crit. Care Med.* **22,** 306–312 (1994).

169. Preiser, J.C. and Vincent, J.L. Nitric oxide involvement in septic shock; Do human beings behave like rodents? In: *1996 Yearbook of Intensive Care and Emergency Medicine.* Ed. Vincent, J.L. Springer-Verlag, Berlin, 1996, pp. 358–365.

170. Ochoa, J.B., Udekwu, A.O., Billiar, T.R., Curran, R.D., Cerra, F.B., Simmons, R.L., and Peitzman, A.B. Nitrogen oxide levels in patients after trauma and during sepsis. *Ann. Surg.* **214,** 621–626 (1991).

171. Evans, T., Carpenter, A., Kinderman, H., and Cohen, J. Evidence of increased nitric oxide production in patients with the sepsis syndrome. *Circ. Shock* **41,** 77–81 (1993).

172. Gomez-Jimenez, J., Salgado, A., Mourelle, M., Martin, M.C., Segura, R.M., Peracaula, R., and Moncada, S. L-Arginine-nitric oxide pathway in endotoxemia and human septic shock. *Crit. Care Med.* **23,** 253–258 (1995).

173. Wong, H.R., Carcillo, J.A., Burckhart, G., Shah, N., and Janosky, J.E. Increased serum nitrite and nitrate concentrations in children with the sepsis syndrome. *Crit. Care Med.* **23,** 835–842 (1995).

174. Jacob, T.D., Ochoa, J.B., Udekwu, A.O., Wilkinson, J., Murray, T., Billiar, T.R., Simmons, R.L., Marion, D.W., and Peitzman, A.B. Nitric oxide production is inhibited in trauma patients. *J. Trauma* **35,** 590–597 (1993).

175. Evans, T.G., Rasmussen, K., Wiebke, G., Hibbs, J.B., Jr. Nitric oxide synthesis in patients with advanced HIV infection. *Clin. Exp. Immunol.* **97,** 83–86 (1994).

176. Goode, H., Howdle, P., Walker, B., and Webster, N. Nitric oxide synthase activity is increased in patients with sepsis syndrome. *Clin. Sci.* **88,** 131–133 (1995).

177. Geller, D.A., Lowenstein, C.J., Shapiro, R.A., Nussler, A.K., Di Silvio, M., and Wang, S.C. Molecular cloning and expression of inducible nitric oxide synthase from human hepatocytes. *Proc. Natl. Acad. Sci. USA* **90,** 3491–3495 (1993).

178. Nicolson, A.G., Haites, N.E., McKay, N.G., Wilson, H.M., MacLeod, A.M., and Benjamin, N. Induction of nitric oxide synthase in human mesangial cells. *Biochem. Biophys. Res. Commun.* **193,** 1269–1274 (1993).

179. Goureau, O., Hicks, D., and Courtois, Y. Human retinal pigmented epithelial cells produce nitric oxide in response to cytokines. *Biochem. Biophys. Res. Commun.* **198,** 120–126 (1994).

180. Robbins, R.A., Barnes, P.J., Springall, D.R., Warren, J.B., Kwon, O.J., Buttery, L.D., Wilson, A.J., Geller, D.A., and Polak, J.M. Expression of inducible nitric oxide synthase in human lung epithelial cells. *Biochem. Biophys. Res. Commun.* **203,** 209–218 (1994).

181. Watson, D., Donaldson, J., Grover, R., Mottola, D., Guntipalli, K., and Vincent, J.L. The cardiopulmonary effects of 546C88 in human septic shock. *Int. Care Med.* **21,** S117 (1995).

182. Gross, S.S., Stuehr, D.J., Aisaka, K., Jaffe, E.A., Levi, R., and Griffith, O.W. Macrophage and endothelial cell nitric oxide synthesis: cell-type selective inhibition by N^G-amino-arginine, N^G-nitro-arginine and N^G-methyl-arginine. *Biochem. Biophys. Res. Commun.* **170,** 96–103 (1990).

15

NO and Glomerulonephritis
Victoria Cattell and H. Terence Cook

1. Introduction

The various forms of human glomerulonephritis are a group of diseases of largely unknown etiology, which account for a significant proportion of human renal disease. Most have an immune pathogenesis and injury results from either immune complex deposition or delayed hypersensitivity-type reactions within the glomerulus [1]. In the majority of cases, evidence from animal models suggests that immune complexes form in situ (deposition of circulating preformed immune complexes is now considered a rare event, with the possible exception of acute serum sickness). In these in situ reactions, the antigens may be fixed intrinsic epitopes on glomerular cells or matrix molecules (e.g., the Heymann antigen on glomerular visceral epithelial cells, or the Goodpasture antigen in the glomerular basement membrane), or they can be exogenous antigens which localize due to some particular affinity for glomerular structures (e.g., components of bacteria). In the former case, the corresponding autoantibodies form through the breakdown of self-tolerance. Recently, it has been appreciated through the development of new experimental models that some types of glomerulonephritis may arise without soluble antibody formation and have characteristics of type IV immune reactions, involving sensitized T lymphocytes.

The effectors of injury are multiple and include attack by terminal complement components, acute neutrophilic inflammation with liberation of reactive oxygen metabolites and proteolytic enzymes, acute macrophage-dependent inflammation, and, possibly, cytotoxic T cells. Most chemical mediators have been implicated at some stage in this reaction, with various roles in cell injury and the ensuing functional derangements. The highly specialized nature of the glomerular capillaries has imparted a certain mystique to these reactions, but, essentially, the cardinal features of inflammation are still at the center of injury causing changes in permeability, hemodynamics, and often permanent structural abnormalities (irreversible scarring).

This combination of the induction of abnormal immune responses and inflammation in a highly vascularized structure, whose function is dependent on actively regulated vasomotor tone, strongly suggests that nitric oxide (NO) could have a major pathogenic role at several different levels. The strong evidence that other radicals (e.g., reactive oxygen metabolites) can be directly injurious in at least some of these diseases, particularly in neutrophil-dependent glomerulonephritis, supports this concept [2]. The induction of NO in pathological situations, including human disease, has recently been reviewed [3].

Although the role of oxygen radicals as critical mediators of neutrophil-dependent glomerulonephritis is unequivocal, the effectors of macrophage-dependent glomerulonephritis have not yet been established. It is clear from experimental models that, in some forms of glomerulonephritis, macrophages are the sole effectors of glomerular injury, and in others, they may be the most significant components of the inflammation [4]. They are usually associated with severe disease and are especially implicated in rapidly progressive types of glomerulonephritis with crescent formation. The macrophages in glomerulonephritis are of activated type, newly differentiated from blood monocytes and synthesizing a wide range of potential mediators, including eicosanoids, procoagulants, cytokines, and growth factors, and expressing high levels of major histocompatibility (MHC) Class II antigens [5]. The discovery that NO is a significant product of activated macrophages, accounting for a major part of their cytostatic and cytotoxic effects [6–8], suggested to us that NO could be the missing effector of macrophage-induced injury in glomerulonephritis, in forms initiated both by immune complexes or by delayed hypersensitivity reactions.

Before describing the recent evidence which demonstrates involvement of NO in glomerulonephritis, evidence for involvement of NO in other forms of inflammation and immune diseases will first be summarized, as this supports a role for NO in the similar pathogenic processes that occur in glomerulonephritis.

2. Nitric Oxide in Immunity and Inflammation

There is considerable evidence that NO is involved in the regulation of the immune response. In the mixed lymphocyte reaction, in vitro stimulation of T cells by alloantigens leads to synthesis of NO and inhibition of NOS enhances lymphocyte proliferation (reviewed in Ref. 9). A similar inhibitory effect is seen in vivo, where NO production in infection with trypanosomes [10] or listeria [11] diminishes T cell proliferation. It has been suggested that the mechanism of these effects is that cytokines, particularly interferon-γ, produced by activated T cells induce inducible nitric oxide synthase (iNOS) synthesis in macrophages and then NO inhibits proliferation of T lymphocytes, possibly by blocking the secretion of IL-2 [12]. Lymphocytes themselves are also a source of NO, and mouse T helper type 1 (Th1) cells can be activated to produce large amounts of NO which then inhibit the secretion of IL-2 and interferon-γ by Th1 cells, thus acting to autoregulate the immune response [13] and possibly alter the balance of Th1 and Th2 cells. Th2 cells could not be induced to produce NO [13]. NO may also regulate macrophages by inhibiting the expression of early growth response-1 protein, which is essential for macrophage differentiation [14], or by inhibiting HLA class II antigen expression, thus interfering with antigen presentation in the generation of the immune response [15]. Th2 cells produce the cytokines interleukin-4 (IL-4) and IL-10, which suppress iNOS activity in macrophages [16,17].

Nitric oxide has been implicated in a number of in vivo models of inflammation. In inflammation induced by local immune complex formation in lung or skin, there is increased formation of nitrite and nitrate, and inhibition of NO synthesis reduces lung and skin permeability and lung hemorrhage [18,19]. In experimental autoimmune encephalomyelitis, there is increased NO generation in the spinal cords of affected mice [20], and aminoguanidine, an inhibitor of NOS with some selectivity for the inducible enzyme, ameliorates the disease. Adjuvant arthritis induced either by *M. tuberculosis* [21] or by bacterial cell-wall fragments [22] is also associated with increased NO synthesis and is improved by NOS inhibition [21]. In a spontaneous autoimmune disease in mice [23] characterized by autoantibody production, arthritis, and glomerulonephritis, NOS inhibition ameliorates the inflammation in glomeruli and joints. There is also evidence of the involvement of NO in human inflammatory disease. Thus, NOS activity is increased in colon biopsies in ulcerative colitis [24], in synovial fluid in inflamed joints [25], and in the bronchial mucosa in asthma [26].

Although these studies provide strong evidence for the role of NO in inflammation, they do not definitively demonstrate the role of the inducible isoform of NOS. Recently, two groups have described mouse strains lacking the iNOS gene [27,28]. These mice are more susceptible to infection with the bacterium Listeria monocytogenes or the protozoan Leishmania major and show a reduced inflammatory response to carrageenin. However, they were more susceptible to lipopolysaccharide-induced mortality, indicating that NO generated from iNOS may have both beneficial and detrimental effects in infection and inflammation.

The evidence for the involvement of nitric oxide in glomerulonephritis will now be reviewed.

3. Nitric Oxide Synthesis in Glomerulonephritis

3.1. Induction of Nitric Oxide Synthesis in Glomerulonephritis

Nitrite Synthesis

The first experiments in which we demonstrated synthesis of NO by nephritic glomeruli were in a rat model of immune complex glomerulonephritis, nephrotoxic nephritis (NTN), where injury is initiated by the deposition of heterologous antiglomerular basement membrane antibodies [29]. The accelerated form of this model is induced by preimmunizing rats with rabbit IgG and Freund's complete adjuvant, followed by an intravenous injection of rabbit antirat glomerular basement membrane antibody 7 days later. Our reason for choosing this model was that it has previously been shown that the injury, as measured by proteinuria, is dependent on macrophage infiltration [30]. The model is characterized by an initial rapid influx of neutrophils and macrophages, but by 24 h, the infiltrating cells are predominantly macrophages and macrophage numbers remain high over the next 4 days [31] (Fig. 15-1).

To detect glomerular NO synthesis, we measured nitrite synthesis by the Griess reaction, a method which has been used widely for studying NO synthesis by activated macrophages [32]. When NO is produced, it is rapidly metabolized to stable end products and, for macrophages in culture, the two major metabolites are nitrite and nitrate, which are generated in an approximate ratio of 3 : 2. Nitrite is stable in tissue culture and can be easily measured by the Griess reaction, in which it forms a purple azochromophore by reacting with a sulphonamide. Isolated glomeruli from rats with accelerated nephrotoxic nephritis were cultured for 48 h, and nitrite which had accumulated in the medium was measured at the end of this time. Because hemoglobin may rapidly catalyze the conversion of nitrite to nitrate [33], kidneys were first perfused with saline. As endotoxin is a potent inducer of iNOS, precautions were taken to ensure that culture conditions were, as far as possible, endotoxin-free: Renal perfusion was carried out with pyrogen-free saline, glassware was baked at 170°C for 4 h, and polymyxin B was added to all buffers used for glomerular isolation. The culture medium was Dulbecco's modified Eagle's MEM (DMEM) without Phenol Red with 10% fetal calf serum (FCS); the FCS contained 0.12 ng/ml endotoxin.

Nitrite was present in the medium from glomeruli isolated at times from 4 h to 21 days after induction of glomerulonephritis with the peak accumulation from glomeruli isolated at 24 h (Fig. 15-1). No nitrite could be detected in the culture medium of normal rat glomeruli. In order to confirm that the nitrite in nephritic glomerular supernatants was derived from the synthesis of NO, we showed with the inhibitor of NOS, N^G-monomethyl-L-arginine (L-NMMA), a dose-dependent inhibition of nitrite generation with almost complete inhibition at a concentration of 300 μM; this inhibition could be reversed by high concentrations of L-arginine. These experiments, therefore, demonstrated that nephritic glomeruli were able to synthesize nitric oxide from L-arginine without further stimulation in culture and were the first demonstration of NO production at a site of immunologically induced inflammation. The amount of nitrite accumulated in the culture medium at 24 h was about half that found at 48 h, suggesting that the rate of synthesis was approximately linear over this period. Addition of lipopolysaccharide at 1 μg/ml led to an increase in nitrite generation. Because, in some instances, nephritic glomeruli synthesize superoxide which might react with NO and possibly interfere

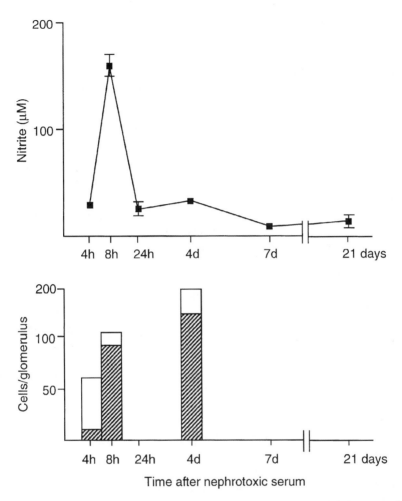

Figure 15-1. Acute accelerated nephrotoxic nephritis. *Upper panel*: Nitrite generation by isolated glomeruli cultured at 2000/ml for 48 h. *Lower panel*: Leukocyte infiltration in this model. Open columns = neutrophils; hatched columns = macrophages. Numbers were not determined at 24 h or 7 and 21 days.

with its detection as nitrite, the incubations were carried out in the presence of superoxide dismutase, but this had no effect on nitrite production.

Since this initial demonstration of NO synthesis by glomeruli in accelerated nephrotoxic nephritis, we have used the same method to demonstrate that there is NO synthesis in a number of other models of acute glomerulonephritis. Nitrite synthesis was substantial in a unilateral model induced in rats preimmunized with human IgG by perfusing the left kidney with cationized human IgG, which localizes to the glomerular capillary wall because of its charge [34]. This immune complex formation leads to a floridly hypercellular model of glomerulonephritis with a major influx of macrophages and neutrophils. We had previously found that macrophages isolated from inflamed glomeruli in this model appear highly activated as assessed by their expression of class II histocompatibility antigens, pattern of eicosanoid synthesis, and ability to synthesize superoxide [5]. In this model, there was basal synthesis of nitrite by glomeruli, which was maximal at day 4. Again, the production of nitrite was approximately linear over 48 h in culture and could be inhibited by L-NMMA; this inhibition was reversible by L-arginine but not by D-arginine. At 7 days, nitrite synthesis was much less than at 4 days in spite of the fact that there was no change in the number of glomerular macrophages during this time. The possible reasons for this decline are discussed later. As this was a unilateral model, the non–nephritic right kidney was available as a control and no significant basal nitrite production was found from day 1 to 4, indicating that there was no systemic induction of NO synthesis but that induction was confined to the nephritic kidney. At 7 days, however, there was a small but measurable synthesis of nitrite by glomeruli from the intact right kidneys.

Nitrite synthesis by glomeruli was also present in a model of mesangial proliferative glomerulonephritis induced by a monoclonal antibody to the Thy 1.1 antigen which is present on the surface of rat glomerular mesangial cells (the Thy 1 model) [35] and in active Heymann nephritis, a rat model of membranous glomerulonephritis [36]. The results in these models will be discussed further in Section 3.2.

iNOS mRNA

The generation of micromolar quantities of NO over many hours in culture strongly suggested that the enzyme involved in glomerulonephritis is the calcium-independent inducible nitric oxide synthase (iNOS). Since our initial demonstration of NO synthesis in nephritic glomeruli, the different types of NOS have been identified and the genes sequenced, making it possible to study the NOS isoform involved by the detection of mRNA and by immunohistochemistry.

To detect the expression of the iNOS gene, we used the reverse transcription polymerase chain reaction (RT-PCR) on RNA extracted from glomeruli in accelerated NTN [37]. Glomeruli were isolated by sieving in RNase-free conditions and RNA was extracted using RNAzol B. RNA was then reverse transcribed with Moloney leukemia virus reverse transcriptase and PCR was carried out using primers corresponding to rat vascular smooth muscle NOS. We initially carried out semiquantitative PCR using primers which amplified a 222 base-pair (bp) product. We could detect this product in activated rat peritoneal macrophages obtained with *Corynebacterium parvum* but not in thioglycollate-elicited macrophages. It was also present in rat neutrophils elicited with oyster glycogen and in cultured rat mesangial cells stimulated with IL-1. The PCR product was detectable in glomeruli from nephritic rats from 6 h to 7 days (Fig. 15-2). Its

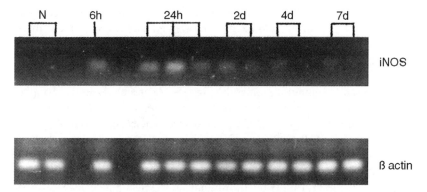

Figure 15-2. Time course of inducible nitric oxide synthase (iNOS) expression in accelerated nephrotoxic nephritis. Polymerase chain reaction was carried out with iNOS and β-actin primers on cDNA from normal glomeruli (N) and glomeruli isolated 6 h, 24 h, and 2, 4 and 7 days after induction of glomerulonephritis. PCR was performed with cDNA corresponding to 178 ng RNA. (From Ref. 37 with permission.)

identity was confirmed by sequencing. It was just detectable in glomeruli from normal rats when the highest concentration of cDNA was amplified. Competitive RT-PCR was used to study the time course of iNOS mRNA. For this, another set of primers were designed which amplified a 735-bp product and then a mutant was made with an internal deletion of 221 base pairs and used as a standard in competitive PCR reactions [38]. There was an approximate 440-fold increase in iNOS mRNA during the first 24 h of accelerated NTN and the levels remained elevated by 40-fold at 7 days.

The presence of glomerular iNOS mRNA has been confirmed in Thy 1 glomerulonephritis using a ribonuclease protection assay by Goto et al. [39]; mRNA was present at 1 h but had declined by 6 h and 12 h and was not detectable by 24 h or at later time points. This study also identified the glomerular expression of endothelial NOS in both control and nephritic glomeruli, with a slight increase in nephritic glomeruli particularly at 1 h and 4 days although, at each time, it was much less than iNOS at 1 h. The source of iNOS was also investigated, as will be discussed (see Section 3.2). In MRL-*lpr/lpr* mice which develop a spontaneous autoimmune glomerulonephritis increased iNOS mRNA in whole kidney accompanied by an increase in iNOS protein detected by Western blotting has been demonstrated [23].

iNOS Immunohistochemistry

We have looked for protein expression by immunohistochemistry [40] using a polyclonal rabbit antiserum raised against a synthetic peptide corresponding to amino acids 47–71 of the mouse iNOS protein. This antiserum reacts with a protein band of 135 kDa in a Western blot of an ADP eluate of induced murine macrophages but does not recognize neuronal or endothelial NOS [26]. Immuno-histochemistry was carried out on formalin-fixed paraffin-embedded sections of unilateral glomerulonephritis induced by cationized IgG. In nephritic kidneys, there was strong staining in glomeruli (Fig. 15-3) particularly in the cytoplasm of mononuclear cells in capillary loops and in cells in Bowman's space in glomeruli with early crescents. iNOS-positive mononuclear cells were also present in the interstitium. No staining was seen in normal kidneys or in control right kidneys from nephritic rats. This distribution of iNOS was confirmed by in situ hybridization using synthetic oligonucleotide probes. Recently, in collaboration with C. Pusey and F. Tam (Royal Postgraduate Medical School) we have been able to demonstrate iNOS induction immunohistochemically in a model of NTN in the Wistar Kyoto (WKY) rat (Fig. 15-4). Immunohistochemical localization of iNOS has also been demonstrated in Thy-1 glomerulonephritis [39], where the main staining appears to be in infiltrating neutrophils.

Thus, there is clear evidence that in several experimental models of immune complex glomerulonephritis there is induction of iNOS and that nephritic glomeruli have the capacity to synthesize NO.

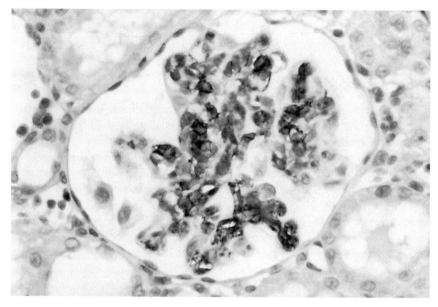

Figure 15-3. Acute unilateral glomerulonephritis induced by cationized IgG—day 4. Immunohistochemistry shows strong glomerular staining for iNOS. Immunoperoxidase/hematoxylin. (From Ref. 40 with permission.) **See plate for color illustration.**

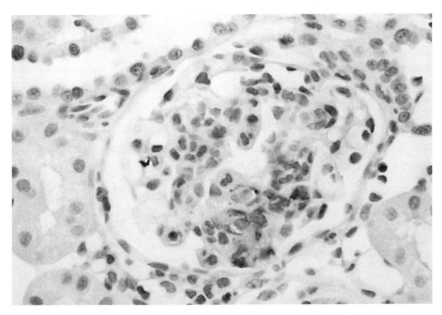

Figure 15-4. Nephrotoxic nephritis in the WKY rat—day 4. Positive immunohistochemistry for iNOS; immunoperoxidase/hematoxylin. **See plate for color illustration.**

3.2. Sources of iNOS

Macrophages

Our next goal was to identify the cellular source of NO production in nephritic glomeruli, as NO can be induced in almost every cell, at least in vitro. Activated macrophages, which are the source of many of the chemical mediators implicated in the pathogenesis of glomerulonephritis [4], can produce large quantities of NO [6–8, 32]. We therefore first studied NO synthesis by macrophages isolated from nephritic glomeruli. We, and others, had previously shown that these macrophages can be isolated by several techniques which do not affect viability or function [5,41]. One problem, however, is which macrophages to use as appropriate controls, as infiltrating macrophages in glomerulonephritis are a mixed population of newly recruited blood monocytes and macrophages in various stages of maturation. We have used syngeneic, elicited or activated, peritoneal macrophages, as these represent a similarly mixed population recently recruited from peripheral blood. Macrophages isolated from nephritic glomeruli produced abundant nitrite under basal conditions; basal production of 30.7 ± 2.8 nmol per 48 h per 10^6 cells was found in accelerated nephrotoxic nephritis [29] and higher levels in the unilateral model induced by cationized IgG. (Thioglycollate-elicited peritoneal macrophages, in contrast, do not produce nitrite spontaneously.) From calculations based on the number of macrophages per glomerulus, macrophages could account for the entire NO synthesis from nephritic glomeruli. This major source was also suggested by the linear relationship between glomerular nitrite levels, and the number of infiltrating macrophages in glomeruli in the different models of glomerulonephritis studied [42] (Fig. 15-5).

We, therefore, sought direct evidence that macrophages are a source of nitrite in glomeruli by studying active Heymann nephritis, where leukocyte-independent injury occurs due to the activation of lytic complement components by in situ immune complex formation on glomerular capillary walls [43]. The disease is induced by immunization with a renal tubular epithelial (RTE) antigen, generating antibodies which react in situ with this epitope expressed on glomerular visceral epithelium. Glomeruli isolated from this chronic model 8 weeks after the onset of proteinuria spontaneously synthesized low levels of nitrite (7.1 ± 1.4 nmol per 2000 gl per 48 h) [36]. Surprisingly, these glomeruli contained a significant

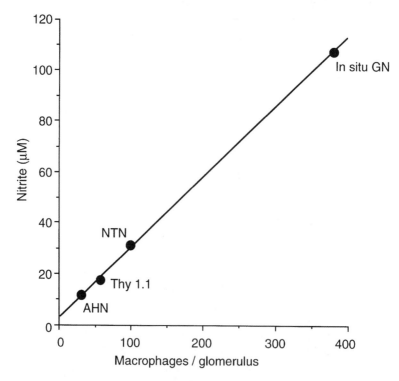

Figure 15-5. Linear relationship between glomerular nitrate generation and macrophage infiltration in four models of glomerulonephritis. AHN = active Heymann nephritis; Thy 1.1 = Thy 1 model; NTN = accelerated nephrotoxic nephritis; in situ gn = acute immune complex glomerulonephritis induced by cationized IgG. (From Ref. 94 with permission.)

macrophage infiltrate (32 ± 6 macrophages/gl, compared with 14 ± 12 macrophages/gl in adjuvant-alone non-nephritic controls), which had not previously been detected in this model. We were able to show that these macrophages were the source of glomerular nitrite by irradiation-depletion of leukocytes.

Mesangial Cells

Cultured mesangial cells possess iNOS (Chapter 10), and nitrite synthesis can be induced by cytokine stimulation, a factor likely to be highly relevant in vivo in glomerulonephritis. We have investigated whether mesangial cells in vivo might be a source of NO production using the Thy 1.1 model of glomerulonephritis [35]. The mesangial proliferation in this model depends on an initial complement-mediated injury to mesangial cells produced by injection of anti-Thy 1 antibody (either polyclonal or monoclonal) which reacts with constitutive Thy 1 on mesangial cells. The model can be broadly divided into an early phase (up to 4 days after administration of antibody where injury with leukocyte infiltration occurs) and a later phase of mesangial proliferation. Our interest was to see whether NO was produced in the proliferative phase, where the mesangial cells in vivo have characteristics thought to resemble those of mesangial cells in culture [44]. Spontaneous glomerular nitrite synthesis occurred in the early inflammatory phase where there was significant macrophage infiltration, but not during the phase of mesangial proliferation [35] (Fig. 15-6). This strongly suggested that in Thy 1 model, as in previous models studied, leukocytes are the major source of NO generation in nephritic glomeruli. One intriguing finding, however, was that although there was no spontaneous NO synthesis by glomeruli with established mesangial proliferation, low levels could be induced by IL-1, a cytokine which induces NO in mesangial cells in vitro [45,46], whereas IL-1 at a similar dose (1 nM) did not induce NO in normal glomeruli. Ex vivo stimulation with IL-1 on day 4, when many mesangial cells are proliferating, produced higher nitrite than on day 7, when hypercellularity is established, a further similarity between cultured and in vivo mesangial cells.

Neutrophils

In many models of glomerulonephritis, neutrophils form a significant component of the leukocyte infiltration in glomeruli, and in some diseases, they are the mediators of injury, through generation of proteases and reactive oxygen metabolites. In the rat, neutrophils isolated from the peritoneal cavity 5 h after intraperitoneal injection of oyster glycogen express iNOS mRNA [37]. These cells are, therefore, a possible source of NO in nephritic glomeruli. In immunohistochemical studies on glomeruli in the unilateral cationized IgG model [40], we did not obtain convincing positive staining in infiltrating neutrophils. However, in a recent study on the initial phase of the Thy 1 model, expression of iNOS

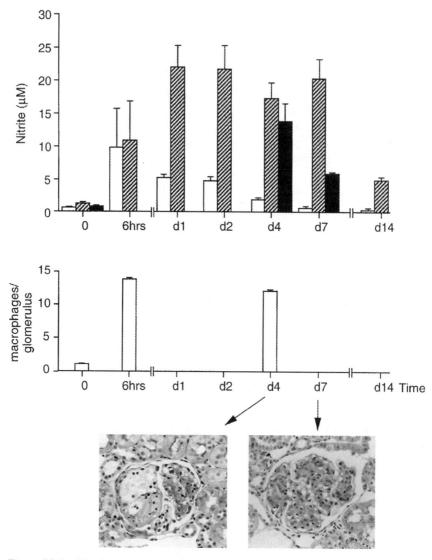

Figure 15-6. Thy 1.1 glomerulonephritis. *Upper panel*: Glomerular nitrite generation at different times after induction of disease. Open columns = basal; hatched columns = LPS stimulated; solid columns = IL-1 stimulated. *Center panel*: Macrophage numbers per glomerulus at times 0, 6 h, and 4 days. *Lower panel*: Left—mesangiolytic lesion at 4 days; right—mesangial proliferation at 7 days.

was identified principally in neutrophils by double-labeling, immunohistochemistry [39] although a few iNOS positive cells were macrophages. iNOS was not detected in mesangial cells through the course of this model, despite examination of the mesangial proliferative lesions on days 4 and 10.

3.3. iNOS in Normal Glomeruli

In normal glomeruli in some rats, we find detectable iNOS mRNA by RT-PCR, which may reflect the extreme sensitivity of this technique. This finding has also been reported by other groups [47,48]. Endothelial constitutive NOS has been identified in rat glomeruli by RT-PCR [49], ribonuclease protection assay [39], and immunohistochemistry [50]. In the Thy 1 model of glomerulonephritis, an increase in endothelial NOS mRNA has been detected [39].

In spite of this iNOS expression in normal glomeruli, nitrite cannot be detected by the Griess reaction in cultured normal glomeruli under either basal conditions or when stimulated with 1-μg/ml LPS. However, glomeruli stimulated with higher levels of LPS (10 or 100 μg) do show L-NMMA inhibitable nitrite production (6.2 ± 0.3 nmol per 2000 gl per 48 h with 100 μg/ml LPS). In the normal rat glomerulus, there is a small population of resident macrophages in the mesangium. Irradiation with renal shielding 4 days before glomerular isolation reduced the numbers of resident glomerular macrophages from 13 ± 0.4 to 2.9 ± 0.3 gl and significantly reduced the nitrite accumulation from 6.2 ± 0.3 to 2.6 ± 0.9 nmol per 2000 gl per 48 h, suggesting that these resident macrophages are the source of LPS-inducible NO.

3.4. Significance and Source of Urinary Nitrite and Nitrate

The urinary excretion of nitrite and nitrate is the net result of dietary intake, endogenous synthesis, and metabolic loss. NO synthesis is the endogenous source of these metabolites, but this contribution can be masked by ingested nitrate; if animals are fed a low-nitrate diet, urinary nitrate excretion then becomes a measure of endogenous NO production [51].

The ratio of nitrite/nitrate in urine is approximately 1 : 1000, and nitrite is almost undetectable in normal rat urine. In our first study of accelerated nephrotoxic nephritis, we found greatly elevated nitrite in the urine, but at that time, nitrate was not measured nor were the animals on a nitrite/nitrate-free diet. We therefore further examined nitrite and nitrate excretion under dietary restriction to determine whether the kidney was the source of these metabolites in glomerulonephritis. Rats on such a diet showed a dramatic drop in urinary nitrate excretion. Preimmunization alone with rabbit IgG and Freund's complete adjuvant caused a rise over 6 days, which would correspond with systemic immune stimulation. Comparing the levels between rats which subsequently received a rabbit nephrotoxic globulin to produce nephritis and those which received equivalent amounts of normal rabbit globulin, there was no difference between the two groups, both showing a second rise in nitrate. Similar results with much lower levels were found for urinary nitrite. These experiments showed that the major source of these metabolites in urine in glomerulonephritis was systemic not glomerular, at least in this model with systemic immunization.

3.5. Regulation of NO Production in Glomerulonephritis

Nitric oxide synthesis is regulated at many levels [52,53], including transcription and translation of NOS, and the availability of arginine and cofactors required for NO synthesis (see Chapter 3). Very little is known about these events in glomerulonephritis. Induction of iNOS occurs rapidly after glomerular immune complex formation. This is almost certainly cytokine mediated, as many of the cytokines known to induce iNOS, (e.g., IL-1 and TNF) have been detected in the early phases of glomerulonephritis [54]. IFN-γ could be involved, but to date, despite the presence of small numbers of T lymphocytes in nephritic glomeruli [55], there is no clear evidence of IFN-γ production in glomerulonephritis.

Preliminary experiments in the Thy 1 model of glomerulonephritis have shown that the message for iNOS appears as rapidly as 1 h after injection of anti-Thy 1 antibody (H. Ebrahim, personal communication). This is preceded by induction of IL-1 mRNA, a cytokine which is known to induce NO synthesis in a wide variety of cell types. We, in collaboration with Pfeilschifter, have tested the ability of administered IL-1 to induce iNOS mRNA in vivo and have found that IL-1 induces iNOS mRNA, detected by RT-PCR 1 h after cytokine administration. These preliminary results therefore suggest that IL-1, together with its many other effects in glomerulonephritis, could cause the induction of NO synthesis.

In the two models of acute leukocyte-dependent glomerulonephritis we have studied, NO synthesis peaks within 1 day of the onset of disease and then declines, despite persistence of the macrophage infiltrate. The macrophages in the lesions therefore behave similarly to those in vitro after γ-IFN/LPS stimulation, which show peak iNOS activity at 12 h and declining activity, until at 72 h when iNOS activity is no longer detectable [56]. The macrophages in nephritic glomeruli are not refractory to induction of iNOS, for ex vivo stimulation with LPS results in high levels of nitrite. Quantitative RT-PCR shows a decline in iNOS mRNA [37]. A similar decline in iNOS in vivo is found in the mononuclear cells infiltrating the renal cortex in the acute allograft reaction in the rat; about 40% of mononuclear cells are positive at the onset of rejection (4 days after transplantation), but almost none by 7 days [57]. These findings suggest that transcription of iNOS mRNA ceases or that the message is destabilized. This could be due to loss of stimulatory cytokines or the production of inhibitory cytokines. Transforming growth factor β (TGFβ) is a cytokine which inhibits NO production through multiple effects, including reducing the stability of iNOS mRNA, and is implicated in glomerulonephritis as a cytokine which promotes glomerular scarring [58].

The inhibition of iNOS message in infiltrating macrophage could be autocrine or due to products of intrinsic glomerular cells, possibly mesangial cells, as activated mesangial cells produce a wide range of cytokines, including TGFβ [59], and we found that supernatants from cultured mesangial cells have a suppressive effect on NO production by peritoneal macrophages [60]. In addition, NO itself is able to modify the expression of iNOS in a biphasic fashion. Thus, at low concentrations, NO augments the expression of iNOS mRNA by LPS and IFN-γ-treated macrophages, but at higher concentrations, it is inhibitory [61]. It is therefore possible that, in vivo, NO released from endothelial cells may potentiate the rapid secretion of NO by newly arrived macrophages, but that once high levels of NO are achieved, the further induction of iNOS is prevented.

4. L-Arginine Metabolism and Arginase Activity in Glomeruli

In addition to acting as a substrate for NOS, L-arginine may also be metabolized to L-ornithine and urea by the enzyme arginase, one of the enzymes of the urea cycle. L-Ornithine is further metabolized to polyamines (putrescine, spermine, and spermidine), which are essential for cell proliferation, and to L-proline, which is required for collagen synthesis. Possible interactions and competition for substrate between these metabolic pathways may, therefore, have significant pathophysiological effects. This is a major pathway of arginine metabolism in the liver where arginase levels are very high. There is high arginase activity in wounds and at other inflammatory sites which leads to marked depletion of L-arginine [62–64]. The levels in wound fluid are well below the effective K_m for macrophage iNOS and so there may be limitation of macrophage NO generation. In an experimental model of wound healing, there is an initial phase of high NOS activity followed after several days by an increase in arginase activity [63]. In view of this possible reciprocal relationship between the NO and arginase pathways, we looked for evidence of arginase activity in glomerulonephritis.

To determine arginase activity, we studied the conversion of ^{14}C-arginine to ^{14}C-urea by glomeruli in culture. Glomeruli were isolated and cultured as for measurement of nitrite synthesis. Normal glomeruli have a low level of constitutive arginase activity and this is increased 5.7-fold in nephritic glomeruli isolated 3 days after induction of accelerated NTN. Addition of LPS to cultured glomeruli increases nitrite generation and reduces urea synthesis, whereas L-NMMA produces an increase in urea synthesis consistent with a reciprocal relationship between the NOS and arginase pathways. In order to study the relationship between the pathways in more detail and also the time course of arginase induction, we studied the unilateral model induced by cationized IgG.

Glomerular urea synthesis was increased at each time point studied (1, 4, and 7 days) and was maximal on day 4, whereas nitrite synthesis was maximal on day 1. L-NMMA increased urea synthesis on days 1 and 4 consistent with

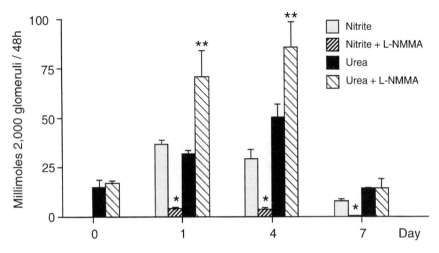

Figure 15-7. Sequential activation of NOS and arginase pathways of L-arginine metabolism in glomeruli in acute in situ immune complex glomerulonephritis in the rat induced by cationized IgG. Nitrite production declines with time, whereas metabolism by arginase increases from day 1 to day 4. Competition between the pathways is demonstrated by the significantly increased urea production in the presence of L-NMMA.*p < .05 versus basal; **p < .037 versus basal. (Modified from Ref. 94 with permission.)

competition between the two pathways for substrate or possibly an inhibitory effect on arginase of an intermediate in the NO synthetic pathway [65] (Fig. 15-7). In order to confirm that there was induction of arginase in vivo, intracellular activity was studied in glomerular lysates at day 4 of nephritis; this was increased threefold compared with control glomeruli (nephritic 47.6 ± 12 nmol urea/min per 10,000 gl; controls 13.7 ± 1.3). What is the likely cellular source of the arginase in glomeruli? We studied the arginase activity of macrophages isolated from nephritic glomeruli and of cultured glomerular mesangial cells. The results showed for the first time that mesangial cells have high arginase activity and were the likely source of the arginase activity in glomeruli, whereas macrophages could entirely account for the NOS activity. In a preliminary report, increased arginase activity has also been shown in experimental mesangial proliferative glomerulonephritis and was associated with increased ornithine decarboxylase activity [66]. These findings suggest that increased arginase activity in nephritic glomeruli could be a factor promoting cell proliferation and scarring, as ornithine will act as a substrate for the synthesis of polyamines and collagen [67].

5. The Role of NO in Glomerulonephritis

5.1. Potential Actions of Nitric Oxide in Glomerulonephritis

The possible roles that NO may play in glomerulonephritis will now be summarized before the effects of manipulating its synthesis are reviewed in the following section. Although the precise role of NO during pathological processes remains unclear, it is likely that it has both damaging and protective roles [3]. First, NO may be cytotoxic, either directly or by reacting with superoxide to form peroxynitrite [68,69] and this cytotoxicity could affect intrinsic glomerular cells or, in an autocrine manner, the infiltrating leukocytes synthesizing NO. The cellular toxicity caused by NO itself is due to its effects on many intracellular targets including iron–sulfur clusters in the mitochondrial electron transport chain [6]. Peroxynitrite can decompose spontaneously to form the highly reactive hydroxyl radical, which may cause peroxidative damage to a wide variety of biological molecules [68]. However, it has also been suggested that, in situations where tissue damage is primarily due to reactive oxygen species, NO may scavenge superoxide or peroxyl radicals [70] and thus limit damage. The interaction of NO with superoxide may also have a role in modulating leukocyte adhesion and migration. In the mesenteric vasculature, inhibition of NO causes a superoxide-dependent increase in the vascular expression of P-selectin [71] and NO itself can reduce direct superoxide induced leukocyte adhesion to mesenteric vessels [72]. Thus, the reaction of NO with superoxide may be toxic or protective, depending on the circumstances, and this, in turn, will depend on the relative concentrations of the two radicals and the concentration of other antioxidants; it is possible that the balance of these effects may change with time during the course of glomerulonephritis.

There are a number of other potentially beneficial effects of NO in glomerulonephritis. NO has a major role in the control of glomerular hemodynamics (discussed in more detail in Chapter 8) which may help to maintain glomerular blood flow and glomerular filtration rate during inflammation. This could partly result from the ability of NO to increase prostaglandin synthesis by stimulating the activity of cyclooxygenase [73]. NO has inhibitory effects of platelets and could therefore inhibit glomerular thrombosis, as has been shown for glomeruli after endotoxin administration [74]. It may also cause downregulation of the expression of pro-inflammatory cytokines [75] during inflammation. Following the initial phase of injury, NO may also affect repair, because, in vitro, it has been shown to inhibit mesangial cell proliferation [76] although others [77] have failed to confirm this effect. It may also cause apoptosis and thus promote resolution of inflammation by removal of inflammatory cells and proliferating intrinsic cells [78,79]. The role of NO in wound healing, a situation analogous to glomerular scarring, is discussed in Chapter 5.

5.2. Manipulation of Nitric Oxide Synthesis in Glomerulonephritis

Nitric oxide has now been implicated as a mediator in many types of inflammatory reaction, including immune complex lesions which have many pathological features in common with glomerulonephritis. So far, published studies manipulating the L-arginine NO pathway in immune complex injury in lung [18,80], skin [18], joint [22], and peritoneum [81] have supported a pro-inflammatory role for L-arginine metabolites. In all, the early formation of edema, measured either by changes in vascular permeability or tissue swelling, has been ameliorated by local or systemic administration of competitive inhibitors of NO synthesis, such as L-NMMA or L-NAME [18,22,80,81], and in some experiments, permeability has been exacerbated by the administration of L-arginine [18]. These findings suggest involvement of NO in initial increases in vascular permeability following immune complex deposition in the microvasculature, although how dependent these effects are on changes in flow rather than permeability per se has not yet been determined. It is, therefore, a reasonable hypothesis that NO inhibition could have similar effects on glomerular proteinuria induced by immune complex deposition. The effects of NO on the inflammatory cell infiltration in the above models also have possible significance for glomerulonephritis. These few studies suggest that macrophage, but not neutrophil, accumulation is reduced by NO inhibition. Thus, in the experiments by Mulligan et al. [18,80] on immune complex lung injury, the neutrophil infiltration induced by the intratracheal installation of IgG antibodies to bovine serum albumin, followed by intravenous bovine serum antigen, was not reduced by coinstallation of L-NMMA [18], but the macrophage response in a similar system using IgA antibodies was [80]. Similarly, in a biphasic model of immune arthritis [22] where streptococcal cell-wall fragments injected intraperitoneally localize in synovium causing a destructive arthritis, the neutrophil infiltration of the early phase of arthritis was not affected by administration of systemic L-NMMA, but the mononuclear cell infiltration in the chronic phase was significantly reduced.

The large quantities of NO generated by nephritic glomeruli, through the inducible NOS, suggest a role in tissue damage. Assessment of the role of NO in glomerulonephritis in vivo is, however, complicated first by the necessity to demonstrate convincingly that NO inhibition has not influenced the amount of immune deposition in glomeruli and, second, by the effects of the currently available NO inhibitors on constitutive NO which changes both systemic and glomerular hemodynamics [82,83] precluding attributing effects to iNOS inhibition. With these reservations, the current evidence for an in vivo role for NO in glomerulonephritis will now be reviewed.

The first in vivo evidence for a role for NO in the pathogenesis of glomerulonephritis came from a study of chronic NO inhibition in spontaneous murine autoimmune disease [23]. MRL-*lpr/lpr* mice spontaneously develop disease characterized by autoantibody production, glomerulonephritis, and other manifestations of autoimmune disorder, including lymphadenopathy, vasculitis, and arthritis. There is strong evidence of iNOS induction in these mice compared with normal strains. Increased urinary nitrite/nitrate excretion begins at 10–12 weeks of age, peritoneal macrophages have enhanced responses to endotoxin and IFN-γ, and large numbers of iNOS-positive cells are present in the spleen. In the kidney and spleen, iNOS mRNA and NO synthase activity are present. The authors of this study proposed that the high levels of cytokine production in macrophages in these mice could be responsible for the widespread induction of iNOS and elevated NO production. Treatment with oral L-NMMA, beginning at 8 weeks of age, was shown to block increased NO production and reduce proteinuria. The histological changes of proliferative glomerulonephritis and arthritis were ameliorated. These results strongly implicate NO in this autoimmune disease. As L-NMMA treatment did not reduce serum anti-DNA antibodies, the authors further suggested that NO had a role as a mediator of inflammation in the affected organs and was not merely a regulator of afferent immune mechanisms.

Another recent study has also implicated NO as a mediator of injury in glomerulonephritis [84]. In the rat Thy 1 model, acute NO inhibition with L-NMMA almost abolished mesangiolysis and significantly reduced proteinuria and the later increases in mesangial matrix. No comment was made on effects on mesangial cell proliferation, which presumably was also reduced, as mesangiolysis did not occur. This remarkable result, achieved by a single injection of a small dose of L-NMMA 60 min prior to administration of cytotoxic antibody, suggested that NO was a major effector of mesangial injury. By minimizing the dose of L-NMMA, there was no hypertension at the time of antibody administration and no detectable difference in the amount of antibody localizing in glomeruli (although this was only assessed by immunofluorescence and not by more sensitive radiolabeled antibody binding studies). It is of some interest that macrophage infiltration was unaffected, as this implies that mesangiolysis per se is not the stimulus to macrophage accumulation and, further, that macrophages are not important for the subsequent increased matrix synthesis, as has previously been suggested. Similar ameliorative results were obtained by feeding rats with a low-arginine diet.

In contrast with the above experimental data, others have found a protective role for NO in glomerulonephritis. Ferrario et al. [85] studied acute NO inhibition in the heterologous phase of rat nephrotoxic nephritis (2 h after nephrotoxic serum) using single nephron puncture in Munich Wistar rats to assess glomerular hemodynamic responses. L-NMMA infusion caused a dramatic augmentation of urinary protein excretion, accompanied by further increases in the elevated glomerular pressure and efferent arteriolar resistance seen in this model. This exacerbation of glomerular vasoconstriction supports a role for induced NO in counteracting the acute vasoconstriction at mesangial and arteriolar level induced by the generation of eicosanoids in acute nephrotoxic injury. Preliminary experiments by the same group using L-NAME inhibition and a milder level of injury also showed a dramatic exacerbation of proteinuria and an increase in neutrophil infiltration, again suggesting a protective role for NO [86]. However, in both these experiments with nonselective NO inhibitors, there were significant increases in mean arterial blood pressure, which complicates interpretation of the results. Similarly, experiments with L-NAME in passive [87] and active [88] Heymann nephritis, which resulted in exacerbation of proteinuria, have been complicated by hypertension. This lack of availability of selective NOS inhibitors has delayed progress in studying the role of NO in acute glomerular injury, for hypertension is known to exacerbate immune complex injury [89] and is a constant feature of inhibition of endothelial-derived NO. There is a preliminary report on the use of aminoguanidine, an arginine analog more selective for iNOS than eNOS, in autoimmune glomerulonephritis induced by mercuric chloride [90]. Aminoguanidine profoundly enhanced proteinuria when administered during the effector phase of the disease but was not effective in the induction phase, supporting this increasing evidence for a NO-protective role in glomerular immune injury.

In an attempt to overcome the adverse effects of competitive NO inhibitors on constitutive NO synthesis, we have performed studies to determine whether a more selective form of NO inhibition can be achieved by acute arginine depletion, as the high output of NO resulting from iNOS, particularly in macrophages, is dependent on L-arginine concentration. We have shown that the enzyme arginase is intimately linked with NO production through the availability of the common substrate L-arginine, for NO inhibition will increase urea production in nephritic glomeruli [91], and exogenous arginase can inhibit NO production in both macrophages [92] and nephritic glomeruli [93]. Injection of bovine liver arginase in rats causes rapid depletion of plasma L-arginine levels (Fig. 15-8, upper panel). In accelerated nephrotoxic nephritis, this was accompanied by exacerbation of proteinuria and reduced glomerular nitrite synthesis; when this was studied in a unilateral form of this model, there was no systemic hypertension, indicating an intact constitutive NOS system, but exacerbation of proteinuria still occurred (Fig. 15-8, lower panel) [93]. These results suggest, as in the competitive NO inhibitor experiments discussed above, that NO also has a protective role in the autologous (macrophage-dependent) phase of nephrotoxic injury. We found no influence of this treatment on leukocyte infiltration, which again suggests that the principal effects are hemodynamic.

Clearly, these early studies on NO manipulation in vivo have produced conflicting results which are far from defining how NO is involved in immune glomerular injury. However, the evidence for NO induction in nephritic glomeruli is now indisputable and major pathological roles are being defined for NO in other tissues. We believe that further research will reveal a significant role for NO in the pathogenesis of glomerulonephritis.

Acknowledgments

The authors acknowledge colleagues in the Department of Histopathology at St. Mary's Hospital Medical School, London and at the Wellcome Research laboratories, Beckenham, Kent and support for this work from the Medical Research Council, The National Kidney Research Fund, and The Wellcome Trust.

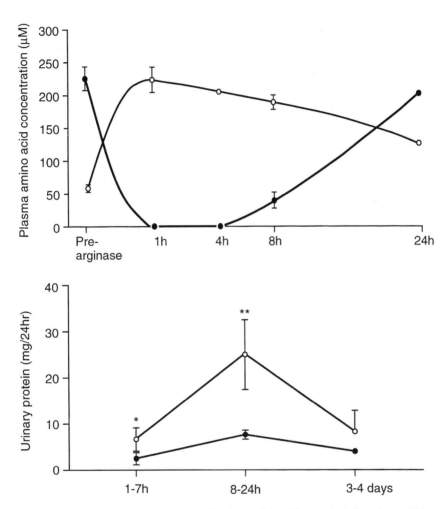

Figure 15-8. Upper panel: Plasma levels of L-arginine (closed circles) and L-ornithine (open circles) following an intravenous dose of bovine liver arginase (2500 units) in the rat. *Lower panel*: Effect of arginase on urinary protein 1–7 h, 8–24 h, and 3–4 days after induction of unilateral accelerated nephrotoxic nephritis. Rats received IV 2500 units arginase (open circles) or saline (closed circles) 1 h before intrarenal nephrotoxic globulin and 1250 units arginase or saline 8 h later. *p < .05, **p < .01.

Note Added in Proof

Sakurai et al (Biochim Biophys Acta 1316: 132–138, 1996) have studied the role of NF-kappa B in the induction of iNOS. In WKY rats with nephrotoxic nephritis there was increased NF-kappa B DNA-binding activity in glomeruli on day 1 with maximal activation at days 3–5. Pyrrolidine dithiocarbamate (PDTC), a potent inhibitor of NF-kappa B activation, inhibited the increase in NF-kappa B DNA-binding activity, inhibited transcription of IL-1 beta, MCP-1 and iNOS and prevented proteinuria.

References

1. McCluskey, R.T. In: *Pathology of the Kidney*, Ed. Heptinstall, R.H. Little, Brown and Company, Boston, 1983, Vol. 3, pp. 335–336.

2. Shah, S.V. *Annu. Rev. Physiol.* **57**, 245–262 (1995).

3. Moncada, S. and Higgs, E.A. *FASEB J.* **9**, 1319–1330 (1995).

4. Cattell, V. *Kidney Int.* **45**, 945–952 (1994).

5. Cook, H.T., Smith, J., Salmon, J.A., and Cattell, V. *Am. J. Pathol.* **134**, 431–437 (1989).

6. Stuehr, D.J. and Nathan, C.F. *J. Exp. Med.* **169**, 1543–1545 (1989).

7. Marletta, M.A., Yoon, P.S., Iyengar, R., Leaf, C.D., and Wishnok, J.S. *Biochemistry* **21**, 8706–8711 (1988).

8. Hibbs, J.B., Taintor, R.R., Vavrin, Z., and Rachlin, E.M. *Biochem. Biophys. Res. Commun.* **157**, 87–94 (1988).

9. Langrehr, J.M., Hoffman, R.A., Lancaster, J.R., Jr., and Simmons, R.L. *Transplantation* **55**, 1205–1212 (1993).

10. Schleifer, K.W. and Mansfield, J.M. *J. Immunol.* **151**, 5492–5503 (1993).

11. Gregory, S.H., Wing, E.J., Hoffman, R.A., and Simmons, R.L. *J. Immunol.* **150**, 2901–2909 (1993).

12. Liew, F.Y. *Curr. Opin. Immunol.* **7**, 396–399 (1995).

13. Taylor-Robinson, A.W., Liew, F.Y., Severn, A., et al. *Eur. J. Immunol.* **24**, 980–984 (1994).

14. Henderson, S.A., Lee, P.H., Aeberhard, E.E., et al. *J. Biol. Chem.* **269**, 25239–25242 (1994).

15. Sicher, S.C., Vazquez, M.A., and Lu, C.Y. *J. Immunol.* **153**, 1293–1300 (1994).

16. Bogdan, C., Vodovotz, Y., Paik, J., Xie, Q., and Nathan, C. *J. Leukocyte Biol.* **55**, 227–233 (1994).

17. Cunha, F.Q., Moncada, S., and Liew, F.Y. *Biochem. Biophys. Res. Commun.* **182**, 1155–1159 (1992).

18. Mulligan, M.S., Hevel, J.M., Marletta, M.A., and Ward, P.A. *Proc. Natl. Acad. Sci. USA* **88**, 6338–6342 (1991).

19. Mulligan, M.S., Moncada, S., and Ward, P.A. *Br. J. Pharmacol.* **107**, 1159–1162 (1992).

20. Lin, R.F., Lin, T.-S., Tilton, R.G., and Cross, A.H. *J. Exp. Med.* **178**, 643–648 (1993).

21. Ialenti, A., Moncada, S., and Di Rosa, M. *Br. J. Pharmacol.* **110**, 701–706 (1993).

22. McCartney-Francis, N., Allen, J.B., Mizel, D.E., et al. *J. Exp. Med.* **178**, 749–754 (1993).

23. Weinberg, J.B., Granger, D.L., Pisetsky, D.S., et al. *J. Exp. Med.* **179**, 651–660 (1994).

24. Boughton-Smith, N.K., Evans, S.M., Hawkey, C.J., et al. *Lancet* **342**, 338–340 (1993).

25. Farrell, A.J., Blake, D.R., Palmer, R.M.J., and Moncada, S. *Ann. Rheum. Dis.* **51**, 1219–1222 (1992).

26. Hamid, Q., Springall, D.R., Riveros-Moreno, V., et al. *Lancet* **342**, 1510–1513 (1993).

27. Wei, X., Charles, I.G., Smith, A., et al. *Nature* **375**, 408–411 (1995).

28. MacMicking, J.D., Nathan, C., Hom, G., et al. *Cell* **81**, 641–650 (1995).

29. Cattell, V., Cook, T., and Moncada, S. *Kidney Int.* **38**, 1056–1060 (1990).

30. Schreiner, G.F., Cotran, R.S., Pardo, V., and Unanue, E.R., *J. Exp. Med.* **147**, 369–384 (1978).

31. Cattell, V., Smith, J., and Cook, H.T. *Clin. Exp. Immunol.* **79**, 260–265 (1990).

32. Ding, A.H., Nathan, C.F., and Stuehr, D.J. *J. Immunol.* **141**, 2407–2412 (1988).

33. Kosaka, H., Imaizumi, K., Imai, K., and Tyuma, I. *Biochim. Biophys. Acta* **581**, 184–188 (1979).

34. Cook, H.T. and Sullivan, R.S. *Am. J. Pathol.* **139**, 1047–1052 (1991).

35. Cattell, V., Lianos, E., Largen, P., and Cook, T. *Exp. Nephrol.* **1**, 36–40 (1993).

36. Cattell, V., Largen, P., de Heer, E., and Cook, T. *Kidney Int.* **40**, 847–851 (1991).

37. Cook, H.T., Ebrahim, H., Jansen, A.S., Foster, G.R., Largen, P., and Cattell, V. *Clin. Exp. Immunol.* **97**, 315–320 (1994).

38. Peten, E.P., Garcia-Perez, A., Terada, Y., et al. *Am. J. Physiol.* **263**, F951–F957 (1992).

39. Goto, S., Yamamoto, T., Feng, L., et al. *Am. J. Pathol.* **147**, 1133–1141 (1995).

40. Jansen, A., Cook, T., Taylor, G.M., et al. *Kidney Int.* **45**, 1215–1219 (1994).

41. Boyce, N.W., Tipping, P.G., and Holdsworth, S.R. *Kidney Int.* **35**, 778–782 (1989).

42. Cattell, V., Largen, P., de Heer, E., and Cook, T. In: *The Biology of Nitric Oxide. Part 2.* Eds. Moncada, S., Marletta, M.A., Hibbs, J.B., Jr., and Higgs, E.A. Portland Press, London, 1992, Vol. 1, pp. 257–261.

43. Couser, W.G. and Abrass, C.K. *Annu. Rev. Med.* **39**, 517–530 (1988).

44. Lovett, D.H. and Sterzel, R.B. *Kidney Int.* **30**, 246–254 (1986).

45. Pfeilschifter, J. and Schwarzenbach, H. *FEBS Lett.* **273**, 185–187 (1990).

46. Pfeilschifter, J., Rob, P., Mulsch, A., Fandrey, J., Vosbeck, K., and Busse, R. *Eur. J. Biochem.* **203**, 251–255 (1992).

47. Morrissey, J.J., McCracken, R., Kaneto, H., Vehaskari, M., Montani, D., and Klahr, S. *Kidney Int.* **45**, 998–1005 (1994).

48. Mohaupt, M.G., Elzie, J.L., Ahn, K.Y., Clapp, W.L., Wilcox, C.S., and Kone, B.C. *Kidney Int.* **46**, 653–665 (1994).

49. Ujiie, K., Yuen, J., Hogarth, L., Danziger, R., and Star, R.A. *Am. J. Physiol. Renal, Fluid Electrolyte Physiol.* **267**, F296–F302 (1994).

50. Bachmann, S., Bosse, H.M., and Mundel, P. *Am. J. Physiol.* **268**, F885–F898 (1995).

51. Granger, D.L., Hibbs, J.B., and Broadnax, L.M. *J. Immunol.* **146**, 1294–1302 (1991).

52. Nathan, C. and Xie, Q. *J. Biol. Chem.* **269**, 13725–13728 (1994).

53. Wang, Y. and Marsden, P.A. *Curr. Opin. Nephrol. Hyperten.* **4**, 12–22 (1995).

54. Noronha, I.L., Niemir, Z., Stein, H., and Waldherr, R. *Nephrol. Dial. Transplant.* **10**, 775–786 (1995).

55. Florquin, S. and Goldman, M. *Springer Semin. Immunopathol.* **16**, 71–80 (1994).

56. Assreuy, J., Cunha, F.Q., Liew, F.Y., and Moncada, S. *Br. J. Pharmacol.* **108**, 833–837 (1993).

57. Cattell, V., Smith, J., Jansen, A., Riveros-Moreno, V., and Moncada, S. *Transplantation* **58**, 1399–1402 (1994).

58. Border, W.A. and Noble, N.A. *Exp. Nephrol.* **2**, 13–17 (1994).

59. Sterzel, R.B., Schulze-Lohoff, E., and Marx, M. *Kidney Int.* **43** (Suppl. 39), S-26–S-31 (1993).

60. Largen, P.J., Tam, F.W.K., Rees, A.J., and Cattell, V. *Exp. Nephrol.* **3**, 34–39 (1995).

61. Sheffler, L.A., Wink, D.A., Melillo, G., and Cox, G.W. *J. Immunol.* **155**, 886–894 (1995).

62. Currie, G.A., Gyure, L., and Cifuentes, L. *Br. J. Cancer* **39**, 613–620 (1979).

63. Albina, J.E., Mills, C.D., Henry, W.L., Jr., and Caldwell, M.D. *J. Immunol.* **144**, 3877–3880 (1990).

64. Albina, J.E., Mills, C.D., Barbul, A., et al. *Am. J. Physiol.* **254**, E459–E467 (1988).

65. Daghigh, F., Fukuto, J.M., and Ash, D.E. *Biochem. Biophys. Res. Commun.* **202**, 174–180 (1994).

66. Ketteler, M., Border, W.A., Brees, D.K., and Noble, N.A. *J. Am. Soc. Nephrol.* **4**, 455 (1993).

67. Ketteler, M., Border, W.A., and Noble, N.A. *Am. J. Physiol. Renal, Fluid Electrolyte Physiol.* **267**, F197–F207 (1994).

68. Beckman, J.S. and Crow, J.P. *Biochem. Soc. Trans.* **21**, 330–334 (1993).

69. Ischiropoulos, H., Zhu, L., and Beckman, J.S. *Arch. Biochem. Biophys.* **298**, 446–451 (1992).

70. Rubbo, H., Radi, R., Trujillo, M., et al. *J. Biol. Chem.* **269**, 26066–26075 (1994).

71. Davenpeck, K.L., Gauthier, T.W., and Lefer, A.M. *Gastroenterology* **107**, 1050–1058 (1994).

72. Gaboury, J., Woodman, R.C., Granger, D.N., Reinhardt, P., and Kubes, P. *Am. J. Physiol. Heart Circ. Physiol.* **265**, H862–H867 (1993).

73. Salvemini, D., Misko, T.P., Masferrer, J.L., Seibert, K., Currie, M.G., and Needleman, P. *Proc. Natl. Acad. Sci. USA* **90**, 7240–7244 (1993).

74. Shultz, P.J. and Raij, L. *J. Clin. Invest.* **90**, 1718–1725 (1992).

75. Florquin, S., Amraoui, Z., Dubois, C., Decuyper, J., and Goldman, M. *J. Exp. Med.* **180**, 1153–1158 (1994).

76. Garg, U.C. and Hassid, A. *Am. J. Physiol.* **257**, F60–F66 (1989).

77. Mohaupt, M., Schoecklmann, H.O., Schulze-Lohoff, E., and Sterzel, R.B. *J. Hyperten.* **12**, 401–408 (1994).

78. Sarih, M., Souvannavong, V., and Adam, A. *Biochem. Biophys. Res. Commun.* **191**, 503–508 (1993).

79. Baker, A.J., Mooney, A., Hughes, J., Lombardi, D., Johnson, R.J., and Savill, J. *J. Clin. Invest.* **94**, 2105–2116 (1994).

80. Mulligan, M.S., Warren, J.S., Smith, C.W., et al. *J. Immunol.* **148**, 3086–3092 (1992).

81. Teixeira, M.M., Fairbairn, S.M., Norman, K.E., Williams, T.J., Rossi, A.G., and Hellewell, P.G. *Br. J. Pharmacol.* **113**, 1363–1371 (1994).

82. Walder, C.E., Thiemermann, C., and Vane, J.R. *Br. J. Pharmacol.* **102**, 967–973 (1991).

83. Radermacher, J., Klanke, B., Schurek, H-J., Stolte, H.F., and Frolich, J.C. *Kidney Int.* **41**, 1549–1559 (1992).

84. Narita, I., Border, W.A., Ketteler, M., and Noble, N.A. *Lab. Invest.* **72**, 17–24 (1995).

85. Ferrario, R., Takahashi, K., Fogo, A., Badr, K.F., and Munger, K.A. *J. Am. Soc. Nephrol.* **4**, 1847–1854 (1994).

86. Munger, K.A., Fogo, A., Nassar, G., and Badr, K.F. *J. Am. Soc. Nephrol.* **5**, 588 (1994).

87. Kaysen, G.A., Martin, V.I., and Jones, H., Jr. *Am. J. Physiol.* **263**, F907–F914 (1992).

88. Tikkanen, I., Uhlenius, N., Tikkanen, T., et al. *J. Am. Soc. Nephrol.* **5**, 594 (1994).

89. Neugarten, J., Feiner, H., Schact, R.G., Gallo, G.R., and Baldwin, D.S. *Kidney Int.* **22**, 257–263 (1982).

90. Van der Meide, P.H., Groenestein, M.J., De Labie, M.C.D.C., Aten, J., and Weening, J.J. Abstracts of the XIII th International Congress of Nephrology 1995, p. 303 (abstract).

91. Cook, H.T., Jansen, A., Lewis, S., et al. *Am. J. Physiol.* **267**, F646–F653 (1994).

92. Hibbs, J.B., Jr., Vavrin, Z., and Taintor, R.R. *J. Immunol.* **138**, 550–565 (1987).

93. Waddington, S., Cook, H.T., Reaveley, D., Jansen, A., and Cattell, V. *Kidney Int.* **49,** 1090–1096 (1996).

94. Cattell, V. and Cook, H.T. *Exp. Nephrol.* **1**, 265–280 (1993).

16

NO in Acute Renal Failure

John D. Conger

1. Introduction

Over a decade ago, it was noted that there was an absence of kidney vasodilator response to intrarenal acetylcholine (ACh) in a rat model of ischemic acute renal failure (ARF) [1]. This was the first evidence suggesting the possibility that renovascular endothelium-derived relaxing factor activity (EDRF) may be abnormal following an ischemic insult to the kidney. Subsequent studies have shown that ischemia, ischemia–reperfusion, and hypoxia cause altered endothelial function in several vital organs, including brain, limb, lung, and heart, as well as the kidney [2–7]. Since its identification as the principal active component of EDRF [8], there has been continuing interest in determining what role nitric oxide (NO) might play in both ischemic and nephrotoxic ARF. What is apparent from these studies of NO in ARF is that the relationship is much more complex than a simple loss of endothelial NO activity. First, basal constitutive nitric oxide synthase (cNOS) and inducible nitric oxide synthase (iNOS) generated NO at the time of ARF induction may have a substantial impact on the subsequent severity of renal injury. Second, endothelial NOS (eNOS) and NO activity in the early and established post-ARF induction phases may vary from reduced to increased and, in either circumstance, have an important effect on associated renal hemodynamics. Finally, whereas the vascular effects of NO in ARF are presumed to be protective, there is recent evidence that NO generated during hypoxia in renal tubules may actually contribute to cell injury and death. Table 16-1 summarizes the reported estimates of NO activity in ARF models. Thus, with our present level of understanding, the role of NO in ARF must be viewed as highly variable, depending on whether one is addressing the vasculature or tubules and at what point in the course of disease one is interested. This review is organized in such a manner as to account for these differing potential effects of NO in ARF.

Table 16-1. Summary of Presumed NO Activity in Acute Renal Failure Models

Model	Phase	Vascular	Tubular
Ischemia	Initiation	↑ or ↓	↑
Ischemia	Early established	↓	↑
Ischemia	Late established	↑	—
Hypoxia	Initiation	—	↑
Oxidant stress	Initiation	—	↑
Cyclosporine A	Initiation	N1 (normal)	—
Cyclosporine A	Established	↓	—

As preliminary information, the various methods of assessing NOS/NO will be discussed, because this is vital to interpreting the results of various studies. Next, data will be presented regarding vascular NOS activity. Its role in the initiation and evolution of ARF will be considered first, followed by evidence of its participation in the hemodynamics of the early and later established phases of this disorder. Finally, the tubular toxicity of NO in hypoxia–reoxygenation and oxidant stress injuries will be examined.

2. Measurement of the NOS/NO System

Nitric oxide is a radical with a molecular life span of less than 10 s. Moreover, its primary route of action, at least as an endothelial-generated substance, is as a paracrine hormone. Thus, NO is difficult to measure directly in in vivo systems. Indirect assessment techniques, therefore, have been developed which take advantage of the defined steps in NO generation and action (i.e., L-arginine→NOS→ NO→cGMP).

L-Arginine can be measured in plasma and tissue to determine adequacy of NOS substrate in NO production [9]. Frequently, L-arginine has been given in both in vivo and in vitro test systems [10]. This maneuver tests for the presence of NOS activity, as well as increases the production of NO. Nitric oxide synthase activity and NO generation can be blocked with a variety of L-arginine analogs (L-NMMA, L-NNA, L-NAME, NLA) [11]. Aminoguanidine, a nucleophilic hydrazine, is a putative selective inhibitor of inducible NOS [12]. Nitric oxide synthase activity can be inferred from the biologic effects or end-product measurements of L-arginine addition. L-citrulline is a by-product in NO generation that can be determined in in vitro systems. Endothelium-dependent vasodilators such as ACh, bradykinin, histamine, and thrombin stimulate endothelial NOS [11]. Immunoblot estimates of tissue content of NOS isoenzymes can be performed in vitro and Northern blots can detect tissue messenger RNA [13–15].

Using in vitro preparation, there is a limited experience with direct NO measurement using electron-sensing microprobes [16,17]. However, the more common, indirect measurements of NO rely on detecting the relatively stable NO metabolites NO_2 and NO_3 [18]. Because the major fraction of excreted NO_2/NO_3 is derived from glomerular filtration and not generated in the kidney and because NO_2/NO_3 are variably reabsorbed by the tubules, urinary measurements may be of value in estimating systemic NO generation but have limited sensitivity in relation to renal NO production [19]. Inhibitors of NO include scavenging agents such as hemoglobin and Methylene Blue [20]. NO life span can be prolonged with superoxide dismutase, which reduces available superoxide radical to react with NO [21]. Superoxide dismutase can be used to increase NO detection by other methods.

Urinary and plasma cGMP levels are increased when there is increased NO stimulation of target cells such as smooth muscle [22]. Like NO_2/NO_3, measurement specificity of cGMP in vivo is not optimal and interpretation of results must be cautious. Nitric oxide donors, such as sodium nitroprusside and glyceryl-trinitrate, provide NO directly and, therefore, test cGMP generation and its cell-dependent effects in smooth muscle or other target cells [23].

Because nitric oxide generation and its cellular effects involve several steps and the measurements of these intermediates largely represent indirect reflections of NO activity, caution is required in interpreting the results. This principle must be kept in mind, particularly when cause and effect relationships are sought or suggested by individual studies.

3. Vascular NO in the Initiation and Evolution of ARF

In all forms of hemodynamically induced and several forms of nephrotoxic acute renal failure, a substantial reduction in renal blood flow is a critical factor in the initiation of renal injury. Intrinsic mechanisms in the renal vasculature protect against these major declines in renal perfusion. The importance of vasodilator prostaglandins in this regard has been well documented [24]. More recently, the paracrine activity of NO generated through constitutive eNOS has been shown to participate in the protective response when renal perfusion pressure and blood flow are compromised. Pretreatment of rats with inhibitors of eNOS aggravates ARF induced by renal artery clamping (RAC) [25,26]. Similarly, when salt-depleted rats are given NOS inhibitors prior to exposure to radiocontrast agents, subsequent declines in renal plasma flow and glomerular filtration rate are significantly greater than in control rats [27]. Glycerol-induced ARF in rats also has been shown to be aggravated when the NOS inhibitor NLA was given at the time of ARF induction [28].

In other experiments, L-arginine treatment has been used in models of ARF. Schramm et al. [29] gave L-arginine 300 mg/kg, L-monoethylarginine (MeArg) 30 mg/kg, and a combination of the two intravenously over 60 min, after 40 min bilateral renal artery occlusion. L-Arginine or a combination of L-arginine and

MeArg improved inulin clearance at 180 min after renal artery clamp release compared to controls or MeArg alone. Similar results were obtained in a rat model of uranyl nitrate compared to control. L-Arginine was shown by Maree et al. [28] to ameliorate glycerol-induced ARF when infused immediately after giving intramuscular glycerol.

The above studies examined the consequences of manipulation of constitutive NO generation by the kidney during induction in ischemic and nephrotoxic ARF models. Although these data are of interest, they do not address the role of native renal NO production and its effect in ARF initiation in a cause and effect manner. Increasing NO generation by supplying L-arginine or blocking NO production with NOS inhibitors have positive and negative effects, respectively, on renal function during and immediately after ARF induction. However, these are not necessarily ARF specific effects. Analogous results have been found when other endogenous vasodilators (e.g., prostaglandins) are inhibited or stimulated during ARF induction [31–33]. In general, treatment with exogenous substances that produce renal vasoconstriction will worsen and those that produce renal vasodilation will improve postinduction kidney function in ARF. Thus, altering renal hemodynamics by changing constitutive NOS activity does not confirm that native NOS/NO have unique and specific roles in ARF induction.

There are other data, albeit indirect, that suggest a vascular injury process as part of ischemically induced ARF may interfere with constitutive eNOS activity and NO generation. In several vascular beds, it has been shown that hypovolemic shock or ischemia–reperfusion impairs subsequent endothelial function. Lieberthal et al. [34] have shown in an isolated perfused kidney model that renovascular resistance is increased immediately following 25 min of ischemia. Endothelium-independent vasodilators such as atrial natriuretic peptide (ANP) and sodium nitroprusside prevented the increase in renovascular resistance (RVR). However, endothelium-dependent vasodilators (i.e., those that stimulate endothelial NO generation, including ACh and A23187) did not prevent the increase in RVR as illustrated in Table 16-2. Similar observations have been made by Cristol et al. in anesthetized rats [35]. The increase in basal vascular tone following ischemia is thought to be due to loss of constitutive eNOS activity which, under normal physiological conditions, modulates vascular tone both directly and by attenuating endogenous vasoconstrictor agonist activity [36]. The precise nature of the early-phase functional endothelial injury in hypovolemic shock and ischemia–reperfusion is not known. There is some evidence that there may be a loss of L-arginine substrate from endothelial cells, as L-arginine treatment can attenuate the increased vascular tone [37]. There are, however, no tissue measurements of L-arginine. Other evidence suggests that superoxide radical formed during reperfusion scavenges endothelial NO [38]. Other possibilities, including decreased NOS expression, have not been adequately explored.

Another mechanism of shock leading to ischemic ARF also may involve NO. This is the somewhat controversial role of NO in the hypotension of endotoxemia.

Table 16-2. Effect of Endothelial-Dependent Vasodilators on Renal Hemodynamics Following Ischemia

	n	Renal Perfusate Flow (ml/min/g)	Renal Vascular Resistance (mm Hg/ml/min/g)	GFR (ml/min/g)
Control	7	9.7±0.05	10.2±0.8	0.51±0.07
Ischemia	8	6.1±0.6*	16.7±1.4*	0.20±0.04*
Control + acetylcholine	3	10.7±1.3	9.8±1.3	0.42±0.14
Ischemia + acetylcholine	5	5.5±0.4**	19.2±1.4**	0.10±0.03**
Control + A23187	3	11.3±0.4	8.9±0.6	0.42±0.10
Ischemia + A23187	5	5.3±0.4***	16.0±1.0***	0.12±0.06***

Note: Values are means ± SE; n = No. of rats; GFR = glomerular filtration rate.

*$p < .005$ compared with control.

**$p < .02$ compared with control + acetylcholine.

***$P < .02$ compared with control + A23187.

Source: Reprinted with permission from Ref. 34.

A number of studies suggest that endotoxic shock is the result of excessive release of NO into the blood with subsequent massive vasorelaxation [39,40]. There is in vitro evidence indicating that lipopolysaccharide (LPS) antigen can stimulate inducible NOS in macrophages and smooth muscle cells through tumor necrosis factor-α or interleukin-1β. In support of a role of NO generated through iNOS in endotoxic shock is the observation that NOS inhibitors such as L-NMMA can limit the hypotension of endotoxic shock [39]. Against a role of NO in endotoxic shock is the time lag of several hours for LPS to induce macrophage and smooth muscle NOS. This would not account for the early hypotension observed after endotoxin exposure [41]. In addition, circulating NO, with its half-life in blood of only 6–10 s, does not seem a likely candidate to produce severe systemic hypotension. This issue is considered in greater detail in Chapters 8 and 14. A related observation by Shultz and Raij [42] suggests that NO production in endotoxic shock, whether generated through iNOS or eNOS, is critical to protection from glomerular thrombosis seen in this disorder [43]. They found that treatment with L-NAME markedly increased thrombi within glomerular capillary loops after LPS exposure, as illustrated in micrographs in Fig. 16-1. Glomerular thrombosis could be attenuated with L-arginine pre-treatment.

In a nephrotoxic form of ARF induced with intravenous cyclosporine A, a renal vasoconstrictor response has been observed. However, a recent study by Conger et al. [44] using intrarenal L-NAME indicated that NO generation was present despite renal vasoconstriction. The author is unaware of other studies of endogenous NOS activity during ARF induction in nephrotoxic models.

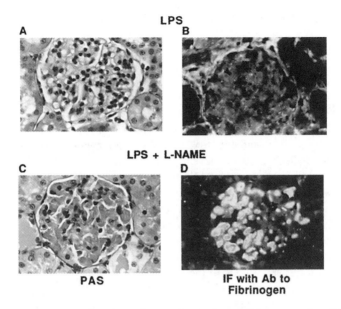

Figure 16-1. Representative photomicrographs of glomeruli from rats given LPS alone (A and B) or LPS + L-NAME (C and D). Kidneys were removed at the end of the 8-h experimental period, and portions were examined under light microscopy with PAD staining (A and C) or processed for immunofluorescence with FITC-labeled antibody to fibrinogen (B and D). Glomeruli from animals given LPS alone showed no significant changes from normal and negative fluorescence, whereas those given LPS + L-NAME showed extensive thrombi within the glomerular capillary loops and markedly positive fluorescent staining (original magnification: 400 ×). (Reprinted with permission from Ref, 42.).

4. Vascular NO in the Established Phase of ARF

The preceding studies in the evolutionary phase of ARF and in ischemia–reperfusion vascular injury in other organs predicted that eNOS/NO activity in the established phase of ARF would be reduced. This was, of course, supported by the repeated observation in experimental models that renal blood is reduced in the early phase of ARF. This has not turned out to be the case.

Our laboratory investigated NOS/NO in the norepinephrine (NE)-induced models of ARF in rats at 48 h [45] and 1 week [46] into the disease course. This model is induced by intrarenal infusion of NE for 90 min during which renal blood flow is reduced to approximately 10% of normal. The animals are returned to metabolic cages. Thereafter, renal function, hemodynamics, and vascular reactivity are determined at various times after ARF induction. Rats that have been similarly treated but received intrarenal saline vehicle rather than NE served as controls.

At 48 h, basal renal blood flow in NE-ARF rats was 40% less than in control rats ($p < .01$). Glomerular filtration rate was 80% less than in controls ($p < .01$). There was an increased vasoconstrictor sensitivity to intrarenal angiotensin II and endothelin-1 in the NE-ARF rats compared to controls ($p < .02$). There was a decreased vasodilator response to intrarenal infusion of endothelium-dependent ACh in the NE-ARF rats ($p < .01$), but a comparable vasorelaxation to that in control rates when endothelium-independent prostacyclin (PGI_2) was infused. Although these findings would have been consistent with a loss of eNOS activity, the response to intrarenal L-NAME was inconsistent with this interpretation. When L-NAME was infused into the kidneys of NE-ARF rats, there was a marked vasoconstrictor response which exceeded that observed in control rats ($p < .01$). This finding indicated that eNOS/NO activity, in fact, was present and was not the pathophysiologic basis for the decreased basal renal blood flow at 48 h. The lack of renal vasodilation to ACh coupled with an exaggerated vasoconstrictor response to L-NAME suggested the possibility that eNOS/NO activity actually was maximal under basal conditions and could not be stimulated further.

Similar studies have been performed in NE-ARF rats at 1 week into the disease course. At this time, basal renal blood flow was similar to or only slightly less than in control animals [46]. However, glomerular filtration rate continued to be reduced at approximately 35% of controls ($p < .001$). As was observed at 48 h, there was a lack of vasodilator responses to ACh and BK (bradykinin), but an increased constrictor response to L-NAME in NE-ARF kidneys compared to controls. Figure 16-2 shows the renal blood flow responses to sequential intrarenal infusions of ACh and L-NAME in NE-ARF kidneys. Vasodilator response to PGI_2 was similar to controls. Aminoguanidine, an iNOS inhibitor, had no effect on renal blood flow when infused into the kidney of NE-ARF rats. When the NO donor sodium nitroprusside was given to NE-ARF kidneys, there was no vasodilator response, unlike that observed in control rat kidneys ($p < .001$). Taken together, these pharmacologic hemodynamic experiments indicated that NOS activity of the constitutive type was present in the later established phase in an experimental model of ischemic ARF. In addition, these experiments suggested that NOS activity was maximal as was the case at 48 h. Not only was it impossible to stimulate eNOS further with endothelium-dependent vasodilator, but it was also not possible to stimulate smooth muscle cells cGMP further with exogenous infused NO donors. In this same study of 1-week postischemic vascular reactivity, direct evidence of eNOS was sought by immunologic methods. Using a mouse monoclonal antibody, immunohistochemical and immunofluorescence detection of cNOS in the arterial vasculature of 1-week control and NE-ARF kidneys were performed. The intensity of endothelial staining or fluorescence for cNOS was at least as strong in NE-ARF as in sham-ARF arterial microvessels, as illustrated in Fig. 16-3 [46].

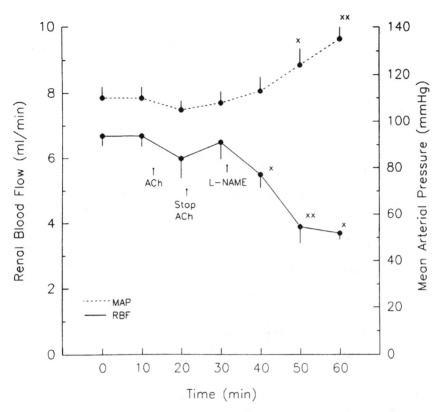

Figure 16-2. Effects of sequential intrarenal infusion of ACh followed by intravenous administration of L-NAME on MAP and RBF in NE-ARF rats. ACh did not have significant effects on either parameter. However, L-NAME caused a significant increase in MAP and a significant decrease in RBF. *Different from baseline at $p > .01$. **Different from baseline and preceding value at $p < .01$. (Reprinted with permission from Ref. 46.)

An interesting aspect of the above observations in established ARF in the norepinephrine model is the pathophysiologic process that underlies basal vascular tone. At 48 h, renovascular resistance is increased by approximately 40%. At the same time, eNOS activity appears to be present, if not maximal. This combination of findings suggests that there is a potent postischemic vasoconstrictor mechanism operating, to which increased eNOS/NO activity may be a protective response. At 1 week, when basal renovascular resistance is only slightly increased, and approaches normal, eNOS/NO activity is still maximal and the observed effect of L-NAME indicated that there remains a strong vasoconstrictor stimulus to which eNOS/NO continues to respond. What is the nature of the postischemic vasoconstrictor mechanism? To date, this question is unanswered. Several possibilities exist, including a smooth muscle cell membrane defect that

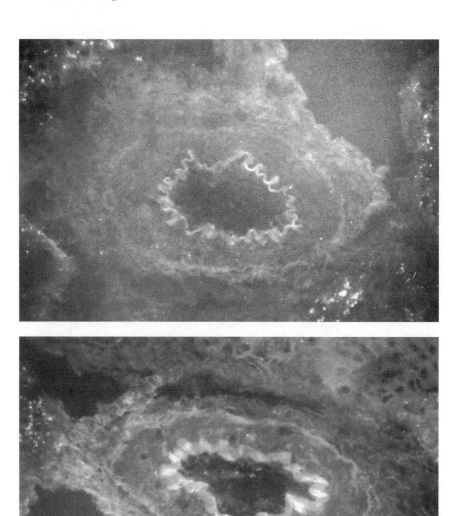

Figure 16-3. Immunofluorescence demonstration of representative sections of distal interlobular arteries incubated with mouse anti-NOS monoclonal antibody. *Top*: Anti-NOS in sham-ARF kidney; *Bottom*: anti-NOS in NE-ARF kidney. The intersity of anti-NOS immunofluorescence was at least similar, if not greater, in the endothelium of NE-ARF compared with sham-ARF kidneys. (Reprinted with permission from Ref. 46.)

allows increased calcium entry or defective calcium extrusion, an altered smooth muscle cell membrane potential, and persistent autocrine or paracrine activity of vasoconstrictor agonists such as endothelin-1, constrictor prostaglandins and other eicosanoids, angiotensin II, or other as yet unidentified humoral substances. Sorting out these possibilities obviously will require further investigation.

There are no data currently regarding vascular NOS/NO activity in the established phase of nephrotoxic models of ARF, with the exception of cyclosporine A (CsA). Even in the studies on CsA nephrotoxicity, it is uncertain whether these models are more representative of acute or chronic CsA renal toxicity. Unlike acute, single-dose CsA treatment in which the vasoconstrictor response was questionably due to decreased eNOS activity, 1-week CsA-exposure studies in animal models have found consistent evidence for impaired endothelium-dependent vasodilation [47–49]. The findings of partial sparing of the endothelium-dependent vasodilator response by administering L-arginine [10,50,51] and the lack of vasoconstrictor response to NOS inhibitors have suggested that continual CsA treatment depresses the endothelial L-arginine/eNOS/NO vasodilator pathway. In support of a primary endothelial abnormality was the study by Gallego et al. [50] in rats given CsA 25 mg/kg for 2 weeks. Like others [10,52], they found a decreased vasodilator response to ACh, in arterial rings, which was partially corrected by L-arginine treatment. Response to the NOS inhibitor NLA was suppressed. However, vasorelaxation to the calcium ionophore A23187, which directly stimulates NOS, and to sodium nitroprusside (SNP) (an NO donor) were normal. Recent experiments carried out in cortical neuron cell cultures have suggested that, in fact, there may be a direct link between CsA and NOS inhibition. Cyclosporine A, coupled with its intracellular receptor cyclophilin, inhibited the serine/threonine phosphatase calcineurin [53]. Calcineurin inhibition by CsA has been shown by Dawson et al. [54] to prevent dephosphorylation activation of brain NOS.

A number of experiments examining the longer-term effects of CsA treatment (2 weeks and longer) on renal vascular resistance have shown significantly different renovascular responses. In addition to an "apparent" decline in constitutive NOS activity, there was a depressed vasodilator response to SNP. Results from the author's laboratory [55] in a rat model given CsA, 5 mg/kg per day, for 2 weeks showed not only a reduced response to ACh and the NOS inhibitor L-NAME but also a significantly attenuated response to SNP as well, despite a normal vasodilator response to PGI_2. Rego et al. [56] found decreased vasodilator response to endothelium-dependent ACh, BK, and A23187, as well as to the endothelium-independent SNP, in the mesenteric vascular bed and thoracic aortic rings from rats treated with CsA, 5–10 mg/kg daily, for 3 weeks. Similar observations were made by Roullet et al. [52] in the mesenteric resistance arterial vessels from rats. Gerkens [47] found a decrease in renal vasodilator response to both ACh and SNP in rats treated for 3 weeks with 5–20 mg/kg CsA daily. Huand et al. [57] reported that SNP-induced vascular relaxation was impaired in the

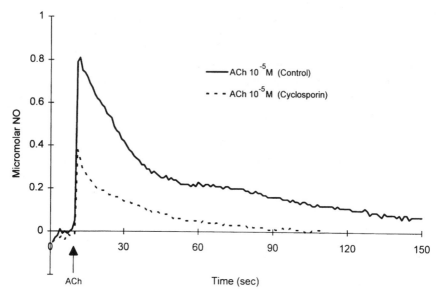

ACh-evoked release of NO from control and cyclosporin-treated rat renal arterioles

Figure 16-4. Nitric oxide response as measure with NO-selective microsensor positioned at the open end of a perfused afferent arteriole. In a series of such vessels, the NO response to ACh was significantly less in arterioles from rats treated with daily cyclosporine A injections. (From Ref. 59 with permission.)

mesenteric vasculature of rabbits given 5–20 mg/kg CsA daily for 7 weeks. Bossaller et al. [58] showed a decreased renal vasodilator response to SNP at 8, but not 3, weeks after continual CsA treatment in rats. Despite the fact that these studies point to a CsA-induced defect in smooth muscle cGMP with time, Conger et al. [59] have shown (Fig. 16-4) that ACh-stimulated NO release, measured with a sensitive microelectrode, is reduced in afferent arterioles isolated from 2-week CsA-treated rats. Collectively, these data suggest that in 1–3-week models of CsA exposure there are systemic and renal defects in vasorelaxation that are more complex than loss of endothelial NOS activity alone and, likely, involve abnormalities of smooth muscle cells as well.

5. NO in the Tubular Injury of ARF

To this point, the discussion in this chapter has focused almost entirely on NOS/NO activity of the vasculature in the events initiating and maintaining ischemic and nephrotoxic forms of ARF. Moreover, the presumed role of NO, with the exception of its participation in the hypotension of endotoxic shock, was considered to be protective. However, recent evidence indicates that NO, in fact, may mediate cell injury in a number of tissues, including the renal tubules. NO has been found to be cytotoxic to rat hepatoma cells [60], neuronal cells [61–63], and to the respiratory epithelium [64]. Furthermore, NO appears to be responsible for brain or heart ischemic damage after cerebral occlusion [65] and myocardial infarction [66,67], respectively. Most recently, NO cytotoxicity has been described in renal tubular epithelial cells. Yu et al. [68] showed that NOS activity is increased during hypoxia in freshly isolated rat proximal tubules. They found that cell membrane damage was prevented by a NOS inhibitor. A further increase in hypoxic injury was observed when the NOS substrate, L-arginine, was added to the hypoxic proximal tubules suspension as shown in Fig. 16-5.

The mechanism of NO-mediated injury during hypoxia is unknown. However, there are a number of possibilities. NO has been shown to cause nuclear DNA fragmentation which, in turn, activated nuclear poly(ADP)ribose synthase (PARS) [69]. Enhanced activation of PARS promotes ribosylation of nuclear and cytoskeletal (F-actin) proteins utilizing NAD as a substrate. Cellular ATP is then shunted to regenerate NAD, an event which may contribute to cell injury because of further decrements in already depleted energy stores during hypoxia [69]. NO also may cause membrane damage either by direct attack or by combining with O_2^- to form the toxic radical, $ONOO^-$, which can attack cellular membranes by lipid peroxidation [70].

Whereas the collective evidence indicates that NO is cytotoxic, the enzymatic form of NOS that mediates NO generation appears to differ depending on the renal cell injury model. Constitutive and inducible forms of NOS have been identified in renal tubular cells other than the vasculature including macula densa cells (cNOS), inner medullary collecting ducts (cNOS and iNOS), and proximal tubules (cNOS and iNOS) [71,72]. In the studies of hypoxic renal tubules, Yagoob et al., [73] presented data consistent with constitutive eNOS and neuronal (n)NOS activation by increased cellular calcium as the mechanism for increased renal tubular NO and cell injury. Aminoguanidine (inhibitor of iNOS) did not prevent hypoxia-induced NO release. Immunoblotting of the tubular tissue showed detectable eNOS and nNOS, but not the macrophage form of iNOS. In contrast, Peresleni et al. [74], using a model of BSC-1 kidney tubular epithelial cell oxidative stress injury, found that antisense oligodeoxynucleotides (AS-ODN) to iNOS prevented generation of NO in response to H_2O_2. The effect was isoform specific, as AS-ODN-pretreated cells retained functional eNOS as judged from the ability to generate NO in response to ionomycin. The potential toxic role of NO in hypoxic and oxidant stress injuries to renal parenchymal cells and the underlying mechanisms of specific enzyme upregulation will require further examination. In addition, their precise relationship to in vivo injury events will have to be determined.

Noiri et al. [75] have utilized AS-ODNs in vivo. Rats pretreated with an AS-ODN construct hybridizing with iNOS RNA/DNA protected the kidneys against

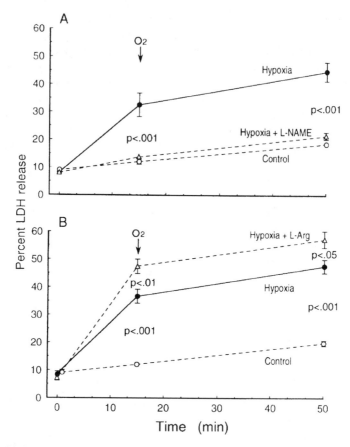

Figure 16-5. (A) Hypoxia for 15 min and reoxygenation for 35 min increased LDH release in H/R tubules (•) compared with control tubules (O). L-NAME completely prevented the H/R injury (△). O$_2$ was reintroduced at 15 min as shown by the arrow. (B) Addition of L-arginine to H/R tubules enhances LDH release (△) as compared with (•) and control (O) tubules. O$_2$ was reintroduced at 15 min as shown by the arrow. Values are mean ± SE. (Reprinted with permission from Ref. 68.)

acute ischemia. The functional protection was associated with reduced morphologic signs of tubular injury, suggesting that the macrophage-specific isoform of iNOS is responsible for the generation of cytotoxic NO.

6. Summary and Conclusions

Much remains uncertain about the role of NO in acute renal failure. What is known at this point comes from animal models and in vitro preparations. Adding to the frustration of understanding how NOS/NO participates in ARF is the short life span of NO and its predominant paracrine and autocrine activity, factors that make direct measurement problematic and indirect assessments difficult to interpret.

In the initiation phase of ischemic ARF, available information suggests that endothelial NOS activity is reduced and may contribute to the severity of ischemic injury. The evidence regarding vascular NO in nephrotoxic ARF is mixed. Although inhibiting eNOS can worsen nephrotoxic forms of ARF, this does not prove that its activity and NO generation are part of the pathophysiologic process of renal injury. The initial vasoconstrictor response that may contribute to CsA-induced ARF does not appear to depend on a decrease in eNOS/NO generation. The role of increased iNOS and NO generation in the hypotension and renal failure of endotoxic shock is controversial. However, eNOS/NO in the renal vasculature appears to protect glomeruli from thrombosis in this disorder.

In the established phase of ischemic ARF, endothelial NOS/NO is likely increased, rather than reduced, as determined by pharmacologic and immunologic studies in an NE-ARF model in rats. The findings in established ischemic ARF suggest that the increased eNOS/NO may be a protective response to an underlying potent local postischemic vasoconstrictor stimulus within the kidney. The only data regarding the established phase of nephrotoxic ARF come from studies of continual short-term exposure to CsA. The findings in this model indicate that the L-arginine/NOS/NO/cGMP activation sequence is impaired. However, it is unclear whether the primary defect resides in the endothelium or smooth muscle cells of the vasculature.

Recent evidence indicates that NO generated in proximal tubular epithelium may actually contribute to the cellular injury of hypoxia and oxidant stress. This interesting observation suggests that renal NO may be contributing to both protection and injury in ARF.

Additional studies need to be carried out to determine the roles of NO in ARF—both protective and toxic. It is important, however, that the relative contributions of these mechanisms ultimately be understood in the context of whole organ pathophysiology so that therapeutic ventures are well founded and do not run the risk of inducing overall negative in vivo effects.

References

1. Kelleher, S.P., Robinette, J.B., and Conger, J.D. *Am J Physiol.* **246**, F379–F386 (1984).
2. Mayhan, W.G., Amundsen, S.M., Faraci, F.M., et al. *Am J Physiol.* **255**, H879–H884 (1988).
3. Sternbergh, W.C., Makhoul, R.G., and Adelman, B. *Surgery* **114**, 960–967 (1993).
4. Adnot, S., Raffestin, B., Eddahibit, S., et al. *J. Clin. Invest.* **87**, 155–162 (1991).
5. VanBenthuysen, K.M., McMurty, I.F., and Horwitz, L.D. *J. Clin. Invest.* **79**, 265–274 (1987).

6. Quillen, J.E., Sellke, F.W., Brooks, L.A., et al. *Circulation* **82**, 586–594 (1990).

7. Conger, J.D., Robinette, J.B., and Schrier R.W. *J. Clin. Invest.* **82**, 532–537 (1988).

8. Palmer, R.M.J., Ferrige, A.G., and Moncada, S. *Nature (London)* **327**, 524–526 (1987).

9. Konings, C.H. *Clin. Chim. Acta* **176**, 185–194 (1988).

10. Gallego, M.J., Farre, A.L., Riesco, A., et al. *Am. J. Physiol.* **265**, H708–H718 (1993).

11. Moncada, S., Palmer, R.M.J., and Higgs, E.H. *Pharmacol. Rev.* **43**, 109–142 (1990).

12. Griffiths, M.J.D., Messent, M., MacAllister, R.J., et al. *Br. J. Pharmacol.* **110**, 963–968 (1993).

13. Pollock, J.S., Nakane, M., Buttery, D.K. et al. *Am. J. Physiol.* **265**, C1379–C1387 (1993).

14. Nakayama, D.K., Geller, D.A., Lowenstein, C.J., et al. *Am. J. Respir. Cell Mol. Biol.* **7**, 471–476 (1992).

15. MacNaul, K.L. and Hutchinson, N.I. *Biochem. Biophys. Res. Commun.* **196**, 1330–1334 (1993).

16. Gerhardt, G.A., Oke, A.F., Nagy, G., et al. *Brain Res.* **290**, 390–395 (1984a).

17. Gerhardt, G., Rose, G., and Hoffer, B.J. *J. Neurochem.* **46**, 842–850 (1986).

18. Shultz, P.J., Tayeh, M.A., Marletta, M.A. et al. *Am. J. Physiol.* **261**, F600–F606 (1991).

19. Suto, T.G., Losonezy, G., Qiu, C., et al. *J. Am. Soc. Nephrol.* **5**, 593A (1994).

20. Martin, W., Villani, G.M., Jothianandan, D., et al. *J. Pharm. Exp. Ther.* **232**, 708–716 (1985).

21. Kennedy T.P., Rao, N.V., Hopkins, C., et al. *J. Clin. Invest.* **83**, 1326–1335 (1989).

22. Chen, Y.P. and Sanders P.W. *J. Clin. Invest.* **88**, 1559–1567 (1991).

23. Ignarro, L.J. *Circ. Res.* **65**, 1–21 (1989).

24. Torsello, G., Schror, K., Szabo, A., et al. *Eur. J. Vasc. Surg.* **3**, 5–10 (1989).

25. Papadimitriou, M., Economidou, D., Vakiania, P., et al. *Nephrol. Dial. Transplant.* **9**, 82–87 (1994).

26. Chintala, M.S., Chiu, P.J., Vemulapalli, S., et al. *Naunyn Schmiedebergs Arch. Pharmacol.* **348**, 305–310 (1993).

27. Schwartz, D., Blum, M., Peer, G., et al. *Am. J. Physiol.* **267**, F374–F379 (1994).

28. Maree, A., Peer, G., Schwartz, D., et al. *Nephrol. Dial. Transplant.* **9**, 78–81 (1994).

29. Schramm, L., Heidbreder, E., Schmitt, A., et al. *Renal Failure* **16**, 555–569 (1994).

30. Schramm, L., Heidbreder, E., Lopau, K., et al. *Nephrol. Dial. Transplant.* **9**, 88–93 (1994).

31. Mauk, R.H., Patak, R.V., Fadem, S.Z., et al. *Kidney Int.* **12**, 122–130 (1977).

32. Lifschitz, M.D., and Barnes, J.L. *Am. J. Physiol.* **274**, F714–F719 (1984).

33. Finn, W.F., Kaj, L.J., and Grossman, S.H. *Kidney Int.* **32**, 479–485 (1987).

34. Lieberthal, W., Wolf, E.F., Rennke, H.G., et al. *Am. J. Physiol.* **256**, F894–F900 (1989).

35. Cristol, J.P., Thiemermann, C., Mitchell, J.A., et al. *Br. J. Pharmacol.* **109**, 188–194 (1993).

36. Conger, J.D. and Weil, J.V. *J. Invest. Med.* **43**, 431–442 (1995).

37. Szabó, C.C., Csáki, A., Benyó, Z., et al. *Circ. Shock* **37**, 307–316 (1992).

38. Moncada, S., Palmer, R.J., and Higgs, E.A. *Hypertension* **12**, 365–372 (1988).

39. Kilbourn, R.G., Jubran, A., Gross, S.S., et al. *Biochem. Biophys. Res. Commun.* **172**, 1132–1138 (1990).

40. Hibbs, J.B., Jr., *Res. Immunol.* **142**, 656–569 (1991).

41. Wright, C.E., Rees, D.D., and Moncada, S. *Cardiovasc. Res.* **26**, 48–57 (1992).

42. Shultz, P.J. and Raij, L. *J. Clin. Invest.* **90**, 1718–1725 (1992).

43. Conger, J.D. and Falk, S.A., *J. Clin. Invest.* **67**, 1334–1342 (1981).

44. Conger, J.D., Kim, G.E., and Robinette, J.B. *Am. J. Physiol.* **267**, F443–F449 (1994).

45. Conger, J.D., Falk, S.A, and Robinette, J.D. *J. Clin. Invest.* (in press).

46. Conger, J.D., Scultz, P., Raij, L., et al. *J. Clin. Invest.* **96**, 631–639 (1995).

47. Gerkens, J.F. *J. Pharmacol. Exp. Ther.* **250**, 1105–1112 (1989).

48. Chan, B.B.K., Kern, J.A., Flanagan, T.L., et al. *Circulation* **86**, SII-295–SI-299 (1992).

49. Rego, A., Vargas, R., Cathapermal S., et al. *J. Pharmacol. Exp. Ther.* **259**, 905–915 (1991).

50. Gallego, M.J., Garcia Villalon, A., Lopez Farre, A., et al. *Circ. Res.* **74**, 477–484 (1994).

51. De Nicola, L., Thomson, S.C., Waed, L.M., et al. *J. Clin. Invest.* **92**, 1859–1865 (1993).

52. Roullet, J.B., Xue, H., McCarron, D.A., et al. *J. Clin. Invest.* **93**, 2244–2250 (1994).

53. Li, W. and Hanschumacher, R.E. *J. Biol. Chem.* **268**, 14040–14044 (1993).

54. Dawson, T.M., Steiner, J.P., Dawson, V.L., et al. *Proc. Natl. Acad. Sci. USA* **90**, 9809–9812 (1993).

55. Conger, J.D. Unpublished results.

56. Rego, A., Vargas, R., Wroblewska, B., et al. *J. Pharmacol. Exp. Ther.* **252**, 165–170 (1990).

57. Huand, H.C., Rego, A., Vargas, R., et al. *Transplant Proc.* **19**, 126–130 (1987).

58. Bossaler, C., Forstermann, U., and Hetel, R. *Eur. J. Pharm.* **165**, 165–169 (1989).

59. Conger, J.D., Falk, S.A., Friedmann, M., et al. *J. Am. Soc. Nephrol.* **6**, 995A (1995).

60. Ioannidis, I. and deGroot, H. *Biochem. J.* **296**, 341–345 (1993).

61. Lafon-Cazal, M., Pietri, S., Culcasi, M., et al. *Nature* **364**, 535–537 (1993).

62. Lipton, S.A., Choi, Y.-B., Pan, Z.-H., et al. *Nature* **364**, 626–632 (1993).

63. Oury, T.D., Piantadosi, C.A., and Crapo, J.D. *J. Biol. Chem.* **268**, 15396–15398 (1993).

64. Heiss, L., Lancaster, J., Corbett, J., et al. *Proc. Natl. Acad. Sci. USA* **91**, 267–270 (1994).

65. Malinsky, T., Bailey, F., Zhang, A., et al. *J. Cerebral Blood Flow Metab.* **13**, 355–358 (1993).

66. Dinerman, J., Lowenstein, C., and Synder, S. *Circ. Res.* **73**, 217–222 (1993).

67. Patel, V.C., Yellon, D.M., Singh, K.J., et al. *Biochem. Biophys. Res. Commun.* **196**, 234–238 (1993).

68. Yu, L., Gengaro, P.E., Niederberger, M., Schrier, R. *Proc. Natl. Acad. Sci. USA* **91**, 1691–1695 (1994).

69. Zhang, J., Dawson, V.L., Dawson, T.W., Snyder, S. *Science* **263**, 687–689 (1994).

70. Beckman, J.S., Beckman, T.W., Chen, J., et al. *Proc. Natl. Acad. Sci. USA* **87**, 1620–1624 (1990).

71. Mohaupt, M.G., Elize, J.L., Ahn, K.Y., et al. *Kidney Int.* **46**, 653–665 (1994).

72. Terada, Y., Tomito, K., Nonoguchi, H., et al. *J. Clin. Invest.* **90**, 659–665 (1992).

73. Yaqoob, M., Edelstein, C., Alkhunaizi, A., Schrier, R. *J. Am. Soc. Nephrol.* **6**, 993A (1995).

74. Peresleni, T., Noiri, E., Bahou, W.F., Goligorsky, M. *Am. J. Physiol.* **270**, F971–F977 (1996).

75. Noiri, E., Peresleni, T., Miller, F., Goligorsky, M. *J. Clin. Invest.* **97**, 2377–2383 (1996).

17

NO in Diabetic Nephropathy

Dennis Diederich

1. Introduction and Overview

The incidence of diabetes mellitus has reached epidemic proportions; some 10–12 million people in the United States alone are afflicted with diabetes mellitus [1]. The vascular complications of diabetes represent major risk factors for cardiovascular morbidity and mortality. Diabetic nephropathy has emerged as the leading cause of end-stage kidney failure in many centers in the United States, and diabetic retinopathy is the leading cause of blindness. The risk of fatal myocardial infarction is increased threefold to fourfold in diabetic individuals. Diabetic nephropathy is essentially a microvascular disease of the kidneys [2]. The natural history of renal involvement in insulin-dependent diabetes mellitus is characterized by a spectrum of abnormalities, beginning with functional hemo-dynamic alterations [3], followed by structural alterations [4], which, in some 30–50% of subjects progress to end-stage renal failure. Renal blood flow and glomerular filtration rate are increased during the initiating phase of diabetes in both humans and in animal models. The recent discovery that nitric oxide (NO) serves as a physiologic modulator of renal blood flow and of glomerular filtration rate prompted investigators to examine the potential role of NO in the mediation of diabetes-induced hyperfiltration. NO-mediated responses appear to be enhanced during the initiating phase of experimental diabetes, but as the disease progresses, responses mediated by NO become impaired. Accelerated free radical production and glycated proteins increase NO inactivation and contribute to inflammatory gene activation. Vasoprotective responses mediated by NO are replaced by cyto-toxic, vasodestructive responses. Preliminary evidence suggests that enhanced, dysfunctional production of NO via the inducible NO synthase (iNOS) pathway may be contributing to the enhanced free radical production and vasodestructive responses which characterize the advanced phase of diabetic vascular disease.

This presentation will focus on the potential roles of alterations of NO homeostasis in producing vascular dysfunction, characteristic of the initiating and early phases of overt nephropathy, and destruction of the kidneys, characteristic of the advanced phase of diabetic nephropathy. Clinical correlates for this arbitrary classification are outlined in Table 17-1.

A brief overview of NO homeostasis in the normal kidney will facilitate understanding alterations in NO production and function which have been described in diabetes. As in other organs, NO in the kidney is produced from L-arginine by a family of NO synthases (NOS) [5]. Four isoforms of NOS have been demonstrated in the rat kidney: (a) endothelial NOS (eNOS) [6,7], (b) neuronal NOS (nNOS) [8], and (c) two isoforms of "inducible" NOS (iNOS), vascular smooth muscle (vsmNOS), and macrophage (macNOS) [9]. Endothelial NOS is localized primarily in glomeruli, afferent arterioles, and arcuate and interlobular arteries, whereas nNOS is localized predominantly in cells of the macula densa and inner medullary collecting duct [10]. Both eNOS and nNOS are constitutive and are activated by mechanisms which increase intracellular calcium, including receptor activation by agonists such as acetylcholine and bradykinin. The small amounts of NO produced by eNOS maintain basal renal blood flow and glomerular filtration rate via cyclic guanosine,3′,5′,monophosphate (cGMP)-mediated vascular smooth muscle and mesangial cell relaxation [11]. The afferent arterioles are exquisitely sensitive to the vasorelaxing effects of basal and activated production of endothelium-derived nitric oxide (EDNO) [12–14]. EDNO modulates (opposes) renal vasoconstriction induced by angiotensin II, norepinephrine, endothelin, and prostaglandins [12,15]. Inhibition of NOS by the intrarenal infusion of subpressor doses of L-arginine analogs leads to abrupt decreases in renal plasma flow, glomerular filtration rate, and sodium excretion [16,17]. Stimulation of EDNO production by infusion of acetylcholine increases renal blood flow, glomerular filtration rate, sodium excretion, and urinary excretion of cGMP [18,19]. Nitric oxide synthesized in glomerular endo-

Table 17-1. Clinical Spectrum of Diabetic Renal Disease

Initiating phase
Hypertrophy
Glomerular hyperfiltration
Microalbuminuria
Overt-phase diabetic nephropathy
Mesangial expansion
Proteinuria
Hypertension
Glomerular sclerosis
Tubulointerstitial disease
Advanced phase
Interstitial fibrosis
End-stage renal failure

thelial cells produces cGMP-mediated responses in adjacent mesangial cells ("cross-talk") [20,21], examples of which are suppression of mesangial cell proliferation and inhibition of protein synthesis [22]. In addition to vasodilatory and antiproliferative effects, NO modulates microvascular permeability [23], inhibits platelet adhesion and aggregation [24], as well as leukocyte adhesion to the vessel wall [25,26]. These observation underscore the important physiologic roles of NO in modulation of renal function and in vasoprotection.

The iNOS isoforms differ in being calcium independent and in requiring transcriptional activation, classically by cytokines, for full expression. Activation of the iNOS pathway leads to sustained, nonmodulated production of high net amount of NO. NO production via the iNOS pathway is generally viewed as providing defensive and cytotoxic functions. However, recent evidence indicates that the original concepts of constitutive and inducible NOS may require some modification. Constitutive eNOS is subject to upregulation by shear stress [27,28] and estrogen [29] as well as downregulation by NO itself [30–32]. Similarly, iNOS has recently been demonstrated to be constitutively present in tubular cells [9] and in glomeruli [33,34]. The responses of vsmNOS and macNOS may differ depending on the activating stimuli [9].

2. Growth Factors, Glomerular Hyperfiltration, and NO

The initial targets of diabetes in the kidney are the endothelial and mesangial cells. Within 36–48 h of development of hyperglycemia and glycosuria, kidney size, glomerular filtration, and renal blood flow increase [35–37]. The enlargement of the kidneys is due to hypertrophy of tubular epithelium and glomerular structures. The increase in glomerular filtration rate (GFR) arises from the combination of increased renal plasma flow and glomerular capillary hydraulic pressure (P_{GC}); the increase in P_{GC} results from preferentially greater reduction in afferent arteriolar resistance [38]. Glomerular hyperfiltration appears to follow enlargement of the kidneys, suggesting that the increase in size of the kidneys is not due to the increased function [36]. The mechanism(s) responsible for the hypertrophic and hemodynamic responses in diabetic kidneys have been the subject of considerable study but remain incompletely understood. Hyperglycemia plays a critical role in the production and maintenance of both glomerular hyperfiltration and renal enlargement in diabetes [39]. Normalization of blood glucose with insulin eliminates glomerular hyperfiltration and decreases the size of the kidneys in both humans and animals with diabetes [40–42]. The close temporal relationship between renal enlargement and hyperfiltration prompted investigators to explore the vasoactive effects of renal growth factors. Increased synthesis of extracellular matrix components, an early finding in kidneys of diabetic animal, is preceded by release of growth factors, including insulinlike growth factor I (IGF-1) growth hormone (GH), transforming growth factor (TGF)-β1 and platelet-derived growth factor (PDGF) [2,43–46]. In addition to their mitogenic effects, IGF-1 and PDGF produce vasodilation mediated by NO [47,48]. Infusion of IGF-1 increases glomerular filtration rate in humans and animals [49–51]. Renal vasodilation produced by IGF-1 infusion in the isolated perfused rat kidney was blocked by inhibition of NOS with N^G-nitro-L-arginine methyl ester (L-NAME), indicating that the response induced by the growth factor was mediated by increased NO production [52]. In addition to diabetes, renal IGF-1 production is increased in two other states characterized by glomerular hyperfiltration, hypersomatotropism, and during compensatory hypertrophy of the kidney [49]. In summary, hyperglycemia stimulates the production of renal growth factors; growth factors contribute to glomerular hyperfiltration by stimulation of vascular NO production.

The potential role of increased production of NO in the pathogenesis of glomerular hyperfiltration in diabetic animals has been addressed in several studies [53–56]. Urinary excretion of nitrite and nitrate, stable metabolites of NO, was increased in diabetic rats [53–55,57]; this finding provides indirect evidence in support of increased NO production. The relationship between glomerular hyperfiltration and increased nitrate + nitrite excretion is not clear, however. In one study, L-arginine supplementation prevented glomerular hyperfiltration in diabetic rats without affecting the elevated urinary nitrite + nitrate excretion [54]. In addition, acute changes in urinary excretion of nitrite + nitrate do not necessarily predict renal vascular NO production [58]. Comparison of the responses of GFR in control and diabetic rats to infusions of either N^ω-nitro-L-arginine (L-NA) or N^G-nitro-L-arginine methyl ester (L-NAME) has been utilized to evaluate the NO-mediated component in basal GFR [16,59]. Changes in GFR in diabetic rats in response to infusion of NOS inhibitors have been quite varied, ranging from no response [56] to incomplete [53,57] or complete elimination of glomerular hyperfiltration [55]. These findings suggest that other factors, in addition to increased NO production, contribute to glomerular hyperfiltration in experimental diabetes [43,60–62].

3. Central Role of Glucose Elevation in EDNO Dysfunction

Vascular relaxation mediated by endothelium-derived nitric oxide (EDNO) is impaired in diabetic subjects. The bulk of evidence indicating impairment in EDNO-mediated arterial relaxations in diabetes is based on decreased relaxation noted in response to acetylcholine. Relaxation in response to acetylcholine is blocked by removal of the endothelial layer and by inhibition of NOS by L-arginine analogs, confirming that EDNO mediates the relaxation [63,64]. Elevations in glucose play a major role in the initiation of diabetes-induced endothelial dysfunction. EDNO-mediated relaxations in both conduit and resistance arteries from normal rats were markedly impaired following 2–4 h exposure of the arteries to elevated concentrations of glucose [65–68]. The impairment of EDNO-mediated relaxations were specific for D-glucose; mannitol and L-glucose at equimolar concentrations exerted no effect on relaxations in response to acetylcholine. The impairment in EDNO-mediated relaxations induced by elevated glucose arise from multiple mechanisms, including increased production of free radicals [67,69,70], vasoconstricting prostaglandins [65,67], and activation of protein kinase C (PKC). Addition of free radical scavengers [67,69,70], indomethacin [65,67], or PKC inhibitors [71] during glucose incubation prevented the impairment in EDNO-mediated relaxations. These findings point to increased destruction, rather than decreased production, of EDNO as the major mechanism by which glucose elevation impairs EDNO responses.

Similar findings were noted in the renal microcirculation; EDNO-mediated relaxations of perfused afferent arterioles from normal rabbits were impaired following a 3-h exposure to 30 mM glucose [72]. Exposure of isolated glomeruli to 30 mM glucose increased production of lipid peroxide radicals (LPO) within 1 h; the increased LPO production was sustained during a 48-h incubation [73]. The hydroxyl radical scavenger, dimethylthiourea (DMTU), blocked the production of LPO by glomeruli throughout the 48-h incubation. Pretreatment of glomeruli with PKC antagonists (H-7 or staurosporine) blocked LPO production during the initial hour but not after 48 h of incubation.

4. Activation of Protein Kinase C by Elevated Glucose

High ambient concentrations of glucose activate PKC in arteries [71,74], isolated glomeruli [75], and in cultures of mesangial [76,77], endothelial [78], vascular smooth muscle [79,80], and tubular epithelial cells [81]. Mechanisms by which elevated glucose activated PKC activity included increased levels of diacylglycerol (DAG) [82], increased intracellular calcium (Ca^{2+}) [83], increased prostaglandin endoperoxides (thromboxane A$_2$/prostaglandin H$_2$), and phospholipids [84]. Activation of PKC has been reported to have a wide range of effects, including inhibition of NO production by all isoforms of NOS [75,85–91], increased permeability of endothelial cell layers [92], activation of mesangial contraction [74], increased mesangial matrix production [76,78,93,94], and mediation of renal vasoconstriction by angiotensin II [95]. Growth factor production may also be mediated by PKC-dependent mechanisms [96–99].

The above findings provide evidence that high ambient concentrations of glucose induce oxidative stress in vascular tissues, including those of the kidney. Free radicals produced by oxidant stress induce reversible endothelial dysfunction by inactivation of EDNO (Fig. 17-1).

5. Endothelial (EDNO) Dysfunction in Early-Phase Diabetes

There is considerable evidence indicating that vascular relaxations mediated by EDNO are impaired in humans and animals with diabetes [100–102]. Relaxation

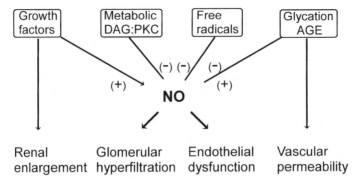

Figure 17-1. Diagram outlining some of the important responses induced by glucose elevation which alter nitric oxide (NO) production and function during the early phases of diabetic vascular disease. Hyperglycemia induces the expression of growth factors; insulinlike growth factor 1 (IGF-1) and platelet derived growth factor (PDGF) contribute to glomerular hyperfiltration by stimulating release of NO. NO production is downregulated by protein kinase C (PKC).Free radicals and advanced glycosylation end products (AGEP) produce endothelial dysfunction by inactivation of NO. During later stages of diabetes, AGEP can stimulate production of NO via cytokine-induced expression of inducible NO synthase. Plus (+) denotes stimulation; minus (−) indicates inhibition or inactivation.

in response to acetylcholine is impaired in both large and small arteries obtained from diabetic animals. The mechanisms responsible for diabetes-induced EDNO dysfunction (Table 17-2) have been explored most extensively in the aorta obtained from diabetic animals [102,103]. Impaired relaxations in response to acetylcholine noted in the aorta from diabetic rats are normalized by pretreatment of the arteries with the free radical scavengers, superoxide dismutase or dimethylthiourea [69,102,103]. These findings point to increased inactivation of EDNO by free radicals as the cause of the faulty response to EDNO in diabetic arteries.

Endothelium-derived NO-mediated relaxation in renal vessels also is impaired in diabetic animals. Renal vasodilation induced by infusion of acetylcholine and production of cGMP by isolated glomeruli is markedly decreased in diabetic rats

Table 17-2. Potential Mechanisms for EDNO Dysfunction

Decreased production of NO
L-Arginine, cofactor deficiency
Downregulation of NOS
Receptor, postreceptor dysfunction
Decrease in transport NO (RSNO)
Increased destruction of NO
Increased free radical production
Quenching of NO by AGEP
Decreased responsiveness to NO
Increased production of NO antagonists

[104]. Both responses were normalized in insulin-treated diabetic animals. The diameter of afferent arterioles measured in situ is increased in diabetic rats [105,106]. Constriction of afferent arterioles in response to L-NAME was found to be blunted in diabetic rats but was normalized following pretreatment with superoxide dismutase [105]. EDNO-mediated relaxations were impaired in isolated perfused interlobar arteries obtained from diabetic rats [107]; the impairment in endothelium-dependent relaxation became progressively worse with duration of the diabetes [108]. Pretreatment with dimethylthiourea normalized relaxations in response to acetylcholine in arteries from diabetic rats but had no effect on arteries obtained from control rats [107]. Collectively, the findings reviewed above demonstrate a consistent pattern of impairment in relaxations mediated by EDNO in large as well as small vessels from diabetic animals. The impaired responses to EDNO in diabetic vessels during the early phase of diabetes (4–6 weeks) are due to an increased production of free radicals which inactivate NO; scavengers of free radicals normalize EDNO-mediated relaxations in diabetic vessels.

Administration of insulin or of dimethylthiourea to diabetic rats (starting 3 days after onset of hyperglycemia) normalized EDNO-mediated relaxations in interlobar arteries from the kidneys of diabetic rats studied 4–6 weeks after induction of the diabetes [109]. These findings underscore the critical roles of sustained glucose elevation and of increased free radical production in the genesis of reversible EDNO dysfunction, a characteristic finding during the early phases of diabetic vascular disease (Figure 17-1). The findings outlined above present a seeming paradox: nitric oxide-mediated relaxations in the renal microcirculation appear to be enhanced in diabetes, whereas ENDO-mediated relaxations in renal arteries obtained from diabetic animals are impaired.

6. EDNO in the Advanced Phase of Diabetic Vascular Disease

Vascular relaxations mediated by EDNO in interlobar arteries from diabetic rats became progressively less with duration of diabetes; after 12 weeks of diabetes, relaxations of diabetic arteries were less than 10% of those observed in arteries from control rats [109]. Pretreatment of the arteries with dimethylthiourea produced negligible improvement in acetylcholine relaxations. Vascular smooth muscle relaxation induced by nitroprusside, mediated by NO, become impaired also during the advanced phases of diabetic vascular disease. The lack of reversibility of EDNO-mediated relaxations and the impairment in NO-mediated vascular smooth muscle relaxations provide evidence for irreversible injury to the diabetic arteries with long-standing diabetes. Mechanisms contributing to the irreversible phases of diabetic vascular disease include the production of advanced glycosylation end products (AGEP), accelerated free radical production, and structural damage to the vasculature. Ligation of AGEP with specific cell membrane receptors triggers free radical production. AGEP and free radicals (oxidant stress) both can induce gene expression, resulting in inflammatory products and iNOS activation (Fig. 17-2).

There is compelling evidence indicating that advanced glycosylation end products (AGEP) are intimately involved in the pathogenesis of diabetic vascular disease; this area has been extensively reviewed [1,110,111]. This discussion will focus upon effects of glycosylation and of AGEP which impact on NO homeostasis. AGEP alter NO-mediated responses in several ways. AGEP inactivate NO by quenching of the molecule [112], thereby interfering with the ability of NO to induce vascular relaxation or to suppress proliferative responses [112,113]. Free radicals are by-products of both the early as well as late glycation reactions [111,114–116]; free radicals increase the rate of NO inactivation. Glycosylation of proteins such as albumin decreases the formation of S-nitroso-albumin [117], a physiologically active, stabilized form of NO [118]. S-Nitroso-albumin exhibits NO-like vasoprotective responses, including vasodilation and antiplatelet and antiproliferative effects [118–124]. In addition, rapid uptake of locally produced NO by reactive sulfhydryl groups may prevent toxic effects of high concentrations of NO [122,125–128]. The principal mechanism through which AGEP exert their cellular effects is via specific cellular receptors for AGE [129–131]; receptors for AGEP (RAGE) are expressed in many cells, including endothelial

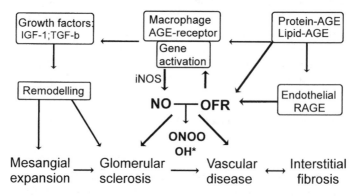

Figure 17-2. Diagram depicting potential mechanisms capable of inducing the expression of inducible NO synthase (iNOS) during the later phases of diabetic vascular disease. Ligation of AGEP to macrophage receptors triggers the production of cytokines (tumor necrosis factor alpha and interleukin-1) and induces gene expression, both of which may induce expression of iNOS. Oxygen free radicals (OFR) produced by glycation reactions and in response to AGEP binding to receptors for AGE (RAGE) also may induce gene expression leading to iNOS expression. Activation of iNOS leads to sustained production of high concentrations of NO. Reaction between NO and OFR may lead to the production of damaging peroxynitrite (ONOO) and hydroxyl radicals (OH⁺). Damage to cells induced by high concentrations of NO or by NO-mediated ONOO and OH* may contribute to glomerular sclerosis, accelerated vascular disease, and tubulointerstitial fibrosis.

[132] and mesangial cells [131]. Ligation of AGEP by RAGE induces oxidant stress with free radical production [132–134]. Free radicals generated by AGEP–receptor interaction may inactive NO. Binding of AGEP to receptors on monocytes and macrophages may activate iNOS via production of tumor necrosis factor alpha (TNF-α) and interleukin-1 (IL-1) [110,111,135,136]. Aminoguanidine, a phenylhydrazine derivative that inhibits advanced glycosylation reactions, has been also reported to prevent early diabetic vascular dysfunction [137], which may be related to inhibition of iNOS by aminoguanidine [138,139]. Administration of early glycosylation products to normal rats induces glomerular hyperfiltration [140,141], mesangial growth and matrix production [140–142], vascular insensitivity to vasodilating agents, and other renal findings which closely simulate those of the native diabetic state [140,141]. AGEP peptides released from local metabolism of AGEP are cleared from the circulation by glomerular filtration [1,143]. AGEP peptides accumulate in the circulation as renal function declines [1,144,145]. The accumulation of AGEP peptides may be a major factor contributing to the accelerated atherosclerosis commonly observed in diabetic subjects with advanced renal insufficiency [142,146–148]. In summary, AGEP alter NO homeostasis initially by inhibiting the vasoprotective effects of EDNO. AGEP also trigger the release of cytokines and growth factors which promote structural alterations, including mesangial expansion and glomerular sclerosis. Cytokines production initiated by AGEP also may induce the expression of iNOS leading to dysfunctional and potentially destructive responses mediated by NO.

Accelerated free radical production during late-phase diabetic vascular disease arises from multiple pathways [110,111,114,149,150]. Free radicals are formed in the early as well as late steps in glycosylation reactions [114–116] and following AGEP interaction with receptors [132–134,151] noted above. Low-density lipoprotein (LDL) oxidation is increased by pathophysiological concentrations of glucose [152], by glycation reactions [149,153], and by free radicals [149,154]. Modified LDL and lipid peroxidation products interfere with both the production and bioactivity of EDNO [155]. Oxidized LDL and AGEP induce the expression of genes regulating the production of inflammatory mediators, including growth factors, adhesion molecules, and iNOS [151,154,156–160] (Fig. 17-3).

7. Evidence for iNOS Expression in Diabetes Mellitus

Schönfelder and colleagues recently provided direct evidence for iNOS expression in humans with gestational diabetes [161]. Employing the techniques of reverse transcription–polymerase chain reaction and Western blot analyses, these investigators demonstrated iNOS messenger RNA and protein expression in placental tissue from patients with gestational diabetes but not in nondiabetic control patients. The authors suggested that increased NO production by iNOS contributes to abnormalities in the intervillous circulation and dilatation of capillaries noted in the placenta of diabetic subjects.

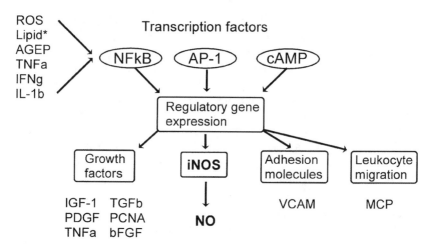

Figure 17-3. Overview of potential mechanisms for iNOS expression or activation in diabetes. Reactive oxygen species (ROS), oxidized lipids, advanced glycosylation end products (AGEP), tumor necrosis alpha (TNFα), interferon gamma (IFN-γ), and interleukin-1 beta (IL-1β) are all capable of activating transcription factors which participate in regulatory gene expression. NFκB refers to nuclear factor κB, AP-1 refers to activator protein-1, and cAMP refers to cyclic adenosine 3′,5′-monophosphate. Expression of regulatory genes leads to expression of various growth factors, iNOS, adhesion molecules, and leukocyte migration factors. Abbreviations: IGF-1 = insulinlike growth factor 1; PDGF = platelet derived growth factor; TNFa = tumor necrosis factor alpha, TGFb = transforming growth factor beta, PCNA = proliferating cell nuclear antigen, bFGF = basic fibroblast growth factor, VCAM = vascular cell adhesion molecule, and MCP = monocyte chemotactic protein.

There is increasing evidence of iNOS expression in experimental diabetes. Cytokine gene expression accompanied by both iNOS and cyclooxygenase expression have been demonstrated in lymphocytes and macrophages infiltrating the pancreas of animals with hereditary diabetes mellitus (NOD mice and bio-breeding BB rats) [162,163]. Islet cell destruction in these animal models is accompanied by increased local production of NO by the activated infiltrating cells [162,164–166]. Macrophages activated by AGEP or by oxidized lipids release TNF-α and IL-1β, cytokines known to induce expression of iNOS [110,167,168]. Glucose elevation has recently been reported to enhance the expression of iNOS induced by lipopolysaccharide plus interferon-γ in cultured murine mesangial cells [169]. Infusion of AGEP into normal animals induces glomerular hyperfiltration [141] and vascular dysfunction characterized by increased permeability and unresponsiveness to vasodilatory agents [140]. Amino-guanidine, a relatively selective inhibitor of iNOS [139], has been reported to inhibit the development of diabetic vascular dysfunction [137,138,140]. Expression of iNOS in renal mesangial cells is controlled by at least two separate signaling pathways, one involving 3',5',-cyclic monophosphate (cAMP) and the other triggered by cytokines such as IL-1β or TNF-α [170]. The two pathways operate in a synergistic fashion. Interleukin-1β- and TNF-α-induced iNOS expression is dependent on activation of the transcription factor, nuclear factor κB (NFκB), whereas the cAMP is not. NFκB is activated by reactive oxygen species and by AGEP [151,158]. Oxidized LDL activates both the cAMP and NFκB signal pathways [158,167].

8. Vasodestructive Implications of iNOS Activation in Diabetes

Activation of iNOS leads to rampant, nonmodulated production of very high concentrations of NO [171]. The simultaneous increased production of NO and of superoxide anion, an important target of NO, produces conditions favorable for the formation of peroxynitrite anion in diabetes [10,172–175]. Peroxynitrite directly oxidizes many important biologic molecules by attacking sulfhydryl and iron–sulfur centers [173,175–177]. Peroxynitrite may spontaneously decompose to form hydroxyl radical and nitrogen dioxide [172]. In the presence of trace metals, peroxynitrite may form a nitronium ion, an intermediate which readily nitrates tyrosine on proteins, forming 3-nitrosotyrosine residues [178,179]. The cellular toxicity caused by high concentrations of NO may arise from NO-mediated production of damaging peroxynitrite anion and hydroxyl radicals. In addition, NO can produce toxicity by attacking iron–sulfur clusters in the mitochondrial electron transport chain [10] and by reacting with critical iron or sulfhydryl groups on various enzymes. Evidence of NO-mediated toxicity in experimental diabetes at the present is largely indirect. The abnormal vasodilation noted in the afferent arteriole in diabetic animals, as well as the insensitivity of the afferent arteriole to vasodilating and vasoconstricting agents may both be reflections of increased NO resulting from iNOS activation in the arteriole. The later development of sclerosing glomerular disease likewise may arise at least in part from cytotoxic effects of NO-mediated free radical production. Continued infusions of AGEP into normal animals induce glomerular sclerosis and albuminuria [142]; as reviewed above, AGEP may induce iNOS expression via multiple pathways. AGEP cross-linking in structural proteins is enhanced by increased free radicals. The synergistic effects of accumulation of AGEP [1,110,146,156] and enhanced NO-mediated radical production may be responsible for the widespread acceleration of atherosclerosis characteristic of end-stage diabetic nephropathy.

9. Intervention Strategies to Prevent Diabetic Vascular Disease

The focus of prevention must be on control of glucose levels in diabetic individuals. The critical role of elevation in glucose in the pathogenesis of diabetic vascular disease cannot be overemphasized. The findings from the recently completed Diabetes Control & Complications Trial offer conclusive proof for the protective effects of aggressive control of blood glucose in diabetic subjects [180]. Adjunctive measures which may afford some protection from the delayed complications of diabetes currently under study include the trials with aminoguanidine [1], with aldose reductase inhibitors [181], and with antioxidants [182]. Discussion of these trials is beyond the scope of this chapter. Intervention strategies directed at iNOS activation in diabetes must await direct demonstration of a pathophysiological role of iNOS in diabetic vascular disease. The recently reported findings that selective knock-down of iNOS in the kidney ameliorates ischemic acute renal failure [183] offers promise that it may be possible to selectively target iNOS activation in other clinical states.

References

1. Bucala, R. and Vlassara, H. *Am. J. Kidney Dis.* **26,** 875–888 (1995).

2. Woolf, A.S., Bosch, R.J., and Fine, L.G. *J. Hypertens.* **10**(Suppl. 1), S11–S16 (1992).

3. Mogensen, C.E., *Contemp. Issues Nephrol.* **20,** 19–49 (1989).

4. Hostetter, T.H. and Daniels, B.S. *Contemp. Issues Nephrol.* **20,** 51–65 (1989).

5. Knowles, R.G. and Moncada, S. *Biochem. J.* **298,** 249–258 (1994).

6. Bachmann, S., Bosse, H.M., and Mundel, P. *Am. J. Physiol.* **268,** F885–F898 (1995).

7. Ujiie, K., Yuen, J., Hogarth, L., Danziger, R., and Star, R.A. *Am. J. Physiol.* **267,** F296–F302 (1994).

8. Tojo, A., Gross, S.S., Zhang, L., et al. *J. Am. Soc. Nephrol.* **4,** 1438–1447 (1994).

9. Mohaupt, M.G., Elzie, J.L., Ahn, K.Y., Clapp, W.L., Wilcox, C.S., and Kone, B.C. *Kidney Int.* **46,** 653–665 (1994).

10. Cook, T. *Contemp. Issues Nephrol.* **30,** 119–142 (1995).

11. Raij, L. and Baylis, C. *Kidney Int.* **48,** 20–32 (1995).

12. Ito, S., Arima, S., Ren, Y.L., Juncos, L.A., and Carretero, O.A. *J. Clin. Invest.* **91,** 2012–2019 (1993).

13. Edwards, R.M. and Trizna, W. *J. Am. Soc. Nephrol.* **4,** 1127–1132 (1993).

14. Deng, A. and Baylis, C. *Am. J. Physiol.* **264,** F212–F215 (1993).

15. Ito, S., Johnson, C.S., and Carretero, O.A. *J. Clin. Invest.* **87,** 1656–1663 (1991).

16. Lahera, V., Salom, M.G., Miranda Guardiola, F., Moncada, S., and Romero, J.C. *Am. J. Physiol.* **261,** F1033–F1037 (1991).

17. Baylis, C., Mitruka, B., and Dend, A. *J. Clin. Invest.* **90,** 278–281 (1992).

18. Tolins, J.P., Palmer, R.M., Moncada, S., and Raij, L. *Am. J. Physiol.* **258,** H655–H662 (1990).

19. Lahera, V., Salom, M.G., Fiksen-Olsen, M.J., Raij, L., and Romero. J.C. *Hypertension* **15,** 659–663 (1990).

20. Shultz, P.J., Schorer, A.E., and Raij, L. *Am. J. Physiol.* **258,** F162–F167 (1990).

21. Marsden, P.A., Brock, T.A., and Ballerman, B.J. *Am. J. Physiol.* **258,** F1295–F1303 (1990).

22. Garg, U. and Hassid, A. *Am. J. Physiol.* **257,** F60–F66 (1989).

23. Kubes, P. and Granger, D.N. *Am. J. Physiol.* **262,** H611–H615 (1992).

24. Radomski, M.W., Palmer, R.M.J., and Moncada, S. *Br. J. Pharmacol.* **92,** 639–646 (1987).

25. Gaboury, J., Woodman, R.C., Granger, D.N., Reinhardt, P., and Kubes, P. *Am. J. Physiol.* **265,** H862–H867 (1993).

26. Kubes, P., Suzuki, M., and Granger, D.N. *Proc. Natl. Acad. Sci. USA* **88,** 4651–4655 (1991).

27. Uematsu, M., Ohara, Y., Navas, J.P., et al. *Am. J. Physiol.* **269,** C1371–C1378 (1995).

28. Riser, B.L., Cortes, P., Zhao, X., Bernstein, J., Dumler, F., and Narins, R.G. *J. Clin. Invest.* **90,** 1932–1943 (1992).

29. Weiner, C.P., Lizasoain, I., Baylis, S.A., Knowles, R.G., Charles, I.G., and Moncada, S. *Proc. Natl. Acad. Sci. USA* **91,** 5212–5216 (1994).

30. Griscavage, J.M., Hobbs, A.J., and Ignarro, L.J. *Adv. Pharmacol.* **34,** 215–234 (1995).

31. Assreuy, J., Cunha, F.Q., Liew, F.Y., and Moncada, S. *Br. J. Pharmacol.* **108,** 833–837 (1993).

32. Buga, G.M., Griscavage, J.M., Rogers, N.E., and Ignarro, L. J. *Circ. Res.* **73,** 808–812 (1993).

33. Morrissey, J.J., McCracken, R., Kaneto, H., Vehaskari, M., Montani, D., and Klahr, S. *Kidney Int.* **45,** 998–1005 (1994).

34. Cook, H.T., Ebrahim, H., Jansen, A.S., Foster, G.R., Largen, P., and Cattell, V. *Clin. Exp. Immunol.* **97,** 315–320 (1994).

35. Seyer-Hansen, K. *Kidney Int.* **23,** 643–646 (1983).

36. Cortes, P., Dumler, F., Goldman, J., and Levin, N.W. *Diabetes* **36,** 80–87 (1987).

37. Mogensen, C.E. and Anderson, M.J.F. *Diabetes* **22,** 706–712 (1973).

38. Hostetter, T.H., Troy, J.L., and Brenner, B.M. *Kidney Int.* **19,** 410–415 (1981).

39. Mogensen, C.E., Christensen, C.K., Pedersen, M.M., et al. *J. Diabet. Complications.* **4,** 159–165 (1990).

40. Stackhouse, S., Miller, P.L., Park, S.K., and Meyer, T.W. *Diabetes* **39,** 989–995 (1990).

41. Christiansen, J.S., Gammelgaard, J., Tronier, B., Svendsen, P.A., and Parving, H.H. *Kidney Int.* **21,** 683–688 (1982).

42. Mogensen, C.E. and Anderson, M.J.F. *Diabetologia* **11,** 221–224 (1975).

43. Blankestijn, P.J., Derkx, F.H., Birkenhager, J.C., et al. *J. Clin. Endocrinol. Metab.* **77,** 498–502 (1993).

44. Nakamura, T., Fukui, M., Ebihara, I., et al. *Diabetes* **42**, 450–456 (1993).

45. Yamamoto, T., Nakamura, T., Noble, N.A., and Ruoslahti, E. *Proc. Natl. Acad. Sci. USA* **90**, 1814–1818 (1993).

46. Knecht, A., Fine, L.G., Rodemann, P., Muller, G., Woo, D.D.L., and Norman, J.T. *Am. J. Physiol.* **261**, F292–F299 (1991).

47. Tsukahara, H., Gordienko, D.V., Tonshoff, B., Gelato, M.C., and Goligorsky, M.S. *Kidney Int.* **45**, 598–604 (1994).

48. Juncos, L.A., Ren, Y.L., Arima, S., and Ito, S. *J. Clin. Invest.* **91**, 1374–1379 (1993).

49. Hammerman, M.R. and Miller, S.B. *Contemp. Issues Nephrol.* **30**, 425–438 (1995).

50. LeRoith, D., Werner, H., Phillip, M., and Roberts, C.T.J. *Am. J. Kidney Dis.* **22**, 722–726 (1993).

51. Hirschberg, R., Brunori, G., Kopple, J.D., and Guler, H.P. *Kidney Int.* **43**, 387–397 (1993).

52. McKillop, I., Haylor, J., and El Nahas, A.M. *Exp. Nephrol.* **3**, 49–57 (1995).

53. Komers, R., Allen, T.J., and Cooper, M.E. *Diabetes* **43**, 1190–1197 (1994).

54. Reyes, A.A., Karl, I.E., Kissane, J., and Klahr, S. *J. Am. Soc. Nephrol.* **4**, 1039–1045 (1993).

55. Tolins, J.P., Shultz, P.J., Raij, L., Brown, D.M., and Mauer, S.M. *Am. J. Physiol.* **265**, F886–F895 (1993).

56. Mathis, K.M. and Banks, R.O. *J. Am. Soc. Nephrol.* **7**, 105–112 (1996).

57. Bank, N. and Aynedjian, H.S. *Kidney Int.* **43**, 1306–1312 (1993).

58. Suto, T., Losonczy, G., Qui, C., et al. *Kidney Int.* **48**, 1272–1277 (1995).

59. Baylis, C., Harton, P., and Engels, K. *J. Am. Soc. Nephrol.* **1**, 875–881 (1990).

60. Hirata, Y., Suzuki, Y., Hayakawa, H., et al. *Clin. Sci.* **88**, 413–419 (1995).

61. Zhang, P.L., Mackenzie, H.S., Troy, J.L., and Brenner, B.M. *J. Am. Soc. Nephrol.* **4**, 1564–1570 (1994).

62. Hoogenberg, K., Sluiter, W.J., and Dullaart, R.P. *Acta Endocrinol. Copenh.* **129**, 151–157 (1993).

63. Furchgott, R.F. and Zawadzki, J.V. *Nature* **288**, 373–376 (1980).

64. Moncada, S., Palmer, R.M., and Higgs, E.A. *Pharmacol. Rev.* **43**, 109–142 (1991).

65. Tesfamariam, B., Brown, M.L., Deykin, D., and Cohen, R.A. *J. Clin. Invest.* **85**, 929–932 (1990).

66. Pieper, G.M., Meier, D.A., and Hager, S.R. *Am. J. Physiol.* **269**, H845–H850 (1995).

67. Taylor, P.D. and Poston, L. *Br. J. Pharmacol.* **113**, 801–808 (1994).

68. Cameron, N.E., Cotter, M.A., Dines, K.C., and Maxfield, E.K. *Diabetologia* **36**, 516–522 (1993).

69. Tesfamariam, B. and Cohen, R.A. *Am. J. Physiol.* **263**, H321–H326 (1992).

70. Bohlen, H.G. and Lash, J.M. *Am. J. Physiol.* **265**, H219–H225 (1993).

71. Tesfamariam, B., Brown, M.L., and Cohen, R.A. *J. Clin. Invest.* **87**, 1643–1648 (1991).

72. Arima, S., Ito, S., Omata, K., Takeuchi, K., and Abe, K. *Kidney Int.* **48,** 683–689 (1995).

73. Ha, H. and Kim, K.H. *Kidney Int.* **48** (Suppl. 51), S18–S21 (1995).

74. Williams, B. *J. Hyperten.* **13,** 477–486 (1995).

75. Craven, P.A., Studer, R.K., and DeRubertis, F.R. *J. Clin. Invest.* **93,** 311–320 (1994).

76. Studer, R.K., Craven, P.A., and DeRubertis, F.R. *Diabetes* **42,** 118–126 (1993).

77. Haneda, M., Kikkawa, R., Sugimoto, T., et al. *J. Diabetes Complications* **9,** 246–248 (1995).

78. DeRubertis F.R. and Craven, P.A. *Diabetes* **43,** 1–8 (1994).

79. Williams, B. and Howard, R.L. *J. Clin. Invest.* **93,** 2623–2631 (1994).

80. McKenna, T.M., Li, S., and Tao, S. *Am. J. Physiol.* **269,** H1891–H1898 (1995).

81. Cole, J.A., Walker, R.E., and Yordy, M.R. *Diabetes* **44,** 446–452 (1995).

82. Xia, P., Inoguchi, T., Kern, T.S., Engerman, R.L., Oates, P.J., and King, G.L. *Diabetes* **43,** 1122–1129 (1994).

83. Xia, P., Kramer, R.M., and King, G.L. *J. Clin. Invest.* **96,** 733–740 (1995).

84. Williams, B. and Schrier, R.W. *J. Clin. Invest.* **92,** 2889–2896 (1993).

85. Craven, P.A., Studer, R.K., and DeRubertis, F.R. *Metabolism* **44,** 695–698 (1995).

86. Gopalakrishna, R., Chen, Z.H., and Gundimeda, U. *J. Biol. Chem.* **268,** 27180–27185 (1993).

87. Hirata, K., Kuroda, R., Sakoda, T., et al. *Hypertension* **25,** 180–185 (1995).

88. Ohara, Y., Sayegh, H.S., Yamin, J.J., and Harrison, D.G. *Hypertension* **25,** 415–420 (1995).

89. Okada, D. *J. Neurochem.* **64,** 1298–1304 (1995).

90. Geng, Y.J., Wu, Q., and Hansson, G.K. *Biochim. Biophys. Acta* **1223,** 125–132 (1994).

91. Bredt, D.S., Ferris, C.D., and Snyder, S.H. *J. Biol. Chem.* **267,** 10976–10981 (1992).

92. Pugliese, G., Tilton, R.G., and Williamson, J.R. *Diabetes Metab. Rev.* **7,** 35–59 (1991).

93. Ziyadeh, F.N., Fumo, P., Rodenberger, C.H., Kuncio, G.S., and Neilson, E.G. *J. Diabetes Complications* **9,** 255–261 (1995).

94. Ayo, S.H., Radnik, R.A., Garoni, J.A., Glass, W.F.I., and Kriesberg, J.I. *Am. J. Pathol.* **136,** 1339–1348 (1990).

95. Williams, B., Tsai, P., and Schrier, R.W. *J. Clin. Invest.* **90,** 1992–1999 (1992).

96. Pfeiffer, A. and Schatz, H. *Exp. Clin. Endocrinol. Diabetes* **103,** 7–14 (1995).

97. Serri, O. and Renier, G. *Metabolism* **44,** 83–90 (1995).

98. Muhl, H. and Pfeilschifter, J. *Biochem. J.* **303,** 607–612 (1994).

99. Schwieger, J. and Fine, L.G. *Semin. Nephrol.* **10,** 242–253 (1990).

100. Elliott, T.G., Cockcroft, J.R., Groop, P.H., Viberti, G.C., and Ritter, J.M. *Clin. Sci.* **85,** 687–693 (1993).

101. Calver, A., Collier, J., and Vallance, P. *J. Clin. Invest.* **90**, 2548–2554 (1992).

102. Cohen, R.A. *Circulation* **87**(Suppl. V), V67–V76 (1993).

103. Langenstroer, P. and Pieper, G.M. *Am. J. Physiol.* **263**, H257–H265 (1992).

104. Wang, Y.X., Brooks, D.P., and Edwards, R.M. *Am. J. Physiol.* **264**, R952–R956 (1993).

105. Ohishi, K. and Carmines, P.K. *J. Am. Soc. Nephrol.* **5**, 1559–1566 (1995).

106. Ohishi, K., Okwueze, M.I., Vari, R.C., and Carmines, P.K. *Am. J. Physiol.* **267**, F99–F105 (1994).

107. Dai, F.X., Diederich, A., Skopec, J., and Diederich, D. *J. Am. Soc. Nephrol.* **4**, 1327–1336 (1993).

108. Diederich, D., Skopec, J., Diederich, A., and Dai, F.X. *Am. J. Physiol.* **266**, H1153–H1161 (1994).

109. Dai, F.X., Skopec, J., Diederich, A., and Diederich, D. *J. Am. Soc. Nephrol.* **6**, 657A (1995).

110. Vlassara, H. *J. Lab. Clin. Med.* **124**, 19–30 (1994).

111. Brownlee, M., Cerami, A., and Vlassara, H. *N. Engl. J. Med.* **318**, 1315–1321 (1988).

112. Bucala, R., Tracey, K.J., and Cerami, A. *J. Clin. Invest.* **87**, 432–438 (1991).

113. Hogan, M., Cerami, A., and Bucala, R. *J. Clin. Invest.* **90**, 1110–1115 (1992).

114. Lyons, T.J. *Am. J. Cardiol.* **71**, 26B–31B (1993).

115. Mullarkey, C.J., Edelstein, D., and Brownlee, M. *Biochem. Biophys. Res. Commun.* **173**, 932–939 (1990).

116. Hunt, J.V., Smith, C.C.T., and Wolff, S.P. *Diabetes* **39**, 1420–1424 (1990).

117. Farkas, J. and Menzel, E.J. *Biochim. Biophys. Acta* **1245**, 305–310 (1995).

118. Stamler, J.S., Jaraki, O., Osborne, J., et al. *Proc. Natl. Acad. Sci. USA* **89**, 7674–7677 (1992).

119. Scharfstein, J.S., Keaney, J.F., Jr., Slivka, A., et al. *J. Clin. Invest.* **94**, 1432–1439 (1994).

120. Simon, D.I., Stamler, J.S., Jaraki, O., et al. *Arterioscler. Thromb.* **13**, 791–799 (1993).

121. Keaney, J.F., Jr., Simon, D.I., Stamler, J.S., et al. *J. Clin. Invest.* **91**, 1582–1589 (1993).

122. Stamler, J.S. *Curr. Top. Microbiol. Immunol.* **196**, 19–36 (1995).

123. Stamler, J.S., Singel, D.J., and Loscalzo, J. *Science* **258**, 1898–1902 (1992).

124. Marks, D.S., Vita, J.A., Folts, J.D., Keaney, J.F.J., Welch, G.N., and Loscalzo, J. *J. Clin. Invest.* **96**, 2630–2638 (1995).

125. Arnelle, D.R. and Stamler, J.S. *Arch. Biochem. Biophys.* **318**, 279–285 (1995).

126. Lipton, S.A., Singel, D.J., and Stamler, J.S. *Prog. Brain Res.* **103**, 359–364 (1994).

127. Jia, L., Bonaventura, C., Bonaventura, J., and Stamler, J.S. *Nature* **380**, 221–226 (1996).

128. Lipton, S.A., Choi, Y.B., Pan, Z.H., et al. *Nature* **364,** 626–632 (1993).

129. Radoff, S., Vlassara, H., and Cerami, A. *Arch. Biochem. Biophys.* **263,** 418–423 (1988).

130. Vlassara, H. *Diabetes* **41**(Suppl. 2), 52–56 (1992).

131. Skolnik, E.Y., Yang, Z., Makita, Z., Radoff, S., Kirstein, M., and Vlassara, H. *J. Exp. Med.* **174,** 931–939 (1991).

132. Wautier, J.L., Zoukourian, C., Chappey, O., et al. *J. Clin. Invest.* **97,** 238–243 (1996).

133. Schmidt, A.M., Hori, O., Brett, J., Yan, S.D., Wautier, J.L., and Stern, D. *Arterioscler. Thromb.* **14,** 1521–1528 (1994).

134. Wautier, J.L., Wautier, M.P., Schmidt, A.M., et al. *Proc. Natl. Acad. Sci. USA* **91,** 7742–7746 (1994).

135. Hasegawa, G., Nakano, K., Sawada, M., et al. *Kidney Int.* **40,** 1007–1012 (1991).

136. Lopes-Virella, M.F. and Virella, G. *Diabetes* **41**(Suppl. 2), 86–91 (1992).

137. Corbett, J.A., Tilton, R.G., Chang, K., et al. *Diabetes* **41,** 552–556 (1992).

138. Tilton, R.G., Chang, K., Hasan, K.S., et al. *Diabetes* **42,** 221–232 (1993).

139. Misko, T.P., Moore, W.M., Kasten, T.P., et al. *Eur. J. Pharmacol.* **233,** 119–125 (1993).

140. Vlassara, H., Fuh, H., Makita, Z., Krungkrai, S., Cerami, A., and Bucala, R. *Proc. Natl. Acad. Sci. USA* **89,** 12043–12047 (1992).

141. Sabbatini, M., Sansone, G., Uccello, F., Giliberti, A., Conte, G., and Andreucci, V.E. *Kidney Int.* **42,** 875–881 (1992).

142. Vlassara, H., Striker, L.J., Teichberg, S., Fuh, H., Li, Y.M., and Steffes, M. *Proc. Natl. Acad. Sci. USA* **91,** 11704–11708 (1994).

143. Makita, Z., Radoff, S., Rayfield, E.J., et al. *N. Engl. J. Med.* **325,** 836–842 (1991).

144. Bucala, R. and Vlassara, H. *Blood Purif.* **13,** 160–170 (1995).

145. Makita, Z., Bucala, R., Rayfield, E.J., et al. *Lancet* **343,** 1519–1522 (1994).

146. Vlassara, H. *Blood Purif.* **12,** 54–59 (1994).

147. Makita, Z., Bucala, R., Rayfield, E.J., et al. *Lancet* **343,** 1519–1522 (1994).

148. Vlassara, H. *Kidney Int.* **48**(Suppl. 51), S43–S44 (1995).

149. Lyons, T.J. *Diabetes* **41**(Suppl. 2), 67–73 (1992).

150. Brownlee, M. *Diabetes Care* **15,** 1835–1843 (1992).

151. Yan, S.D., Schmidt, A.M., Anderson, G.M., et al. *J. Biol. Chem.* **269,** 9889–9897 (1994).

152. Kawamura, M., Heinecke, J.W., and Chait, A. *J. Clin. Invest.* **94,** 771–778 (1994).

153. Bucala, R., Makita, Z., Vega, G., et al. *Proc. Natl. Acad. Sci. USA* **91,** 9441–9445 (1994).

154. Griendling, K.K. and Alexander, R.W. *FASEB J.* **10,** 283–292 (1996).

155. Myers, P.R., Wright, T.F., Tanner, M.A., and Ostlund, R.E. *J. Lab. Clin. Med.* **124,** 672–683 (1994).

156. Schmidt, A.M., Hori, O., Chen, J.X., et al. *J. Clin. Invest.* **96**, 1395–1403 (1995).

157. Yang, C.W., Vlassara, H., Peten, E.P., He, C.J., Striker, G.E., and Striker, L.J. *Proc. Natl. Acad. Sci. USA* **91**, 9436–9440 (1994).

158. Sen, C.K. and Packer, L. *FASEB J.* **10**, 709–720 (1996).

159. Kirstein, M., Brett, J., Radoff, S., Stern, D., and Vlassara, H. *Proc. Natl. Acad. Sci. USA* **87**, 9010–9014 (1990).

160. Kirstein, M., Aston, C., Hinz, R., and Vlassara, H. *J. Clin. Invest.* **90**, 439–446 (1992).

161. Schonfelder, G., John, M., Hopp, H., Fuhr, N., van der Giet, M., and Paul, M. *FASEB J.* **10**, 777–784 (1996).

162. McDaniel, M.L., Kwon, G., Hill, J.R., Marshall, C.A., and Corbett, J.A. *Proc. Soc. Exp. Biol. Med.* **211**, 24–32 (1996).

163. Rabinovitch, A., Suarez-Pinzon, W., El-Sheikh, A., and Power, R.F. *Diabetes* **45**, 749–754 (1996).

164. Corbett, J.A., Kwon, G., Turk, J., and McDaniel, M.L. *Biochemistry* **32**, 13767–13770 (1993).

165. Corbett, J.A. and McDaniel, M.L. *Diabetes* **41**, 897–903 (1992).

166. Corbett, J.A. and McDaniel, M.L. *J. Exp. Med.* **181**, 559–568 (1995).

167. Parhami, F., Fang, Z.T., Fogelman, A.M., Andalabi, A., Territo, M.C., and Berliner, J.A. *J. Clin. Invest.* **92**, 471–478 (1993).

168. Vlassara, H., Brownlee, M., Monogue, K., Dinarello, C.A., and Pasagian, A. *Science* **240**, 1546–1548 (1988).

169. Sharma, K., Danoff, T.M., DePiero, A., and Ziyadeh, F.N. *Biochem. Biophys. Res. Commun.* **207**, 80–88 (1995).

170. Pfeilschifter, J. *Kidney Int.* **48**(Suppl. 51), S50–S60 (1995).

171. Nathan, C. and Xie, Q. *Cell* **78**, 915–918 (1994).

172. Beckman, J.S., Beckman, T.W., Chen, J., Marshall, P.A., and Freeman, B.A. *Proc. Natl. Acad. Sci. USA* **87**, 1620–1624 (1990).

173. Radi, R., Beckman, J.S., Bush, K.M., and Freeman, B.A. *Arch. Biochem. Biophys.* **288**, 481–487 (1991).

174. Beckman, J.S. *J. Dev. Physiol.* **15**, 53–59 (1991).

175. Radi, R., Beckman, J.S., Bush, K.M., and Freeman, B.A. *J. Biol. Chem.* **266**, 4244–4250 (1991).

176. Kooy, N.W., Royall, J.A., Ye, Y.Z., Kelly, D.R., and Beckman, J.S. *Am. J. Respir. Crit. Care Med.* **151**, 1250–1254 (1995).

177. Radi, R., Beckman, J.S., Bush, K.M., and Freeman, B.A. *J. Biol. Chem.* **266**, 4244–4250 (1991).

178. Ischiropoulous, H., Zhu, L., Chen, J., et al. *Arch. Biochem. Biophys.* **298**, 431–437 (1992).

179. Koppenol, W.H., Moreno, J.J., Pryor, W.A., Ischiropoulos, H., and Beckman, J.S. *Chem. Res. Toxicol.* **5**, 834–842 (1992).

180. DCCT Trial Res. Group, *N. Engl. J. Med.* **329,** 977–986 (1993).

181. Boel, E., Selmer, J., Flodgaard, H.J., and Jensen, T. *J. Diabetes Complications* **9,** 104–129 (1995).

182. Ting, H.H., Timimi, F.K., Boles, K.S., Creager, S.J., Ganz, P., and Creager, M.A. *J. Clin. Invest.* **97,** 22–28 (1996).

183. Noiri, E., Peresleni, T., Miller, F., and Goligorsky, M.S. *J. Clin. Invest.* **97,** 2377–2383 (1996).

18

NO and Hypertension

Eduardo Nava, Georg Noll, and Thomas F. Lüscher

Although nitric oxide (NO) release seems to be influenced by changes in blood pressure, it remains controversial whether its production is or is not affected in hypertension. Recent research work is unraveling profound differences in the role of NO in different forms of hypertension. This role seems to vary depending on the stage of the disease and model studied. In human hypertension, pharmacological experiments revealed an impaired NO dilator mechanism. In spontaneous and renovascular hypertension in the rat, the production of NO is increased, probably as a compensatory mechanism. However, in genetic hypertension, NO is ineffective in performing its biological functions, presumably because of increased inactivation by oxidative radicals. In this form of hypertension, an increased production of contractile factors and/or a decreased release of hyperpolarizing factors seem to be involved. NO also plays a crucial role in the kidney, where it contributes to the regulation of blood pressure. A reduced renal production of NO, even with a normal overall endothelial function, could contribute to the genesis of hypertension. In salt-dependent hypertension (Dahl- and DOCA salt-sensitive), NO production is impaired, probably due to a deficiency of the substrate for NO synthase. In pulmonary hypertension, the use of NO gas inhalation has been proposed as a future therapy for this condition. The above issues will be the focus of this chapter.

1. Nitric Oxide in the Cardiovascular System

Nitric oxide has quickly become a well-established mediator of cell communication implicated in many physiological and pathophysiological processes, including a variety of cardiovascular diseases [1]. NO is synthesized from L-arginine by NO synthases (Fig. 18-1) [2], a family of enzymes including a calcium-dependent isoform which is constitutively present in endothelial cells and platelets (NOS III) [2]. This enzyme is modulated by shear stress and a variety of receptor-activated agonists as well as hormones and Chapter 8 [3–5]. Nitric oxide synthesized by endothelial cells plays a crucial role in the regulation of blood pressure, vascular tone, and platelet aggregation [6–8]. There is also a second form of constitutive calcium-dependent enzyme present in neuronal cells (NOS I). This enzyme can be found within the nerve endings surrounding the vessels [2]. Another type of NO synthase is calcium independent and inducible by immunological stimuli (NOS II; Fig. 18-1) [2]. This enzyme is present in a number of

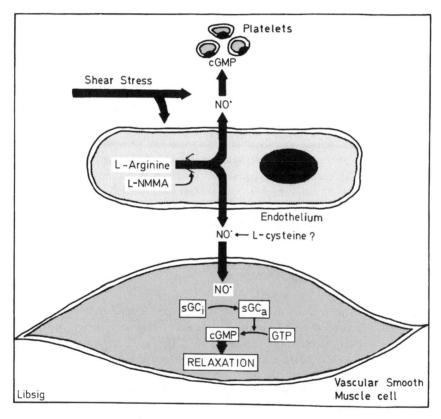

Figure 18-1. The L-arginine/NO pathway in the blood vessel wall. sGC = soluble guanylate cyclase; cGMP = cyclic 3′,5′-guanosine monophosphate; LPS = lipopolysaccharide; TNF = tumor necrosis factor; IL-1 = interleukin 1b; cNOS/iNOS = constitutive/inducible NO synthase; ADMA = asymmetrical dimethylarginine; L-NMMA = N^G-monomethyl-L-arginine; L-NAME = N^G-nitro-L-arginine methylester.

tissues, including smooth muscle cells and macrophages, and takes part in several immunopathological processes [2].

2. Nitric Oxide and Blood Pressure

Research carried out by the authors has shown that brief pharmacologically induced elevations in blood pressure are followed by an increased release of NO to the circulation, which can be detected by measuring small variations in plasma nitrate. On the contrary, drops in pressure cause a decreased production of NO (Fig. 18-2) [9]. We have also observed that the production of NO and the activity of constitutive NOS are higher in a genetic model of hypertension compared to normotensive controls (Fig. 18-3) [10,11]. These findings strongly suggest that high blood pressure upregulates NO production. The mechanism by which elevated blood pressure leads to an increased production of NO is not clear yet. Extensive work has been done on the effect of blood flow, shear stress, and other related mechanical stimuli on the production of NO and the expression of endothelial NOS [3,12–15]. Kelm et at. proposed that only variations in blood flow, but not blood pressure, are able to stimulate the endothelial release of NO [16]. Although cardiac output remains constant in spontaneous hypertension [17,18], it is likely that the shear stress against the vessel wall is increased because of the generalized vasoconstriction seen in this condition. Also, shear stress may be higher due to the increased viscosity of the blood occurring in hypertension [19]. In addition, it is possible that the availability of NO is diminished because of increased oxidative degradation of NO (see Section 5). The overexpression of NOS could be ascribed to cessation of downregulation by NO [20].

3. Nitric Oxide and Hypertension

The role of the endothelium and NO in systemic hypertension is still controversial. We initially proposed that an impaired release of relaxing factors may partly underlie the pathogenesis of hypertension [21]. However, later it appeared that endothelium-dependent relaxation is nonuniformly affected in this condition. In some vascular beds of spontaneously hypertensive rats, such as the aorta, mesenteric, carotid, and cerebral vessels, endothelium-dependent relaxation is impaired [21–23]. In contrast, in coronary and renal arteries of spontaneously hypertensive rats, endothelial function does not seem to be affected by high blood pressure [23,24].

Research carried out so far reveals that the role of NO in hypertension varies widely, depending on the situation and animal model studied (see below for specific details). In some models, NO production seems to be diminished and this might be partly responsible for the high blood pressure. In other situations,

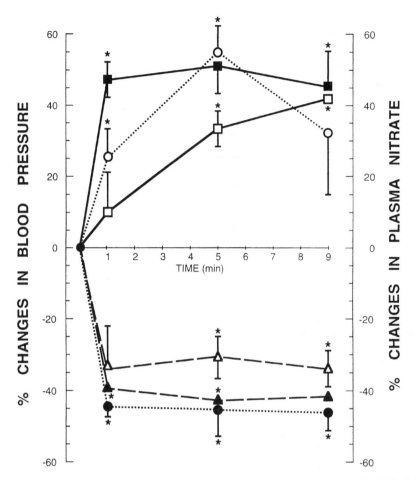

Figure 18-2. Effects of substance P (3 μg/kg/min, dotted lines), prostacyclin (0.3–0.6 mg/kg/min, dashed lines) and angiotensin II (3 μg/kg/min, solid lines) on blood pressure (closed symbols) and plasma nitrate (open symbols). Plasma nitrate concentration follows the changes in blood pressure caused by angiotensin II and prostacyclin. For substance P, however, there is an inverse relation between blood pressure and plasma nitrate. *Statistically significant ($p<.05$) compared to basal levels. (From Ref. 9, by permission.)

the cause of hypertension is not related to the NO pathway and the production of NO is activated as a compensatory mechanism (Fig. 18-4) [25].

4. Nitric Oxide in Human Hypertension

Experiments in humans, have demonstrated a diminished basal and stimulated NO production [26–30]. The decrease in forearm blood flow induced by N_G-

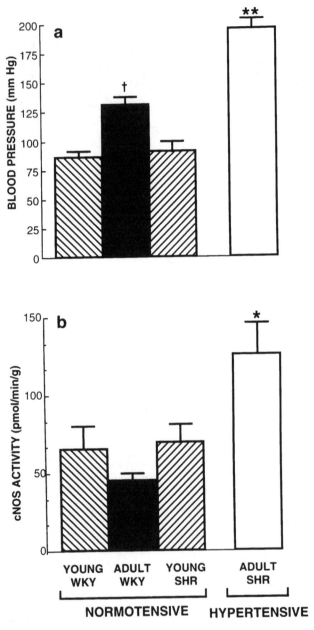

Figure 18-3. Blood pressure (a) and activity of constitutive NO synthase (cNOS; b) in hearts of spontaneously hypertensive (SHR) and normotensive Wistar–Kyoto rats (WKY) at age 3 weeks (young) and 18 weeks (adult). * $P < .05$ and ** $P < .01$ versus normotensives; †$p < .01$ versus young animals. (From ref 10, by permission.)

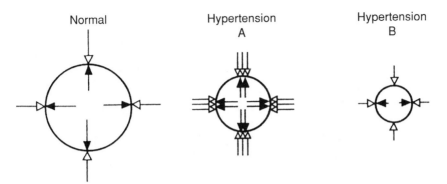

Figure 18-4. Moncada's interpretation of the biological events observed in different experimental situations of hypertension. The production of NO by the vessel wall may vary in three ways. In a normal artery, the vasoconstrictor tone (open-headed arrows) is counterbalanced by the NO-dependent vasodilatory tone (closed-headed arrows). In hypertensive situation A (e.g., spontaneous or renovascular hypertension), the vasoconstrictor tone is primarily increased and NO production appears higher than normal in an attempt to equilibrate blood pressure. In situation B (e.g., salt-sensitive hypertension), the primary cause is a diminished production of NO. The usual vasoconstrictor tone is enough to produce hypertension. (From Ref. 25, by permission.)

monomethyl-L-arginine (L-NMMA), is smaller in hypertensive than in normotensive patients [26] and endothelium-dependent vasodilations to acetylcholine appear to be reduced in patients with primary or secondary hypertension (Fig. 18-5 [27–29]). The impaired endothelial response in hypertensives can be enhanced by indomethacin, suggesting that vasoconstrictor prostanoids also contribute to the impaired endothelium-dependent relaxation in hypertensive patients [29]. Most of these studies are based, unfortunately, on pharmacological experiments, and few analytical measurements of NO have been performed (except for urinary nitrate which seems to be lower in hypertensive patients [30]). Tracing the real production of NO from endothelial cells in human hypertension is an endeavor which still remains to be undertaken. New horizons have been opened in this field by the recent use of porphyrinic microprobes in human vessels [31].

5. Nitric Oxide in Spontaneous Hypertension

The spontaneously hypertensive rat (SHR) is a genetic model of hypertension which we have investigated for many years. In the mid-1980s, we demonstrated that the impaired relaxations of aortic segments from SHR are not caused by a deterred production of endothelium-derived relaxing factor (EDRF) but instead, by an increased release of prostanoid contractile factors (Fig. 18-6) [21,32,33]. With the advent of the NO age and using new methodologies, we have confirmed

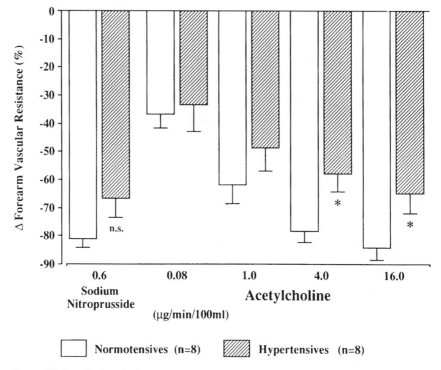

Figure 18-5. Changes in forearm vascular resistance caused by acetylcholine and sodium nitroprusside in normotensives and hypertensives. The response to acetylcholine, but not that to nitroprusside, was impaired in hypertensive patients. * p<.05. (From Ref. 27, by permission.)

the initial observations on the release of EDRF in this form of hypertension (Fig. 18-2) [10,11]. The activity of constitutive NOS, as assessed by the conversion of [^{14}C]-L-arginine to [^{14}C]-L-citrulline is higher in the heart and in mesenteric resistance arteries obtained from SHR compared to age-matched normotensive rats [10,11]. Moreover, the concentration of the oxidative product of NO, nitrate, measured by high-performance liquid chromatography and capillary electrophoresis, is higher in hypertensive rats as compared to normotensive controls [11]. In contrast, prehypertensive young SHR exhibit similar nitrate levels as the normotensive. These observations demonstrate that the basal release of NO is increased in rats with spontaneous hypertension and that this increased production is directly related to the high blood pressure of the animals. Interestingly, Hirata's group has found that, although the vasoactive effects of acetylcholine in the kidney are indeed abnormal, the release of NO from renal vessels is not diminished but is even slightly higher in the SHR (Fig. 18-7) [33]. Treatment with inhibitors of the endothelium-derived hyperpolarizing factor (EDHF) brings the vasoactive

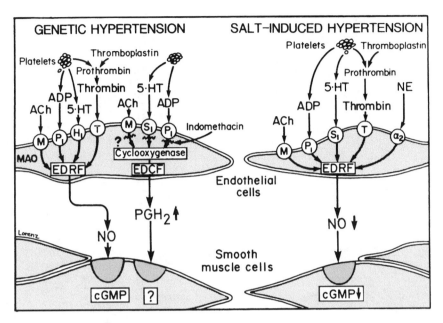

Figure 18-6. Different mechanisms of endothelial dysfunction in genetic (i.e., spontaneous; left) and salt-induced hypertension of the rat (right). ACh = acetylcholine; ADP = adenosine diphosphate; cGMP = cyclic 3′,5′-guanosine monophosphate; NE = noradrenaline; 5-HT = 5-hydroxytryptamine (serotonin); PGH_2 = prostaglandin H_2; EDCF = endothelium-derived contracting factor. (Modified from Ref. 44, by permission.)

effects of acetylcholine to a similar level in normotensive and hypertensive rats, suggesting that, in this form of hypertension, a decreased EDHF production, but not NO, is taking place [33].

Further studies carried out in our laboratory have demonstrated that the accumulation of cyclic GMP in mesenteric resistance arteries is similar in WKY and SHR [11]. Moreover, the NO-dependent vasodilator tone, assessed by the blood pressure effects of L-NAME, is not higher in hypertensive rats, as would be expected in a situation in which the production of NO is increased. The capacity of vascular smooth muscle cells of hypertensive rats to respond to NO, on the other hand, must be fully maintained as organic nitrates lower blood pressure in a similar fashion in both strains of rats and relaxations to sodium nitroprusside are enhanced in this condition [32]. Altogether, these findings suggest that endogenously produced NO, though increased in SHR, is unable to properly raise cyclic GMP levels in the vascular smooth muscle cells of these animals. Hence, it appears that an additional event takes place that prevents NO from accomplishing normal hemodynamic functions. The hypertrophied and fibrotic intimal layer of hypertensive vessels [34] may represent a physical barrier for NO, accounting

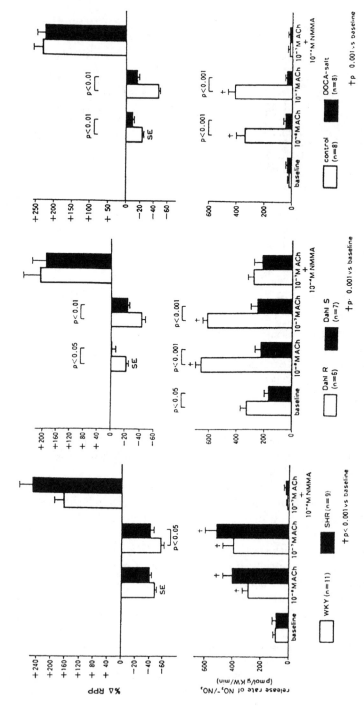

Figure 18-7. Effects of acetylcholine (Ach) and N^G-monomethyl-L-arginine (L-NMMA) on renal perfusion pressure (RPP) and release rates of nitrite and nitrate ($NO_2^- - NO_3^-$) into the perfusate in spontaneously hypertensive rats (SHR) and Wistar–Kyoto rats (WKY). KW, kidney weight (From Ref. 33, by permission.)

for the blunted hemodynamic actions of NO. Also the chemical environment that NO encounters, such as oxicative stress [35] can determine its fate.

6. Nitric Oxide and the Kidney in Hypertension

Nitric oxide plays an important role in renal function. Indeed, the kidney is extremely sensitive to NO inhibition as very low doses of L-arginine analogs, which do not affect blood pressure, diminish diuresis and natriuresis [36,37]. Research carried out by Salazar and co-workers demonstrated the importance of NO in the kidney, particularly when the balance of sodium is altered or extracellular volume expansion occurs [37–39]. Indeed, the activity of constitutive NOS is higher in the renal medulla (where the regulation of pressure-induced natriuresis takes place) than in any other tissue related to the control of blood pressure [40]. It is also noteworthy that the kidney displays the highest increases in vascular resistance upon treatment with NO synthesis inhibitors [41]. It is plausible that in certain circumstances, minimal alterations in the production of NO occur. In such cases, the cardiovascular system might not be affected, but systemic hypertension emerges due to a deterrence of kidney function. It has been recently shown that renal failure is also associated with an accumulation of an endogenous inhibitor of NO synthesis, the asymmetrical dimethylarginine [42], which could also explain the increase in peripheral resistance and hypertension observed in this condition.

In renovascular hypertension endothelium-dependent vascular relaxation is impaired, suggesting a diminished production of EDRF [43]. However, Sigmon and Beierwaltes have demonstrated in a model of renovascular hypertension that treatment with an inhibitor of NO synthesis similarly affects clipped and nonclipped kidneys as well as normotensive controls [41]. Thus, in this model of hypertension, the endothelium is not disfunctional but is a critical component in the adaptation to increased blood pressure. These authors also observed that the clipped kidney lacks the NO-mediated adaptive mechanism and suggested that this may contribute to the pathogenesis of hypertension in this model [41].

7. Nitric Oxide in Salt-Sensitive Hypertension

We reported that Dahl salt-sensitive rats show impaired endothelium-dependent relaxations (Fig. 18-6) [44], but no release of vasoconstrictor prostanoids could be demonstrated [44,45]. This suggests that a decreased NO production may contribute to the pathogenesis of this form of hypertension. Hayakawa et al. recently demonstrated that Dahl salt-sensitive rats display a reduced response to acetylcholine in renal blood vessels which is accompanied by a decreased release of NO (Fig. 18-7) [33]. Chen and Sanders showed interesting data supporting this idea [46,47]. Inhibition of NO synthesis causes a higher increase in blood pressure in normotensive rats (Sprague–Dawley and Dahl salt-resistant) than in Dahl salt-sensitive rats, suggesting that the synthesis of NO is lower in the latter group of animals [46,47]. These authors also demonstrated that NO production, as assessed by the pharmacological effects of NO inhibitors, improves by feeding NaCl to salt-resistant Dahl rats [46]. Moreover, administration of L-arginine was effective in lowering blood pressure in the salt-sensitive strain, but not the SHR [46]. These findings provide new insights into the pathogenesis of salt-dependent hypertension and emphasize the profound differences in the role of NO in several forms of hypertension.

Blood vessels from DOCA salt-sensitive hypertensive rats elicit impaired endothelium-dependent relaxations [48], and cGMP accumulation is diminished in these vessels [49]. Thus, a decreased production of NO may be involved. Hayakawa and co-workers have demonstrated that these impaired relaxations are paralleled by a diminished release of NO from perfused kidney vessels (Fig. 18-7) [33]. Beneficial effects of treatment with L-arginine on endothelial function have also been shown [50,51], suggesting, once again, a compromised availability of the substrate for NOS as a possible pathogenesis of this form of hypertension.

8. Nitric Oxide in the Hypertensive Heart

Hypertension leads to left ventricular hypertrophy, myocardial damage and heart failure [52]. Recent studies have shown that in spontaneous hypertension, the production of NO is increased in the heart. Kelm et al. have shown that the release of NO from isolated coronary vessels is augmented in the spontaneously hypertensive rat (SHR [53]). Moreover, experiments performed in our laboratory demonstrate that adult SHR possess a higher activity of constitutive NOS in the heart endothelial cells than their normotensive counterparts [10]. Very young prehypertensive SHR have, in contrast, lower enzymatic activity than the normotensive animals, indicating that the increased activity of NOS in these cells is indeed related to hypertension (Fig. 18-3) [10]. The augmented activity of constitutive NOS in the hypertensive heart develops at the expense of the left ventricle, where the highest differential pressure in the cardiovascular system is found [10]. These observations further support the concept that high blood pressure upregulates the endothelial NOS and, hence, the production of NO. It appears that cardiac cNOS activity remains unchanged within the normotensive blood pressure range; however, there exists a pressure threshold above which upregulation of the enzyme takes place.

An enhanced production of NO in the hypertensive heart probably acts as a compensatory mechanism by decreasing myocardial contractility and causing vascular dilatation. Nitric oxide in the hypertensive heart may also protect it from hypertrophy. High blood pressure causes cardiac hypertrophy and fibrosis which often leads to left ventricular failure [54]. Nitric oxide, a potent inhibitor of smooth muscle cell growth and migration [54], might protect the heart from these damaging effects of hypertension.

9. Nitric Oxide and Pulmonary Hypertension

Nitric oxide is present in exhaled air of animals [55] and humans [56]. It is an inhibitory modulator of the pulmonary hypoxic pressor response [57]. Furthermore, NO plays a role in the pulmonary circulation and in the normal oxygenation of the blood through regulation of ventilation–perfusion match [57]. It is likely that the NO involved in the physiological regulation of the pulmonary circulation is synthesized by the constitutive NO synthase [55]. On the basis of these findings, inhaled NO has been used for the treatment of various forms of pulmonary hypertension [58], such as the pulmonary hypertension of the newborn [59], pulmonary hypertension of congenital heart disease [60], idiopathic pulmonary hypertension [61], acute pneumonia [62], and severe obstructive airways disease [63]. It is important to realize that inhaled NO is still at the experimental stage and the potential hazards of this treatment have not been fully explored. If further data confirm the early promise of this novel treatment, it will not be long before NO delivery and monitoring machines are in routine use in neonatal and intensive care units.

Acknowledgments

Original work was supported by grants of the Swiss National Research Foundation (# 32-32541.91 and 32-32655.91) and the Wilhelm Bitter Foundation and the Fondo de Investigaciones Sanitarias de la Seguridad Social (#95/1763).

References

1. Moncada, S. and Higgs, A. Mechanisms of disease. The L-arginine nitric oxide pathway. *N. Engl. J. Med.* **329**(27), 2002–2012 (1993).

2. Moncada, S. The L arginine: Nitric oxide pathway. *Acta Physiol. Scand.* **145**, 201–227 (1992).

3. Nishida, K., Harrison, D.G., Navas, J.P., Fisher, A.A., Dockery, S.P., Uematsu, M., Nerem, R.M., Alexander, W., and Murphy T.J. Molecular cloning and characterization of the constitutive bovine aortic endothelial cell nitric oxide synthase. *J. Clin. Invest.* **90**, 2092–2096 (1992).

4. Schmidt, H.H.H.W., Zernikow, B., Baeblich, S., and Böhme, E. Basal and stimulated formation and release of L arginine derived nitrogen oxides from cultured endothelial cells. *J. Pharmacol. Exp. Ther.* **254**(2), 591–597 (1990).

5. Weiner, C.P., Lizasoain, I., Baylis, S.A., Knowles, R.G., Charles, I.C., and Moncada, S. Induction of calcium dependent nitric oxide synthase by sex hormones. *Proc. Natl. Acad. Sci. USA* **91**, 5212–5216 (1994).

6. Rees, D.D., Palmer, R.M.J., and Moncada, S. Role of endothelium derived nitric oxide in the regulation of blood pressure. *Proc. Natl. Acad. Sci. USA* **86**, 3375–3378 (1989).

7. Vallance, P., Collier, J., and Moncada, S. Effects of endothelium derived nitric oxide on peripheral arteriolar tone in man. *Lancet* **ii**, 997–1000 (1989).

8. Radomski, M.W., Palmer, R.M.J., and Moncada, S. An ʟ arginine to nitric oxide pathway in human platelets regulates aggregation. *Proc. Natl. Acad. Sci. USA* **87**, 5193–5197 (1990).

9. Nava, E., Leone, A.M., Wiklund, N.P., and Moncada, S. Detection of release of nitric oxide by vasoactive substances in the anaesthesized rat. In: *The Biology of Nitric Oxide;* Eds. Feelisch, M., Busse, R., and Moncada, S. Portland Press, London, 1994, pp. 179–181.

10. Nava, E., Noll, G., and Lüscher, T.F. Increased activity of constitutive nitric oxide synthase in cardiac endothelium in spontaneous hypertension. *Circulation* **91**, 2310–2313 (1995).

11. Nava, E., Moreau, P., Lüscher, T.F. Basal production of nitric oxide is increased, but inefficacious in spontaneous hypertension. *Circulation* **92**(8), I 347 (1995).

12. Hutcheson, R. and Griffith, T.M. Release of endothelium derived relaxing factor is modulated both by frequency and amplitude of pulsatile flow. *Am. J. Physiol.* **30**, H257–H262 (1991).

13. Kanai, A.J., Strauss, H.C., Truskey, G.A., Crews, A.L., Grunfeld, S., and Malinski, T. Shear stress induces ATP independent transient nitric oxide release from vascular endothelial cells, measured directly with a porphyrinic microsensor. *Cir. Res.* **77**, 284–293 (1995).

14. Pohl, U., Herlan, K., Huang, A., and Bassenge, E. EDRF mediated shear induced dilation opposes myogenic vasoconstriction in small rabbit arteries. *Am. J. Physiol.* **261**, H2016–H2023 (1991).

15. Awolesi, M.A., Sessa, W.C., and Sumpio, B.E. Cyclic strain upregulates nitric oxide synthase in cultured bovine aortic endothelial cells. *J. Clin. Invest.* **96**, 1449–1454 (1995).

16. Kelm, M., Feelisch, M., Deussen, A., Strauer, B.E., and Scharader, J. Release of endothelium derived nitric oxide in relation to pressure and flow. *Cardiovasc. Res.* **25**, 831–836 (1991).

17. Smith, T.L. and Hutchings, P.M. Central hemodynamics in the developmental stage of spontaneous hypertension in the unanesthetized rat. *Hypertension* **1**(5), 508–517 (1979).

18. Nishiyama, K., Nishiyama, A., and Frohlich, E.D. Regional blood flow in normotensive and spontaneously hypertensive rats. *Am. J. Physiol.* **230**(3), 691–698 (1976).

19. De Clerck, F., Beerens, M., Van Gorp, L., and Xhonneux R. Blood hyperviscosity in spontaneously hypertensive rats. *Throm. Res.* **18**, 291–295 (1980).

20. Assreuy, J., Cunha, F.Q., Liew, F.Y., and Moncada, S. Feedback inhibition of nitric oxide synthase activity by nitric oxide. *Br. J. Pharmacol.* **108**, 833–837 (1993).

21. Lüscher, T.F. and Vanhoutte, P.M. Endothelium dependent contractions to acetylcholine in the aorta of the spontaneously hypertensive rat. *Hypertension* **8**, 344–348 (1986).

22. Dohi, Y., Thiel, M.A., Buhler, F.R., and Lüscher, T.F. Activation of the endothelial L arginine pathway in pressurized mesenteric resistance arteries: effect of age and hypertension. *Hypertension* **15**, 170–179 (1990).

23. Lüscher, T.F. Endothelium derived nitric oxide: The endogenous nitrovasodilator in the human cardiovascular system. *Eur. Heart. J.* **12**(Suppl. E), 2–11 (1991).

24. Tschudi, M.R., Criscione, L., and Lüscher, T.F. Effect of aging and hypertension on endothelial function of rat coronary arteries. *J. Hyperten.* **9**, 164–165 (1991).

25. Moncada, S. Nitric oxide. *J. Hyperten.* **12**(Suppl. 10), S35–S39 (1994).

26. Calver, A., Collier, J., Moncada, S., and Vallance, P. Effect of local intra arterial N^G-monomethyl-L-arginine in patients with hypertension: The nitric oxide dilator mechanism appears abnormal. *J. Hyperten.* **10**, 1025–1031 (1992).

27. Linder, L., Kiowski, W., Bühler, F.R., and Lüscher, T.F. Indirect evidence for the release of endothelium derived relaxing factor in human forearm circulation in vivo: Blunted response in essential hypertension. *Circulation* **81**, 1762–1768 (1990).

28. Panza, J.A., Casino, P.R., Kilcoyne, C.M., and Quyyumi, A.A. Role of endothelium dependent vascular relaxation of patients with essential hypertension. *Circulation* **87**, 1468–1474 (1993).

29. Taddei, S., Virdis, A., Mattei, P., and Salvetti, A. Vasodilation to acetylcholine in primary and secondary forms of human hypertension. *Hypertension* **21**, 929–933 (1993).

30. Benjamin, N., Copland, M., and Smith, L.M. Reduced nitric oxide synthesis in essential hypertension. *Endothelium,* **3**, s94 (1995).

31. Vallance, P., Patton, S., Bhagat, K., Macalister, R., Radomski, M., Moncada, S., Malinski, T. Direct measurement of nitric oxide in humans. *Endothelium* **3**, s81 (1995).

32. Diederich, D., Yang, Z., Bühler, F.R., and Lüscher, T.F. Impaired endothelium dependent relaxations in hypertensive resistance arteries involve cyclooxigenase pathway. *Am. J. Physiol.* **258**, H445–H451 (1990).

33. Hayakawa, H., Hirata, Y., Suzuki, E., Sugimoto, T., Matsuoka, H., Kikuchi, K., Nagano, T., Hirobe, M., Sugimoto, T. Mechanisms for altered endothelium dependent vasorelaxation in isolated kidneys from experimental hypertensive rats. *Am. J. Physiol.* **264**, H1535–H1541 (1993).

34. Lindop, G.B.M. *Textbook of Hypertension.* Ed. Swales, J.D. Blackwell Scientific Publications, London, 1994, pp. 663–669.

35. Grunfeld, S., Hamilton, C.A., Mesaros, S., McClain, S.W., Dominiczak, A.F., Bohr, D.F., and Malinski, T. Role of superoxide in the depressed nitric oxide production by the endothelium of genetically hypertensive rats. *Hypertension* **26**(Part 1), 854–857 (1995).

36. Lahera, V., Salom, M.G., Miranda Guardiola, F., Moncada, S., and Romero, J.C. Effects of *N*-nitro-L-arginine methyl ester on renal function and blood pressure. *Am. J. Physiol.* **261,** F1033–F1037 (1991).

37. Salazar, F.J., Pinilla, J.M., López, F., Romero, J.C., and Quesada, T. Renal effects of prolonged synthesis inhibition of endothgeliun derived nitric oxide. *Hypertension* **22,** 49–55 (1992).

38. Salazar, F.J., Alberola, A., Pinilla, J.M., Romero, J.C., Quesada, T. Salt induced increase in arterial pressure during nitric oxide synthesis inhibition. *Hypertension* **20,** 113–117 (1993).

39. Alberola, A., Pinilla, J.M., Quesda, T., Romero, J.C., Salom, M.G., and Salazar, F.J. Role of nitric oxide in mediating renal response to volume expansion. *Hypertension* **19,** 780–784 (1992).

40. Nava, E., Gonzalez, J.D., Llinas, M.T., and Salazar, F.J. Relative importance of constitutive NO synthase in the rat kidney. *Eur. J. Clin. Invest.* (in press).

41. Sigmon, D.H. and Beierwaltes, W.H. Nitric oxide influences blood flow distribution in renovascular hypertension. *Hypertension* **23**(1), I-34–I-39 (1994).

42. Vallance, P., Leone, A., Calver, A., Collier, J., and Moncada, S. Accumulation of an endogenous inhibitor of nitric oxide synthesis in chronic renal failure. *Lancet* **329,** 572–575 (1992).

43. Lockette, W., Otsuka, Y., and Carretero, O.A. The loss of endothelium-dependent vascular relaxation in hypertension. *Hypertension* **8**(Suppl. II), II-61–II-66 (1986).

44. Lüscher, T.F. and Vanhoutte, P.M. Mechanisms of altered endothelium dependent responses in hypertensive blood vessels. In: *Relaxing and Contracting Factors:* Humana Press Clifton New Jersey 1988 Ed.

45. Lüscher, T.F., Raij, L., and Vanhoutte, P.M. Endothelium dependent vascular responses in normotensive and hypertensive Dahl rats. *Hypertension* **9,** 157–163 (1987).

46. Chen, P.Y. and Sanders, P.W. L Arginine abrogates salt sensitive hypertension in Dahl/Rapp rats. *J. Clin. Invest.* **88,** 1559–1567 (1991).

47. Chen, P.Y. and Sanders, P.W. Role of nitric oxide synthesis in salt sensitive hypertension in Dahl/Rapp rats. *Hypertension* **22,** 812–818 (1993).

48. Voorde, J.V. and Leusen, I. Endothelium dependent and independent relaxation of aortic rings from hypertensive rats. *Am. J. Physiol.* **250,** H711–H717 (1986).

49. Otsuka, Y., Dipiero, A., Hirt, E., Brennaman, B., and Lockette, W. Vascular relaxation and cGMP in hypertension. *Am. J. Physiol* **254,** H163–H169 (1988).

50. Hayakawa, H., Hirata, Y., Suzuki, E., Kimura, K., Kikuchi, K., Nagano, T., Hirobe, M., and Omata, M. Long term administration of L-arginine improves nitric oxide release from kikney in deoxycorticosterone acetate salt hypertensive rats. *Hypertension* **23**(1), 752–756 (1994).

51. Laurant, P., Demolombe, B., Berthelot, A. Dietary L-arginine attenuates blood pressure in mineralocorticoid salt hypertensive rats. *Clin. Exp. Hyperten.* **17**(7), 1009–1024 (1995).

52. Anversa, P., Peng, L.I., Malhotra, A., Zhang, X., Herman, M.V., and Capasso, J. Effects of hypertension and coronary constriction on cardiac function, morphology, and contractile proteins in rats. *Am. J. Physiol.* **265**, 713–725 (1993).

53. Kelm, M., Feelisch, M., Krebber, T., Motz, W., and Strauer, B.E. The role of nitric oxide in the regulation of coronary vascular resistance in arterial hypertension: Comparison of normotensive and spontaneously hypertensive rats. *J. Cardiovasc. Pharmacol.* **20**, 183–186 (1992).

54. Dubey, R.K., Jackson, E.K., and Lüscher, T.F. Nitric oxide inhibits angiotensin-II-induced migration of rat aortic smooth muscle cell. *J. Clin. Invest.* **96**, 141–149 (1995).

55. Gustafsson, S.E., Leone, A.M., Persson, M.G., Wiklund, N.P., and Moncada, S. Endogenous nitric oxide is present in the exhaled air of rabbits, guinea pigs and humans. *Biochem. Biophys. Res. Commun.* **181**, 852–857 (1991).

56. Leone, A.M., Gustafsson, S.E., Francis, P.L., Persson, M.G., Wiklund, N.P., and Moncada, S. Nitric oxide is present in exhaled breath in humans: direct GC MS confirmation. *Biochem. Biophys. Res. Commun.* **201**, 883–887 (1994).

57. Wiklund, N.P., Persson, M.G., Gustafsson, L.E., Moncada, S., and Hedqvist, P. Modulatory role of endogenous nitric oxide in pulmonary circulation *in vivo*. *Eur. J. Pharmacol.* **185**, 123–124 (1990).

58. Landzberg, M.J., Graydon, E., Fernandes, S.M., Hare, J.M., Body, S.C., Atz, A.M., and Wessel, D.L. Inhaled nitric oxide in primary pulmonary hypertension the perfect pulmonary vasodilator. *Circulation* **92**(8), I241 (1995).

59. Roberts, J.D., Polanger, D.M., Lang, P., and Zapol, W.M. Inhaled nitric oxide in persistent pulmonary hypertension of the newborn. *Lancet* **340**, 818–819 (1992).

60. Roberts, J.D., Lang, P., Bigatello, L.M., Vlahakes, G.J., and Zapol, W.M. Inhaled nitric oxide in congenital heart disease. *Circulation* **87**, 447–453 (1993).

61. Pepke-Zaba, J., Higenbottam, T.X., Tuan Dinh-Xuan, A., Stone, D., and Wallwork, J. Inhaled nitric oxide as a cause of selective pulmonary vasodilatation in pulmonary hypertension. *Lancet* **338**, 1173–1174 (1991).

62. Blomqvist, H., Wickerts, C.J., Andreen, M., Ullberg, U., Ortqvist, A., and Frostell, C. Enhanced pneumonia resolution by inhalation of nitric oxide. *Acta Anaesthesiol. Scand.* **37**, 110–114 (1993).

63. Adatia, I., Thomson, J., Landzberg, M., and Wessel, D. Inhaled nitric oxide in chronic obstructive lung disease. *Lancet* **341**, 307–308 (1993).

19

NO in Normal and Preeclamptic Pregnancy

Chris Baylis

Normal pregnancy involves dramatic hemodynamic maternal changes which are suppressed in preeclampsia, leading to serious complications for mother and baby. The mechanisms of the physiologic adaptations are unknown, as is the precise cause of preeclampsia, and there are no adequate methods of prevention or reliable early markers of the disease. As discussed in detail in this chapter, it is possible that the normal hemodynamic physiologic adaptations are due to increased nitric oxide (NO) production and that the systemic manifestations of preeclampsia are the result of a net NO-deficient synthesis.

1. Renal and Systemic Hemodynamics in Normal Pregnancy

Large increases occur in plasma volume, stroke volume, and heart rate, which together produce an ~40% rise in cardiac output [1,2]. Despite the increase in cardiac output, blood pressure (BP) falls [1,3] due to a reduction in total peripheral vascular resistance (TPVR) [1]. As well as arteriolar vasodilation, normal pregnancy is accompanied by vascular refractoriness to the pressor action of administered vasoconstrictors including angiotensin II (AII) and arginine vasopressin [2,4,5]. In addition to changes in systemic hemodynamics, there are marked changes in kidney function indicated by the fall in serum creatinine, due to a rise in glomerular filtration rate (GFR), maintained throughout most of the gestational period [6,7]. The time course and relative magnitude of these hemodynamic responses to pregnancy are shown in Fig. 19-1. The rise in GFR is correlated with increases in renal plasma flow (RPF) [7], and all of these systemic and renal responses are seen in the rat as well as in women. Studies by the author have shown that in the rat, the rise in GFR is entirely due to increases in RPF, secondary to renal vasodilation [8]. The renal vasodilation precedes the general volume expansion and involves parallel relaxation of both preglomerular and

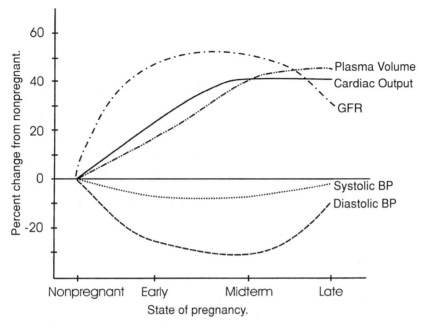

Figure 19-1. Diagram depicting the time course and magnitude of the changes in blood pressure (BP), GFR, plasma volume, and cardiac output in normal pregnant women.

postglomerular arteriolar resistances (R_A and R_E). An important consequence of the parallel falls in R_A and R_E is that no change occurs in glomerular blood pressure, P_{GC} during pregnancy [8].

The renal and systemic hemodynamic changes are initiated by maternal factors, as they also occur in pseudopregnant rats [8]. However, the exact causes of the renal and peripheral vasodilation of pregnancy are unknown although various hormone systems have been implicated as potential mediators. Increased production of prostaglandins occur in pregnancy [8,9], but studies by the author suggest that these vasodilatory agents do not directly mediate the gestational rise in GFR or falls in BP [10,11]. The renin–angiotensin II system is modified in pregnancy with increased plasma renin activity (PRA) and plasma AII levels and decreased responsivity to administered AII [4,5]. We have shown that AII does not control BP or renal function in the normal (physiologic) nonpregnant state [12]; thus, it is unlikely that gestational changes in the AII system cause the fall in BP or increased GFR. Plasma atrial natriuretic peptide (ANP) increases moderately during pregnancy, although there is substantial variability [13,14]. These increases in ANP are not responsible for the gestational fall in BP because plasma ANP remains elevated long after delivery, when BP has returned to prepregnancy values [14] and Masilamani and Baylis have shown that the depressor response to ANP is unaltered in pregnant rats [15]. Despite much investigative effort, the cause of the vasodilation and plasma volume expansion of normal pregnancy remains elusive. The possible role of the nitric oxide (NO) system is discussed in section 4.

2. Nitric Oxide: Control of Systemic and Renal Hemodynamics in the Nongravid State

Many cell types, including vascular endothelial cells, make the vasodilatory NO synthesized enzymatically from L-arginine, which acts predominantly via the second messenger, cyclic GMP (cGMP) [16]. Vascular endothelial cells contain the constitutive, endothelial NO synthase (eNOS) that continually produces NO, which controls TPVR and BP by direct vasodilatory actions and by blunting the responsiveness to vasoconstrictors [16,17]. A second constitutively expressed neuronal or "brain-type" NOS (bNOS) influences BP and kidney function by central and peripheral neural activity [18] and possibly by influencing renal epithelial cell transport in various locations throughout the tubule, including the macula densa [19,20]. A third inducible NOS isoform (iNOS) may influence BP and/or kidney function under some circumstances [21,22]. Nonselective inhibition of NO production from the various NOS isoforms can be produced by systemic administration of substituted L-arginine analogs which compete with the endogenous substrate and inhibit NOS [16]. Studies in animals have shown that acute NOS inhibition leads to dose-dependent increases in BP, to a maximum of ~30–40% above basal values [12,16,23]. Pressor doses of NO inhibitors also produce large increase in renal vascular resistance (RVR), decline in RPF, and small reduction in GFR [12,23]. Local inhibition of NO production in the human forearm circulation produces vasoconstriction [24,25] and systemic NOS inhibition increases BP and TPVR [26], showing that tonically produced NO vasodilates the periphery in normal man. Infusion of the native substrate L-arginine stimulates NO production in some settings, and studies in normal man have shown that L-arginine lowers BP and increases RPF [27]. Thus, tonically produced NO plays an important role in control of BP and renal function in the normal, nonpregnant human, as discussed in more detail elsewhere in this volume (Chapters 8 and 18).

The author and others have shown that in animals, chronic NOS inhibition produces a dose-dependent, chronic systemic hypertension, renal vasoconstriction with substantial elevations in glomerular blood pressure, proteinuria, and glomerular injury [28,29]. Whereas partial NOS inhibition for 8 weeks produces a moderate stable hypertension [28], near-complete NOS inhibition elicits a higher mean BP over a shorter (5–6 week) period, with some rats developing malignant hypertension [29]. Although the role of NO deficiency in human hypertension has yet to be clearly defined, several clinical studies have suggested that NO deficiency occurs in some essential hypertensives, and acute L-arginine (NO substrate) infusion is an effective antihypertensive agent in essential and secondary hypertension in nonpregnant humans [24,30]. This important area is discussed elsewhere in this volume (Chapter 18).

3. Methods of Assessment of Activity of NO

Plasma or tissue levels of the NO substrate arginine give a measure of substrate availability, because NO is enzymatically synthesized from L-arginine [16]. It is generally believed that L-arginine is not rate limiting for normal constitutive NO synthesis, but the substrate can become rate limiting when NO production rates are high due to iNOS; whether increased levels of constitutively generated NO also become substrate limited is not known [16,31,32]. There are also endogenous, substituted L-arginine analogs such as asymmetric dimethylarginine (ADMA) which functions as an inhibitor of NO synthesis, by competing with the native substrate, L-arginine, for the enzyme(s) that synthesize NO (the NO synthases, NOS) [33,34]. Because endogenous levels of these L-arginine analogs are present in concentrations which can inhibit NO synthesis, plasma or tissue levels of these compounds may be informative about overall NO synthesis. In vivo, orotic acid (a precursor of pyrimidine synthesis) is increased as a marker of arginine deficiency, which could reflect arginine depletion secondary to excessive NO synthesis or NO deficiency secondary to inadequate arginine availability [35,36 and Chapter 6].

The type, location, and abundance of the NOS is also an important determinant of NO production. So far, three NOS isoforms have been sequenced and cloned; NOS I is the constitutively expressed brain-type NOS (bNOS) first found in the cerebellum; NOS II is the inducible NOS (iNOS) activated by immune and possibly other stimuli and first found in the macrophage; NOS III is the endothelial constitutively expressed NOS (eNOS), and a second iNOS isoform has recently been detected in the rat kidney [20,21]. The activity of these various NOS isoforms can be measured by arginine to citrulline conversion, localized with immunohistochemistry, and NOS mRNA expression can be quantitated using specific cDNA probes [16,19,20,21,22,37]. Several cofactors are required for NOS activity, such as calcium in the case of the constitutively expressed NOS, and FAD, NADPH, and tetrahydrobiopterin for all NOS [16,38]; however, these are ubiquitous and it is unlikely that cofactor levels will be useful as an index of activity of the NO system.

Nitric oxide is extremely labile and cannot be measured directly because it rapidly undergoes oxidation to inorganic NO_2 and $NO_3 = NO_x$ [16]. The amount of NO_x present in body fluids represents the production of NO when corrected for the NO_x ingested in the diet. The best way to assess NO production is from 24-h urine NO_x excretion rates, corrected for intake. We have found that acutely $U_{NO_x b}V$ does *not* reliably reflect NO production because there is extensive renal epithelial handling of filtered NO_x and when tubular reabsorption is altered,

$U_{NO_x}V$ will change independently of the NO system [39]. Plasma and 24-h urine values of the NO second messenger cGMP are informative but may be misleading when viewed in isolation, as cGMP is also the second messenger of atrial natriuretic peptide.

In addition to measuring the activity and/or levels of the various components of the NO system, acute interruption of NO synthesis, either locally or systemically and either in vivo or in vitro, can also give important insights into the functional activity of the endogenous NO system, as discussed above. In vivo and in vitro responses to certain agonists, such as acetylcholine (Ach) and bradykinin (Bk) as well as the physiologic stimulus of shear stress, have provided another means of assessing the activity of the NO system [16]. Finally, the responsiveness to NO donors (such as sodium nitroprusside) and the distribution and isoform of the NO receptor (predominantly soluble guanylate cyclase) are also informative [40,41].

Many of these approaches have been employed to investigate the role of NO in normal and hypertensive pregnancy, as discussed in the next section.

4. NO in Normal Pregnancy: Control of Maternal Hemodynamics

4.1. Experimental

A number of studies have investigated whether enhanced production and/or sensitivity to NO occurs in normal pregnancy. Plasma and urinary levels of cGMP (the second messenger of NO) increase during pregnancy in rats and urinary cGMP increases in pseudopregnant rats [42]. This probably reflects increased tissue production of cGMP, as metabolic clearance rate is unaffected [42]. Marked increases in 24-h urinary NO_x excretion have been reported during pregnancy in the normal rat which cannot be accounted for by increased dietary intake [43,44]. In addition, increases in plasma NO_x were also observed during pregnancy, and plasma and urinary NO_x correlate well with increases in cGMP (Fig. 19-2 [43]).

Nitric oxide production in pregnant rats is relatively resistant to chronic NO synthesis inhibition [44], suggesting enhanced basal production, although one study failed to show an exaggerated pressor response to acute NO synthesis inhibition in conscious pregnant versus virgin rats [45]. In vivo studies suggest that NO is responsible for the pregnancy-associated refractoriness to the pressor action of administered vasoconstrictors [46,47]. Data from in vitro studies have also supported the concept that NO activity is increased in some blood vessels during pregnancy although the findings have been variable. Much work has been done in guinea pigs, where agonist-stimulated NO release by Ach from carotid, uterine, and mesenteric arteries, but not renal arteries, is enhanced in pregnancy [48,49]. However, these same workers failed to show a consistent gestational

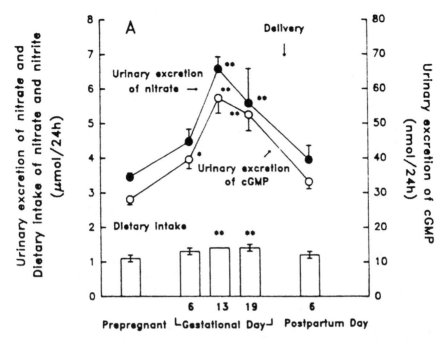

Figure 19-2. Summary of the 24-h urinary excretion of nitrite + nitrate (●) and cGMP (○) and the dietary intake of nitrate in a group of nine rats before, during, and after pregnancy. * and ** denotes *p*<.05 and *p*<.01, respectively, compared to prepregnant values. (Reproduced with permission from Ref. 43.)

blunting of the carotid artery response to a range of vasoconstrictors which argues against tonic basal NO production in this conduit vessel [50–52]. In the rat thoracic aorta, there is no evidence of increased basal or agonist stimulated NO in late pregnancy [53]. Although uterine blood vessels are partially dependent on NO to attenuate the effect of some vasoconstrictors, cyclooxygenase products and other unidentified agents also play a role [51,52]. In the guinea pig, there is only a small increase in RPF during pregnancy, and in this species, pregnancy has little impact on NO release from the renal artery [49]. None of these studies in the guinea pig have addressed the issue of vascular reactivity in resistance-sized vessels. One study in isolated rat renal interlobar arteries suggests blunted basal and enhanced agonist-stimulated NO release in late pregnancy [54]. Because RVR is returning to nonpregnant values close to term, the more likely time to see increased basal renal NO release would be at midterm, when the kidney is maximally vasodilated [8].

Recent studies have shown that pregnancy increases the activity of calcium-dependent (constitutive) NOS (by measuring arginine to citrulline conversion) in uterine artery, kidney, heart, and other tissues in both early and late pregnant guinea pigs [37,55]. Increased expression of mRNA levels for both constitutive NOS isoforms have been observed in a variety of locations in late pregnant guinea pig and estrogen apparently provides the primary stimulus, as estrogen receptor antagonism with tamoxifen prevents the gestational stimulation to NOS expression [37]. Increased eNOS has also recently been reported in the aorta of pregnant rat [56] and the author has shown a midterm increase in renal nNOS expression (by Northern blot analysis), at a time when the kidney exhibits the maximal gestational vasodilation in the rat (Baylis and Huffman, unpublished data). There are additional data to suggest that estrogens enhance NO-dependent vascular relaxation in pigs and rabbits [57,58]. However, in pregnant rats, estrogen levels are very low until just before term [59], although increased NO production and vasodilation are seen early; therefore, the precise role of estrogens in mediating the increased NOS expression of pregnancy in the rat remains to be determined.

Normal pregnant rats exhibit signs of relative arginine deficiency because orotic acid excretion increases progressively during pregnancy [60] and plasma arginine levels are ~40% reduced in the basal state [61]. This reduction in maternal plasma L-arginine presumably reflects increased utilization of substrate in response to the increased demand for NO and increasing fetal requirements, as well as increased urea production. Chronic dietary arginine deprivation, in the rat, leads to low fetal birth weight and high perinatal mortality [62]. These observations, in aggregate, suggest that NO plays a role in the systemic cardiovascular responses to normal pregnancy.

In addition to a general increase in vascular NO production, which contributes to the gestational vascular refractoriness to constricting agents and fall in BP and is particularly pronounced close to term, increased NO production apparently plays a role in the renal vasodilation of pregnancy. It is clear that NO is an important physiologic renal vasodilator [63] and recent work in conscious pregnant rats suggests that the renal vasodilation is due to increased NO production [64]. As shown in Figure 19-3, low-dose NOS inhibition abolishes the midterm rise in GFR by selectively reversing the gestational fall in renal vascular resistance, without changing RVR in virgins. Normal late pregnant rats given lipopolysaccharide (LPS, which normally stimulates a large burst of NO production) develop glomerular thrombosis due to inadequate NO generation secondary to substrate limitation, and plasma arginine levels fall to near zero [61]. This same dose of LPS in nonpregnant rats produces only ~50% fall in plasma L-arginine, greater NO generation, and no glomerular thrombosis; thus, pregnant females are particularly susceptible to development of L-arginine deficiency [59]. Presumably, this is partly the result of increased utilization of L-arginine due to increased basal NO synthesis. Finally, chronic NOS inhibition during pregnancy leads to suppression of the normal peripheral and renal vasodilation [65] and produces a pattern that closely resembles the symptoms of preeclampsia (see Section 6). Thus, increased

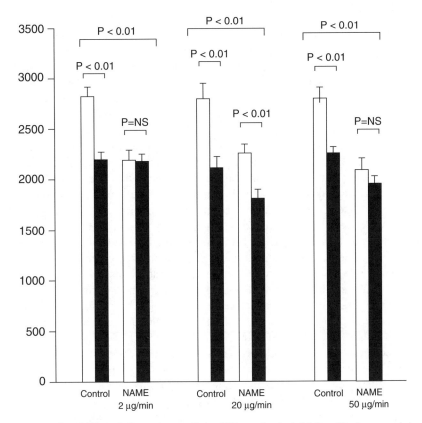

Figure 19-3. Effect of three doses of the NO synthesis inhibitor N-nitro-L-arginine methyl ester (NAME) on GFR in midterm pregnant (open columns) and virgin (closed columns) rats. (Reproduced with permission from Ref. 64.)

NO production seems to be responsible for the renal vasodilation in the normal pregnant rat.

4.2. Clinical

There are specific problems with the clinical assessment of NO activity using NO_x values, which contribute to the lack of consensus in the current literature on the effect of pregnancy on the NO system in women. These include the variability of dietary NO_x intake, which is high in some processed foods and vegetables, as well as the potential impact of other social and environmental influences such as cigarette smoking, alcohol consumption, atmospheric pollution, and exercise [66]. In addition, our work in rats suggests strongly that measurements on spot urines are *uninformative* about the status of the NO system and are more likely to reflect renal tubular handling of NO_3 [39]. Thus, bacteria-free 24-h urine samples are necessary, as 24-h urine excretion of NO_x, when corrected for NO_x intake, provides a measure of NO production. Further, in our experience, human urine samples sometimes contain interfering factors (Samsell and Baylis, unpublished data) which cause the urinary NO_x to "disappear" during assay. One example of an interfering factor is vitamin C [67], which is particularly worrisome in the context of measurements in pregnancy, as pregnant women are usually receiving vitamin supplementation.

The available clinical data are summarized in Table 19-1. The 24-h urinary excretion of cGMP is increased in normal pregnancy [68,69] and Seligman et al. report elevated plasma NO_x in normal pregnancy [70]. In contrast, work by Curtis et al. reports no difference in plasma NO_x values between nonpregnant and normal late pregnant women [71] and Brown et al. report similarity of 24-h urinary NO_x excretion between late pregnant and nonpregnant women [72]. Of note, none of these studies employed a controlled low-NO_x diet or discussed quality control issues regarding urinary NO_x measurements. Preliminary studies in women on a controlled low-NO_x intake unexpectedly show a *reduction* in 24-h $U_{NO_x}V$ in the first trimester of normal pregnancy and no difference later in pregnancy versus the nonpregnant value [73]. These observations, which await confirmation, demonstrate dissociation between NO and cGMP production in normal pregnant women and argue against a role for a widespread increase in NO generation throughout the peripheral vasculature. Strategic, local increases in NO production may, however, have an important impact on maternal BP without being detectable as a change in total NO production.

Because NO is produced by the fetoplacental unit, an overall increase in NO production should occur in the second part of pregnancy, irrespective of whether NO production increases in the maternal resistance vessels. Immunohistochemistry and molecular biology techniques have demonstrated that the normal placenta

Table 19-1. Summary of Clinical Measures of Activity of the NO System in Normal and Preeclamptic Pregnancy

Study [Ref.]	Normal Late Pregnant Versus Nonpregnant	Preeclamptic Versus Normal Late Pregnant
Kopp et al., 1977 [68]	Increased 24-h urinary cGMP excretions	—
Barton et al., 1992 [69]	—	Decreased 24-h urinary cGMP excretion
Cameron et al., 1993 [98]	—	No difference in spot urinary NO_x
Seligman et al., 1994 [70]	Increased plasma NO_x	Decreased plasma NO_x
Curtis et al., 1995 [71]	No difference in plasma NO_x	No difference in plasma NO_x
Brown et al., 1995 [72]	No difference in 24-h urine NO_x	No difference in 24-h urine NO_x
Conrad and Mosher, 1995 [73]	NO difference in plasma NO_x or 24-h $U_{NO_x,b}V$	No difference in plasma NO_x, fall in 24-h $U_{NO_x,b}V$

contains abundant constitutively expressed endothelial NOS in the resistance vasculature [74,75] and, in vitro, the normal placental vasculature makes and vasodilates to NO [76]; in fact, NO is more important than prostacyclin in control of placental vessel tone [77]. Whether this increase in NO_x from fetoplacental sources is sufficiently detectable in maternal blood or urine is not known.

There is some evidence that a relative arginine deficiency develops in normal pregnant women because the 24-h urinary excretion of arginine and plasma arginine levels fall although levels in cord blood are maintained [78–80] and orotic acid excretion increases markedly [81]. In addition, the circulating endogenous NOS inhibitor asymmetric dimethyl arginine (ADMA) falls in normal pregnancy, suggesting a "permissive" increase in NO synthesis [82]. There have been few studies on vascular reactivity in maternal resistance vessels during normal pregnancy, although McCarthy and colleagues could not detect any change in Ach-dependent or NO-donor-induced relaxation of small arteries from fat [83]. In contrast, exaggerated vasoconstriction to an NO synthesis inhibitor occurs in the hand of normal early and late pregnant versus nonpregnant women, suggesting increased tonic NO release in the skin circulation in pregnancy [84].

5. Hypertension in Pregnancy

Hypertensive complications are relatively frequent in pregnancy and can have serious consequences to both mother and baby. Of the various categories of hypertension in pregnancy, preeclampsia is the most serious. This disease is caused by pregnancy and is clinically silent during the first half of the gestation. There are no reliable, simple early indicators. Risk factors include previous preeclamptic pregnancy, multiple pregnancy, extremes of maternal age, family history, possibly low socioeconomic status/poor nutrition, essential hypertension, and renal disease. The overall incidence of essential hypertension, and renal disease. The overall incidence of preeclampsia is 5–7% [85].

The symptoms of preeclampsia are extensive and result from suppression of the normal physiologic responses to pregnancy. They include generalized vasoconstriction, increased responsiveness to vasoconstrictors, increased BP, increased capillary permeability, severe edema, falls in plasma volume, increased intravascular coagulation, reduction in organ perfusion, fall in GFR, proteinuria, widespread vascular endothelial damage, including glomerular endotheliosis, and intrauterine growth retardation (IUGR) [85]. The primary defect in preeclampsia may involve reduced placental perfusion, leading to fetoplacental ischemia. There is some evidence that the ischemic fetoplacental unit produces a circulating agent that causes widespread dysfunction of the maternal vascular endothelium [86].

Research into preeclampsia has been hampered because this disease does not occur spontaneously except in primates and there are no good experimentally induced animal models [87]. Pregnancy is usually antihypertensive when superimposed on rats with established hypertension, whether genetic or experimentally induced [8,87–89]. In the case of the spontaneously hypertensive rat, increased NO production appears to play an important role in the antihypertensive effect of pregnancy [90,91].

Because of the recent evidence, discussed above, which suggests that NO is an important physiologic vasodilator in normal pregnancy, the evidence on a possible NO deficiency in preeclampsia is considered in the next section.

6. NO Deficiency in Preeclamptic Pregnancy

6.1. Experimental

Unlike other models of hypertension, where pregnancy is uniformly antihypertensive [8,88–90], chronic NOS inhibition produces a maintained, dose-dependent hypertension and the normal late fall in BP is suppressed while GFR falls, as shown in Fig. 19-4. In addition, these rats develop proteinuria, suppression of the normal volume expansion, and increased maternal and fetal morbidity and mortality [65] in a pattern that resembles the symptoms of preeclampsia. These observations have subsequently been confirmed by other groups [92,93], although

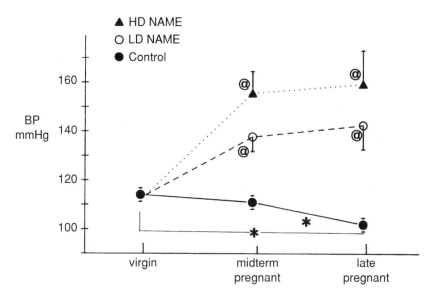

Figure 19-4. Mean arterial blood pressure (BP) in the conscious chronically catheterized rat before and during a normal pregnancy (●) and during pregnancy with low dose (○) and high dose (▲) chronic NO synthesis inhibitor during pregnancy. * denotes a difference in the normal pregnant rats versus the virgin state. @ denotes a difference, during pregnancy, between the NO inhibited and control rats. (Reproduced with permission from Ref. 65.)

a recent report challenges these findings, claiming that fetal malformations and IUGR occur in the absence of maternal hypertension during chronic NOS inhibition [94]. The author has recently confirmed and expanded earlier observations with chronic NOS inhibition during pregnancy [65] and have shown, using micropuncture technique, that R_A and R_E as well as glomerular blood pressure remain extremely high in late pregnant rats with chronic NO inhibition [95]. Furthermore, we have found that, although pregnant rats are somewhat resistant to the actions of NO synthesis inhibitors, a level of NO production that is appropriate for normal BP control in the nonpregnant rat is associated with marked hypertension in pregnancy [44]. Therefore, the hemodynamic adaptations of pregnancy require a large increase in NO production in order for BP to remain low. Finally, studies by Podjarny and colleagues have shown that the symptoms of preeclampsia produced by superimposition of pregnancy in rats with subclinical adriamycin nephropathy can be reversed with chronic L-arginine treatment, suggesting that an NO deficiency is pathogenic in this model [96]. Thus, at least in the rat, relative NO deficiency is associated with compromised pregnancy and a preeclamptic-like disease.

6.2. Clinical

The available clinical data is summarized in Table 19-1. The 24-h urinary excretion of cGMP tends to fall in preeclampsia and, interestingly, treatment of preeclamptics with $MgSO_4$ restores the cGMP excretion to that of a normal late pregnant woman [68]. These observations are supported by recent studies in the rat, which suggest that the antihypertensive action of $MgSO_4$ are mediated by NO [97]. In contrast, a study by Cameron et al. shows no difference in spot urine content of NO_x in normotensive versus preeclamptic late pregnant women [98] although, as discussed above, spot urinary NO_x has little predictive value of the activity of the NO system [39]. Seligman et al. report reduction in plasma NO_x in preeclampsia [70], whereas Curtis et al. show no difference in plasma NO_x values between normal late pregnant and preeclamptic women [71] and Brown et al. report similarity of 24-h urinary NO_x excretion between these same groups [72]. In contrast, the first report with rigorous dietary control does show a significant reduction in 24-h $U_{NO_x,b}V$ in preeclamptic versus normal third trimester women, suggesting diminished NO production in preeclampsia [73].

An in vitro study on vascular reactivity in maternal resistance vessels in preeclamptics shows no abnormality in the NO component of Ach-induced vasodilation or in NO-donor-induced vasodilation [99]. However, the critical measure of the response to NOS inhibition in preeclampsia, reflecting basal NO production, has not yet been reported. The circulating endogenous NOS inhibitor asymmetric dimethyl arginine (ADMA) is increased in preeclampsia [82] and a recent case report shows a highly beneficial effect of an NO donor in treatment of a severe postpartum preeclamptic who developed HELLP syndrome [100]. This agrees with in vitro data showing that platelets from preeclamptics are more sensitive to the inhibitory effects of NO than normal pregnant and nonpregnant women [101]. Overall, it remains unclear whether preeclampsia is associated with NO deficiency in the maternal vasculature. Some of the confusion undoubtedly originates from the difficulty in directly assessing activity of the NO system in the clinical setting.

There are also some conflicting data on human placental NO production in preeclampsia. A recent report by Kovacs et al. demonstrates low cGMP levels in vivo in venous umbilical plasma in preeclamptics versus normal [102]. Pinto and colleagues reported that the high level of agonist-stimulated NO production seen in the "in vitro" normal placenta was profoundly depressed in the preeclamptic placenta [103], but later studies report no difference [104] or increased [105] NO production from preeclamptic versus normal placentas. Recently, immunohistochemical studies have documented alterations in the distribution of endothelial NOS in syncytiotrophoblast in preeclamptic versus normal placentas [106] which may either cause or result from the reduction in placental blood flow that occurs in preeclampsia.

7. Summary and Conclusions

The animal data are reasonably consistent in indicating that NO synthesis increases during gestation and that this increase is required for the physiologic hemodynamic adaptations that are seen in normal pregnancy. The clinical data are contradictory, both regarding normal and preeclamptic pregnancies, and reasons for the variability in the literature have been identified above. There are a number of studies suggesting that an NO deficiency may occur in preeclampsia, which may cause some of the manifestations of the disease, but more, carefully designed clinical studies are required. Two recent review articles provide different perspectives on NO in pregnancy and also conclude that more information is necessary before definitive statements can be made regarding the clinical relevance of NO in normal and hypertensive pregnancy [107,108].

References

1. De Swiet, M. In: *Handbook of Hypertension, Vol. 10: Hypertension in Pregnancy,* Ed. Rubin, P.C. Elsevier Science Publishers, New York, 1988, pp. 1–9.

2. Lindheimer, M.D. and Katz, A.I. *Fluid, Electrolyte and Acid-Base Disorders.* Churchill Livingstone, Inc. New York, 1985, pp. 1041–1080.

3. Chesley, L.C. *Hypertensive disorders in pregnancy.* Appleton-Century-Crofts, 1978, pp. 119–154.

4. Bay, W.H. and Ferris, T.F. *Hypertension* **1**, 410–415 (1979).

5. Gant, N.F., Whalley, P.J., Everett, R., et al. *Am. J. Kidney Dis.* **9**, 303–307 (1987).

6. Davison, J.M. and Hytten, F.E. *J. Obstet. Gynecol.* **81**, 588–595 (1974).

7. Davison, J.M. and Dunlop, W. *Kidney Int.* **18**, 152–161 (1980).

8. Baylis, C. *Baillieres Clinical Obstetrics and Gynecology,* 2nd ed., Bailliere Tindall, London, 1994, pp. 235–264.

9. Ferris, T.F. In: *Handbook of Hypertension, Vol. 10: Hypertension in Pregnancy.* Ed. Rubin, P.C. Elsevier Science Publishers, New York, 1988, pp. 102–117.

10. Conrad, K.P. and Colpoys, M.C. *J. Clin. Invest.* **77**, 236–245 (1986).

11. Baylis, C. *Am. J. Physiol.* **253**, F158–F163 (1987).

12. Baylis, C., Engels, K., Harton, P., et al. *Am. J. Physiol.* **264**, F74–F78 (1993).

13. Fournier, A., ElEsper, G.N., Lalau, J.D., et al. *Can. J. Physiol. Pharmacol.* **69**, 1601–1608 (1991).

14. Gregoire, I., El-Esper, N., Gondry, J., et al. *Am. J. Obstet. Gynecol.* **162**, 71–76 (1990).

15. Masilamani, S. and Baylis, C. *Am. J. Physiol.* **267**, R1611–R1616 (1994).

16. Moncada, S., Palmer, R.M.J., and Higgs, E.A. *Pharm. Rev.* **43**, 109–142 (1991).

17. Conrad, K.P. and Whittemore, S.L. *Am. J. Physiol.* **262**, R1137–R1144 (1992).

18. Cabrera, C. and Bohr, D. *Biochem. Biophys. Res. Commun.* **206**, 77–81 (1995).

19. Tojo, A., Gross, S.S., Zhang, L., et al. *J. Am. Soc. Nephrol.* **4**, 1438–1447 (1994).

20. Bachmann, S. and Mundel, P. *Am. J. Kidney Dis.* **24**, 112–129 (1994).

21. Mohaupt, M.G., Elzie, J.L., Ahn, K.Y., et al. *Kidney Int.* **46**, 653–665 (1994).

22. Morrissey, J.J., Mccracken, R., Kaneto, H., et al. *Kidney Int.* **45**, 998–1005 (1994).

23. Baylis, C., Harton, P., and Engels, K. *J. Am. Soc. Nephrol.* **1**, 875–881 (1990).

24. Panza, J.A., Quyyumi, A.A., Brush, J.E. Jr., et al. *N. Engl. J. Med.* **323**, 22–27 (1990).

25. Vallance, P., Collier, J., and Moncada, S. *Lancet* **2**, 997–1000 (1989).

26. Haynes, W.G., Noon, J.P., Walker, B.R., et al. *J. Hyperten.* **11**, 1375–1380 (1993).

27. Kanno, K., Hirata, Y., Emori, T., et al. *Clin. Exp. Pharmacol. Physiol.* **19**, 619–625 (1992).

28. Baylis, C., Mitruka, B., and Deng, A. *J. Clin. Invest.* **90**, 278–281 (1992).

29. Ribeiro, M.O., Antunes, E., DeNucci, G., et al. *Hypertension* **20**, 298–303 (1992).

30. Hishikawa, K., Nakaki, T., Suzuki, H., et al. *J. Hyperten.* **11**, 639–645 (1993).

31. Reyes, A.A., Karl, I.E., and Klahr, S. *Am. J. Physiol.* **267**, F331–F346 (1994).

32. Morris, S.M. Jr. and Billiar, T.R. *Am. J. Physiol.* **266**, E1–E11 (1994).

33. Vallance, P., Leone, A., Calver, A., et al. *J. Cardiovasc. Pharmacol.* **20**, S60–S62 (1986).

34. Ueno, S.I., Sano, A., Kotani, K., et al. *J. Neurochem.* **59**, 2012–2016 (1992).

35. Visek, W.J. *J. Nutr.* **116**, 36–46 (1986).

36. Milner, J.A. *J. Nutr.* **115**, 516–523 (1985).

37. Weiner, C.P., Lizasoain, I., Baylis, S.A., et al. *Proc. Natl. Acad. Sci. USA* **91**, 5212–5216 (1994).

38. Nathan, C. and Xie, Q. *J. Biol. Chem.* **269**, 13725–13728 (1994).

39. Suto, T., Losonczy, G., Qiu, C., et al. *Kidney Int.* **48**, 1272–1277 (1995).

40. Wong, S.K-F. and Garbers, D.L. *J. Clin. Invest.* **90**, 299–305 (1992).

41. Ujiie, K., Drewett, J.G., Yuen, P.S.T., et al. *J. Clin. Invest.* **91**, 730–734 (1993).

42. Conrad, K.P. and Vernier, K.A. *Am. J. Physiol.* **257**, R847–R853 (1989).

43. Conrad, K.P., Joffe, G.M., Kruszyna, H., et al. *FASEB J.* **7**, 566–571 (1993).

44. Engels, K., Deng, A., Samsell, L., et al. *J. Am. Soc. Nephrol.* **4**, 548A (1993).

45. Umans, J.G., Lindheimer, M.D., and Barron, W.M. *Am. J. Physiol.* **259**, F293–F296 (1990).

46. Allen, R., Castro, L., Arora, C., et al. *Obstet. Gynecol.* **83**, 92–96 (1994).

47. Molnar, M. and Hertelendy, F. *Am. J. Obstet. Gynecol.* **166**, 1560–1567 (1992).

48. Weiner, C., Martinez, E., Zhu, L.K., et al. *Am. J. Obstet. Gynecol.* **161**, 1599–1605 (1989).

49. Kim, T.H., Weiner, C.P., and Thompson, L.P. *Am. J. Physiol.* **267**, H41–H47 (1994).

50. Weiner, C.P., Martinez, E., Chestnut, D.H., et al. *Am. J. Obstet. Gynecol.* **161**, 1605–1610 (1989).

51. Weiner, C.P., Thompson, L.P., Liu, K.Z., et al. *Am. J. Physiol.* **263**, H1764–H1769 (1992).

52. Weiner, C.P., Thompson, L.P., Liu, K.Z., et al. *Am. J. Obstet. Gynecol.* **166**, 1171–1181 (1992).

53. St-Louis, J. and Sicotte, B. *Am. J. Obstet. Gynecol.* **166**, 684–692 (1992).

54. Griggs, K.C., Conrad, K.P., Mackey, K., et al. *Am. J. Physiol.* **265**, F309–F315 (1993).

55. Weiner, C.P., Knowles, R.G., and Moncada, S. *Am. J. Obstet. Gynecol.* **171**, 838–843 (1994).

56. Goetz, R.M., Morano, I., Calovini, T., et al. *Biochem. Biophys. Res. Commun.* **205**, 905–910 (1994).

57. Bell, D.R., Rensberger, H.J., Koritnik, D.R., et al. *Am. J. Physiol.* **268**, H377–H383 (1995).

58. Gisclard, V., Miller, V.M., and Vanhoutte, P.M. *J. Pharmacol. Exp. Ther.* **244**, 19–22 (1988).

59. Garland, H.O., Atherton, J.C., Baylis, C., et al. *J. Endocrinol.* **113,** 435–44 (1987).

60. Milner, J.A. and Visek, W.J. *J. Nutr.* **108,** 1281–1288 (1978).

61. Raij, L. *Kidney Int.* **45,** 775–781 (1994).

62. Pau, M.Y. and Milner, J.A. *J. Nutr.* **111,** 184–193 (1981).

63. Raij, L. and Baylis, C. *Kidney Int.* **48,** 20–32 (1995).

64. Danielson, L.A. and Conrad, K.P. *J. Clin. Invest.* **96,** 482–490 (1995).

65. Baylis, C. and Engels, K. *Clin. Exp. Hyperten.* **B11,** 117–129 (1992).

66. Committee on Nitrate and Alternative Curing Agents in Food, National Academy of Sciences. National Academy Press, Washington, DC, 1981.

67. Doerr, R.C., Rox, J.B., Jr., and Lakritz, L., et al. *Anal. Chem.* **53,** 381–384 (1981).

68. Kopp, L., Paradiz, G., and Tucci, J.R. *J. Clin. Endocrin. Metab.* **44,** 590–594 (1977).

69. Barton, J.R., Sibai, B.M., Ahokas, R.S., et al. *Am. J. Obstet. Gynecol.* **167,** 931–934 (1992).

70. Seligman, S.P., Buyon, J.P., Clancy, R.M., et al. *Am. J. Obstet. Gynecol.* **171,** 944–948 (1994).

71. Curtis, N.E., Gude, N.M., King, R.G., et al. *Hyperten. Pregnancy* **14,** 339–349 (1995).

72. Brown, M.A., Tibben, E., Zammit, V.C., et al. *Hyperten. Pregnancy* **14,** 319–326 (1995).

73. Conrad, K.P. and Mosher, M.D. *J. Am. Soc. Nephrol.* **6,** 657 (1995).

74. Myatt, L., Brockman, D.E., Eis, A.L.W., et al. *Placenta* **14,** 487–495 (1993).

75. Garvey, E.P., Tuttle, J.V., Covington, K., et al. *Arch. Biochem. Biophys.* **311,** 235–241 (1994).

76. Myatt, L., Brewer A., and Brockman, D.E. *Am. J. Obstet. Gynecol.* **164,** 687–692 (1991).

77. Chaudhuri, G., Cuevas, J., Bugo, G.M., et al. *Am. J. Physiol.* **265,** H2036–H2043 (1993).

78. Hytten, F.E. and Cheyne, G.A. *Br. J. Obstet. Gynecol.* **79,** 429–432 (1992).

79. Domenech, M., Gruppuso, P.A., Nishino, V.T., et al. *Pediatr. Res.* **20,** 1071–1076 (1986).

80. Ghadimi, H. and Pecora, P. *Pediatrics* 500–506 (1964).

81. Wood, M.H. and O'Sullivan, W.J. *Am. J. Obstet. Gynecol.* **116,** 57–61 (1973).

82. Fickling, S.A., Williams, D., Vallance, P., et al. *Lancet* **342,** 242–243 (1993).

83. McCarthy, A.L., Taylor, P., Graves, J., et al. *Am. J. Obstet Gynecol* **171,** 1309–1315 (1994).

84. Williams, D.J., Vallance, P.J.T., Neild G.H., et al. *Am. J. Physiol.* **272,** H748–H752 (1997).

85. National High Blood Pressure Education Program Working Group. Report on high blood pressure in pregnancy. *Am. J. Obstet. Gynecol.* **163,** 1689–1712 (1990).

86. Roberts, J.M., Taylor, R.N., and Goldfien, A. *Am. J. Hyperten.* **4,** 700–708 (1991).

87. Phippard, A.F. and Horvath, J.S. *Hyperten. Pregnancy* **10,** 168–185 (1988).

88. Takeda, T. *Jpn. Circ. J.* **28,** 49–54 (1964).

89. Deng, A. and Baylis, C. *Kidney Int.* **48,** 39–44 (1995).

90. Ahokas, R.A., Mercer, B.M., and Sibai, B. *Am. J. Obstet. Gynecol.* **165,** 801–807 (1991).

91. Chu, Z.M. and Beilin, L.J. *Br. J. Pharmacol.* **110,** 1184–1188 (1993).

92. Yallampalli, C. and Garfield, R.E. *Am. J. Obstet. Gynecol.* **169,** 1316–1320 (1993).

93. Molnar, M., Suto, T., Toth, T., et al. *Am. J. Obstet. Gynecol.* **170,** 1458–1466 (1994).

94. Diket, A.L., Pierce, M.R., Munshi, U.K., et al. *Am. J. Obstet. Gynecol.* **171,** 1243–1250 (1994).

95. Deng, A., Engels, K., and Baylis, C. *Kidney Int.* (in press).

96. Podjarny, E., Pomeranz, A., Rathaus, M., et al. *Hyperten. Pregnancy* **12,** 517–524 (1993).

97. Kemp, P.A., Gardiner, S.M., Bennett, T., et al. *Clin. Sci.* **85,** 175–181 (1993).

98. Cameron, I.T., Van Papendorp, C.L., Palmer, R.M.J., et al. *Hyperten. Pregnancy* **12,** 85–92 (1993).

99. McCarthy, A.L., Woolfson, R.G., Raju, S.K., et al. *Am. J. Obstet. Gynecol.* **168,** 1323–1330 (1993).

100. DeBelder, A.J., Lees, C., Martin, J., et al. *Lancet* **345,** 124–125 (1995).

101. Hardy, E., Rubin, P.C., and Horn, E.H. *Clin. Sci.* **85,** 195–202 (1994).

102. Kovacs, G.A., Makary, A., Peto, J., et al. *Hyperten. Pregnancy* **13,** 163–169 (1994).

103. Pinto, A., Sorrentino, R., Sorrentino, P., et al. *Am. J. Obstet. Gynecol.* **164,** 507–513 (1991).

104. Wang, Y., Walsh, S.W., Parnell, R., et al. *Hyperten. Pregnancy* **13,** 171–178 (1994).

105. Lyall, F., Young, A., and Greer, I.A. *Am. J. Obstet. Gynecol.* **173,** 714–718 (1995).

106. Ghabour, M.S., Eis, A.L.W., Brockman, D.E., et al. *Am. J. Obstet. Gynecol.* **173,** 687–694 (1995).

107. Poston, L., McCarthy, A.L., and Ritter, J.M. *Pharm. Ther.* **65,** 215–239 (1995).

108. Morris, N.H., Eaton, B.M., and Dekker, G. *Br. J. Obstet. Gynecol.* **103,** 4–15 (1996).

20

Sickle Cell Disease and NO

Norman Bank, Suzette Y. Osei, and Rexford S. Ahima

1. Background

A wide spectrum of functional and anatomical abnormalities occurs in the kidneys of patients with sickle cell disease (SS). In children and young adults, with either SS or SA - sickle cell trait disease, there is a renal concentrating defect which is reversible with blood transfusion [1]. Beyond the age of 15 however, blood transfusions no longer correct the defect. Occlusion of the vasa recta by sickled red blood cells (RBC), leading to tubular atrophy, interstitial scarring, and microinfarcts is thought to underlie the progressive nature of the concentrating defect [1,2]. More subtle functional disturbances in tubular transport are common and include the inability to sustain a maximum $[H^+]$ gradient and impaired K^+ secretion unrelated to the renin/aldosterone axis [3–5]. The transport defects do not usually translate into clinically apparent electrolyte disturbances unless some intervening event tips the scales, such as sepsis, excess dietary potassium intake, or a potassium-sparing drug is prescribed [6].

Of a more serious nature in terms of long-term prognosis for renal function is the presence of hyperfiltration that occurs in young patients [7,8]. Both glomerular filtration rate (GFR) and effective renal plasma flow (ERPF) have been found to be elevated, indicating reduced renal vascular resistance to blood flow [7,8]. Most likely resulting from the increased blood flow, there is a marked hypertrophy of the glomeruli [9,10], the glomerular area averaging almost seven times that in normal control patients without advanced disease [10]. Closely associated with this remarkable hypertrophy is the presence of focal glomerular sclerosis [11–14] and albuminuria, which often progresses to the nephrotic range. In such patients, renal function usually declines, leading eventually to end-stage renal failure [15,16]. Hemo or peritoneal dialysis is most often prescribed, as renal transplantation can exacerbate painful crises [17] and patients are at risk for recurrence of the nephropathy in the transplanted kidney [18].

Clinical investigation into the mechanisms of hyperfiltration and the tubular transport abnormalities has been somewhat limited. It was shown, however, that urinary vasodilatory prostaglandins are abnormal in sickle cell patients [19] and that treatment with nonsteroidal anti-inflammatory drugs leads to a fall in GFR [20]. Allon et al. found that indomethacin reduced GFR from 119 to 100 ml/min in sickle cell patients but had no significant effect in normal controls [20]. Indomethacin also caused a significant reduction in the clearance of osmotically-free water C_{H_2O} in the sickle cell patients but not in the controls [20]. Urine osmolality (U_{osm}) rose after indomethacin administration in water-loaded sickle cell patients, but again not in the controls [20]. Overall fractional sodium excretion (FE_{Na}) fell by 42% in the sickle cell patients given indomethacin, but a much smaller decrease occurred in the controls. The authors concluded that hyperfiltration is prostaglandin-mediated and that there is a prostaglandin-dependent decrease in salt reabsorption in the medullary thick ascending limb and in the diluting segment of the nephron. De Jong and co-workers [2] suggested that renal synthesis of vasodilatory prostaglandins is increased in sickle cell patients, and this is causally related to afferent arteriolar dilation and glomerular hyperfiltration. More recently, Falk et al. [10] found that treatment of patients with sickle cell nephropathy and proteinuria with an angiotensin-converting enzyme inhibitor led to a decrease in proteinuria without any measurable effect on GFR, ERPF, or filtration fraction. The mechanism of the reduction in proteinuria was not clear, but the absence of any effect on hemodynamic parameters argues against a role for selective efferent arteriolar constriction due to angiotensin II mediating the hyperfiltration.

2. Transgenic Mouse Models of Sickle Cell Disease

In order to study sickle cell disease at a more basic level, several transgenic mouse strains have been developed which express a number of the abnormalities found in patients [21–29]. Different gene strategies have been used in creating these transgenic strains, and the mice vary considerably in their expression of functional and pathological abnormalities. For example, the transgenic SAD mouse manifests microvascular occlusions, occasional thrombi in the lungs, kidney, penis, and myocardium, and a glomerulopathy affecting 75% of the animals in their first year [28]. In contrast, the $\alpha^H\beta^S[\beta^{MDD}]$ transgenic mouse manifests organomegaly but relative few pathologic abnormalities under ambient room air conditions [26,27]. In all strains of sickle cell mice studied, sickling is exacerbated under hypoxic conditions and organ pathology becomes more severe [24,27,28].

The $\alpha^H\beta^S[\beta^{MDD}]$ mouse strain was created by the simultaneous microinjection and cointegration of LCR-β^S and LCR-α^H (LCR, locus control region; α^H, human α-globin) constructs on a normal mouse background [26]. Higher levels of human β^S expression were achieved by breeding the transgenic mice with mice bearing a deletion of the mouse β^{major}-globin gene. When the β^{major} deletion was bred to homozygosity, expression of β^S averaged 72.7% [26]. The designation $\alpha^H\beta^S[\beta^{MDD}]$ indicates the double deletion of mouse β major-globin. The mice manifest a compensated hemolytic anemia, an elevated reticulocyte count of 6.7% (3.3% in controls), and a normal Hct of 47.5% [26]. Oxygen affinity is higher than normal, which could result in lower tissue oxygenation [26]. Examination of the organs reveal that the kidneys, spleen, and lungs are significantly increased in size, the ratio of kidney weight/body weight averaging 1.19% in the sickle cell mice, as compared with 0.87% in the normal controls ($p < .01$) [27]. The kidney weight/body weight ratio increases with age, consistent with progressive vascular engorgement or hypertrophy. Fibrosis is found in the kidneys of some animals, presumably due to healed infarcts [27].

In order to determine whether the mice manifest renal functional abnormalities mimicking those found in sickle cell patients, studies were carried out to measure their ability to maximally concentrate their urine after 18 h of water deprivation [27]. Maximal Urine Osmolality (U_{max}) was not significantly different than in normal control mice. However, when U_{max} was reevaluated after the mice had been housed in environmental chambers filled with a 10% O_2 gas mixture for 1 week, U_{max} showed a 30% fall (Fig. 20-1). This abnormality was reversible upon returning the mice to room air, and then recurred when they were again exposed to hypoxia. Kidney histology revealed severe clumping and sickling of RBC in the vasa rectae with occlusion of blood vessels in the renal medulla and papilla (Fig. 20-2), findings resembling those seen in humans with homozygous sickle cell disease.

Glomerular filtration rate (GFR) was measured in the mice by means of ^{14}C-inulin clearance. The observations are shown in Table 20-1. GFR was found to be significantly higher in the sickle cell mice than in normal controls whether expressed in absolute terms or normalized for body weight (BW). Kidney weight was also increased in the $\alpha^H\beta^S[\beta^{MDD}]$ mice, and the correlation between GFR and kidney weight for both groups was highly significant ($p < .01$) (Fig. 20-3). Whether these two abnormalities are causally related or not is unclear. When GFR was expressed per gram kidney weight, there was no statistical difference between the two groups. Because mean arterial blood pressure was the same in the two groups of animals, renal vascular resistance was presumably lower in the $\alpha^H\beta^S[\beta^{MDD}]$ mice. Hyperfiltration in patients with sickle cell disease has often been attributed to anemia. However, because the $\alpha^H\beta^S[\beta^{MDD}]$ mice are not anemic, other mechanisms seem to be responsible. Overproduction of vasodilatory prostaglandins [19,20] and other endogenous vasodilators seem to be likely candidates.

3. Studies of NO and NOS in Transgenic Mice

Several clinical characteristics of sickle cell patients suggest the possibility that nitric oxide (NO) production may be increased. Hyperfiltration is one of them, as

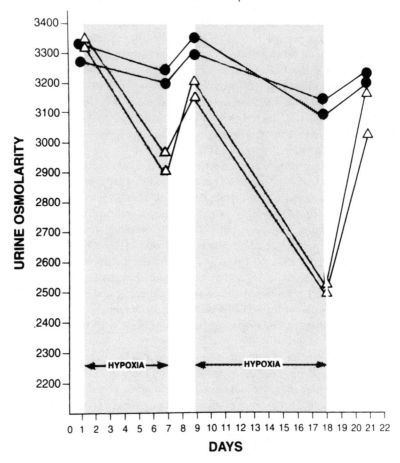

EFFECT OF TWO CYCLES OF HYPOXIA ON
URINE OSMOLARITY IN βS TRANSGENIC MICE

Figure 20-1. Fall in maximum urine osmolality (U_{max}) in $\alpha^H\beta^S[\beta^{MDD}]$ mice after exposure to 10% O_2.

NO has been shown to be an important renal vasodilator, acting on preglomerular arterioles [30] and relaxing mesangial cells [31]. As a group, sickle cell patients have lower blood pressure than controls [32,33], a phenomenon that could conceivably be due to overproduction of NO. Certainly, hypotension and general peripheral vasodilatation has been shown to be associated with high levels of NO production [34]. A third clinical clue is the increased incidence of priapism in sickle cell patients [35,36]. One or more episodes of priapism occurs in approximately 40% of sickle cell patients [36]. Contrary to previously held views, priapism has been shown to be a high-flow condition rather than low flow [37].

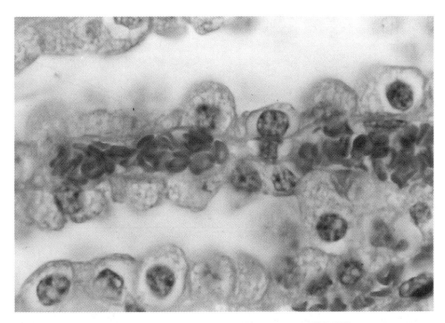

Figure 20-2. Vasoocclusion of medullary blood vessels in $\alpha^H\beta^S[\beta^{MDD}]$ mouse exposed to 10% O_2 for 5 days. Note vacuolization of the cytoplasm and swelling of collecting duct cells.

Evidence for both endothelial and neuronal nitric oxide synthases have been found in rat and rabbit corpus cavernosum, and substantial NO synthesis has been demonstrated [38]. NO synthesis appears to be essential for normal penile erection in experimental animals [38]. Finally, Enwonwu et al. [39] has reported that blood L-arginine levels and urinary excretion of L-arginine are low in patients with sickle cell anemia. This finding could be interpreted to suggest that L-arginine utilization by the NOS pathway is increased in sickle cell patients, perhaps predisposing to depletion of precursor L-arginine and, hence, failure of NO synthesis during periods of stress.

Because of these associations, nitric oxide production was studied in the

Table 20-1. GFR in Control and $\alpha^H\beta^S[\beta^{MDD}]$ Mice

Group	μl/min	ml/min per kg of Body Weight
C57B1/6J (9)	267	10.01
	±20	±0.43
$\alpha^H\beta^S[\beta^{MDD}]$ (12)	402*	12.39
	±40	±0.56
p	<.01	<.01

*p < 0.01 number of animals is shown in brackets [ ]

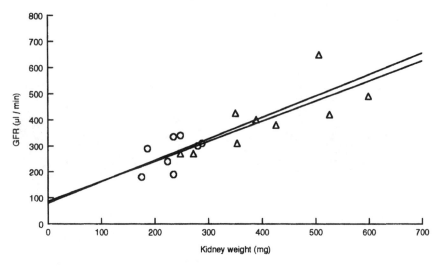

Figure 20-3. Correlation between glomerular filtration rate (GFR) and kidney weight in control ($\bigcirc$) and $\alpha^H\beta^S[\beta^{MDD}]$ mice ($\triangle$).

$\alpha^H\beta^S[\beta^{MDD}]$ mice by measuring urinary nitrite excretion and two of the three isoforms of nitric oxide synthase (NOS) by Western blot and immunohistochemistry. The urinary nitrite/creatinine ratios for control and sickle cell mice are shown in Fig. 20-4. The ratio was significantly higher in the latter group, indicating that nitrite excretion was greater in the $\alpha^H\beta^S[\beta^{MDD}]$ mice than in controls after correction for differences in urine flow rate. Because nitrite is a stable metabolic product of NO, the observation suggests that NO synthesis is increased in the sickle cell mice. The effect of inhibiting NO production was tested in three $\alpha^H\beta^S[\beta^{MDD}]$ mice by administration of nitro-L-arginine (NLA) in the drinking water for 5 days. The correlation between GFR and 24-h urinary nitrite excretion in these mice as well as in untreated $\alpha^H\beta^S[\beta^{MDD}]$ and control mice is shown in Fig. 20-5. A positive correlation ($r=0.54$) was found over the range of nitrite excretion, with the NLA-treated animals having the lowest GFR, and nitrite excretion and the untreated $\alpha^H\beta^S[\beta^{MDD}]$ mice the highest. The observations are consistent with a role for increased NO production in the hyperfiltration of the sickle cell mice but do not constitute proof.

Further studies measured the abundance of endothelial cell nitric oxide synthase (NOS III) and macrophage inducible nitric oxide synthase (NOS II) in protein extracts of the kidneys, using monoclonal antibodies and Western blot. Figure 20-6 shows a Western blot of NOS III. The purified enzyme was applied to lane 5 (positive control). Lanes 1 and 2 are kidney extracts from control mice and lanes 3 and 4 are from $\alpha^H\beta^S[\beta^{MDD}]$ mice. A 140-kDa band, corresponding to the molecular weight of NOS III [40,41] was present in the control mice. Also evident is a higher-molecular-weight band in the kidney extracts. In the $\alpha^H\beta^S[\beta^{MDD}]$

Urine Nitrite/Creatinine Ratios

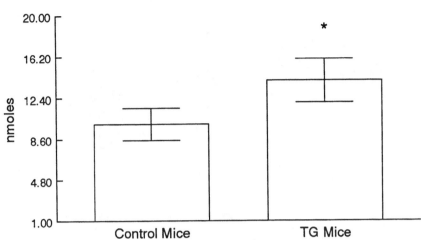

Figure 20-4. Urine nitrite/creatinine ratios in control and transgenic (TG) sickle cell mice. *$p < .05$.

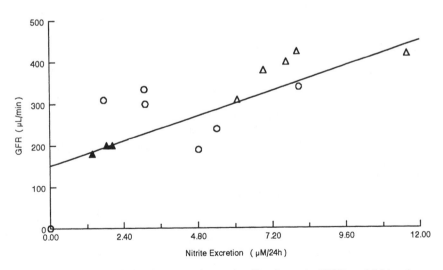

Figure 20-5. P-Correlation between glomerular filtration rate (GFR) and 24-h urinary nitrite excretion. ○: Control mice; △: transgenic sickle cell mice; ▲: transgenic mice + NLA.

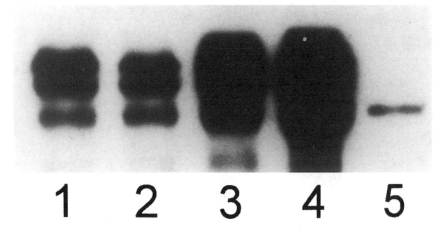

Figure 20-6. Western blot of mouse kidney protein extracts, using a monoclonal antibody directed against human constitutive endothelial cell nitric oxide synthase (NOS III). Lane 1: Control mouse maintained in room air; lane 2: control mouse exposed to 10% O_2; lane 3: sickle cell mouse maintained in room air; lane 4: sickle cell mouse exposed to 10% O_2; lane 5: purified NOS III.

mice, the 140-kDa bands are clearly in greater abundance, and additional higher molecular bands are also evident.

Figure 20-7 is a Western blot in which a monoclonal antibody directed against activated rat macrophage nitric oxide synthase (NOS II) was used. Lanes 1 and 2 are from control mice and lanes 3 and 4 are from $\alpha^H\beta^S[\beta^{MDD}]$ mice. A prominent band is seen at 130 kDa, corresponding to the molecular weight of monomeric NOS II [34,42]. It is clear that the abundance of the 130-kDa protein is much greater in the $\alpha^H\beta^S[\beta^{MDD}]$ mouse. The active forms of NOS II, and perhaps NOS III, are present in vivo as dimers [43]. In both Figs. 20-6 and 20-7, higher-molecular-weight and lower-molecular-weight bands that reacted with the mono-clonal antibodies were observed. These additional bands presumably represent dimeric forms and digestion products of the dimeric enzymes.

Because of the $\alpha^H\beta^S[\beta^{MDD}]$ sickle cell mice manifest relatively mild hematological and pathological abnormalities when they are maintained under room-air conditions [27], they were restudied after exposure to a low-oxygen environment for 4–5 days in order to exaggerate in vivo sickling. It had previously been demonstrated in this transgenic strain [27] as well as in other transgenic strains of sickle cell mice [24,28] that hypoxia increases the number of reticulocytes in the circulation and HbS polymerization, events which lead to vasoocclusive lesions. The resulting effect of chronic hypoxia on NOS III and NOS II is shown in the Western blots (Figs. 20-6 and 20-7). In the control mice, hypoxia had no significant effect on NOS III abundance (lane 2 versus lane 1), but there was a pronounced increase in several molecular-weight bands reacting with the monoclonal antibody

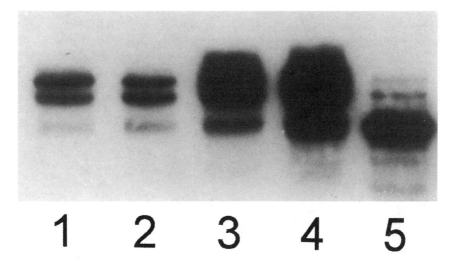

Figure 20-7. Western blot of mouse kidney protein extracts, using a monoclonal antibody directed against rat activated macrophage nitric oxide synthase (NOS II). Lane identification same as in Fig. 20-6.

in the $\alpha^H \beta^S [\beta^{MDD}]$ mice (lane 4 versus lane 3, Fig. 20-6). In the case of NOS II (Fig. 20-7), chronic hypoxia resulted in a small but significant increase in the control mice (lane 2 versus lane 1) and a marked increase in the $\alpha^H \beta^S [\beta^{MDD}]$ mice (lane 4 versus lane 3). Thus, enhancement of the in vivo HbS polymerization due to hypoxia was associated with markedly increased abundance of both NOS III and NOS II in the kidneys of the $\alpha^H \beta^S [\beta^{MDD}]$ mice.

Immunohistochemistry in similarly prepared control and $\alpha^H \beta^S [\beta^{MDD}]$ transgenic mice localized the sites in the kidney of these two isoforms of NOS. NOS III was expressed in the proximal convoluted tubules throughout the cortex of the normal control mice (Fig. 20-8), but was not observed in the glomeruli or in the medulla. In the $\alpha^H \beta^S [\beta^{MDD}]$ mice, NOS III was found in the same sites, but the intensity of staining was greatly increased. NOS II immunoperoxidase staining was not observed in the control mice but was seen in glomeruli, distal convoluted tubules, and collecting ducts in the cortex and medulla of the $\alpha^H \beta^S [\beta^{MDD}]$ mice. Hypoxia resulted in de novo appearance of immunoperoxidase staining of NOS II in distal nephron segments of control mice and increased intensity of NOS II staining in the $\alpha^H \beta^S [\beta^{MDD}]$ mice.

It is of interest that nitric oxide is an important moderator of liver pathophysiology in endotoxin and ischemic types of liver insult [44–48], where it appears to serve a protective function for hepatocytes [49,50]. Sickle cell patients are at risk for developing a wide variety of liver functional and pathological abnormalities [51,52], due in part to intrahepatic vasoocclusion and ischemia. Immunoperoxidase staining of the liver of the $\alpha^H \beta^S [\beta^{MDD}]$ transgenic mice was carried out and

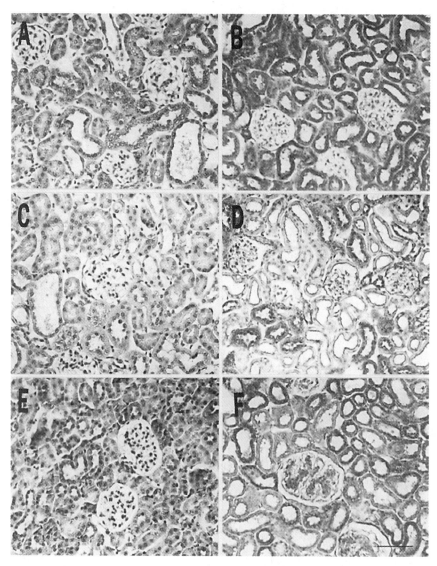

Figure 20-8. Immunohistochemistry of NOS isoforms in the renal cortex under room air (A–D) and hypoxia (E–F). (A) NOS III in cortical tubules of normal mouse. (B) Note the increased NOS III immunostaining in cortical tubules of $\alpha^H\beta^S[\beta^{MDD}]$ mouse. (C) Absence of NOS II in cortex of normal mouse. Nonspecific staining resulting from endogenous peroxidase was also observed in control sections incubated with normal rabbit serum. (D) Strong NOS II immunostaining in distal/collecting tubules of $\alpha^H\beta^S[\beta^{MDD}]$ mouse. (E) Moderate NOS II immunostaining in cortical tubules of normal mouse under hypoxia. (F) Increased NOS II immunostaining in cortical tubules and glomerular mesangium of $\alpha^H\beta^S[\beta^{MDD}]$ mouse. Scale bar = 100 µm. **See plate for color illustration.**

showed NOS II in a rim of hepatocytes lining the central veins, whereas no NOS II staining was present in livers of control mice. Exposure of the $\alpha^H\beta^S[\beta^{MDD}]$ mice to hypoxia led to multifocal areas of liver necrosis, but the rim of hepatocytes surrounding the central veins was free of cellular damage, corresponding to the region where NOS II staining was identified. Thus, NOS II induction occurred in the liver as well as the kidney of the $\alpha^H\beta^S[\beta^{MDD}]$ mice, indicating a systemic mechanism of NOS II induction rather than one confined to the kidney.

4. Mechanisms of Increased NOS Abundance in Sickle Cell Mice

The precise mechanism(s) responsible for the increased NOS II and III immunore-activity in the $\alpha^H\beta^S[\beta^{MDD}]$ mice is presently unknown. NOS II is induced in many disease states that are characterized by chronic inflammatory processes, such as arthritis, inflammatory bowel disease, and encephalitis [53]. Indeed, acute inflammation per se appears to activate NOS II via a number of cytokines [53]. In addition, hypoxia has been shown to induce NOS II mRNA, gene product, and NO production in isolated renal tubular epithelial cells [54]. The observation that normal mice exposed to hypoxia manifest an increase in renal NOS II abundance supports the view that hypoxia at the tissue level may be the mediator in the $\alpha^H\beta^S[\beta^{MDD}]$ mice. A large body of experimental observations has shown that young red blood cells (reticulocytes) from sickle cell patients adhere to the vascular endothelium of small blood vessels, primarily in postcapillary venules [55–58], thereby narrowing the lumen and causing intermittent obstruction [56,58]. This is thought to lead to tissue hypoxia [56]. It seems a reasonable hypothesis, therefore, that local hypoxia is the proximate cause of induction of NOS II, although how hypoxia leads to increased transcription of NOS II is not yet clear.

Much less information is available about factors that regulate long-term changes in constitutive endothelial NOS. Increasing levels of shear stress have been best studied and are known to cause prolonged elevations of NOS III mRNA and protein levels in cultured endothelial cells [59,60]. Shear stress has also been shown to increase NO release from the endothelium in physiological experiments [61–63]. Shear stress is a function of both blood flow rate and viscosity [63]. The kidney in sickle cell disease may suffer from increased shear stress caused by adhesion of RBC to the endothelium via $\alpha_4\beta_1$-integrin expression on sickle reticulocytes to endothelial intercellular adhesion molecule-1 [64]. This could increase the velocity of plasma flow through small vessels due to a narrowed lumen. In addition, clumping of RBC would tend to raise the local hematocrit where adhesion was occurring, and this, in turn, would increase viscosity. These events could conceivably lead to induction of NOS III in endothelial cells in the kidney. However, the immunohistochemistry observations show that NOS III is expressed in cortical tubular epithelial cells and it is this expression which was increased in the sickle cell mice. A number of studies have found evidence for NOS III in renal epithelial cells [65–70], but its function there and its regulation have not been well defined. It is possible that shear stress in the closely adjacent capillaries of the sickle cell mice plays a role in induction of NOS III in the epithelial cells. It has been demonstrated that shear stress invokes a wide array of responses in gene regulation and changes in the cytoskeleton and the adhesion properties of the abluminal surface of endothelial cells [63,71]. Thus, signals originating in endothelial cells could conceivably be transmitted to the adjacent tubular epithelial cells resulting in induction of NOS III at this location. This is highly speculative, however, and it must be said that the cause of the abundant NOS III in the cortical tubules of the $\alpha^H\beta^S[\beta^{MDD}]$ mice is presently unknown.

5. Role of NO in Functional Abnormalities in Sickle Cell Disease

Localization of NOS III and NOS II in the kidneys of the $\alpha^H\beta^S[\beta^{MDD}]$ mice and the increased abundance of these enzymes on Western blots raise the possibility that there is increased NO synthesis at specific sites within the kidney that may be affecting function. For example, NOS II was prominent in the glomerular mesangium of the $\alpha^H\beta^S[\beta^{MDD}]$ mice but was not seen in this location in control mice. NO has been shown to have a direct effect to relax mesangial cells and to increase K_f [31], actions which might contribute to the hyperfiltration. NOS II was expressed in distal convoluted tubules and collecting ducts of the $\alpha^H\beta^S[\beta^{MDD}]$ mice but not in control mice. Patients with sickle cell anemia manifest a defect in urinary acidification and decreased H^+ excretion in response to ammonium chloride loading [72]. Ho Ping Kong and Alleyne [72] reported that urine pH fell to 5.38 in sickle cell patients loaded with ammonium chloride, as compared to 4.83 in controls. Oster et al. [73] confirmed these findings and found that titratable acid excretion and net acid excretion were also reduced. During bicarbonate loading, sickle cell patients could not generate a normal urine/blood pCO_2 gradient, taken to be a sign of impaired distal tubular H^+ secretion [73]. These authors, as well as others [74], concluded that sickle cell patients manifest an incomplete distal renal tubular acidosis. It is of interest, therefore, that Tojo et al. [75] described inhibition of H^+-ATPase activity by nitric oxide in cortical collecting ducts. Thus, induction of NOS II in this location in the sickle cell mice raises the possibility that increased NO synthesis in the cortical collecting ducts might play a role in distal renal tubular acidosis in sickle cell patients. Sickle cell patients also manifest a defect in potassium excretion. Battle et al. [6] and De Fronzo et al. [5] found decreased fractional potassium excretion in sickle cell patients with impaired renal function, as compared with non-sickle-cell patients with comparable reductions in renal function. Hyperkalemia is a common finding in patients with sickle cell anemia. Recent studies have shown that NO can inhibit Na^+/K^+ ATPase in mouse proximal tubular cells [76] and in cortical collecting duct cells [77]. Stoos et al. [77] found that release of NO from activated endothelial cells caused a 50% reduction in short-circuit current in a cocultured line of cells derived from mouse cortical collecting ducts. In addition, cyclic GMP in the collecting duct cells increased markedly under these experimental conditions [77]. These effects were blocked by nitroarginine. Because NOS II is expressed in the cortical collecting ducts of the $\alpha^H\beta^S[\beta^{MDD}]$ mice, the possibility exists that increased NO synthesis at this site may be implicated in reduced sodium reabsorption and, hence, potassium secretion in this segment of the nephron.

As discussed earlier, it has been shown that inhibition of cyclooxygenase by indomethacin causes a significant correction of hyperfiltration in sickle cell patients [2,20,78]. This has been attributed to the expected fall in renal prostaglandin synthesis. However, more recent studies indicate that nonsteroidal anti-inflammatory drugs inhibit NOS II mRNA and gene product [79]. In addition, NOS inhibitors have been found to reduce prostaglandin E_2 synthesis [80]. In a model of renal inflammation, Salvemini et al. [81] found that the marked increase in PGE_2 synthesis in this model was directly related to release of NO from activated infiltrating macrophages. These recent observations, as well as others [82], indicate that there is an important interaction between the inducible forms of NOS (NOS II) and cyclooxygenase. These new findings raise the question of whether the GFR correction by indomethacin in sickle cell patients is due to diminished prostaglandin synthesis, to inhibition of NO synthesis, or both. Clearly, much additional information is needed to understand the role of prostaglandins and NO in sickle cell renal disease and the interrelationship between NOS II and cyclooxygenase.

Acknowledgments

We are appreciative for the contributions of Hagop S. Aynedjian and Judy Qiu to the studies carried out in our laboratory, and for the participation of our hematology colleagues, Drs. Mary E. Fabry and Ronald L. Nagel of the Albert Einstein College of Medicine.

References

1. Statius van Eps, L.W., Pinedo-Veels, C., De Vries, C.H., et al. *Lancet* **1**, 450–452 (1970).

2. de Jong, P.E., and Statius van Eps, L.W. *Kidney Int.* **27**, 711–717 (1985).

3. Goossens, J.R., Statius van Eps, L.W., Schouten, H., et al. *Clin. Chim. Acta* **41**, 149–156 (1972).

4. Ho Ping Kong, H. and Alleyne, G.A.O. *Clin. Sci.* **41**, 505–518 (1971).

5. De Fronzo, R.A., Taufield, P.A., Black, H., et al. *Ann. Intern. Med.* **90**, 310–316 (1979).

6. Battle, D., Itsarayoungyeun, I., Arruda, J.A.L., et al. *Am. J. Med.* **72,** 188–192 (1982).

7. Etteldorf, J.N., Tuttle, A.H., and Clayton, H.W. *Am. J. Dis. Child.* **83,** 185–191 (1952).

8. Hatch, F.E., Jr., Azar, S.H., Ainsworth, T.E., Nardo, J.M., and Culbertson, J.W. *J. Lab. Clin. Med.* **76,** 632–640 (1970).

9. Bernstein, J. and Whitten, C.F. *Arch. Pathol.* **70,** 407–418 (1960).

10. Falk, R.J., Scheinman, J., Phillips, G., Orringer, E., Johnson, A., and Jennette, J.C. *N. Engl. J. Med.* **326,** 910–915 (1992).

11. Bhathena, D.B. and Sandheimer, J.H. *J. Am. Soc. Nephrol.* **1,** 1241–1252 (1991).

12. Elfenbein, I.B., Patchefsky, A., Schwartz, W., and Weinstein, A.G. *Am. J. Pathol.* **77,** 357–374 (1974).

13. Pitcock, J.A., Muirhead, E.E., Hatch, F.E., Johnson, J.G., and Kelly, B.J. *Arch. Pathol.* **90,** 403–410 (1970).

14. Walker, B.R., Alexander, F., Birdsall, T.R., and Warren, R.L. *J. Am. Med. Assoc.* **215,** 437–440 (1971).

15. Bakir, A.A., Hathiwala, S.C., Ainis, H., Hryhorczuk, D.O., Rhee, H.L., Levy, P.S., and Dunea, G. *Am. J. Nephrol.* **7,** 110–115 (1987).

16. Powars, D.R., Elliott-Mills, D.D., Chan, L., Niland, J., Hiti, A.L., Opas, L.M., and Johnson, C. *Ann. Intern. Med.* **115,** 614–620 (1991).

17. Spector, D., Zachery, J.B., Sterioff, S., and Millan, J. *Am. J. Med.* **64,** 835–839 (1978).

18. Miner, D.J., Jorkasky, D.K., Perloff, L.J., Grossman, R.A., and Tomaszewski, J.E. *Am. J. Kidney Dis.* **10,** 306–313 (1987).

19. De Jong, P.R., Saleh, A.W., De Zeeuw, D., et al. *Clin. Nephrol.* **22,** 212–213 (1984).

20. Allon, M., Lawson, L., Eckman, J.R., Delaney, V., and Bourke, E. *Kidney Int.* **34,** 500–506 (1988).

21. Rubin, E.M., Lu, R., Cooper, S., Mohandas, N., and Kan, Y.W. *Am. J. Human Genet.* **42,** 585–591 (1988).

22. Greaves, D.R., Fraser, P., Vidal, M.A., Hedges, M.J., Ropers, D., Luzzatto, L., and Grosveld, F. *Nature* **343,** 183–185 (1990).

23. Ryan, T.M., Townes, T.M., Reilly, M.P., Asakura, T., Palmiter, R.D., Brinster, R.L., and Behringer, R.R. *Science* **247,** 566–568 (1990).

24. Rubin, E.M., Witkowska, H.E., Spangler, E., Curtin, P., Lubin, B.H., Mohandas, N., and Clift, S.M. *J. Clin. Invest.* **87,** 639–647 (1991).

25. Trudel, M., Saadane, N., Garel, M.-C., Bardakdjian-Michau, J., Blouquit, Y., Guerquin-Kern, J.L., Rouyer-Fessard, P., Vidaud, D., Pachnis, A., Romeo, P.-H., Beuzard, Y., and Costantini, F. *EMBO J.* **10,** 3157–3165 (1991).

26. Fabry, M.E., Nagel, R.L., Pachnis, A., Suzuka, S.M., and Costantini, F. *Proc. Natl. Acad. Sci. USA* **89,** 12150–12154 (1992).

27. Fabry, M.E., Costantini, F.D., Pachnis, A., Suzuka, S.M., Bank, N., Aynedjian, H.S., Factor, S., and Nagel, R.L. *Proc. Natl. Acad. Sci. USA* **89,** 12155–12159 (1992).

28. Trudel, M., De Paepe, M.E., Chretien, N., Saadane, N., Jacmain, J., Sorette, M., Hoang, T., and Beuzard, Y. *Blood* **84**, 3189–3197 (1994).

29. Fabry, M.E., Sengupta, A., Suzuka, S.M., Costantini, F.D., Rubin, E.M., Hofrichter, J., Christoph, G., Manci, E., Culberson, D., Factor, S.M., and Nagel, R.L. *Blood* **86**, 2419–2428 (1995).

30. Deng, A. and Baylis, C. *Am. J. Physiol.* **264** (Renal, Fluid, Electrolyte Physiol. **33**, F212–F215 (1993).

31. Shultz, P.J., Schorer, A.E., and Raij, L. *Am. J. Physiol.* **258**, F162–F167 (1990).

32. Johnson, C.S. and Giorgio, A.J. *Arch. Intern. Med.* **141**, 891–893 (1981).

33. Rodgers, G.P., Walker, E.C., and Podgor, M.J. *Am. J. Med. Sci.* **305**, 150–156 (1993).

34. Xie, Q.-W., and Nathan, C.J. *Leukoc. Biol.* **56**, 576–582 (1994).

35. Emond, A.M., Holman, R., Hayes, R.J., and Sergeant, G.R. *Arch. Intern. Med.* **140**, 1434–1437 (1980).

36. Fowler, J.E., Jr., Koshy, M., Strub, M., and Chinn, S.K. *J. Urol.* **145**, 65–68 (1991).

37. Ramos, C.E., Park, J.S., Ritchey, M.L., and Benson, G.S. *J. Urol.* **153**, 1619–1621 (1995).

38. Andersson, K.E. and Wagner, G. *Physiol. Rev.* **75**, 191–236 (1995).

39. Enwonwu, C.O., Xu, X.-X., and Turner, E. *Am. J. Med. Sci.* **300**, 366–371 (1990).

40. Sessa, W.C., Harrison, J.K., Barber, C.M., Zeng, D., Durieux, M.E., D'Angelo, D.D., Lynch, K.R., and Peach, M.J. *J. Biol. Chem.* **267**, 15274–15276 (1992).

41. Nishida, K., Harrison, D.G., Navas, J.P., Fisher, A.A., Dockery, S.P., Uematsu, R., Nerem, M., Alexander, R.W., and Murphy, T.J. *J. Clin. Invest.* **90**, 2092–2096 (1992).

42. Hevel, J.M., White, K.A., and Marletta, M.A. *J. Biol. Chem.* **266**, 22789–22791 (1991).

43. Marletta, M.A. *J. Biol. Chem.* **268**, 12231–12234 (1993).

44. Curran, R.D., Billiar, T.R., Stuehr, D.J., Ochoa, J.B., Harbreecht, B.G., Flint, S.G., and Simmons, R.L. *Ann. Surg.* **212**, 462–471 (1990).

45. Nussler, A.K., DiSilvio, M., Billiar, T., Hoffman, R.A., Geller, D.A., Selby, R., Madariaga, J., and Simmons, R.L. *J. Exp. Med.* **176**, 261–264 (1992).

46. Geller, D.A., Nussler, A.K., DiSilvio, M., Lowenstein, C.J., Shapiro, R.A., Wang, S.C., Simmons, R.L., and Billiar, T.R. *Proc. Natl. Acad. Sci. USA* **90**, 522–526 (1993).

47. Wood, E.R., Berger, H., Jr., Sherman, P.A., and Lapentina, E.G. *Biochem. Biophys. Res. Commun.* **191**, 767–774 (1993).

48. Buttery, L.D.K., Evans, T.J., Springall, D.R., Carpenter, A., Cohen, J., and Polak, J.M. *Lab. Invest.* **71**, 755–764 (1994).

49. Harbrecht, B.G., Billiar, T.R., Stadler, J., Demetris, A.J., Ochoa, J.B., Curran, R.D., et al. *Crit. Care Med.* **20**, 1568–1574 (1992).

50. Billiar, T.R. *Gastroenterology* **180**, 603–605 (1995).

51. Bauer, T.W., Moore, G.W., and Hutchins, G.M. *Am. J. Med.* **69**, 833–837 (1980).

52. Johnson, C.S., Omata, M., Tong, M.J., Simmons, J.F., Weiner, J., and Tatter, D. *Medicine* **64**, 349–356 (1985).

53. Kroncke, K.-D., Fehsel, K., and Kolb-Bachofen, V. *Biol. Chem. Hoppe–Seyler* **376**, 327–343 (1995).

54. Yu, L., Gengaro, P.E. Mederberger, M., Burke, T.J., and Schrier, R.W. *Proc. Natl. Acad. Sci. USA* **91**, 1691–1695 (1994).

55. Hebbel, R.P., Schwartz, R.S., and Mohandas, N. *Clin. Haematol.* **14**, 141–161 (1985).

56. Kaul, D.K., Fabry, M.E., and Nagel, R.L. *Proc. Natl. Acad. Sci. USA* **86**, 3356–3360 (1989).

57. Mohandas, N. and Evans, E. *Ann. NY Acad. Sci.* **565**, 327–337 (1989).

58. Kaul, D.K., Chen, D., and Zhan, J. *Blood* **83**, 3006–3017 (1994).

59. Kuckan, M.J., Hanjoong, J., and Frangos, J.A. *Am. J. Physiol.* **267**, C753–C758 (1994).

60. Ranjan, V., Xiao, Z., and Diamond, S.L. *Am. J. Physiol.* **269**, H550–H555 (1995).

61. Koller, A., Sun, D., and Kaley, G. *Circ. Res.* **72**, 1276–1284 (1993).

62. Macarthur, H.M., Hecker, R., Busse, R. and Vane, J.R. *Br. J. Pharmacol.* **108**, 100–105 (1993).

63. Davis, P.F. *Physiol. Rev.* **75**, 519–560 (1995).

64. Swerlick, R.A., Eckman, J.R., Kumar, A., Jeitler, M., and Wick, T.M. *Blood* **82**, 1891–1899 (1993).

65. Inoue, N., Venema, R.C., Sayegh, H.S., Ohara, Y., Murphy, T.J., and Harrison, D.G. *Arterioscler. Thromb. Vasc. Biol.* **15**, 1255–1261 (1995).

66. Iskii, K., Chang, B.J.F., Kerwin, F.L., Jr., Wagenaar, F.L., Huang, Z.J., and Murad, F. *J. Pharmacol. Exp. Ther.* **256**, 38–43 (1991).

67. Tracey, W.R., Pollock, J.S., Murad, F., Nakane, M., and Forstermann, U. *Am. J. Physiol.* **266**, C22–C28 (1994).

68. Terada, Y., Tomita, K., Nonoguchi, H., and Marumo, F. *J. Clin. Invest.* **90**, 659–665 (1992).

69. Ujiie, K., Yuen, J., Hogarth, L., Danziger, R., and Star, R.A. *Am. J. Physiol.* **267**, F296–F302 (1994).

70. Bachmann, S. and Mundel, P. *Am. J. Kidney Dis.* **24**, 112–129 (1994).

71. Resnick, N. and Gimbrone, M.A., Jr. *FASEB J.* **9**, 874–882 (1995).

72. Ho Ping Kong, H. and Alleyne, G.A.O. *Clin. Sci.* **41**, 505–518 (1971).

73. Oster, J.R., Lespier, L.E., Lee, S.M., Pellegrini, E.L., and Vaamonde, C.A. *J. Lab. Clin. Med.* **88**, 389–401 (1976).

74. Goossens, J.R., Statius van Eps, L.W., Schouten, H., et al. *Clin. Chim. Acta* **41**, 149–156 (1972).

75. Tojo, A., Guzman, N.J., Garg, L.C., Tisher, C.C., and Madsen, K.M. *Am. J. Physiol.* **267**, F509–F515 (1994).

76. Guzman, N.J., Fang, M.-Z., Tang, S.-S., Ingelfinger, J.R., and Garg, L.C. *J. Clin. Invest.* **95**, 2083–2088 (1995).

77. Stoos, B.A., Carretero, O.A., Farhy, R.D., Scicli, G., and Garvin, I.L. *J. Clin. Invest.* **89**, 761–765 (1992).

78. De Jong, P.E., De Jong-Van den Berg, L.T.W., De Zeeuw, D., Donker, A.J.M., Schouten, H., and Statius van Eps, L.W. *Clin. Sci.* **63**, 53–58 (1982).

79. Aeberhard, E.E., Henderson, S.A., Arabolos, N.S., Griscavage, J.M., Castro, F.E., Barrett, C.T., and Ignarro, L.J. *Biochem. Biophys. Res. Commun.* **208**, 1053–1059 (1995).

80. Salvemini, D., Manning, P.T., Zweifel, B.S., Seibert, K., Connor, J., Currie, M.G., Needleman, P., and Masferrer, J.L. *J. Clin. Invest.* **96**, 301–308 (1995).

81. Salvemini, D., Seibert, K., Masferrer, J.L., Misko, T.P., Currie, M.G., and Needleman, P. *J. Clin. Invest.* **93**, 1940–1947 (1994).

82. Swierkosz, T.A., Mitchell, J.A., Warner, T.D., Botting, R.M., and Vane, J.R. *Br. J. Pharm.* **114**, 1335–1342 (1995).

21

A Putative Role of NO and Oxidant Injury in the Pathogenesis of Hemolytic Uremic Syndrome

Marina Noris and Giuseppe Remuzzi

1. Introduction

Hemolytic uremic syndrome (HUS) is a disease of nonimmune hemolytic anemia, thrombocytopenia, and renal failure due to platelet thrombi in the microcirculation of the kidney. It mainly affects infants and small children, although older children and adults may also suffer. The characteristic lesion, thrombotic microangiopathy (TMA) [1], is unique to this syndrome and consists of vessel wall thickening (capillaries and arterioles), with swelling and detachment of the endothelial cells from the basement membrane and accumulation of fluffy material in the subendothelium (Fig. 21-1). As a result of these changes and occasional thrombi, the glomerular capillary lumina may become occluded. Glomerular thrombi consist of fibrin and platelets, with platelets being preponderant in the early lesions. Extrarenal lesions have also been found, which include microthrombotic lesions in the colon, heart, brain, and pancreas [1]. Typical HUS of children begins with diarrhea, sometimes bloody, and other gastrointestinal symptoms. Classic HUS forms have a remarkably better prognosis than atypical (sporadic or D–) HUS that manifests without prodromal diarrhea in children of all ages and in adults, and accounts for about 5–10% of all cases [2]. The latter usually occurs in families and has a very poor outcome, with end-stage renal failure or death in more than 50% of cases [2].

Available evidence points toward vascular endothelial dysfunction as an important step in the sequence of events leading to the development of microangiopathic lesions in HUS. Consistent with this possibility are findings that all the proposed causative agents for HUS (i.e., bacterial endotoxins, verotoxins, antibodies and immunocomplexes, and certain drugs) are toxic to vascular endothelium in vitro.

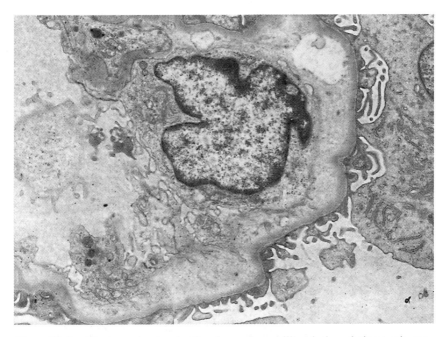

Figure 21-1. Electron micrograph of a glomerular capillary in hemolytic uremic syndrome. The endothelium is detached from the glomerular basement membrane; the subendothelial space is widened and occupied by electron-lucent fluffy material and cell debris. The capillary lumen is narrowed. (Magnification: 3000 ×).

2. The Pivotal Role of Endothelial Cell Injury in the Pathogenesis of HUS

A number of endothelial cell abnormalities have been reported in HUS and its related syndrome, thrombotic thrombocytopenic purpura (TTP). In both syndromes, the vessels form less prostacyclin (PGI_2) than normal [3] and serum binding capacity for PGI_2 is reduced [4]. A recent study showed that in vivo PGI_2 synthesis is reduced in children with HUS, as indicated by lower than normal urinary excretion of renal PGI_2 metabolites in the acute phase of the disease, which normalizes in remission [5]. Unusually large von Willebrand factor (vWF) multimers in plasma of patients with HUS or TTP can be taken as additional evidence of vascular endothelial cell damage [1,6,7]. These multimers do not normally circulate but are stored in the endothelial cell Weibel-Palade bodies and their presence in the peripheral blood may reflect endothelial cell damage.

Experimental evidence indicate that all the factors known to cause HUS may damage vascular endothelium. Most cases of HUS are due to an infection with *Escherichia coli* 0157:H7 [8,9] that produces two toxins, verotoxin-1 and verotoxin-2, also called Shiga-like toxin due to their identity to the cytotoxin of

Shigella dysenteriae type 1 [10,11]. Verotoxins are toxic to human endothelial cells in culture [12,13] and reduce the synthesis of the natural platelet aggregation inhibitor prostacyclin from rat aortic tissue [14]. As recently documented [15], human renal cortex and medulla express specific glycolipid receptors (Gb3 receptors) which bind both verotoxin-1 and verotoxin-2.

Endotoxin can combine with verotoxin to amplify its cytotoxic potential. Louise and Obrig [16] found that this synergism was dose dependent in the umbilical vein and renal endothelial cells in culture and was maximal after preincubating the cells with endotoxin. Endotoxin alone transiently inhibited protein synthesis by umbilical vein endothelial cells and enhanced the protein synthesis-inhibiting activity of verotoxin. A similar synergism has been demonstrated between verotoxin and tumor necrosis factor-α (TNF-α) [17]. Preincubation of umbilical vein endothelial cells with TNF-α increased specific binding sites for ^{125}I verotoxin-1 10–100-fold, and upregulated cell-surface-associated verotoxin receptors [18]. Given the frequent finding of endotoxemia in HUS [19], it is conceivable that endothelial damage results from the toxic effect of verotoxin added to the effects of endotoxin and endotoxin-elicited TNF-α.

More recent studies focused on the possibility that endothelial damage could derive from neutrophil activation with the consequent excessive release of toxic products [20]. In shigellosis, neutrophilia, endotoxemia and complement activation correlate with the development of HUS [21]. Plasma concentration of elastase, a neutrophil-derived proteolytic enzyme, is increased in patients with diarrhea-associated HUS [22]. Moreover, polymorphonuclear cells from HUS patients avidly adhere to endothelial cells in culture and induce endothelial injury by degrading fibronectin [23].

The possibility that activated neutrophils contribute to microvascular injury in HUS is consistent with findings that the cytotoxic activity of endotoxin in vitro [24] and in vivo [25,26] mainly results from its ability to promote neutrophil adhesion to endothelium [27] and to stimulate cytokines [28] and oxygen radical [29] formation. Recently, we have demonstrated that verotoxin-1 increases leukocyte adhesion to cultured endothelial cells under physiologic flow conditions [30] by upregulating adhesive proteins on endothelial cell surface. This finding is relevant if one considers that leukocyte adherence and/or emigration are key events for the increased vascular leakage and microvascular dysfunction associated with acute inflammation.

3. NO and Peroxynitrite as Mediators of Endothelial Damage: Evidence from Studies in Vitro and in Experimental Models of Endothelial Cell Injury

Bacterial and viral products as well as leukocyte-derived cytokines [31–35], by upregulating a calcium-independent isoform of NO synthase, may induce endothelial cells to release NO. Overproduction of NO is part of a defense mechanism against invading microorganisms [35,36], but it can exert undesired cytotoxicity on the cells that produce it [37] and on neighboring cells [38], as documented by data that the NO antagonist *N*-monomethyl-L-arginine (L-N-MMA) protects experimental animals from immune-mediated vascular injury [39,40]. Proposed mechanisms for NO toxicity include binding to iron in the prosthetic groups of a variety of enzymes, *S*-nitrosation of proteins, and mono- or poly(ADP)ribosylation leading to posttranslational modification of proteins and also to NAD depletion [35,37,41,42]. Autocytotoxicity of NO formed from the inducible enzyme has been recently reported upon endothelial cell stimulation with TNF-α [33]. Thus, TNF-α induced a time- and dose-dependent release of cytoplasmic enzyme lactate dehydrogenase (LDH) in cultured bovine aortic endothelial cells, which was preceded by increased production of NO and was prevented by the NO synthase inhibitor *N*-iminoethyl-L-ornithine and by dexamethasone, which blocks [43] expression of the inducible NO synthase in the endothelium. In two recent articles, NO liberated from various NO-donating agents caused cytolysis in rat liver endothelial cells [44] and in human umbilical vein endothelial cells (HUVEC) [45], confirming a direct toxic effect of NO.

The production of large amounts of NO has been implicated in the cell and organ damage associated with endotoxic shock [46,47 and Chapter 14]. Endotoxin causes intravascular NO production through the induction of NO synthase either in medial-layer smooth muscle cells or in endothelial cells [43,48–50]. Excessive NO release likely mediates endotoxin-induced vascular hyporeactivity, as documented by data that both N_G-monomethyl-L-arginine (L-NMMA) and N_W-nitro-L-arginine (L-NAME) were able to restore in vivo responsiveness to norepinephrine after intravenous infusion of endotoxin in rats [51]. In another study, L-NMMA administration in dogs with endotoxin-mediated shock fully restored vascular resistance and systemic arterial pressure [52]. In rats, administration of L-NMMA concurrently with endotoxin enhanced the leakage of albumin in the gastrointestinal tract. By contrast, when L-NMMA was injected 3 h after endotoxin (i.e., at the time of the expression of an inducible NO synthase enzyme), a dose-dependent reduction in the vascular damage was observed [53]. Similarly, in another study in rabbits, pretreatment with dexamethasone had no effect on the initial hemodynamic changes following endotoxin but prevented the subsequent fall in mean arterial pressure observed in animals treated with endotoxin alone [54]. This suggests that even though the role of constitutively produced NO might be protective in maintaining local organ perfusion in endotoxemia [55], the large quantities generated by the inducible enzyme contribute to tissue damage.

There is also evidence that NO can rapidly react with leukocyte-derived superoxide anion (O_2^-) to form peroxynitrite ($ONOO^-$) [56,57], a potent oxidizing agent known to initiate lipid peroxidation in biological membranes, hydroxylation and nitration of aromatic amino acid residues and sulfhydryl oxidation of proteins [58,59]. Furthermore, once protonated, peroxynitrite decays to the very reactive and toxic hydroxyl radical (OH•) and nitrogen dioxide radical NO_2• [56]. This pathway is of particular interest during acute inflammatory reaction, when excess oxygen radicals are produced by activated leukocytes. Actually, formation of peroxynitrite has been recently detected, by measuring the immunoreactivity of nitrotyrosine, in the aorta of rats 6 h after injection of endotoxin, which was prevented by L-NMMA administration [60]. Similarly, Wizemann and co-workers [47] found evidence for the presence of nitrotyrosine residues in sections of lungs 48 h after endotoxin administration to rats, which correlated with histological evidence of cellular damage [47].

4. Increased NO Formation in HUS: A Possible Mediator of Microvascular Injury

We recently explored the possibility that in HUS and the related disease TTP, an exaggerated formation of NO could increase in vivo lipid peroxidation, possibly as a consequence of NO interacting with neutrophil-derived products [61]. Plasma concentrations of the NO metabolites nitrites/nitrates (NO_2^-/NO_3^-) in patients with recurrent HUS/TTP studied during the acute phase of the disease were elevated as compared to healthy controls, indicating an increased NO synthesis in vivo. At recovery, NO_2^-/NO_3^- plasma concentrations decreased in all patients. To investigate whether substances which increase vascular NO synthesis were present in the circulation of patients with HUS/TTP, human umbilical vein endothelial cells (HUVEC) were exposed for 24 h in vitro to serum from patients or controls, and NO synthesis was evaluated as conversion of [^{3}H]-L-arginine to [^{3}H]-L-citrulline. Serum from patients with acute HUS/TTP induced NO synthesis in cultured endothelial cells more than normal serum. Enhanced stimulatory activity was no longer found in the recovery phase. Similar results have been obtained recently in four patients with typical HUS and diarrhea prodrome (Table 21-1, unpublished data). Of note, in the latter patients, differences between patients and controls were even greater than in recurrent HUS/TTP.

Altogether these data suggest that the high NO levels measured in vivo in acute HUS/TTP depend on NO synthesis induction, most probably in vascular endothelial cells. Activated inflammatory cells are also possible sources of NO [36]. In humans, there is indirect evidence of NO formation by macrophages [62] and peripheral monocytes [63]; however, a conclusive proof is still lacking [64].

TNF-α is a likely candidate for NO synthase induction in HUS/TTP due to the fact that this cytokine is one of the most potent inducers of the inducible NO synthase enzyme in various cell system, including endothelial cells [33,34]. Actually, we found a higher plasma TNF-α concentration in patients with acute HUS/TTP, confirming previous data in the literature [65]. In addition, Siegler et al. [66] found elevated levels of TNF-α in the urine of children with HUS, very likely reflecting renal overproduction of the cytokine. Apart from TNF-α, other

Table 21-1. Nitric Oxide and Oxygen Radical Synthesis in Patients with Typical HUS

	O_2^- release by PMNs (nmol/10^6 cells/30 min)	Plasma NO_2^- NO_3^- (nmol/ml)	[^{3}H]-L-citrulline[a] (pmol/10^5 cells)	Plasma MDA (nmol/ml)
HUS				
Acute	11.6±3.5*.**	121.8±4.2*.**	182±92*.**	0.549±0.373*.**
Remission	4.6±0.4	85.8±15.4	98±50	0.313±0.081
CTR	3.2±0.4	43.1±14.3	58±11	0.146±0.034

Note: Values are mean ± S.D.

[a]HUVEC were incubated for 24 h with serum (diluted 1 : 2 in Phosphate-buffered saline (PBS) from patients and controls in the presence of [^{3}H]-L-arginine. HUS, patients with typical, diarrhea associated hemolytic uremic syndrome (n=4), CTR, healthy volunteers (n=4).

*p < .01 versus CTR.

**p < .05 versus remission.

substances that accumulate in the blood of patients with HUS/TTP, such as bacterial cytotoxins [31,32,43] and other cytokines (i.e., interleukin-1β [31]), could theoretically contribute to endothelial NO synthesis induction in this disease.

In turn, NO itself releases TNF-α and interleukin-1 [67,68] from human leukocytes, thus promoting secondary activation that may amplify inflammatory damage [69]. Actually, recent evidence has indicated (see above) that neutrophils contribute to microvascular injury in HUS. Because activated neutrophils release reactive oxygen radicals, we measured ex vivo O_2^- production by polymorphonuclear cells from patients with acute HUS/TTP. Polymorphonuclear cell-derived O_2^- was remarkably higher than normal in patients with acute forms of either typical HUS (Table 21-1) or recurrent HUS/TTP [61] but decreased in the recovery phase. It is possible that this excess O_2^- reacts with NO, yielding peroxynitrite, a strong cell toxic due to its ability to induce oxidant injury and initiate lipid peroxidation. The latter possibility is supported by data of higher malondialdehyde (MDA) and conjugated diene concentrations in the circulation of patients with acute HUS/TTP ([61] and Table 21-1). This is not a completely new finding, as previous studies showed increased lipid peroxidation products in the serum of an infant [70] and children with HUS [71]. Moreover, increased levels of MDA and decreased antioxidant enzyme activity have been found in red blood cells of children with diarrhea-associated HUS [72]. It is possible that oxidative stress utilizes and depletes the intracellular defense mechanism of red blood cells in these patients, inducing a pro-oxidant/antioxidant imbalance. Interestingly, in our patients with HUS/TTP, either O_2^- release from polymorphonuclear cells or NO metabolite plasma levels positively correlated [61] with plasma LDH, the best available marker of disease activity, further confirming a role of these toxic radicals in the microangiopathic process.

A preliminary report by Westberg et al. [73] has recently documented increased

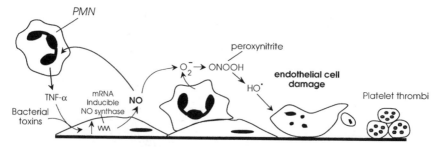

Figure 21-2. Proposed mechanism for nitric oxide and oxygen radical-mediated endothelial damage in hemolytic uremic syndrome. PMN: polymorphonuclear cell; TNF-α: tumor necrosis factor-α; NO: nitric oxide; O_2^-: superoxide anion; HO•: hydroxyl radical.

NO_3^- plasma levels in patients with active TMA. The authors suggested that increased NO synthesis in TMA may be an adaptive response to prevent platelet aggregation in this disease. Given the inhibitory properties of NO on platelet activation [35], this possibility cannot theoretically be excluded. However, the very rapid interaction of NO with neutrophil-derived O_2^- likely prevents NO from exerting its beneficial effects in this syndrome.

5. Hypothesis

A schematic rendition of the hypothesis on a possible novel pathway of injury that could account for microvascular damage in HUS/TTP is depicted in Fig. 21-2. The initial insult (verotoxin, endotoxin, immune complexes, cytokines) induces NO synthesis by endothelial cells and polymorphonuclear cell activation. Activated polymorphonuclear cells adhere to the endothelium and release cytokines and O_2^-: the former amplify inflammatory response, the latter can interact with NO to form ONOO⁻, thus contributing to microvascular damage. NO, in turn, induces the release of cytokines by leukocytes. Thus, once activated, the inflammatory process sustains itself.

Acknowledgments

This work has been supported by Italian Telethon Grant No. E 353.

References

1. Remuzzi, G., Ruggenenti, P., and Bertani, T. In: *Renal Pathology with Clinical and Functional Correlations.* Eds. Tisher, C.G. and Brenner, B.M. Lippincott, Philadelphia, 1994, pp. 1154–1184.

2. Remuzzi, G. and Ruggenenti, P. *Kidney Int.* **47,** 2–19 (1995).

3. Remuzzi, G., Misiani, R., Marchesi, D. et al. *Lancet* **2,** 871–872 (1978).

4. Wu, K.K., Hall, E.R., Rossi, E.C., and Papp, A.C. *J. Clin. Invest.* **75,** 168–174 (1985).

5. Noris, M., Benigni, A., Siegler, R.L., Gaspari, F., Casiraghi, F., Mancini, M., and Remuzzi, G. *Am. J. Kidney Dis.* **20,** 144–149 (1992).

6. Galbusera, M., Ruggenenti, P., Noris, M., Burnouf-Radosevich, M., Benigni, A., Mannucci, P.M., and Remuzzi, G. *Lancet* **345,** 224–225 (1995).

7. Mannucci, P.M., Lombardi, R., Lattuada, A. et al. *Blood* **74,** 978–983 (1989).

8. Riley, L.W., Remis, R.S., Helgerson, S.D. et al. *N. Engl. J. Med.* **308,** 681–685 (1983).

9. Karmali, M.A., Steele, B.T., Petric, M., and Lim, C. *Lancet* **1,** 619–620 (1983).

10. O' Brien, A.D., Lively, T.A., Chen, M.E., Rothman, S., and Formal, F.B. *Lancet* **1,** 702 (1983).

11. O' Brien, A.D. and Holmes, R.K. *Microbiol. Rev.* **51,** 206–220 (1987).

12. Obrig, T.G., Del Vecchio, P.J., Karmali, M.A., Petric, M., Moran, T.P., and Judge, T.K. *Lancet* **2,** 687 (1987).

13. Kavi, J., Chant, I., Maris, M., and Rose, P.E. *Lancet* **2,** 1035 (1987).

14. Karch, H., Bitzan, M., Pietsch, R. et al. *Microb. Pathol.* **5,** 215–221 (1988).

15. Boyd, B. and Lingwood, C. *Nephron* **51,** 207–210 (1989).

16. Louise, C.B. and Obrig, T.G. *Infect. Immunol.* **60,** 1536–1543 (1992).

17. Louise, C.B. and Obrig, T.G. *Infect. Immunol.* **59,** 4173–4179 (1991).

18. van de Kar, N.C.A.J., Monnens, L.A.H., Karmali, M.A., and van Hinsbergh, V.W.M. *Blood* **80,** 2755–2764 (1992).

19. Koster, F., Boonpucknavig, V., Suiaho, S., Gilman, R., and Rahaman, M. *Clin. Nephrol.* **21,** 126–133 (1984).

20. Henson, P.M. and Johnston, R.B. Jr. *J. Clin. Invest.* **79,** 669–674 (1987).

21. Koster, K., Levin, J., Walker, L. et al. *N. Engl. J. Med.* **298,** 927 (1978).

22. Milford, D.V., Staten, J., MacGreggor, I., Dawes, J., Taylor, C.M., and Hill, F.G. *Nephrol. Dial. Transplant.* **6,** 232–237 (1991).

23. Forsyth, K.D., Simpson, A.C., Fitzpatrick, M.M., Barratt, T.M., and Levinsky, R.J. *Lancet* **2,** 411–414 (1989).

24. Raghu, G., Striker, L.J., and Striker, G.E. *Clin. Immunol. Immunopathol.* **38,** 275–281 (1986).

25. Butler, T., Rahaman, H., Al-Mahmua, K.A. et al. *Br. J. Exp. Pathol.* **66,** 7–15 (1985).

26. Bohn, E. and Muller-Berghaus, G. *Am. J. Pathol.* **84,** 239–258 (1976).

27. Thomas, P.D., Hampson, F.W., Casale, J.M., and Hunninghake, G.W. *J. Lab. Clin. Med.* **111,** 286–292 (1988).

28. Firestein, G.S., Boyle, D., Bullough, D.A., et al. *J. Immunol.* **152,** 5853–5859 (1994).

29. Taylor, C.M. and Powell, H.R. In: *Hemolytic Uremic Syndrome and Thrombotic Thrombocytopenic Purpura.* Eds. Kaplan, B.S., Trompeter, R.S., and Moake, J.L. Marcel Dekker, Inc., New York, 1992, pp. 335–372.

30. Morigi, M., Micheletti, G., Figliuzzi, M. et al. *Blood* **86,** 4556–4558 (1995).

31. Kilbourn, R.G. and Belloni, P. *J. Natl. Cancer Inst.* **82,** 772–776 (1990).

32. Noris, M., Benigni, A., Boccardo, P. et al. *Kidney Int.* **44,** 445–450 (1993).

33. Estrada, C., Gomez, C., Martin, C., Moncada, S., and Gonzalez C. *Biochem. Biophys. Res. Commun.* **186,** 475–482 (1992).

34. Lamas, S., Michel, T., Brenner, B.M., and Marsden, P.A. *Am. J. Physiol.* **261,** C634–C641 (1991).

35. Moncada, S. and Higgs, E.A. *FASEB J.* **9,** 1319–1330 (1995).

36. Marletta, M.A., Yoon, P.S., Iyengar, R., Leaf, C.D., and Wishnok, J.S. *Biochemistry* **27,** 8706–8711 (1988).

37. Drapier, J.C. and Hibbs, J.B., Jr. *J. Immunol.* **140,** 2829–2838 (1988).

38. Seekamp, A., Mulligan, M.S., Till, G.O., and Ward, P. *Am. J. Pathol.* **142,** 1217–1226 (1993).

39. Mulligan, M.S., Moncada, S., and Ward, P.A. *Br. J. Pharmacol.* **107,** 1159–1162 (1992).

40. Mulligan, M.S., Hevel, J.M., Marletta, M.A., and Ward, P.A. *Proc. Natl. Acad. Sci. USA* **88,** 6338–6342 (1991).

41. Kwon, N.S., Stuehr, D.H., and Nathan, C.F. *J. Exp. Med.* **174,** 761–767 (1991).

42. Stadler, J., Billiar, T.R., Curran, R.D., Stuehr, D.J., Ochoa, J.B., and Simmons, R.L. *Am. J. Physiol.* **260,** C910–C916 (1991).

43. Radomski, M.W., Palmer, R.M.J., and Moncada, S. *Proc. Natl. Acad. Sci. USA* **87,** 10043–10047 (1990).

44. Volk, T., Ioannidis, I., Hensel, M., deGroot, H., and Kox, W.J. *Biochem. Biophys. Res. Commun.* **213,** 196–203 (1995).

45. Bratt, J. and Gyllenhammar, H. *Arthritis Rheum.* **38,** 768–776 (1995).

46. Stoclet, J.C., Fleming, I., Gray, G. et al. *Circulation* **87** (Suppl. V), V77–V80 (1993).

47. Wizemann, T.M., Gardner, C.R., Laskin, J.D. et al. *J. Leuk. Biol.* **56,** 759–768 (1994).

48. Rees, D.D., Cellek, S, Palmer, R.M.J., and Moncada, S. *Biochem. Biophys. Res. Commun.* **173,** 541–547 (1990).

49. Knowles, R.G., Salter, M., Brooks, S.L., and Moncada, S. *Biochem. Biophys. Res. Commun.* **172,** 1042–1048 (1990).

50. Weigert, A.L., Higa, E.M.S., Niederberger, M., McMurtry, I.F., Raynolds, M., and Schrier, R.W. *J. Am. Soc. Nephrol.* **5,** 2067–2072 (1995).

51. Julou-Schaeffer, G., Gray, G.A., Fleming, I., Schott, C., Parratt, J.R., and Stoclet, J.-C. *Am. J. Physiol.* **259,** H1038–H1043 (1990).

52. Kilbourn, R.G., Jubran, A., Gross, S.S. et al. *Biochem. Biophys. Res. Commun.* **172,** 1132–1138 (1990).

53. Laszlo, F., Whittle, B.J., and Moncada, S. *Br. J. Pharmacol.* **111,** 1309–1315 (1994).

54. Wright, C.E., Rees, D.D., and Moncada, S. *Cardiovasc. Res.* **26,** 48–57 (1992).

55. Mulder, M.F., Van Lambalgen, A.A., Huisman, E., Visser, J.J., Van Den Bos, G.C., and Thijs, L.G. *Am. J. Physiol.* **266,** H1558–H1564 (1994).

56. Pryor, W.A. and Squadrito, G.L. *Am. J. Physiol.* **268,** L699–L722 (1995).

57. Beckman, J.S., Beckman, T.W., Chen, J., Marshall, P.A., and Freeman, B.A. *Proc. Natl. Acad. Sci. USA* **87,** 1620–1624 (1990).

58. Radi, R., Beckman, J.S., Bush, K.M., and Freeman, B.A. *J. Biol. Chem.* **266,** 4244–4250 (1991).

59. Radi, R., Beckman, J.S., Bush, K.M., and Freeman, B.A. *Arch. Biochem. Biophys.* **288,** 484–487 (1991).

60. Szabo, C., Salzman, A.L., and Ischiropoulos, H. *FEBS Lett.* **363,** 235–238 (1995).

61. Noris, M., Ruggenenti, P., Todeschini, M. et al. *Am. J. Kidney Dis.* **27** (1996).

62. Denis, J. *J. Leukoc. Biol.* **48,** 380–387 (1991).

63. Hunt, N.C.A. and Goldin, R.D. *J. Hepatol.* **14,** 146–150 (1992).

64. Moncada, S. and Higgs, A. *N. Engl. J. Med.* **329,** 2002–2011 (1993).

65. Wada, H., Kaneko, T., Ohiwa, M., et al. *Am. J. Hematol.* **40,** 167–170 (1992).

66. Siegler, R.L., Edwin, S.S., Christofferson, R.D., and Mitchell, M.D. *J. Am. Soc. Nephrol.* **2,** 274 (1991) (Abstract).

67. Lander, H.M., Sehajpal, P., Levine, D.M., and Novogrodsky, A. *J. Immunol.* **150,** 1509–1516 (1993).

68. Magrinat, G., Mason, S.N., Shami, P.J., and Weinberg, J.B. *Blood* **80,** 1880–1884 (1992).

69. Larrick, J.W., Graham, D., Toy, K., Lin, L.S., Senyk, G., and Fendly, B.M. *Blood* **69,** 640–644 (1987).

70. Brown, R.E., Alade, A.L., Knight, J.A., and Evans B.J. *Clin. Chem.* **34,** 2382–2384 (1988).

71. Situnayake, R.D., Crump, B.J., Thurman, D.I., and Taylor, C.M. *Pediatr. Nephrol.* **5,** 387–392 (1991).

72. Turi, S., Nemeth, I., Vargha, I., and Matkovic, B. *Pediatr. Nephrol.* **8,** 26–29 (1994).

73. Westberg, G., Herlitz, H., Petersson, A., Sigstrom, L., and Wennmalm, A. *J. Am. Soc. Nephrol.* **6,** 458 (1995) (Abstract).

22

Role of NO in Urinary Tract Infections

Robert M. Weiss, Marcia A. Wheeler, and
Shannon D. Smith

Nitric oxide (NO) has been shown to have a pathophysiologic role in several disease states including sepsis [1], allograft rejection [2], rheumatoid arthritis [3,4], glomerulonephritis [5], and glomerulosclerosis [6,7]. Inducible nitric oxide synthase (iNOS) has been identified in rodent monocytes/macrophages, and iNOS mRNA has been detected in purified human monocyte-derived macrophages induced with LPS and interferon-γ [8]. As urine from patients with urinary tract infections (UTIs) contains the components involved in the induction of NOS, including bacterial products, cytokines, and neutrophils, infected urine has provided a model for the molecular and enzymatic characterization of iNOS in human inflammatory cells. The definitive documentation of iNOS in human inflammatory cells previously had been lacking.

There are at least three major types of nitric oxide synthase (NOS) enzymes that convert L-arginine to NO and L-citrulline [9]. One is a soluble constitutive calcium/calmodulin-dependent enzyme (nNOS) that has been localized primarily to neuronal tissues [10–12]. The activity of this form of NOS is regulated by changes in intracellular calcium concentration and a nNOS has been characterized in nonadrenergic, noncholinergic (NANC) innervated tissues [13–15]. The second type of NOS, endothelial NOS (eNOS), has similar properties to nNOS in that it also is a constitutive enzyme that releases NO for short periods of time and requires NADPH, tetrahydrobiopterin, FAD, and FMN as cofactors (see Chapter 2). eNOS, however, is primarily membrane bound due to posttranslational modifications [16,17] and has been identified in endothelial cells [18]. The other major type of NOS, an inducible NOS (iNOS), is induced in cultured rabbit bladder smooth muscle cells [19,20], rodent macrophages [21–24], human hepatocytes [25], and vascular smooth muscle cells. iNOS is induced in cells treated with lipopolysaccharides (LPS) [21] and cytokines [26,27]. Once activated, iNOS synthesizes NO for long periods of time and can exert pro-inflammatory, bacteriocidal, and antiviral actions [28–30]. (see Chapter 14).

During UTIs, there is a complex interaction between the host defense mechanisms and the invading bacteria (Fig. 22-1). Bacteria may attach to the urothelium and induce cytokines [31,32]. Interleukin-6 (IL-6) is found in the urine of mice within minutes of the intravesical instillation of *Escherichia coli* or isolated *P. fimbriae* [33]. Urinary tract infections in children infected with *P. fimbriated E. coli* are associated with higher IL-6 levels than infections occurring in children with other *E. coli* strains [34]. Both IL-6 and IL-8 are found in urine after colonization of the human urinary tract with nonvirulent strains of *E. coli* [31,32]. *E. coli* also has been shown to stimulate the production of IL-8 in normal bladder epithelial cells, urinary tract epithelial cell lines, and neutrophils, and IL-8 is an important neutrophil chemoattractant [35]. Both IL-6 and IL-8 appear to be secreted by uroepithelial cells in the bladder and initiate the influx of neutrophils. When cytokines are induced, neutrophil numbers can rise to 10^3–10^8 cells/ml urine in UTIs.

Most neutrophils present in infected urine are viable and are potentially capable of phagocytosis [36–38]. The ability of leukocytes to phagocytose bacteria may derive from their ability to produce hydrogen peroxide or reactive nitrogen

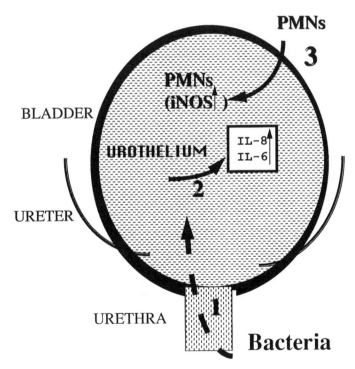

Figure 22-1. Schematic representation of a UTI. 1. Bacteria travel through the urethra and attach to the bladder urothelium. 2. The urothelium is induced to produce cytokines. 3. Neutrophils (PMNs) invade the bladder and NOS is induced.

intermediates, including NO [39]. Combinations of cytokines and LPS which generate large amounts of NO in rodent monocyte/macrophages elicit none or very small amounts of NO in human neutrophils isolated from blood [40,41]. Studies with nontraditional inducers of NOS in human monocyte/macrophages indicate that these cells can produce NO following infection with human immuno-deficiency virus type I [42] and after cross-linking of the surface receptors CD69 [43] or FcεII/CD23 [44,45].

Reactive nitrogen intermediates produced by leukocyte cytoplasts can kill bacteria [46] and neutrophils in freshly voided urines from symptomatic bacteri-uric patients also can phagocytoze bacteria. Therefore, although phagocytosis may be compromised by low urinary pH and adverse osmolality [38], the increased NOS activity seen in urinary leukocytes from patients with UTIs may facilitate bacterial killing.

In the urine from patients with UTIs, the conditions are present for the induction of nitric oxide synthase (NOS), that is,

1. Bacterial products including LPS

2. Cytokines

3. Large numbers of neutrophils

This, coupled with the presence of reactive nitrogen intermediates, including nitrites, led to our studies on infected urine as a means of verifying iNOS activity in human inflammatory cells.

Urine nitrite levels are increased in patients with UTIs [47] and detection of urinary nitrites serves as a rapid indirect method for detecting bacteriuria [48]. The elevated urinary nitrite levels with UTIs are based on the bacterial reduction of endogenous nitrates (NO_3^-) to nitrites (NO_2^-).

$$NO_3^- \xrightarrow{\text{bacteria}} NO_2^-$$

Another possible source of elevated urinary nitrite levels is related to the oxidation of nitric oxide (NO) to nitrates and nitrites with further bacterial reduction of nitrates to nitrites.

Smith et al. [49] have shown that with UTIs in humans, in addition to elevated urinary nitrite levels, there is increased NOS activity in the particulate fraction of the urine pellet. Furthermore, the elevated urinary nitrite levels in infected urine, increased with time when the urine was incubated at 37°C (Fig. 22-2). Nitrite levels in noninfected urine did not increase with incubation. The increase in nitrite levels with time in the infected specimens could be related to both the bacterial conversion of nitrate to nitrite and to the oxidation of NO to nitrate and nitrite. The dual mechanism for the increase in urinary nitrite levels is supported by the finding that the addition of L-arginine (1 mM) to the infected urine during

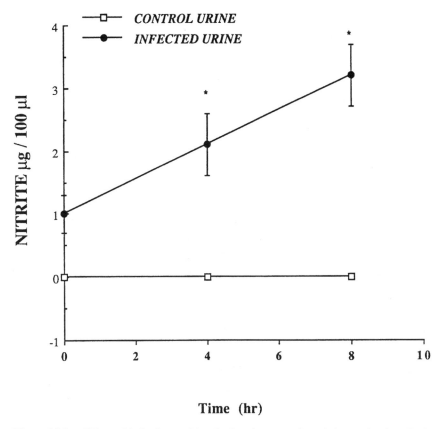

Time (hr)

Figure 22-2. Effect of infection and incubation time on urine nitrite production. Each data point is the mean ± SEM of the urinary nitrite levels from at least five different donors. Asterisk represents a 95% significance level between control and infected urine. [Reprinted with permission from *Kidney International,* **45** (1994), pp. 586–591.]

a 4-h incubation period caused a further 102±27% increase in urinary nitrite production. In bacteria-enriched filtrates of infected urine, NOS activity was not measurable and L-arginine did not increase nitrite production, suggesting that the primary NOS activity responsible for the increased nitrite production in infected urine is related to the cellular fraction rather than to the bacteria. Furthermore, NOS activity was not detected in samples of *E. coli*. The addition of sodium nitrate (NaNO$_3$; 1 m*M*) to infected urine during a 4-h incubation period caused a 124±35% increase in urinary nitrite production and caused a 33±11% increase in nitrite levels in bacteria-enriched filtrates of infected urine. Neither L-arginine nor NaNO$_3$ altered nitrite levels in noninfected urine during incubation.

Nitric oxide synthase activity, as measured by the conversion of [^{14}C]-L-arginine to [^{14}C]-L-citrulline, was significantly elevated in homogenates isolated from urine pellets of patients with UTIs compared to controls. Production of [^{14}C]-citrulline was confirmed by thin-layer chromatography. The increase in NOS activity was evident in both soluble and particulate fractions of the urine pellet, although the soluble activity was less than 10% of the particulate activity. NOS activity in the particulate fraction of the urine pellet from women with noninfected urine was 18 times greater than that in the particulate fraction of the urine pellet from men with noninfected urine, 14±3 and 0.8±0.2 pmol citrulline formed per minute per milligram of protein, respectively. NOS activity was increased significantly in the particulate fraction of the urine pellet in both women and men with infected urine, 121±25 and 49±9 pmol citrulline formed per minute per milligram of protein, respectively (Fig. 22-3). The K_m for enzyme activity, using 3.0–200 μM arginine as substrate, in urine pellet particulate fractions was 18±5 μM arginine, which is consistent with published results for purified iNOS [49].

Nitric oxide synthase activity in the particulate fraction isolated from the urine pellet of patients with UTIs depends on NADPH but not on calcium. The average NOS activity in particulate fractions of the urine pellet of infected urine obtained

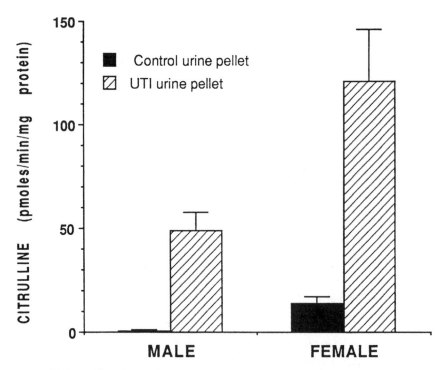

Figure 22-3. Difference in urine particulate NOS activity between men and women with and without urinary tract infections. Each value represents determinations from at least five patients.

from males and females in the presence of 1 mM CaCl$_2$ is 81±14 pmol arginine metabolized per minute per milligram of protein. NOS activity is 75±19 pmol arginine metabolized per minute per milligram of protein in the nominally calcium-free medium. Similarly, particulate NOS activity in uninfected urine from both men and women is not affected by removal of calcium, whereas it decreases on the removal of NADPH.

Nitric oxide synthase activity in the particulate fraction of the urine pellet from infected urine was inhibited 39±3%, 30±4%, and 5±2% by 100 μM L-canavanine, 100 μM N^G-monomethyl-L-arginine (NMA) and 100 μM N^G-nitro-L-arginine (NNA), respectively (Fig. 22-4). The cofactor and inhibitor profile of the urine pellet NOS activity provide further support that the NOS activity measured in the particulate fraction of the urine pellet is an inducible isoform, an

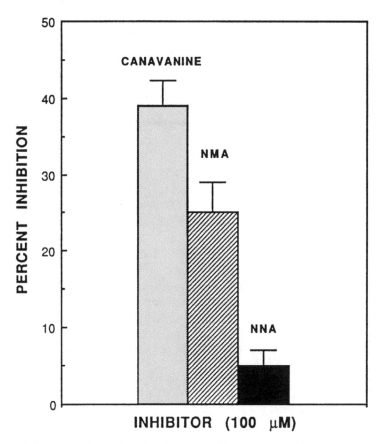

Figure 22-4.   Percent inhibition of particulate NOS by L-NNA, L-NMMA, and L-canavanine. Results are reported as percent inhibition ± SEM. [Reprinted with permission from *Kidney International,* **45** (1994), pp. 586–591.]

iNOS, and is similar to the activity described in rodent macrophages, inflammatory neutrophils, and carcinoma cells [50] and to that induced by microbes, microbial products, tumor cells, and cytokines [51]. NOS inhibitors blunt citrulline formation from arginine in a concentration-dependent manner (i.e., 1000 μM L-canavanine inhibits 68±4% of the NOS activity. An explanation for the inability of these inhibitors to suppress all of the citrulline production is that some of the citrulline may be generated by a NOS-independent pathway. In murine macrophages and bovine epithelial cells, arginine is metabolized to ornithine by arginase, and the ornithine, in turn, is converted to citrulline by a transcarbamoylase [52,53]. A similar pathway may account for the production of some of the citrulline in the urine pellet.

To characterize the cell type containing iNOS, Ficoll gradient centrifugation was performed on leukocyte-positive urines from patients with and without UTIs. Ficoll gradient fractionation of cell pellets from the urine of patients with UTIs shows that the majority of the NOS activity is found in neutrophil-enriched fractions and that the NOS activity is 43-fold higher in neutrophil-enriched fractions obtained from patients with UTIs than from controls. When NOS activity is expressed as NADPH-dependent activity, 23.5±13.8 pmol citrulline per minute per milligram of protein is measured in neutrophil-enriched fractions obtained from patients with UTIs compared to 0.55±0.16 pmol citrulline per minute per milligram of protein measured in neutrophil-enriched fractions isolated from controls.

Inhibition profile of leukocyte NOS activity in leukocyte-enriched fractions isolated from the urine with a series of NOS inhibitors is consistent with the pattern seen with human recombinant iNOS [54] and with iNOS in activated rat peritoneal neutrophils [55]. The rank order of NOS-inhibitor potency on NADPH-dependent neutrophil iNOS activity is L-N^5-(1-iminoethyl) ornithine (L-NIO) > L-thiocitrulline > L-NMMA > L-canavanine > aminoguanidine > N^G-nitro-L-arginine methyl ester (L-NAME) = (L-NNA) (Table 22-1). The iNOS in human neutrophils is primarily membrane associated, which is similar to the membrane partitioning of eNOS [56]. Although human neutrophil iNOS is calcium independent, it is inhibited by the calmodulin, antagonists, trifluoperazine (TFP), and N-(6-aminohexyl)-5-chloro-1-naphthalene-sulfonamide (W-7) at effective concentrations of 10 μM and 39 μM, respectively (Table 22-1). This suggests that calmodulin is not bound as tightly to the human neutrophil iNOS as to the rodent form.

A reverse transcriptase–polymerase chain reaction (RT-PCR) on cDNA prepared from leukocyte-enriched total RNA reveals a 413-bp (base-pair) fragment which is consistently amplified with human iNOS primers. DNA sequencing of the PCR products reveals 99.9% nucleotide sequence identity with their respective cloned cDNAs.

Using a specific C-terminal human iNOS polyclonal antibody [57], Western blot analysis of total lysates or 2′5′-ADP-sepharose-purified samples isolated

Table 22-1.   NOS Inhibitors and Calmodulin
Antagonists Inhibit Leukocyte NOS Activity

	IC_{50}^a (μM)
NOS inhibitor	
L-NIO	3.0
L-Thiocitrulline	4.1
L-NMMA	7.1
L-Canavanine	12.1
Aminoguanidine	22.0
L-NAME	116
L-NNA	>200
Calmodulin antagonist	
Trifluoperizine	10
W-7	39

[a]IC_{50} values are determined using a computer-assisted
log–logit plot for two to six experiments for each inhibitor, each done in duplicate.

from leukocyte fractions obtained from patients with UTIs identified iNOS protein at approximately 130 kDA, identical in molecular weight to the protein product seen in HEK 293 cells stably transfected with the human hepatocyte iNOS cDNA. Collectively, these data provide definitive molecular proof of an iNOS in human neutrophils whose activity is increased with UTIs.

It is of note that iNOS activity which is 10.7±2.9 pmol citrulline per minute per milligram of protein at the time of UTI diagnosis remains elevated at 23.5±8.6 pmol citrulline per milligram of protein after 2–4 days of antibiotic therapy but decreases significantly to 1.5±0.5 pmol citrulline per milligram of protein after 6–10 days of antibiotic treatment. (Fig. 22-5) Thus, with antibiotic treatment of a UTI, NOS activity remains elevated at a time when pyuria is significantly decreased and may be involved in the inflammatory response seen in UTIs.

Nitric oxide causes nitrosation of the heme moiety of the soluble guanylyl cyclase enzyme and this leads to its activation. Guanylyl cyclase converts guanosine triphosphate (GTP) to cyclic GMP [58] and many of the actions of NO are mediated via cyclic GMP [59]. Therefore, we measured cGMP levels in the urine from patients with UTIs ($n = 56$) and age-matched controls ($n = 50$) (Fig. 22-6). The predominant organisms identified in the infected urine samples are *Escherichia coli* (69%), *Klebsiella* (12%), *Pseudomonas* (7%), *Staphylococcus* (5%), and others (7%). In addition to the increased NOS activity in infected urine, there is a fourfold to fivefold increase in cyclic GMP levels in the supernatants of infected urine as determined with an ^{125}I radioimmunoassay [60]. Cyclic GMP levels are 0.78±0.08 and 3.05±0.44 μmol cyclic GMP per gram of creatinine in uninfected and infected urine, respectively.

In urinary tract infections, cytokines induced in the bladder epithelial layer cause a rapid influx of neutrophils. These neutrophils have been implicated in

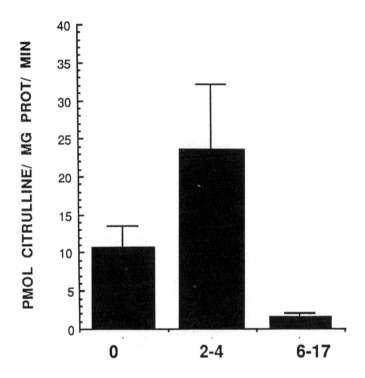

Figure 22-5. Activity of iNOS after antibiotic treatment of UTIs. In patients with UTIs ($n=5$), NOS activity is measured at the time of diagnosis, and at 2–4 days and 6–17 days after initiation of antibiotic treatment. Results are expressed as mean ± SEM, and NOS activity is significantly decreased 6–17 days after initiation of antibiotic treatment.

phagocytotic and inflammatory processes. Because of the presence of both cytokines and bacterial products in the urine, these neutrophils were examined for the presence of nitric oxide synthase. These results can be summarized as follows:

1. The NOS found in the pellet fraction of urine is increased in patients with UTIs and is an NADPH-dependent, CA^{2+}-independent enzyme, suggesting that it is an iNOS. The majority of neutrophil NOS activity is in the particulate fraction. NOS isolated from rat polymorphonuclear neutrophils [24] and iNOS isolated from transformed macrophages are primarily soluble proteins [23]. Only about 40% of the iNOS isolated from primary mouse peritoneal macrophages [61] and 25% of the iNOS in the cloned murine macrophage cell line RAW264.7 [62] are membrane associated. In contrast, greater than 90% of the human neutrophil iNOS is tightly membrane associated, which is similar to the membrane partition of eNOS. Because the

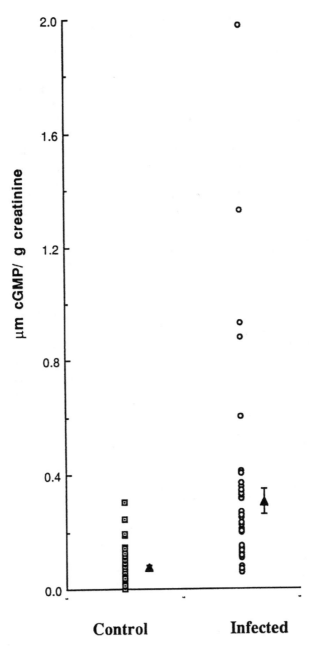

Figure 22-6. Cyclic GMP levels are significantly elevated in the urine of patients with UTIs compared to controls. Closed triangles represent the mean ± SEM of cGMP values for individual patients with UTIs (*n*=56) and their age and sex matched controls (*n*=50).

primary structure of the NOSs are similar, and posttranslational modifications cause the targeting of eNOS to membranes, human neutrophil iNOS also may be modified after translation, perhaps by ubiquitinylation [61] or by fatty acid modification.

2. The inhibitor profile of neutrophil NOS also suggests that it is an iNOS. L-NIO, L-NMMA, and canavanine have a lower IC_{50} than L-NAME and L-NNA. The IC_{50}s for L-NIO and L-NMMA for neutrophil iNOS are similar to those reported for human recombinant iNOS [54] and iNOS in activated rat neutrophils [55]. Unlike iNOS isolated from rodent sources, human neutrophil iNOS is inhibited by calmodulin antagonists, TFP and W-7, suggesting that calmodulin is bound less tightly to the human enzyme than to the rodent form. Irrespective of the potential differences in calmodulin binding affinity to human and rodent iNOS, both enzymes are clearly Ca^{2+}-independent [63].

3. RT-PCR amplified iNOS and DNA sequencing confirmed the PCR results, indicating that human neutrophils contained iNOS message. These results were further confirmed by identification of an immunoreactive iNOS protein at approximately 130 kDa, using a specific C-terminal human iNOS polyclonal antibody. iNOS immunoreactivity present in neutrophils isolated from patients with bacterially infected urine is decreased after antibiotic treatment and is not detected in the urine from patients without UTIs.

4. Cyclic GMP levels in the urine of patients with urinary tract infections are significantly increased compared to cyclic GMP levels in control urines [60]. Because NO increases cGMP levels, NO may cause an increase in cGMP during UTIs.

Although identification of iNOS has been problematic in human blood-borne inflammatory cells [40,41], iNOS has been definitively identified in urinary tract infections [64]. Because UTIs represent an ongoing infectious process with interplay of bacteria and host defense mechanisms, the characterization of iNOS in neutrophils from infected urine suggests an important role for NO in inflammatory processes.

Acknowledgment

Work in authors laboratory was supported by NIH grants DK 38311 and DK 09049.

References

1. Ochoa, J.B., Udekwu, A.O., Billiar, T.R., Curran, R.D., Cerra, F.B., Simmons, R.L., and Peitzman, A.B. Nitrogen oxide levels in patients after trauma and during sepsis. *Ann. Surg.* **214,** 621 (1991).
2. Smith, S.S., Wheeler, M.A., Zhang, R., Weiss, E.D., Lorber, M.I., Sessa, W.C., and Weiss, R.M. Nitric oxide synthase induction with renal transplant rejection or infection. *Kid. Int.* **50,** 2088 (1996).

3. McCartney-Francis, N., Allen, J.B., Mizel, D.E., Albina, J.E., Xie, Q.W., Nathan, C.F., and Wahl, S.M. Suppression of arthritis by an inhibitor of nitric oxide synthase. *J. Exp. Med.* **178,** 749 (1993).

4. Farrell, A.J., Blake, D.R., Palmer, R.M.J., and Moncada, S. Increased concentration of nitrite in synovial fluid and serum samples suggest increased nitric oxide synthesis in rheumatic diseases. *Ann. Rheum. Dis.* **51,** 1219 (1992).

5. Sever, R., Cook, T., and Cattell, V. Urinary excretion of nitrite and nitrate in experimental glomerulonephritis reflects systemic immune activation and not glomerular synthesis. *Clin. Exp. Immunol.* **90,** 326 (1992).

6. Reyes, A.A., Porras, B.H., Chasalow, F.L., and Klahr, S. L-Arginine decreases the infiltration of the kidney by macrophages in obstructive nephrophathy and puromycin-induced nephrosis. *Kidney Int.* **45,** 1346 (1994).

7. Raij, L. and Baylis, C. Glomerular actions of nitric oxide. *Kidney Int.* **48,** 20 (1995).

8. Reiling, N., Ulmer A.J., Duchrow, M., Ernst, M., Flad, H-D., and Hauschildt, S. Nitric oxide synthase: mRNA expression of different isoforms in human monocytes/macrophages. *Eur. J. Immunol.* **24,** 1941–1944 (1994).

9. Sessa, W.C. The nitric oxide synthase family of proteins. *J. Vas. Res.* **31,** 131 (1994).

10. Bredt, D.S. and Snyder, S.H. Isolation of nitric oxide synthase, a calmodulin-requiring enzyme. *Proc. Natl. Acad. Sci. USA* **87,** 682–685 (1990).

11. Förstermann, U., Ishii, K., Gorsky, L.D., and Murad, F. The cyosol of NIE-115 neuroblastoma cells synthesizes an EDRF-like substance that relaxes rabbit aorta. *Naunyn–Schmiedeberg's Arch. Pharmacol.* **340,** 771 (1989).

12. Förstermann, U., Gorsky, L.D., Pollock, J.S., Schmidt, H.H.H.W., Heller, M., and Murad, F. Regional distribution of EDRF/NO-synthesizing enzyme(s) in rat brain. *Biochim. Biophys. Res. Commun.* **168,** 727 (1990).

13. Mitchell, J.A., Shen, G.H., Förstermann, U., and Murad, F. Characterization of nitric oxide synthases in non-adrenergic, non-cholinergic nerve containing tissue from the rat anococcygeus muscle. *Br. J. Pharmacol.* **104,** 289 (1991).

14. Dokita, S., Morgan, W.R., Wheeler, M.A., Yoshida, M., Latifpour, J., and Weiss, R.M. N^G-nitro-L-arginine inhibits non-adrenergic, non-cholinergic relaxation in rabbit urethral smooth muscle. *Life Sci.* **48,** 2429 (1991).

15. Dokita, S., Smith, S.D., Nishimoto, T., Wheeler, M.A., and Weiss, R.M. Involvement of nitric oxide and cyclic GMP in rabbit urethral relaxation. *Eur. J. Pharmacol.* **269,** 269–275 (1994).

16. Liu, J. and Sessa, W.C. Identification of covalently bound amino-terminal myristic acid in endothelial nitric oxide synthase. *J. Biol. Chem.* **269,** 11691–11694 (1994).

17. Sessa, W.C., Garcia-Cardeña, G., Liu, J., Keh, A., Pollock, J.S., Bradley, J., Thiru, S., Braverman, I.M., and Desai, K.M. The Golgi association of endothelial nitric oxide synthase is necessary for the efficient synthesis of nitric oxide. *J. Biol. Chem.* **270,** 17641–17644 (1995).

18. Förstermann, U., Pollock, J.S., Schmidt, J.H.H.W., Heller, M., and Murad, F. Calmodulin-dependent endothelium-derived relaxing factor/nitric oxide synthase activity is present in the particulate cytosolic fractions of bovine aortic endothelial cells. *Proc. Natl. Acad. Sci. USA* **88,** 1788 (1991).

19. Nangia, A.K., Smith, S.D., Wheeler, M.A., and Weiss, R.M. Characterization of nitric oxide synthase (NOS) activity in rabbit urinary bladder. *FASEB J.* **7,** A259 (1993).

20. Weiss, R.M., Nangia, A.K., Smith, S.D., and Wheeler, M.A. Nitric oxide synthase (NOS) activity in urethra, bladder, and bladder smooth muscle. *Neurourol. Urodyn.* **13,** 397–398 (1994).

21. Stuehr, D.J. and Marletta, M.A. Mammalian nitrate biosynthesis: Mouse macrophages produce nitrite and nitrate in response to *Escherichia coli* lipopolysaccharide. *Proc. Natl. Acad. Sci. USA* **82,** 7738–7742 (1985).

22. Marletta, M.A., Yoon, P.S., Iyengar, R., Leaf, C.D., and Wishnok, J.S. Macrophage oxidation of L-Arginine to nitrite and nitrate: Nitric oxide is an intermediate. *Biochemistry* **27,** 8706–8711 (1988).

23. Stuehr, D.J., Cho, H.J., Kwon, N.S., Weise, M.F., and Nathan, C.F. Purification and characterization of the cytokine-induced macrophage nitric oxide synthase. *Proc. Natl. Acad Sci. USA* **88,** 7773 (1991).

24. Yui, Y., Hattori, R., Kosuga, K., Elizawa, H., Hiki, K., and Kawai, C. Purification of nitric oxide synthase from rat macrophages. *J. Biol. Chem.* **266,** 12544 (1991).

25. Geller, D.A., Lowenstein, C.J., Shapiro, R.A., Nussler, A.K., Di Silvio, M., Wang, S.C., Nakayama, D.K., Simmons, R.L., Snyder, S.H., and Billiar, T.R. Molecular cloning and expression of inducible nitric oxide synthase from human hepatocytes. *Proc. Natl. Acad. Sci. USA* **90,** 3491 (1993).

26. Stuehr, D.J. and Marletta, M.A. Induction of nitrate/nitrate synthesis in murine macrophages by BCG infection, lymphokines, or interferon-gamma. *J. Immunol.* **139,** 518 (1987).

27. Busse, R. and Mulsch, A. Induction of nitric oxide synthase by cytokines in vascular smooth muscle cells. *FEBS. Lett.* **275,** 87 (1990).

28. Nathan, C. and Xie, Q.-W. Nitric oxide synthases: roles, tolls and controls. *Cell* **78,** 915–918 (1994).

29. Gross, S.S. and Wolin, M.S. Nitric oxide: Pathophysiological mechanisms. *Annu. Rev. Physiol.* **57,** 737–769 (1995).

30. MacMicking J.D., Nathan, C., Hom, G., Chartrain, N., Fletcher, D.S., Trumbauer, M., Stevens, K., Xie, Q.-W., Sokol, K., and Hutchinson, N. Altered responses to bacterial infection and endotoxic shock in mice lacking inducible nitric oxide synthase. *Cell* **81,** 641–650 (1995).

31. Hedges, S., Anderson, P., Liden-Janson, G., de Man, P., and Svanborg., C. Interleukin-6 response to deliberate Gram-negative colonization of the human urinary tract. *Infect. Immun.* **59,** 421–427 (1991).

32. Agace, W.W., Hedges, S., Ceska, M., and Svanborg, C. Interleukin-8 and the neutrophil response to mucosal gram-negative infection. *J. Clin. Invest.* **92,** 780–785 (1993).

33. DeMan, P., Aarden, L., Engberg, I., Linder, H., Svanborg-Edén, C., and Van Kooten, C. Interleukin-6 induced at mucosal surfaces by Gram-negative bacterial infection. *Infect. Immun.* **57,** 3383 (1989).

34. Benson, M., Jodal, U., Andreasson, A., Karlsson, Å., Rydberg, J., and Svanborg, C. Interleukin-6 response to urinary tract infection in childhood. *Pediatr. Infect. Dis. J.* **13,** 612 (1994).

35. Svanborg, C., Agace, W., Hedges, S., Lindstedt, R., and Svensson, M.L. Bacterial adherence and mucosal cytokine production. *Ann. NY Acad. Sci.* **730,** 162 (1994).

36. Maeda, S., Deguchi, T., Kanimoto, Y., Kuriyama, M., Kawada, Y., and Nishiura, T. Studies on the phagocytic function of urinary leukocytes. *J. Urol.* **129,** 427 (1983).

37. Katoh, S., Orikasa, S., Toyota, S., Itoh, S., Oikawa, K., Fukushi, Y., and Suzuki, Y. Anti-bacterial defense mechanism of the urinary tract constructed from intestinal segments. Studies on cell population and phagocytotic activity of urinary leukocytes and bacterial growth in urine. *Jpn. J. Urol.* **82,** 1436 (1991).

38. Gargan R.A., Hamilton-Miller, J.M.T., and Brumfitt, W. Effect of pH and osmolality on in vitro phagocytosis and killing by neutrophils in urine. *J. Urol.* **152,** 1615 (1994).

39. Klebanoff, S.J. Reactive nitrogen intermediates and antimicrobial activity: Role of nitrite. *Free Radical Biol. Med.* **14,** 351 (1993).

40. Klebanoff, S.J. and Nathan, C.F. Nitrite production by stimulated human polymorphonuclear leukocytes supplemented with azide and catalase. *Biochem. Biophys. Res. Commun.* **197,** 192 (1993).

41. Yan, L., Vandivier, R.W., Suffredini, A.F., and Danner, R.L. Human polymorphonuclear leukocytes lack detectable nitric oxide synthase activity *J. Immunol.* **153,** 1825 (1994).

42. Bukrinsky, M.I., Nottet, J.S., Schmidtmayerova, H., Dubrovsky, L., Flanagan, C.R., Mullins, M.E., Lipton S.A., and Gendelman, H.E. Regulation of nitric oxide synthase activity in human immunodeficiency virus type 1 (HIV-1)-infected monocytes: Implication for HIV-associated neurological diseases. *J. Exp. Med.* **181,** 735 (1995).

43. De Maria, R., Cifone, M.G., Trotta, R., Rippo, M.R., Festuccia, C., Santoni, A., and Testi, R. Triggering of human monocyte activation through CD69, a member of the natural killer cell gene complex family of signal transducing receptors. *J. Exp. Med.* **180,** 1999 (1994).

44. Mossalayi, M.D., Paul-Eugene, N., Ouaaz, R., Arock, M., Kolb, J.P., Kilchherr, E., Debre, P., and Dugas, B. Involvement of Fc epsilon RII/CD23 and L-arginine-dependent pathway in IgE-mediated stimulation of human monocyte functions. *Int. Immunol.* **6,** 931 (1994).

45. Vouldoukis, I., Riveros-Moreno, V., Dugas, B., Ouaaz, R., Becherel, P., Debre, P., Moncada, S., and Mossalayi, M.D. The killing of Leishmania major by human macrophages is mediated by nitric oxide induced after ligation of the Fc epsilon RII/CD23 surface antigen. *Proc. Natl. Acad. Sci. USA* **92,** 7804 (1995).

46. Malawista, S.E., Montgomery, R.R., and Van Blaricom, G. Evidence for reactive nitrogen intermediates in killing of staphylococci by human neutrophil cytoplasts. A new microbicidal pathway for polymorphonuclear leukocytes. *J. Clin. Invest.* **90,** 631–636 (1992).

47. Cruickshank, J. and Moyes, J.M. The presence and significance of nitrites in urine. *Br. Med. J.* **2,** 712–713 (1914).

48. Griess, P. Bermerkungen zu der Abhandlung der HH. Wesley and Benedikt Ueber einige Azoverbindungen. *Ber. Deutsch. Chem. Gen.* **12,** 426 (1879).

49. Smith, S.D., Wheeler, M.A., and Weiss, R.M. Nitric oxide synthase: An endogenous source of elevated nitrite in infected urine. *Kidney Int.* **45,** 586 (1994).

50. Nathan, C. Nitric oxide as a secretory product of mammalian cells. *FASEB J.* **6,** 3051 (1992).

51. Nussler, A.K. and Billiar, T.R. Inflammation, immunoregulation and inducible nitric oxide synthase. *J. Leuk. Biol.* **15,** 171 (1993).

52. Robbins, R.A., Hamel, F.G., Floreani, A.A., Gossman, G.L., Nelson, K.J., Belenky, S., and Rubinstein, I. Bovine bronchial epithelial cells metabolize L-arginine to L-citrulline: Possible role of nitric oxide synthase. *Life Sci.* **52,** 709 (1993).

53. Schoedon, G., Schneemann, M., Hofer, S., Guerrero, L., Blau, N., and Schaffner, A. Regulation of the L-arginine-dependent and tetrahydrobiopterin-dependent biosynthesis of nitric oxide in murine macrophages. *Eur. J. Biochem.* **213,** 833 (1993).

54. Nakane, M., Pollock, J.S., Klinghofer, V., Basha, F., Marsden, P.A., Hokari, A., Ogura, T., Esumi, H., and Carter, G.W. Functional expression of three isoforms of human nitric oxide synthase in baculovirus-infected insect cells. *Biochem. Biophys. Res. Commun.* **206,** 511 (1995).

55. McCall, T.B., Boughton-Smith, N.K., Palmer, R.M.J., Whittle, B.J.R., and Moncada, S. Synthesis of nitric oxide from L-arginine. Release and interaction with superoxide anion. *Biochem. J.* **261,** 293 (1989).

56. Pollock, J.S., Förstermann, U., Michell, J.A., Warner, T.D., Schmidt, H.H.H.W., Nakane, M., and Murad, F. Purification and characterization of particulate endothelium-derived relaxing factor synthase from cultured and native bovine aortic endothelial cells. *Proc. Natl. Acad. Sci. USA* **88,** 10480 (1991).

57. Maciejewski, J.P., Selleri, C., Sato, T., Cho, H.J., Keefer, L.K., Nathan, C.F., and Young, N.S. Nitric oxide suppression of human hematopoiesis in vitro. Contribution to inhibitory action of interferon-gamma and tumor necrosis factor-alpha. *J. Clin. Invest.* **96,** 1085 (1995).

58. Schmidt, H.H.H.W. and Walter, U. NO at work. *Cell* **78,** 919 (1994).

59. Feelish, M. and Noack, E.A. Correlation between nitric oxide formation during degradation of organic nitrates and activation of guanylate cyclase. *Eur. J. Pharmacol.* **139,** 19 (1987).

60. Smith, S.D., Wheeler, M.A., Foster, H.E., Jr., and Weiss, R.M. Urinary nitric oxide synthase activity and cyclic GMP levels are decreased with interstitial cystitis and increased with urinary tract infections. *J. Urol.* **155,** 1432 (1996).

61. Vodovotz, Y., Russell, D., Xie, Q.-W., Bogdan, C., and Nathan, C. Vesicle membrane association of nitric oxide synthase in primary mouse macrophages. *J. Immunol.* **154,** 2914 (1995).

62. Schmidt, H.H.H.W., Warner, T.D., Nakane, M., Förstermann, U., and Murad, F. Regulation and subcellular location of nitrogen oxide synthases in RAW264.7 macrophages. *Mol. Pharmacol.* **41,** 615 (1992).

63. Bredt, D.S. and Snyder, S.H. Nitric oxide: A physiologic messenger molecule. *Annu. Rev. Biochem.* **63,** 175 (1994).

64. Wheeler, M.A., Smith, S.D., García-Cardeña, G., Nathan, C.F., Weiss, R.M., and Sessa, W.C. Bacterial infection induces nitric oxide synthase in human neutrophils. *J. Clin. Invest.* **99,** 110 (1997).

23

Role of the NO System in Urinary Tract Obstruction

David A. Schulsinger, Steven S. Gross, and E. Darracott Vaughan, Jr.

1. Introduction

Obstructive uropathy with subsequent nephron injury is a major cause of renal damage and occurs in a significant number of urological diseases. Obstruction to the urinary tract can come from either intrinsic disease or external compression to the genitourinary system. The pathophysiology of urinary tract obstruction has been a subject of a number of recent reviews [1–3]. In a large autopsy study of patients of all ages, frequency of urinary tract obstruction was 3.1%. Obstruction in the younger patient is usually due to a congenital anomaly; during young adulthood, the most common etiology is urinary stone disease, and later in life, the male can develop hydronephrosis secondary to prostatic disease. The terminology is confusing, with the term *hydronephrosis* being an anatomical description of a dilated ureter and collecting system. Better terms are either *obstructive uropathy* or *obstructive nephropathy*, which imply that there is actual damage to the urinary tract. In the latter setting, there is not only an anatomic change but a reduction in renal function due to the obstructive injury. The mechanism of injury to the nephron during obstruction remains an area of intense controversy. It is clear that this is not a pure pressure phenomenon although the ureteral pressure initially rises following obstruction. Subsequently, it falls and there is renal vasoconstriction which is injurious to the nephron. During the course of either complete or partial obstruction, there appears to be a complex interaction between renal vasoconstrictors and renal vasodilators. Accordingly, with the advent of knowledge concerning nitric oxide, studies have been initiated to define the role of nitric oxide system in obstructive injury. Therefore, in this chapter we will review the current information available concerning this system in experimental models of urinary tract obstruction.

2. Renal Hemodynamics in Ureteral Occlusion

Renal hemodynamic regulation during complete unilateral ureteral obstruction (UUO) is described by a triphasic relationship in renal blood flow (RBF) and ureteral pressure (UP) [4–6]. In Phase I (0–90 min), there is a rise in both RBF and UP, primarily as a result of afferent arteriole vasodilation. During Phase II (90 min–5 h), there is a decline in RBF and a rise in UP associated with efferent arteriolar constriction. Finally, in Phase III (5–18 h), afferent arteriolar constriction results in a decline in both RBF and UP. The triphasic relationship characterizing these changes in RBF and UP is shown in Fig. 23-1.

In bilateral ureteral obstruction (BUO), the changes in RBF and UP are not

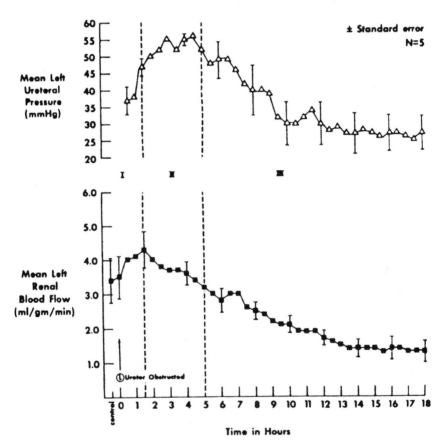

Figure 23-1. The triphasic relationship between ipsilateral renal blood flow and left ureteral pressure during 18 h of left ureteral occlusion. (From Vaughan, E.D. Jr., Sorenson, E.J. Gillenwater, J.Y. The renal hemodynamic response to chronic unilateral complete ureteral occlusion. Invest. Urol. 1970;8:78. Used by permission.)

entirely the same as the patterns seen in the triphasic model of UUO. Similar to UUO, there is an increase in both RBF and UP during the first 90 min after BUO [7]. Between 90 min and 7 h, however, the decrease in RBF during BUO is significantly less than the decline in RBF seen in UUO. By 24 h, the decrease in RBF is similar in UUO and BUO. During the first 4.5 h, the progressive rise in UP follows a similar pattern for both UUO and BUO. In contrast to UUO, after 4.5 h, the UP remains elevated until at least 24 h. Unlike the triphasic UUO model of preglomerular vasodilatation, postglomerular vasoconstriction followed by preglomerular vasoconstriction, BUO passes through a phase of preglomerular vasodilatation and, finally, postglomerular vasoconstriction and remains in that phase. In contrast to chronic UUO, increased preglomerular vasoconstriction does not appear to occur in BUO.

Although the exact mechanisms of renal hemodynamic changes during acute and late phases of UUO are not known, various modulators participate in the effect. Several vasodilators, including prostaglandins [8], bradykinins and nitric oxide (NO) result in dilatation of the afferent arteriole. In vascular endothelial cells, nitric oxide is derived exclusively from the oxidation of the terminal guanidino nitrogen of L-arginine by the enzyme, nitric oxide synthase (NOS) [9]. Nitric oxide, in turn, activates soluble guanylate cyclase and increases the production of guanosine $3'5'$-cyclic monophosphate (cGMP), which mediates vascular relaxation.

The vasoactive compounds potentially responsible for constriction of efferent arteriole during UUO, include thromboxane A_2 [10], ET-1 [11,12], angiotensin II [13], platelet-activating factor [14], and leukotriene B_4 [15]. The focus of this chapter will be the contribution of nitric oxide during acute and late UUO and BUO.

3. Nitric Oxide Synthase and the Urinary Tract

Localizing NO formation at a cellular level throughout the urinary tract was greatly facilitated by immunohistochemical identification of NOS. Several reports document the presence of NOS in the normal kidney, ureter, bladder, and the Leydig cells of the testicle. In pathophysiological conditions, including glomerulonephritis, renal cell carcinoma, and hypoxia, the presence of NOS was identified (see Chapter 15 and 20).

In the kidney, all NOS isoforms have been previously described. The constitutive, "neuronal" isoform, NOS I, is localized to the cells of the macula densa (MD) of rat and mouse kidney [16–19]. Histochemically, only the macula densa and parts of the glomerular capsule were NOS I labeled in rat and human kidney. In guinea pig, selective immunohistochemistry of the postglomerular arteriole was seen. By reverse transcription polymerase chain reaction (RT-PCR), NOS I mRNA was also localized to the inner medullary collecting duct and, to a lesser extent, to the glomerulus, inner medullary thin limb, and cortical and outer medullary collecting ducts as well as in parts of the renal vasculature [20].

The second constitutive-type NOS, NOS III, also termed "endothelial NOS" (eNOS) is present in the endothelium of renal arterioles by NADPH–diaphorase reaction (see Chapter 7 for details).

Finally, the inducible form of NOS (iNOS) or NOS II was localized to the preglomerular portion of the afferent arteriole, including the granular cells [21]. In normal rat [22] and rabbit kidney [23], NO release requires a functional endothelium, as its removal by a detergent abolished its release. This suggested an endothelium-dependent mechanism for NO formation. However, in 3-day UUO, ex vivo detergent-treated endothelium failed to block the release of NO_2^-, suggesting a nonendothelial location for iNOS production of NO [23]. Likewise, we have identified iNOS by immunohistochemistry in the renal tubules after unilateral ureteral obstruction [24]. Figure 23-2 shows the localization of iNOS between 3 and 28 days after UUO. It is well known that L-arginine, the precursor for NO synthesis, is synthesized from citrulline [25] and cortical tubules are the most active site of its synthesis within the kidney [26]. It appears that the renal tubules are the cells responsible for L-arginine production and NO synthesis. This may suggest a central role for the kidney to manufacture NO in response to UUO.

Figure 23-3 shows localization of inducible NOS (NOS II) to the MD in obstructed rat kidney after 28 days. These findings suggest that the MD may produce considerable amounts of NO which may be a mediator substance in signal transfer from the distal tubular fluid to the glomerular arterioles [17]. Wilcox et al. [16] suggest that neuronal NOS in the MD produces NO responsible for mediating a vasodilator arm of the tubuloglomerular feedback response. We propose that NO produced in the MD cells by UUO-induced NOS diffuses across the avascular space of the extraglomerular mesangium. In smooth muscle cells of the afferent arteriole, NO stimulates soluble guanylate cyclase, generating cGMP from GTP to influence the vascular tone of the afferent arteriole during UUO (Fig. 23-4).

4. Nitric Oxide and Acute Unilateral Ureteral Obstruction

During acute UUO, afferent arteriole vasodilatation is responsible for the increase in RBF and UP. A number of studies have examined the role of NO during acute UUO. Lanzone et al. [27] showed that intrarenal infusion of N_G-monomethyl-L-arginine (L-NMMA) attenuated the increase in RBF after acute UUO. In addition, these effects of L-NMMA were abolished by intrarenal L-arginine infusion. Schulsinger et al. [28] demonstrated that intrarenal L-arginine infusion between 90 and 140 min after UUO, in the presence of meclofenamate, a cyclooxygenase (COX) inhibitor, restored the RBF and UP to control UUO values. In the absence of UUO (sham), however, administration of L-arginine had no effect on RBF.

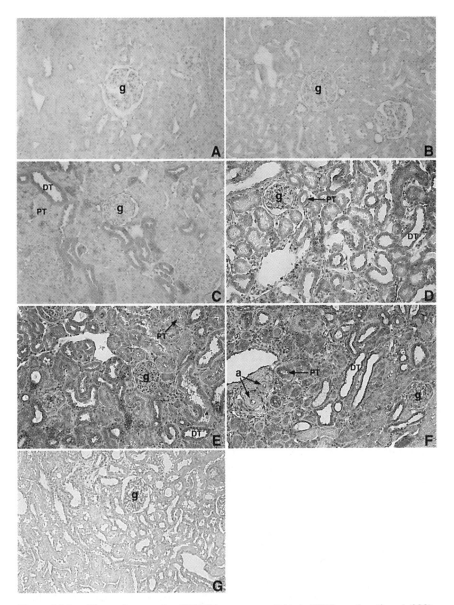

Figure 23-2. Photomicrographs of iNOS immunoreactivity in UUO renal sections (×200). (A) No staining for iNOS in the sham or (B) 1.5-h obstructed kidney. (C) iNOS is localized to the distal tubules of a 3-day obstructed kidney and (D) after 7 days of UUO. (E) iNOS staining of distal tubules, but not in the glomerulus after 21 days of UUO. (F) After 28 days of UUO, staining is visualized in the distal tubule and proximal tubule but not in the glomerulus or arteries. (G) No staining for iNOS in a 14-day steroid treated obstructed kidney (PT, proximal tubule; DT, distal tubule; g, glomerulus: a, artery).

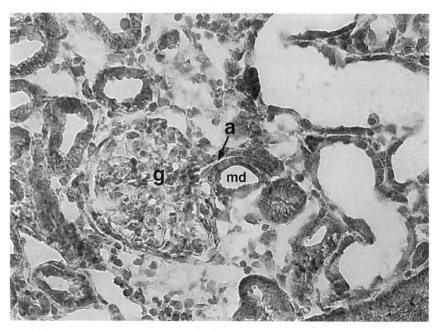

Figure 23-3. Photomicrograph of immunohistochemical localization of iNOS in the macula densa of rat kidney after 28 days of UUO (×400). Note the section through the glomerular capillary tuft and the adjacent tubular structures that show dense localization of iNOS within the cells of macula densa. In contrast, the glomerular cells and the endothelium were not recognized (a, arteriole; g, glomerulus; md, macula densa).

Thus, exogenous L-arginine in the absence of UUO was not necessary in maintaining normal renal hemodynamics.

The activation of the NO system following acute UUO may be due to altered synthesis of L-arginine or increased utilization of L-arginine. It can be argued that there is an increased utilization of endogenous L-arginine upon ureteral occlusion, thereby limiting the production of NO, and this situation is reversed with exogenous L-arginine infusion [28]. Once exogenous L-arginine is added, the augmented NOS activity facilitates the production of NO, promptly reducing renal vascular resistance (RVR) with a subsequent rise in RBF and UP.

The mechanism whereby UUO enhances NOS is not clear. Marsden et al. showed that sheer stress will activate the EDRF system [29]. Therefore, the high intraglomerular hydrostatic pressure may contribute to the production of NO. RT-PCR studies have shown upregulation of mRNA NOS II in obstructed rat kidneys as early as 90 min, but no induction in the contralateral or sham-operated rat kidneys [30]. Thus, iNOS upregulation following UUO may represent the rate-limiting enzyme essential for NO formation.

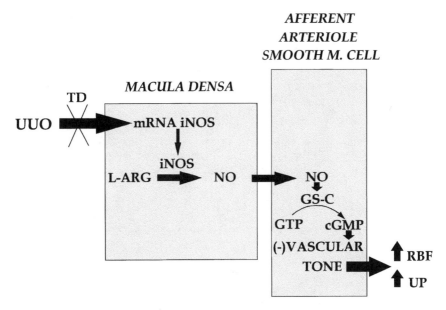

Figure 23-4. Proposed model of UUO-induced NOS in the macula densa and its role in regulating glomerular afferent arteriole smooth muscle tone. (UUO, unilateral ureteral obstruction; TD, triamcinolone diacetate; RBF, renal blood flow; UP, ureteral pressure)

5. Nitric Oxide in Late and Chronic Unilateral Ureteral Obstruction

As described above, Phase III (5–18 h), or late obstruction, is characterized by a decline in both RBF and UP. This effect is explained by vasoconstriction of the afferent arteriole, resulting in decreased GFR and urine production. Unlike the increase in RBF and UP in acute UUO, the vasoconstrictors angiotensin II, thomboxane A_2, and endothelin may counteract the vasodilator effect of NO and other vasodilators during late UUO. Recent studies suggest that decreased NOS activity contributes to renal vasoconstriction observed in rats with bilateral ureteral obstruction [31–33]. During late obstruction, either an altered synthesis of L-arginine or increased utilization of it results in decreased serum levels of L-arginine available for NO production [23,31]. The latter explanation is consistent with our findings. Schulsinger et al. [34] showed that the infusion of L-arginine after 18 h of UUO increased RBF significantly toward baseline (preobstructed) values (Fig. 23-5). Likewise, UP rose sharply after L-arginine during the same time period (Fig. 23-6). However, in the absence of obstruction (sham operated), there was no change in RBF after exogenous L-arginine infusion. Likewise, NOS synthesis inhibition in 24-h UUO rats caused a greater decrease in RBF in the obstructed kidney compared to the sham-operated kidney [35]. Others have shown a similar effect in sham

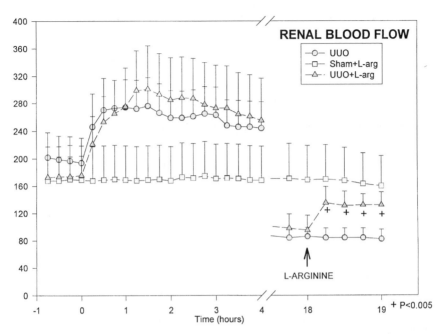

Figure 23-5. RBF before and after UUO (time 0). L-Arginine was infused after 18 h UUO in obstructed (UUO+ L-arg) and sham-operated animals (Sham+ L-arg), but not control obstructed animals (UUO). [+]$p < .005$ compared with preinfusion of L-arginine (18 h).

operated rats [31]. This is consistent with our data showing that iNOS is upregulated after 90 min to 28 days of UUO [24]. The substrate, L-arginine, becomes rate limiting to NO production during this period of obstruction. Thus, the substrate necessary to fuel the renal hyperemic effect during Phase I UUO may be exhausted and not available during late UUO. This, in turn, shifts the delicate balance to favor a vasoconstrictor, rather than a vasodilator effect on the afferent arteriole. In summary, the vasodilator, NO, is able to modulate renal vasoconstriction during late UUO, resulting in an increase in RBF and UP.

In chronic ureteral obstruction, there are few reports on the role of NOS in UUO. In 6-week UUO rats, there was a significant decrease in RBF following inhibition of NO synthesis [36]. Chen et al. [36] suggest that NO is an important modulator of both preglomerular and postglomerular tone in the chronically obstructed kidney.

6. Nitric Oxide and Bilateral Ureteral Obstruction

As mentioned earlier, from 0 to 4.5 h post-BUO there is period of afferent arteriole vasodilatation followed by a phase of efferent arteriole vasoconstriction.

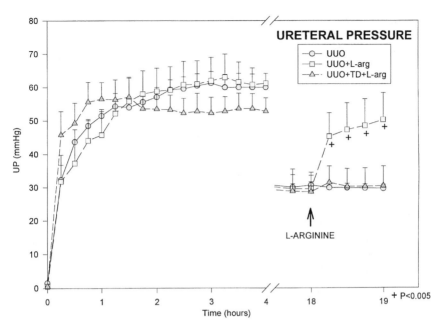

Figure 23-6. UP between 0 and 19 h after UUO. L-Arginine was infused after 18 h UUO in obstructed (UUO+ L-arg) and steroid-treated animals (UUO+ L-arg+TD), but not control obstructed animals (UUO). $^{+}p < .005$ compared with preinfusion of L-arginine (18 h).

Several studies suggest that decreased NO activity contributes to renal vasoconstriction in rats with BUO [31–33]. Reyes et al. showed a decreased availability of arginine for NO synthesis in rats with BUO for 24 h [31]. In rats with BUO, administration of N_W-nitro-L-arginine (L-NAME) further decreased RBF in a dose-dependent manner, suggesting a sustained basal release of NO that counteracts the effect of vasoconstriction which tends to decrease GFR and effective renal plasma flow (ERPF). Subsequent administration of L-arginine significantly increased the GFR and ERPF. L-Arginine given to rats with BUO, or pretreatment with L-arginine, significantly decreased RVR and increased GFR and ERPF. Subsequent administration of an L-arginine antagonist resulted in an increase RVR and a decrease in GFR and ERPF to values comparable to basal conditions. These studies suggest that BUO results in either an alteration in synthesis of L-arginine or an increased utilization of it, resulting in decreased serum levels of the substrate for NO synthesis. Rats obstructed for 24 h have one-third the plasma level of L-arginine compared to serum values prior to obstruction or sham-operated [31]. In summary, the mechanism by which BUO decreases the available endogenous L-arginine for NO synthesis may be due to a decline in renal function or an increase in L-arginine utilization.

7. Regulation of Nitric Oxide During UUO

Various modulators of NOS may effect its expression, synthesis, and activity. Cytokines are known to induce the expression of NOS activity at the level of transcription. Interleukin-6 (IL-6), a macrophage product, causes an upregulation of NOS mRNA. Conversely, steroids such as dexamethasone can inhibit the induction of NOS in vitro [37] and in vivo without affecting the constitutive form of the enzyme [38]. Ex vivo perfusion of dexamethasone or cycloheximide abolished the time-dependent release of NO_2^-, an oxidation product of NO, by the hydronephrotic kidney [23]. PCR and immunoperoxidase staining techniques demonstrated that iNOS mRNA and protein expression were upregulated in renal tubules of 14-day obstructed kidneys [30], and triamcinolone diacetate-pretreatment (TD) prevented iNOS induction. In a dog model, the rise in UP (Fig. 23-6) and RBF (Fig. 23-7) following L-arginine infusion 18 h post-UUO was abolished after TD pretreatment [34]. In contrast, sham-obstructed animals did not show changes in RBF after L-arginine administration. These findings suggest that UUO-induced iNOS mRNA is inhibited by TD at the level of transcription and that excess substrate did not alter the RBF and UP (See fig. 23-4).

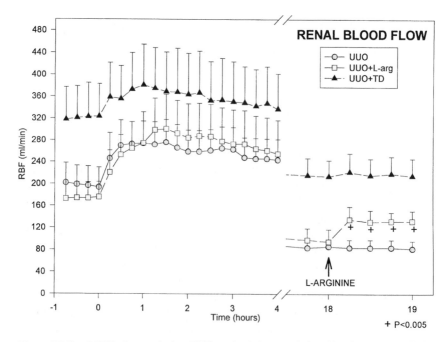

Figure 23-7. RBF before and after UUO. L-Arginine was infused in obstructed (UUO+ L-arg) and steroid-treated animals (UUO+TD) but not control obstructed animals (UUO). $^+p < .005$ compared with preinfusion of L-arginine (18 h).

Selective and nonselective inhibitors NOS isoforms have been described. Thus, L-NMMA blocks the activity of both cNOS and iNOS [39], whereas aminoguanidine (AG) has been shown to be a relatively selective inhibitor of iNOS [40–42]. It is notable that the effects of various inhibitors of NOS activity have been examined during acute unilateral ureteral obstruction [27,43]. Lanzone et al. [27] showed that L-NMMA infused into the renal artery inhibited RBF before UUO. In contrast, after UUO, RBF was not significantly altered. However, when L-NMMA was stopped, RBF increased significantly. Likewise, when animals received L-arginine and L-NMMA simultaneously, there was no change in the RBF compared to the control group. In late-phase UUO, we showed that continuous 18-h ipsilateral renal artery infusion of L-NMMA significantly decreased RBF at all time points prior to and following obstruction [34]. This was reversed when L-arginine was coinfused with L-NMMA after 18-h post-UUO, as both RBF (Fig. 23-8) and UP (Fig. 23-9) were significantly increased. These findings suggest that L-arginine infusion abolished the effects of L-NMMA on NOS during acute and late-phase UUO. In addition, Inman et al. [43] demonstrated that RBF was reduced by 47% and 67% when L-NMMA was infused after and prior to obstruction, respectively. Others showed that the potentiating effect of L-arginine was abrogated by coadministration of L-NMMA or AG [23]. These findings

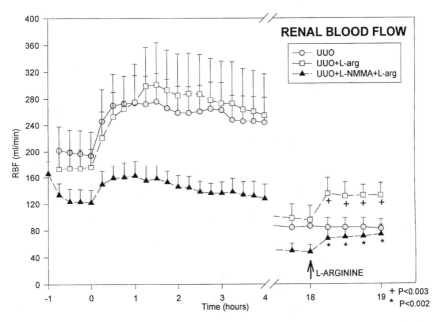

Figure 23-8. RBF before and after UUO. L-Arginine was infused in obstructed (UUO+L-arg) and L-NMMA-treated animals (UUO+ L-NMMA+L-arg) but not control obstructed animals (UUO). ⁺$p < .003$ and *$p < .002$ compared with preinfusion of L-arginine (18 h).

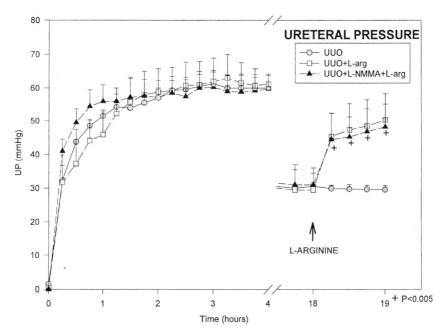

Figure 23-9. UP between 0 and 19 h after UUO. L-Arginine was infused after 18 h UUO in obstructed (UUO+ L-arg) and L-NMMA-treated animals (UUO+ L-NMMA) but not control obstructed animals (UUO). $^+p < .005$ compared with preinfusion of L-arginine (18 h).

support the view that inhibition of NOS prevents the production of the NO which is needed to maintain renal hemodynamics during acute UUO.

8. Interaction of NO and Prostaglandins During UUO

The relationship between NO and the prostanoid system has been investigated. In acute UUO, NO and PG are the major vasoactive regulators in early obstruction. Salvemini et al. [23] demonstrated a relationship between the NO system and the PG system. In ex vivo perfused rabbit hydronephrotic kidneys, production of NO by iNOS causes activation of the inducible COX, resulting in a marked increase in PG synthesis. In the contralateral kidney, production of NO by a cNOS activates constitutive COX, resulting in prostaglandin E_2 release. These in vivo studies may be explained by the in vitro demonstration that NO directly activates purified constitutive and inducible COX isoforms, in association with the formation of an NO complex [44]. Inhibition of PG synthesis entirely prevents the increase in both RBF and UP after UUO [28,45]. Intrarenal infusion of L-arginine promptly increased RBF and UP to levels normally observed initially with UUO and without PG synthesis inhibition [28]. In contrast, L-arginine infusion in the absence of ureteral obstruction does not alter RBF, regardless of whether PG synthesis is inhibited or not. Taken together, these findings suggest that during UUO, the enhanced renal production of PG is largely responsible for the decrease in RVR and that the NO system is also activated, contributing to increased RBF and UP. In summary, activation of the NO system by UUO is capable of superseding the PG system while maintaining high RBF and UP.

9. CONCLUSION

This chapter has attempted to define the role of the nitric oxide system following unilateral and bilateral ureteral obstruction. Clearly, the nitric oxide system plays an important role in the regulation of renal hemodynamics following ureteral obstruction. At the cellular level, UUO results in the upregulation of iNOS mRNA and protein expression and an enhanced synthesis of NO, which serves to maintain renal hemodynamics during UUO. However, the NOS substrate L-arginine apparently becomes rate limiting to NO synthesis in this setting, favoring vasoconstriction of the afferent arteriole. The decreased synthesis of NO during UUO and BUO may contribute significantly to the reduction in RBF and UP after ureteral obstruction.

References

1. Gulmi, F.A., Felson, B., and Vaughan, E.D., Jr. Pathophysiology of urinary tract obstruction. In: *Campbell's Urology*, 7th ed. Eds. Walsh, P.C., Retik, A.B., Stamey, Wein, A.J., and Vaughan, E.D., Jr. W.B. Saunders, Philadelphia, 1997, Chap. 9.

2. Curhan, G.C., and Zeidel, M.L. Urinary tract obstruction. In: *The Kidney*, 5th ed. Ed. Brenner, B.M. W.B. Saunders, Philadelphia, 1996, Chap. 41.

3. Gillenwater, J.Y. *Hydronephrosis in Adult and Pediatric Urology*, 3rd ed. Eds. Gillenwater, J.Y., Grayhack, J.T., Howards, S.S., and Enduckett, J.W. Mosby, St. Louis, MO, 1996, Chap. 21.

4. Vaughan, E.D., Jr., Shenasky, J.H., and Gillenwater, J.Y. Mechanism of acute hemodynamic response to ureteral occlusion. *Invest. Urol.* **9,** 109–118 (1971).

5. Moody, T.E., Vaughan, E.D., Jr., and Gillenwater, J.Y. Relationship between renal blood flow and ureteral pressure during 18 hours of total unilateral ureteral occlusion. *Invest. Urol.* **13,** 246 (1975).

6. Loo, M.H., Felsen, D., Weisman, S., Marion, D.N., and Vaughan, E.D., Jr. Pathophysiology of obstructive nephropathy. *World J. Urol.* **6,** 53 (1988).

7. Moody, T.E., Vaughan, E.D., Jr., and Gillenwater, J.Y. Comparison of the renal hemodynamic response to unilateral and bilateral ureteral occlusion. *Invest. Urol.* **14,** 455 (1977).

8. Yarger, W.E., Schocker, D.D., and Harris, R.H. Obstructive nephropathy in the rat: Possible roles for the renin-angiotensin system, prostaglandins and thromboxanes in postobstructive renal function. *J. Clin. Invest.* **65,** 400 (1981).

9. Palmer, R.M.J., Ashton, D.S., and Moncada, S. Vascular endothelial cells synthesize nitric oxide from L-arginine. *Nature (London)* **333,** 664 (1988).

10. Klahr, S., Harris, K., and Purkerson, M.L. Effect of obstruction on renal functions. *Pediatr. Nephrol.* **2,** 34 (1988).

11. Kahn, S.A., Gulmi, F.A., Chou, S.-Y., Mooppan, U.M., and Kim, H. Contribution of endothelin to renal vasoconstriction in unilateral ureteral obstruction: Reversal by verapamil. *J. Urol.* **153,** 411A (1995).

12. Kelleher, J.P., Shah, V., Godley, M.L., Wakefield, A.J., Gordon, I., Ransley, P.G., Snell, M.E., and Risdon, R.A. Urinary endothelin (ET1) in complete ureteric obstruction in the miniature pig. *Urol. Res.* 63–65 (1992).

13. Purkerson, M.L. and Klahr, S. Prior inhibition of vasoconstrictors normalizes GFR in postobstructed kidneys. *Kidney Int.* **35,** 1306 (1989).

14. Reyes, A.A., and Klahr, S. Role of platelet-activating factor in renal function in normal rats and rats with obstructive nephropathy. *Proc. Soc. Exp. Biol. Med.* **198,** 572 (1991).

15. Reyes, A.A., Lefkowith, J., Pippin, J., and Klahr, S. The role of the 5-lipooxygenase pathway in obstructive nephropathy. *Kidney Int.* **41,** 100 (1992).

16. Wilcox, C.S., Welch, W.J., Murad, F., Gross, S.S., Taylor, G., Levi, R., and Schmidt, H.H.H.W. Nitric oxide synthase in macula densa regulates glomerular capillary pressure. *Proc. Natl. Acad. Sci. USA* **89,** 11993 (1992).

17. Mundell, P., Bachmann, S., Bader, M., Fischer, A., Kummer, W., Mayer, B., and Kriz, W. Expression of nitric oxide synthase in kidney macula densa cells. *Kidney Int.* **42,** 1017 (1992).

18. Bachmann, S., Mundel, P., and Kriz, W. Distribution of nitric oxide synthase (NOS) in the kidney. *J. Am. Soc. Nephrol.* **3,** 540 (1992) (Abstract).

19. Schmidt, H.H.H.W., Gagne, G.D., Nakane, M., Pollock, J.M., Miller, M.F., and Murad, F. Mapping of neuronal nitric oxide synthase in the rat suggests frequent co-localization with NADPH-diaphorase but not with soluble guanylyl cylcase, and novel pareneural function for nitrinergic signal transduction. *J. Histochem. Cytochem.* **40,** 1439 (1992).

20. Terada, Y., Tomita, K., Nonoguchi, H., and Mammo, F. Polymerase chain reaction localization of constitutive nitric oxide synthase and soluble guanylyl messenger RNAs in microdissected rat nephron segments. *J. Clin. Invest.* **90,** 659 (1992).

21. Tojo, A., Gross, S.S., Zhang, L., Tisher, C.C., Schmidt, H.H.H.W., Wilcox, C.S., and Madsen, K. Immunocytochemical localization of distinct isoforms of nitric oxide synthase in the juxtaglomerular apparatus of the normal kidney. *J. Am. Soc. Nephrol.* **4,** 1438 (1994).

22. Bhardwaj, R. and Moore, P.K. The effects of arginine and nitric oxide on resistance blood vessels of the perfused rat kidney. *Br. J. Pharmacol.* **97,** 739–744 (1989).

23. Salvemini, D., Seibert, K., Masferrer, J.L., Misko, T.P., Currie, M.G., and Needleman, P. Endogenous nitric oxide enhances prostaglandin production in a model of renal inflammation. *J. Clin. Invest.* **93,** 1940–1947 (1994).

24. Schulsinger, D.A., Felsen, D., Gulmi, F.A., Gross, S., and Vaughan, E.D., Jr. The expression and localization of inducible nitric oxide synthase (iNOS) following unilateral ureteral obstruction (UUO) in rat kidney. *J. Urol.* **155,** 561A (1996).

25. Borsook, H. and Dubnoff, J.W. The conversion of citrulline to arginine in the kidney. *J. Biol. Chem.* **141,** 717 (1941).

26. Dhanakoti, S.N., Brosnan, J.T., Herzberg, G.R., and Brosnan, M.E. Renal arginine synthesis: Studies in vitro and in vivo. *Am. J. Physiol.* **259,** E437 (1990).

27. Lanzone, J.A., Gulmi, F.A., Chou, S., Mooppan, U.M.M., and Kim, H. Renal hemodynamics in acute unilateral ureteral obstruction: Contribution of endothelium-derived relaxing factor. *J. Urol.* **153,** 2055 (1995).

28. Schulsinger, D.A., Gulmi, F.A., Chou, S., Mootan, U.M.M., and Kim, H. Activation of the endothelium-derived relaxing factor system in acute unilateral ureteral obstruction (UUO). *J. Urol.* **149,** 500A (1993).

29. Marsden, P.A., and Brenner, B.M. Nitric oxide and endothelins: Novel autocrine/paracrine regulators of the circulation. *Semin. Nephrol.* **11,** 169–185 (1991).

30. Schulsinger, D.A., Felsen, D.F., Gulmi, F.A., Gross, S., and Vaughan, E.D., Jr. Expression, localization and inhibition of inducible nitric oxide synthase (iNOS) following unilateral ureteral obstruction (UUO) in rat kidney. *Surgical Forum* **47,** 800–802 (1996).

31. Reyes, A.A., Martin, D., Settle, S., and Klahr, S. EDRF role in renal function and blood pressure of normal rats and rats with obstructive uropathy. *Kidney Int.* **41,** 403 (1992).

32. Reyes, A.A., Karl, I.E., Yates, J., and Klahr, S. Low plasma and renal tissue levels of L-arginine in rats with obstructive nephropathy. *Kidney Int.* **45,** 782 (1994).

33. Reyes, A.A., Porras, B.H., Chasalow, F.I., and Klahr, S. L-Arginine decreases the infiltration of the kidney by macrophages in obstructive nephropathy and puromycin-induced nephrosis. *Kidney Int.* **45,** 1346 (1994).

34. Schulsinger, D., Marion, D.N., Kim, F., Felsen, D.F., Gross, S.S., and Vaughan, E.D., Jr. The nitric oxide system is activated during late unilateral ureteral obstruction (UUO). *J. of Urol.* **157,** 209A (1997).

35. Chevalier, R.L., Thornhill, B.A., and Gomez, R.A. EDRF modulates renal hemodynamics during unilateral obstruction in the rat. *Kidney Int.* **42,** 400–406 (1992).

36. Chen, R.N., Inman, S.R., Stowe, N.T., and Novick, A. Role of endothelium-derived relaxing factor in the maintenance of renal blood flow in a rodent model of chronic hydronephrosis. *Urology* **46**(3), 438–442 (1995).

37. Kunz, D., Walker, G., and Pfeilschkfter, J. Dexamethasone differentially affects interleukin 1β- and cyclic AMP-induced nitric oxide synthase mRNA expression in renal mesangial cells. *Biochem. J.* **304,** 337–340 (1994).

38. Radomski, M.W., Palmer, R.M.J., and Moncada, S. Glucocorticoids inhibit the expression of an inducible but not the constitutive nitric oxide synthase in vascular endothelial cells. *Proc. Natl. Acad. Sci. USA* **87,** 10043–10047 (1990).

39. Moncada, S., Palmer, R.M.J., and Higgs, E.A. Nitric oxide: Physiology, pathophysiology and pharmacology. *Pharmacol. Rev.* **43,** 109–141 (1991).

40. Corbett, J.A., Tilton, R.G., Chang, K., Hasan, K.S., Ido, Y., Wang, J.L., Sweetland, M.A., Lancaster, J.R., Jr., Williamson, J.R., and McDaniel, M.L. Aminoguanidine, a novel inhibitor of nitric oxide formation, prevents diabetic vascular dysfunction. *Diabetes* **41,** 552–556 (1992).

41. Misko, T.P., Moore, W.M., Kasten, T.P., Nickols, G.A., Corbett, J.A., Tilton, R.G., McDaniel, M.L., Williamson, J.R., and Currie, M.G. Selective inhibition of the inducible nitric oxide synthase by aminoguanidine. *Eur. J. Pharmacol.* **233,** 119–125 (1993).

42. Griffith, M.J.D., Messent, M., MacAllister, R.J., and Evans, T.W. Aminoguanidine selectively inhibits inducible nitric oxide synthase. *Br. J. Pharmacol.* **110,** 963–968 (1993).

43. Inman, S.R., Chen, R.N., Stowe, N.T., and Novick, A.C. The influence of endothelium derived relaxing factor (EDRF) on renal blood flow (RBF) during acute ureteral obstruction. *J. Urol.* **153,** 411A (1995).

44. Salvemini, D., Misko, T.P., Masferrer, J., Seibert, K., Currie, M.G., and Needleman, P. Nitric oxide activates cyclooxygenase enzymes. *Proc. Natl. Acad. Sci. USA* **90,** 7240–7244 (1993).

45. Allen, J.T., Vaughan, E.D., Jr., and Gillenwater, J.Y. The effect of indomethacin on renal blood flow and ureteral pressure in unilateral ureteral obstruction in awake dogs. *Invest. Urol.* **15,** 324–327 (1978).

Index

DATE DUE

JUL 07 1998	